ORGANIC PHOSPHORUS COMPOUNDS

ORGANIC PHOSPHORUS COMPOUNDS

COMPOUNDS

Volume 2

G. M. KOSOLAPOFF
Auburn University

and

L. MAIER
Monsanto Research S. A.

WILEY-INTERSCIENCE, a Division of John Wiley & Sons, Inc.
New York • London • Sydney • Toronto

CHEMISTRY

Library of Congress Cataloging in Publication Data

Kosolapoff, Gennady M
Organic phosphorous compounds.

1950 ed. published under title: Organophosphorus
compounds.
Includes bibliographies.
1. Organophosphorus compounds.   I. Maier, L.,
joint author. II. Title.

QD412.P1K55 1972   547'.07   72–1359
ISBN 0-471-50441-6 (v. 2)

Printed in the United States of America

10 9 8 7 6 5 4 3 2

Contents

J. G. VERKADE and K. J. COSKRAN, Depart-
ment of Chemistry, Iowa State University,
Ames, Iowa

PETER BECK, Organisch-Chemisches
Institut der Universität, Mainz,
Germany

v

ORGANIC PHOSPHORUS COMPOUNDS

Chapter 3B.   Phosphite, Phosphonite, Phosphinite, and
              Aminophosphine Complexes

J. G. VERKADE

Iowa State University, Ames, Iowa

K. J. COSKRAN

Michigan State University, East Lansing,
Michigan

The range of ligands considered in this chapter includes representatives of the classes $P(OR)_xR_{3-x}$,
$P(SR)_xR_{3-x}$, $P(NR_2)_xR_{3-x}$, $P(OR)_xX_{3-x}$, $P(NR_2)_xX_{3-x}$ and
$PX(NR_2)R$ where R = substituted or unsubstituted alkyl or
aryl, X = halogen, hydroxyl, or pseudohalogen, and x =
1-3. Although PX_3 compounds might be considered a natu-
ral extension of the series, only complexes where X is a
pseudohalogen (e.g., CN) will be discussed since such
ligands contain carbon and therefore fall within the scope
of this review. It might be noted that PF_3 complexes have
been reviewed recently.[311b] In addition to metal complexes, adducts of Lewis acids such as those formed with
boron acceptors will also be covered. Although boron is
generally considered to be a nonmetal, the justification
for its inclusion is that its electronegativity is actu-
ally substantially less than that of many of the metals
to be discussed (e.g., 2.2-2.4 for the platinum metals).
The ligands of interest in this chapter have been
found to form complexes with all of the transition metals
except those of the scandium and titanium subgroups and
niobium and tantalum in the vanadium subgroup. The
scarcity of technetium is undoubtedly responsible for the
single reported complex with this metal. Except for boron and aluminum adducts, there are only a few scattered
accounts of addition compounds involving other typical
elements. Addition compounds of the latter type are included for completeness.
The distinction between a complex (metallic or non-
metallic) and a conventional compound (if there is such a
substance) has proven over the years to be traditional
rather than chemical. Thus, for instance, $H_3BP(OR)_3$ and
$OP(OR)_3$ formed from B_2H_6 or O_2 and $P(OR)_3$ are generally
considered to be "adduct" (or "complex") and "compound", respectively. Yet the acceptors are isoelectronic and both
normally exist as dimers in the "uncomplexed" state.
Systems which can be considered formally to arise from
the ligands of interest in this chapter and acceptors
such as a carbonium ion (Chapter 4), two OR radicals
(Chapter 5B), or a chalcogen atom (Chapters 14-16, 18, 19)
are treated in the chapters indicated.

A. SYNTHETIC ROUTES

The variety of complexes described in the literature is
extremely wide and their history relatively long. About

half of the more than 25 elements functioning as ligand
sites possess two or more oxidation states in which complex formation can take place. Octahedral and square
planar arrangements of various groups around the metals
complicate matters structurally by allowing the formation
of geometrical isomers upon introducing one of more ligands into the system. To record here a specific synthesis for each compound listed in the compilation at the
end of the chapter would exceed space limitations. Moreover, several routes to a compound are often possible.
An attempt has therefore been made to organize the synthetic methods into categories within which are listed
reactions applicable to making the complexes which are
arranged according to the appropriate groups in the periodic table. The reactions given have been selected main-
ly for their variety and many of those which are very
similar from one periodic group to the next are not repeated unless they are widely used synthetic pathways.
Category I deals with direct reactions of L with the acceptor substrate wherein the oxidation state of the metal
is preserved in the product while category II is concerned with oxidation, reduction, or disproportionation
of the metal substrate in the presence of L. Category
III is a treatment of syntheses wherein changes in a complex of L are brought about which do not involve L.
Changes in complexes of L which alter or eliminate L are
the subject of category IV.

In some cases it was possible to express the synthesis
of many compounds in a single general reaction while in
others only a single compound was reported by a given
route. The latter were included since their narrow ap-
plicability may be only temporary if further work on these
synthetic approaches is forthcoming. Because descriptions
of syntheses began to appear in 1870,[472-477] the reliability of many of the earlier routes and the formulations of the products may be highly questionable especially in cases where reinvestigations have not as yet confirmed the earlier results. This is not to disparage the
earlier work for there are few investigators even today
who have never been misled in reaching conclusions concerning a new compound. The symbol L appearing in a reaction represents a ligand in the classes considered. A
general formula of the type $P(OR)_3$, $P(NR_2)_3$, $P(NR_2)_xR_{3-x}$,
etc., at the right of such reactions designates the ligand
type which has been found to function in the synthesis.
In these general formulas, R may represent substituted or
unsubstituted alkyl or aryl groups, X may be one of the
halogens and x = 1 or 2. It will be indicated when X is
a pseudohalogen. A glance at the synthetic routes which
have been employed shows that ligands of the type $P(OR)_3$
have been most widely investigated and this is very likely

due to the decreased ready availability of ligands from
the other classes. In fact, syntheses for a substantial
number of the rarer ligands have been only recently reported and can be found elsewhere in the appropriate
chapters.

Many of the reactants as well as the products of the
synthetic reactions possess particular stereochemistry.
None of these have been indicated in this section since
the subject of structure and geometrical isomerism is
treated generally in the section following.

Throughout this section we shall use the term "acceptor" in the broad sense that it is a species which functions as a point of attachment for the ligands to be
discussed. No implications concerning the nature of the
acceptor-ligand bond are intended since this link is certainly more complex than a simple electron pair acceptor-
donor interaction in the majority of complexes. Moreover,
no conclusions should be drawn as to the particular atom
in the ligand which is used for bonding to the acceptor
site. These points of interest will be amplified in the
next section.

The abbreviation Cp denotes the pi-coordinated cyclo-
pentadienide ion. The abbreviation chel denotes a poly-
dentate chelating ligand which is further specified for
the given reaction.

I. REACTION OF L WITH THE ACCEPTOR SUBSTRATE

1. SUBSTITUTION BY L OF GROUPS ON THE ACCEPTOR SYS-
TEM. This synthetic process involves displacement of
groups which are normally only very slightly dissociated
from the Lewis acid in the medium employed:

$$-MZ_x + x\ L \longrightarrow -ML_x + x\ Z$$

Historically, substitution of coordinated groups such as
CO, PX_3, PR_3, RCN, NR_3, X^- and chelating agents such as
arenes, polyolefins, diamines, dithianes, and diethers by
one or more L ligand species has received the widest at-
tention and will be considered first. Inasmuch as acceptor compounds which are polymeric generally have bridging
ligands of some type (e.g., H, X, etc.) or are joined by
acceptor-acceptor bonds, cleavage of such links by L will
also be viewed as substitution of a coordinated group.

It should be borne in mind in the stepwise substitutions depicted that not all the classes of ligands indicated to the right of the reaction are necessarily capable
of completing the sequence. To assess the ability of
specific ligands from a given class to substitute, the
compilation of compounds should be consulted. To avoid
repetition, it is assumed in each reaction that L appears

on the left and the displaced species on the right of the arrow.

a. CO Substitution. Carbon monoxide substitution by L can take place under various conditions. Where L is a liquid or low melting solid, reactions in the absence of solvent can take place anywhere from dry ice temperature to about 200°C depending on the reactivity of the parent carbonyl and the desired degree of substitution. The gamut of organic solvents can also be employed and are generally chosen in accordance with the solubilities and reaction temperatures required. Most reactions which are aided by ultraviolet radiation take place near room temperature.

$\underline{V}$ $\quad$ $CpV(CO)_4 \longrightarrow CpV(CO)_3L$ $\quad$ $P(OR)_3, P(NR_2)_3$

$\underline{Cr, Mo,}$ $\quad$ $M(CO)_6 \longrightarrow M(CO)_5L \longrightarrow$ $\quad$ $P(OR)_3, P(SR)_3,$
$\underline{W(=M)}$ $\quad$ $M(CO)_4L_2 \longrightarrow M(CO)_3L_3 \longrightarrow$ $\quad$ $P(NR_2)_3, P(OR)_xX_{3-x},$
$\quad$ $M(CO)_2L_4$ $\quad$ $P(OR)_xR_{3-x}, P(NR_2)_x-X_{3-x},$
$\quad\quad\quad P(NR_2)_xR_{3-x}$

Mixed ligand complexes containing two ligands (L and L') have also been made by stepwise substitution:

$M(CO)_6 \longrightarrow M(CO)_5L \longrightarrow$ $\quad$ $P(OR)_3, P(NR_2)_3$
$M(CO)_4LL'$ $\quad\quad\quad\quad\quad\quad$ $P(OR)_2CN, P(NR_2)_2CN$
$chelM(CO)_4 \longrightarrow$ $\quad\quad\quad\quad$ $P(OR)_3$
$chelM(CO)_3L \longrightarrow$
$chelM(CO)_2L_2$

chel = a derivative of an α, ω-alkane diamine or diphosphine or an o-phenan-throline compound

$[CpMo(CO)_3]_2 \longrightarrow$ $\quad\quad\quad$ $P(OR)_3$
$Cp_2Mo_2(CO)_5L \longrightarrow$
$[CpMo(CO)_2L]_2$
$[CpMo(CO)_3]_2Hg \longrightarrow$ $\quad\quad$ $P(OR)_3$
$[CpMo(CO)_2L]_2Hg$
$CpMo(CO)_3X \longrightarrow$ $\quad\quad\quad$ $P(OR)_3$
$CpMo(CO)_2LX$

X = halogen, SnR_3, HgX, or H

$C_7H_8M(CO)_2X \longrightarrow$ $\quad\quad$ $P(NR_2)_3$
$C_7H_8M(CO)LX$

$\underline{Mn}$ $\quad$ $[Mn(CO)_5]_2 \longrightarrow$ $\quad\quad\quad$ $P(OR)_3, P(OR)_xR_{3-x}$
$\quad$ $Mn_2(CO)_9L \longrightarrow$
$\quad$ $Mn_2(CO)_8L_2$
$\quad$ $[CpMn(CO)_3]_2 \longrightarrow$ $\quad\quad$ $P(OR)_3, P(OR)_xR_{3-x}$
$\quad$ $Cp_2Mn_2(CO)_5L$
$\quad$ $Mn(CO)_5Y \longrightarrow Mn(CO)_4LY \longrightarrow$ $\quad$ $P(OR)_3, P(OR)_xR_{3-x}$
$\quad$ $Mn(CO)_3L_2Y$

Y = alkyl, aryl, halide, or H

<u>Tc</u>   Tc(CO)$_5$Cl $\longrightarrow$ P(OR)$_3$
            Tc(CO)$_3$L$_2$Cl

<u>Fe,</u>   Fe(CO)$_5$ $\longrightarrow$ Fe(CO)$_4$L $\longrightarrow$ P(OR)$_3$, P(NR$_2$)$_3$,
<u>Os</u>   Fe(CO)$_3$L$_2$ $\longrightarrow$ Fe(CO)$_2$L$_3$ P(OR)$_3$X$_{3-x}$
            [Fe(CO)$_3$SPh]$_2$ $\longrightarrow$ P(OR)$_3$
            Fe$_2$(CO)$_5$L(SPh)$_2$

In the reaction of the difunctional ligand P(OCH$_2$)$_3$P, mono- and disubstitution can be achieved. Linkage isomerism in these compounds is discussed in the next section.

              Fe$_3$(CO)$_{12}$ $\longrightarrow$ P(OR)$_3$
              Fe$_3$(CO)$_{11}$L
              Fe(NO)$_2$(CO)$_2$ $\longrightarrow$ P(OR)$_3$
              Fe(NO)$_2$L$_2$
              diolefinFe(CO)$_3$ $\longrightarrow$ P(OR)$_3$
              diolefinFe(CO)$_2$L
       diolefin = 1,5 cyclooctadiene
              Fe(CO)$_4$I$_2$ $\longrightarrow$ P(OR)$_3$
              Fe(CO)$_3$LI$_2$
              CpFe(CO)$_2$X $\longrightarrow$ P(OR)$_3$
              CpFe(CO)LX $\longrightarrow$
              CpFeL$_2$X
       X = halogen, SnR$_3$, SiR$_3$, or R

The only case of CO substitution of an osmium compound also proceeds with addition of L:

              Os(CO)$_3$X$_2$ $\longrightarrow$ P(OR)$_3$
              Os(CO)$_2$L$_2$X$_2$

<u>Co, Rh,</u>   [Co(CO)$_4$]$_2$ $\longrightarrow$ P(OR)$_3$, P(NR$_2$)$_3$
<u>Ir</u>   [Co(CO)$_3$L]$_2$
            CoNO(CO)$_3$ $\longrightarrow$ P(OR)$_3$
            CoNO(CO)$_2$L $\longrightarrow$
            CoNO(CO)L$_2$
            [Co(CO)$_2$L$_3$]$^+$ $\longrightarrow$ P(OR)$_3$
            [Co(CO)L$_4$]$^+$ $\longrightarrow$ [CoL$_5$]$^+$
            Rh(CO)L$_2$Cl $\longrightarrow$ RhL$_3$Cl   P(OR)$_3$, P(NR$_2$)$_x$X$_{3-x}$

The preceding reaction proceeds with substitution of CO, and with the addition of L.

The following two reactions involve displacement of neutral groups and CO:

              Rh(CO)(PR$_3$)$_2$Cl $\longrightarrow$ P(OR)$_3$
              RhL$_3$Cl

$$Ir(CO)_2(NR_3)X \longrightarrow$$
$$Ir(CO)L_2X$$

$P(OR)_3$

__Ni__

$Ni(CO)_4 \longrightarrow Ni(CO)_3L \longrightarrow$
$Ni(CO)_2L_2 \longrightarrow$
$Ni(CO)L_3 \longrightarrow NiL_4$

$P(OR)_3$, $P(OR)_xX_{3-x}$,
$P(NR_2)_xX_{3-x}$,
$P(SR)_xX_{3-x}$, $PRXNR_2$,
$P(NR_2)_3$,
$P(NCO)_3$, $P(NCS)_3$

__B__

$H_8B_4CO \longrightarrow H_8B_4L$

$P(NR_2)_xX_{3-x}$

b. PX_3, PR_3, X^-, R^-, CN^-, NR_3, OR_2, and Olefin
Substitution. Other monodentate groups which can be displaced are PX_3, PR_3, X^-, R^-, CN^-, NR_3, OR_2, and olefin.
Such reactions are generally carried out in organic sol-
vents and occasionally in water at temperatures ranging
from room temperature to reflux.

__Mo, W(=M)__ $M(CO)_5NR_3 \longrightarrow M(CO)_5L$

$P(OR)_3$, $P(NR_2)_3$,
$P(SR)_3$

$W(CO)_3(NCR)_3 \longrightarrow$
$W(CO)_3L_3$

$P(OR)_3$

__Mn__

$[Mn(CO)_4I_2]^- \longrightarrow$
$[Mn(CO)_4LI]$

$P(OR)_3$

__Fe, Ru,__
__Os__

$CpFe(CO)_2I \longrightarrow$
$[CpFe(CO)_2L]^+$

$P(NR_2)_3$

The following reaction involves further addition:

$M(PR_3)_3X_2 \longrightarrow ML_4X_2$
M = Ru or Os

$P(OR)_3$

__Rh, Ir__

$RhL_3Cl \longrightarrow [RhL_4]^+ \longrightarrow$
$[Rh(CO)_2(NCS)_2]^- \longrightarrow$
$RhL_3(NCS)$

$P(OR)_3$

The following two reactions all involve addition as
well as substitution.

$HRh(CO)(PR_3)_2 \longrightarrow HRhL_4$
$Ir(CO)_2(PR_3)_2Cl \longrightarrow$
$[Ir(CO)L_2(PR_3)_2]^+$

$P(OR)_3$
$P(OR)_3$

__Ni, Pd,__
__Pt__

$NiL_3(PR_3) \longrightarrow NiL_4$
$M(PR_3)_4 \longrightarrow ML_3(PR_3)$
ML_4
M = Pd or Pt

$P(OR)_3$
$P(OR)_3$

Addition also is found to take place in the following reaction:

$$Pt(PR_3)_3 \longrightarrow PtL_4 \qquad\qquad P(OR)_3$$
$$PdL_2(NCR)_2 \longrightarrow \qquad\qquad P(OR)_3$$
$$PdL_3(NCR) \longrightarrow PdL_4$$
$$Pd(NCR)_2Cl_2 \longrightarrow PdL_2Cl_2 \qquad P(OR)_3, \; P(NR_2)_3$$
$$PdL_2Cl_2 \longrightarrow [PdL_4]^{2+} \qquad P(OR)_3$$
$$YPt(PR_3)_2Cl \longrightarrow \qquad\qquad P(OR)_3$$
$$[YPtL(PR_3)_2]^{+}$$

Y = aryl or H

$$PtX_4{}^{2-} \longrightarrow \qquad\qquad P(OR)_2OH$$
$$PtL_2[P(O)(OR)_2]_2$$
$$PtZ_2X_2 \longrightarrow \qquad\qquad P(OR)_2OH$$
$$PtZ_2[P(O)(OR)_2]_2$$

Z = NR$_3$ or PR$_3$

The preceding two reactions involve substitution with conversion of L to a coordinated anion.

With $R_2NP(Ph)NR_2$ only one ligand is incorporated:

$$PtX_4{}^{2-} \longrightarrow PtLX_2 \qquad\qquad R_2NP(Ph)NR_2$$

Au

$$Au(PCl_3)Cl \longrightarrow AuLCl \qquad P(OR)_3$$

B

$$H_{12}B_{10}Z_2 \longrightarrow H_{12}B_{10}L_2 \qquad P(OR)_3, \; P(OR)_xR_{3-x}$$

Z = SEt$_2$ or NCMe

$$H_3BNMe_3 \longrightarrow H_3BL \qquad\qquad P(NR_2)_3$$
$$H_4B^- \longrightarrow H_3BL \qquad\qquad P(OR)_3, \; P(OR)_xR_{3-x}$$

c. Chelates. Arenes, polyolefins, and other chelates are displaced in organic solvents by L from room temperature to about 100°C. Ultraviolet radiation is generally unnecessary.

Cr, Mo,
$$areneM(CO)_3 \longrightarrow M(CO)_3L_3 \qquad P(OR)_3, \; P(OR)_xX_{3-x}$$
W(=M)   arene = benzene derivative

$$triolefinM(CO)_3 \longrightarrow \qquad P(OR)_3, \; P(OR)_xR_{3-x},$$
$$M(CO)_3L_3 \qquad\qquad P(NR_2)_xX_{3-x}, \; PXR(NR_2)$$

triolefin = 1,5-cycloheptatriene

$$diolefinM(CO)_4 \longrightarrow \qquad P(OR)_3, \; P(NR_2)_3,$$
$$M(CO)_4L_2 \qquad\qquad P(NR_2)_xX_{3-x}$$

diolefin = norbornadiene or 1,5-cyclooctadiene

$$chelM(CO)_4 \longrightarrow \qquad\qquad P(OR)_3$$
$$M(CO)_4L_2 \longrightarrow M(CO)_3L_3$$

chel = bipyridyl, o-phenanthroline, an α,ω-alkane diamine, or dithiane derivative

Fe $diolefinFe(CO)_3 \longrightarrow$ $P(OR)_3$
           $Fe(CO)_3L_2$
    $diolefin = 1,5-cyclooctadiene$

Co, Rh $diolefin(C_8H_{13})Co \longrightarrow$ $P(OR)_3, \; P(OR)_xR_{3-x}$
           $C_8H_{13}CoL_2 \longrightarrow HCoL_4$

In the last step of the above reaction the overall oxidation state is unchanged inasmuch as C_8H_{13} is an anionic group as well as the hydride.

       $[diolefinRhX]_2 \longrightarrow$ $P(OR)_3, \; P(NR_2)_xX_{3-x}$
       $diolefinRhX_2RhL_2 \longrightarrow$
       $[RhL_2X]_2 \longrightarrow RhL_3X \xrightarrow{\;\;BPh_4^-\;\;}$
       $[RhL_5]^+$

Ni $diolefinNiYC(CF_3)_2 \longrightarrow$ $P(OR)_3$
          $L_2NiYC(CF_3)_2$
    $Y = S$ or O
        $[(chel)_2NiH]^+ \longrightarrow$ $P(OR)_3$
        $[chelNiL_2H]^+$
   $chel = Ph_2PCH_2CH_2PPh_2$

Substitutions accompanied by other changes in the complex make up most of the remainder of the synthetic routes in this category. Upon displacement by L, a CO group may insert between the metal and a coordinated R group to form a coordinated acyl group. Many metal substrates are polymeric being connected by acceptor-acceptor bonds or CO, R, or X bridging links. Here the breaking of an acceptor-ligand bond also separates the units of the polymer. Substitution of a coordinated group may also induce a fundamental change in the incoming ligand.

    d. CO Substitution with Insertion. Carbon monoxide substitution followed by insertion when it does take place generally proceeds in organic solvents from room temperature to 60°C.

Mo $CpMo(CO)_3R \longrightarrow$ $P(OR)_3, \; P(NR_2)_3$
          $CpNo(CO)_2L(COR)$

Mn $Mn(CO)_5R \longrightarrow$ $P(OR)_3, \; P(OR)_xR_{3-x}$
          $Mn(CO)_4L(COR) \longrightarrow$
          $Mn(CO)_3L_2(COR) \longrightarrow$
          $Mn(CO)_2L_3(COR)$

A most unusual reaction is:

$$Mn(CO)_5SiPh_3 + P(OR)_3 \longrightarrow$$
$$Mn(CO)_3L_2(COR)$$

Here an R group from the ligand contributes to the formation of the acyl group. A Michaelis-Arbuzov reaction is probably taking place in which the R group in the cation $[Ph_3SiP(OR)_3]^+$ is attacked by an $[Mn(CO)_5]^-$ anion to form an intermediate $Mn(CO)_5R$ species. Excess ligand could then afford the final product in the usual manner.

Fe $CpFe(CO)_2R \longrightarrow$
            $CpFe(CO)L(COR)$

e. Substitutions with Alteration of L. Substitutions involving alterations in the incoming ligand are relatively few. Except for the first reaction which is induced by ultraviolet radiation, heat seems to be the only requirement.

Mo $[CpMo(CO)_3]_2 + P(OPh)_3 \longrightarrow$
            $Cp(OC)_2Mo[P(OPh)_2]_2Mo(CO)_2Cp$
            $chelMo(CO)_4 + HP(O)(OR)_2 \longrightarrow$
            $chelMo(CO)_3P(OH)(OR)_2 \longrightarrow$
            $chelMo(CO)_2[P(OH)(OR)_2]_2$
      chel = o-phenanthroline
      or bipyridyl

Pd, $MCl_4^{2-} + HP(O)(OR)_2 \longrightarrow M[P(OH)(OR)_2]_2Cl_2$
Pt (=M)

            $M[P(OH)(OR)_2]_2[P(O)(OR)_2]Cl$
                        $M[P(OH)(OR)_2]_2[P(O)(OR)_2]_2$

f. Polymer Cleavage. Cleavage of polymers to form complexes of L occurs from temperatures near 0°C to reflux temperatures of the common organic solvents used.

Mo $[CpMo(NO)I_2]_2 \longrightarrow$ $P(OR)_3$
            $2\ CpMoNOLI_2$

Fe, Ru, $Fe_3(CO)_{12} \longrightarrow 3\ Fe(CO)_4L$ $P(OR)_3,\ P(NR_2)_3$
Os $[FeS_4C_4(CF_3)_4]_2 \longrightarrow$ $P(OR)_3$
            $2\ Fe[S_4C_4(CF_3)_4]L$
            $[Ru(CO)_2X_2]_n \longrightarrow$ $P(OR)_3$
            $2\ Ru(CO)_2L_2X_2$
            $[Os(CO)_3Cl_2]_2 \longrightarrow$ $P(OR)_3$
            $Os(CO)_2L_2Cl_2$

Co, Rh $[Co(NO)_2X]_2$ ⟶ $P(OR)_3$, $P(SR)_3$
          $Co(NO)_2LX$
          $[RhL_2X]_2$ ⟶ RhL_3X $P(OR)_3$

Ni, Pd,   Ni ⟶ NiL_4 $P(NR_2)_xX_{3-x}$
Pt   Here the nickel metal was produced by
          heating NiC_2O_4.
          $[Ni(NO)X]_4$ ⟶ $P(OR)_3$
          $[Ni(NO)LX]_2$ ⟶
          $Ni(NO)L_2X$
          $(PtX_2)_x$ ⟶ PtL_2X_2 $P(OR)_3$
          $[Pd(NCR)_2]_x$ ⟶ $P(OR)_3$
          $PdL_2(NCR)_2$

Cu, Au $[CuL'X]_3$ ⟶ CuL_2X $P(OR)_3$

   In the above reaction L' is a $P(OR)_3$ ligand which is
substituted by one with different R groups.

          $(AuCCPh)_x$ ⟶ $AuL(CCPh)$ $P(OR)_3$

B, Al B_2H_6 ⟶ $2 H_3BL$ ⟶ $P(OR)_3$, $P(NR_2)_3$,
          $(H_3B)_2L$ $PR(OR)(NR_2)$
          B_5H_{11} ⟶ $P(NR_2)_xX_{3-x}$
          $H_8B_4L + 1/2B_2H_6$
          $(AlR_3)_2$ ⟶ $2 R_3AlL$ $P(OR)_3$, $P(NR_2)_xR_{3-x}$,
                                          $P(OR)_xR_{3-x}$

Ge $(GeI_2)_x$ ⟶ $GeLI_2$ $P(NR_2)_3$

   2. REACTION OF L WITH SOLVATED ACCEPTOR SPECIES.
Synthetic reactions of this type generally involve solvated metal salts. Such systems are not ordinarily considered as coordination complexes since the species associated with the metal are only loosely held. Employed
in the syntheses are either a nonsolvated metal salt in
a polar solvating solvent such as acetone, water, or an
alcohol, or a solvated metal salt in a nonpolar hydrocarbon solvent. In many instances hydrated salts in polar
solvents are used. If the anion (generally a halide) from
the metal salt is found to be coordinated in the product,
the salt is written as a halide. If the anion is of the
noncoordinating type such as BF_4^- and ClO_4^-, only the metal
ion is shown. It must be admitted that the distinction in
several instances between polymer cleavage by L and reaction of a monomeric solvated substrate with L is hazy at
best. Reactions of the type discussed here in some cases
lead to redox phenomena and these are discussed in category II.

Mn, Fe M^{2+} + a(See Zn, Cd, Hg)
          M = Mn or Fe

<u>Co, Rh</u> $CoX_2 \longrightarrow CoL_2X_2$ $P(NR_2)_xR_{3-x}$
         X = halogen or NCS
            $CoX_2 \longrightarrow CoL_xX_2$ $P(OR)_3$
      X = 2, 3, or 4
            $RhCl_3 \cdot xH_2O \longrightarrow$ $P(OR)_3$
            $[RhL_4Cl_2]Cl$

<u>Ni</u> $C_3H_5NiBr \longrightarrow C_3H_5NiLBr$ $P(OR)_3$, $P(NR_2)_3$
            $NiX_2 \cdot xH_2O \longrightarrow NiL_2X_2$ $P(OR)_3$, $P(NR_2)_xR_{3-x}$
      X = halogen or NCS
            $Ni(CN)_2 \longrightarrow$ $P(OR)_3$, $P(OR)_xR_{3-x}$
            $NiL_2(CN)_2 \longrightarrow$
            $NiL_3(CN)_2$
            $NiX_2 \longrightarrow NiL_3X_2 \longrightarrow$ $P(OR)_3$
            NiL_4X_2
            $Ni \cdot x\ EtOH^{2+} + Z \longrightarrow$ $P(OR)_3$
            $[NiLZ]^{2+}$
      $Z = P(o\text{-}MeSC_6H_4)_3$
            $[Ni \cdot x\ H_2O]^{2+} \longrightarrow$ $P(OR)_3$
            $[NiL_5]^{2+}$

<u>Cu, Ag</u> $CuX \longrightarrow (CuLX)_{3,4} \longrightarrow$ $P(OR)_3$, $P(NR_2)_xX_{3-x}$,
            $[CuXL_2]_{1,2,4}$ $P(OR)_xR_{3-x}$,
                                                                    $P(NR_2)_xR_{3-x}$

This reaction may well involve depolymerization in the first step. The degree of polymerization in the products is dependent on L, the solvent, and X where X = halogen or CN.

$$Cu \cdot x\ H_2O^{2+} +$$

(a)

or

(b)

$$\{Cu[P(OCH)_3(CH_2)_3]_4\}^{1+} + (x + 4)\ H_2O$$

This unusual synthesis in which dehydration of either isomer of L occurs is also a redox reaction (see category II). Silver ion also allows the dehydration of these phosphorus ester isomers (a) and (b).

$$Ag^+ + 4\ b \longrightarrow$$
$$\{Ag[P(OC\overline{H})_3(CH_2)_3]_4\}^+ +$$
$$4\ H_2O$$
$$Ag^+ \longrightarrow AgL_4^+ \qquad\qquad P(OR)_3$$

Zn, Cd, Hg

The unusual dehydration cited above does not always occur and isomer (a) can be coordinated by rearrangement:

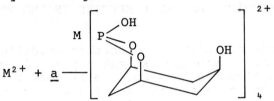

M = Zn or Cd

$$CdI_2 \longrightarrow CdLI_2 \longrightarrow \qquad P(NR_2)_3,\ P(NR_2)_xR_{3-x}$$
$$CdL_2I_2$$
$$HgX_2 \longrightarrow [HgX_2]_2L \longrightarrow \qquad P(OR)_3,\ P(NR_2)_xR_{3-x},$$
$$[HgLX_2]_2 \longrightarrow HgL_2X_2 \qquad\qquad P(NR_2)_3$$

The second and third steps in the above are polymer cleavages.

$$HgI_2 + [MeP(NR_2)_xR_{3-x}]I \longrightarrow [MeP(NR_2)_xR_{3-x}]-$$
$$[HgI_3]\ \text{or}\ \{Hg[MeP(NR_2)_xR_{3-x}]I_3\}$$

3. SIMPLE ADDUCT FORMATION. Although BR_3, BX_3, NMe_3, PF_5, and SO_3 are not strongly polymerized or solvated under the conditions employed, adduct formation is accompanied by reorganization of the electronic environment which stabilizes the Lewis acid. In the case of BR_3 and BX_3 adducts this process is known to involve also a change from planarity in the Lewis acid to tetrahedrality in the adduct. Adduct formation of the type shown below is generally carried out in an organic solvent or neat below 0°C.

B

$$BR_3 \longrightarrow R_3BL \qquad\qquad P(OR)_3,\ P(NR_2)_3,$$
$$\qquad\qquad\qquad\qquad P(NR_2)_xR_{3-x}$$
$$BX_3 \longrightarrow X_3BL \qquad\qquad P(OR)_3,\ P(NR_2)_xX_{3-x},$$
$$\qquad\qquad\qquad\qquad P(NR_2)_xR_{3-x}$$

X = F, Cl

It is interesting that when $HP(O)(OR)_2$ is allowed to interact with BX_3, ligand rearrangement is not observed in contrast to the reactions with metals cited above:

$$BX_3 + OP(H)(OR)_2 \longrightarrow$$
$$X_3BOP(H)(OR)_2$$

N

$$N_2F_4 \longrightarrow N_2F_4L \qquad\qquad P(OR)_3$$

$\underline{P}$ $PF_5 \longrightarrow F_5PL$ $P(OR)_3,\ P(NR_2)_3,$
                                              $P(NR_2)_xR_{3-x}$

$\underline{O}$ $O_3 \longrightarrow O_3L$ $P(OR)_3$

$\underline{S}$ $SO_3 \longrightarrow SO_3L$ $P(OR)_3$

II. REDOX REACTIONS OF METAL SPECIES IN THE PRESENCE OF L

There is little question as to whether oxidation or reduction of the metal species has taken place in most of the synthetic reactions in this section as long as groups which are bonded to the acceptor are clearly recognizable as being more electronegative or electropositive than the acceptor. In the case of coordinated groups such as R, I, and H, however, a more arbitrary approach is necessary since the electronegativities involved (2.5 for C and I, and 2.1 for H) are close to or within the range of the acceptor atoms we are considering (1.5-2.4). For consistency we shall assume addition of R and I groups to the metal to be an oxidation and that they have some degree of anionic character relative to the acceptor atom. Rather than considering the electronegativity for an acceptor atom relative to hydrogen in each redox reaction in this and the remaining categories, we shall adopt the somewhat arbitrary convention that hydrogen bonded to the acceptor is hydridic and therefore oxidizes the metal. This is not unreasonable when it is recognized that the high-field NMR chemical shifts of such hydrogens at least in part reflect an electronic shielding of the nucleus far greater than that for protonic hydrogens attached to the more electronegative elements. It should be pointed out, however, that "hydride" complexes can behave as strong acids in some instances.

Though some oxidations and reductions in this category have been effected in the presence of a specific oxidizing or reducing agent, reductions have been observed to take place spontaneously in the presence of L. In the latter syntheses, the available evidence indicates that the solvent or a group such as H_2O, X, or R associated with the acceptor substrate functions as the reducing agent. In some cases, part of the driving force could well be the high thermodynamic stability of the P=O bond:

$$M^{2+} + L + H_2O \longrightarrow$$
$$M^0 + O{=}L + 2\ H^+$$
$$MX_2 + P(OR)_3 \longrightarrow$$
$$M^0 + O{=}PX(OR)_2 + RX$$

If L is incapable of entering into a Michaelis-Arbuzov reaction, oxidation of L to the appropriate X_2L derivative
of pentavalent phosphorus is possible. Where R groups are
eliminated, it is not unreasonable to suppose that R_2 is
formed. In a few instances, disproportionation of the
metal substrate has been observed in which part of the
metal is found in a higher and some is found in a lower
oxidation state. All of the reactions in this category
are generally done in organic solvents at temperatures in
the room temperature to reflux range. The abbreviations
over the arrows denote the classification of the reaction.

$\underline{Mo}$ $[CpMo(CO)_3]_2 \xrightarrow{\text{disprop.}}$ $P(OR)_3$
$[CpMo(CO)_2L_2]^+ +$
$[CpMo(CO)_3]^-$

$\underline{Re}$ $K_2ReI_6 \xrightarrow{\text{red.}} ReL_3I_3$ $P(OR)_3$
$(chel)_2ReH_3 \xrightarrow{\text{red.}}$
$(chel)_2ReLH$

$\underline{Fe}$ $CpFe(CO)_2Z \xrightarrow{\text{red.}}$ $P(OR)_3$
$[CpFeL_2]_2$
$Z =$ alkyl or Me_3Si

$[FeNO(CO)_3]_2Hg \xrightarrow{\text{disprop.}}$ $P(NR_2)_3$
$[FeNO(CO)_2L_2]^+ +$
$[FeNO(CO)_3]^-$

$Ru(CO)_3L_2 + HSiCl_3 \xrightarrow{\text{oxid.}}$ $P(OR)_3$
$Ru(CO)_2L_2Cl_2$

$OsL(PR_3)_2Cl_3 + NaBH_4 \xrightarrow{\text{oxid.}}$ $P(OR)_3$
$OsL(PR_3)_2H_4$

$Os(PR_3)_2Cl_4 \xrightarrow{\text{red.}}$ $P(OR)_xR_{3-x}$
$OsL(PR_3)_2Cl_3$

$\underline{Co, Ir}$ $Co^{2+} \xrightarrow{\text{disprop.}} [CoL_5]^+ +$ $P(OR)_3$
$[CoL_6]^{3+}$

This disproportionation reaction is unique in that it
is the only one in which L appears in both products.

$[Co(NO)_2Cl]_2 + L \xrightarrow{\text{disprop.}}$ $P(OR)_3$
$Co(NO)L_3 + Co(O=L)_2Cl_2 + O=L + N_2$

$$Co^{2+} + NaBH_4 \text{ or } H_2 \xrightarrow{\text{red.}} P(OR)_3$$
$$HCoL_4$$

$$IrX_6{}^{3-} \xrightarrow[CO]{\text{red.}} Ir(CO)L_2X \qquad P(OR)_3$$

The above reaction also involves substitution and elimination of coordinated groups.

Ni, Pd
$$NiX_2 \xrightarrow{\text{red.}} NiL_4 \qquad P(OR)_3$$

$$NiX_2 + NR_3 \text{(aqueous)} \xrightarrow{\text{red.}} P(OR)_3$$
$$NiL_4$$

$$NiL_5{}^{2+} + \qquad\qquad\qquad P(OR)_3$$
$$HCO_3{}^-\text{(aqueous)} \xrightarrow{\text{red.}} NiL_4$$

$$Cp_2Ni \xrightarrow{\text{red.}} NiL_4 \qquad P(OR)_3, \; P(NR_2)_xX_{3-x}$$

$$Ni(PR_3)_3X \xrightarrow{\text{disprop.}} NiL_2X_2 + \qquad P(OR)_3$$
$$NiL_4$$

$$PdX_4{}^{2-} + NR_3 \xrightarrow{\text{red.}} PdL_4 \qquad P(OR)_3, \; P(OR)_xR_{3-x}$$

$$PtL_2X_2 + H_2NNH_2 \xrightarrow{\text{red.}} PtL_4 \quad P(OR)_3$$

Cu, Au
$$CuX_2 \xrightarrow{\text{red.}} [CuL_4]X \qquad P(OR)_xR_{3-x}$$

$$HAuCl_4 \xrightarrow{\text{red.}} AuLCl \qquad P(OR)_3$$

B
$$B_{10}H_{14} \xrightarrow{\text{oxid.}} H_{12}B_{10}L_2 \qquad P(OR)_3$$

III. ALTERATIONS IN THE COMPLEX WHICH DO NOT INVOLVE COORDINATED L

It has often been found useful to synthesize a complex from an intermediate in which L has already been bonded to the acceptor. Besides oxidation and reduction, various other types of reactions which have been utilized for this purpose include substitution, removal or hydrolysis of groups on the metal other than L, cleavage of polymers, or attachment of additional ligands. All of these reactions are most often carried out in organic solvents from room temperature to solvent reflux temperatures.

1.   REDUCTION.

<u>Mo, W</u>
<u>(=M)</u>
$[CpM(CO)_2L]_2 + Na(Hg) \longrightarrow$ $P(OR)_3$
$[CpM(CO)_2L]^-$
$[CpM(CO)_2L]_2Hg +$ $P(OR)_3$
$Na(Hg) \longrightarrow$
$[CpMo(CO)_2L]^-$

The above reaction can be considered a reduction since the nearly equal electronegativities of Hg and M renders M essentially zerovalent in the Hg derivative.

<u>Mn</u>
$[Mn_2(CO)_8L_2] + Na(Hg) \longrightarrow$ $P(OR)_3$
$[Mn(CO)_4L]^-$

<u>Co</u>
                    butadiene
$HCo(CO)_2L_2 \xrightarrow{\hspace{2cm}}$ $P(OR)_3, P(NR_2)_3$
$Co_2(CO)_5L_3 \longrightarrow$
$Co_2(CO)_4L_4$
$CoL_xX + H_2 \longrightarrow [CoL_4]_2$ $P(OR)_3$
$x = 3$ or 4
$Co(CO)_2L_2Cl + Na \longrightarrow$ $P(OR)_3$
$[Co(CO)_2L_2]^- + Na^+$

<u>Pt</u>
$PtL_2Cl_2 + H_2NNH_2 \longrightarrow$ $P(OR)_3$
$PtL_2(C_2Cl_4)$

2.   OXIDATION.

<u>Mo, W</u>
<u>(=M)</u>
$[CpMo(CO)_2L]_2 + I_2 \longrightarrow$ $P(OR)_3$
$CpMo(CO)_2LI$
$[CpM(CO)_2L]^- + HY \longrightarrow$ $P(OR)_3$
$CpM(CO)_2LH$
$Y =$ organic acid anion
$[CpM(CO)_2L]^- +$ $P(OR)_3$
$RX \longrightarrow CpM(CO)_2LR$
$ \longrightarrow CpM(CO)_2LH +$ olefin $P(OR)_3$

<u>Mn</u>
$[Mn(CO)_4L]^- + RI \longrightarrow$ $P(OR)_3$
$RMn(CO)_4L$
$[Mn(CO)_3L_2] + H^+ \longrightarrow$ $P(OR)_3$
$Mn(CO)_3L_2H$

<u>Co, Rh,</u>
<u>Ir</u>
$[Co(CO)_3L]^- + RH \longrightarrow$ $P(OR)_3$
$[Co(CO)_3LR]$
$[Ir(CO)_2L]_2 + X_2 \longrightarrow$ $P(OR)_3$
$[Ir(CO)LX_3]_2$
$[RhL_4]^+ + H_2 \longrightarrow RhL_4H_2$ $P(OR)_3$

<u>Fe, Ru</u>
$Fe(CO)_3L_2 + 2 HgCl_2 \longrightarrow$ $P(OR)_3$
$[ClHgFe(CO)_3L_2][HgCl_3]$

$$Ru(CO)_3L_2 + I_2 \longrightarrow \qquad P(OR)_3$$
$$Ru(CO)_2L_2I_2$$
$$RuL_4X_2 + NOCl \longrightarrow \qquad P(OR)_3$$
$$RuL_2(NO)Cl_3$$

__Ni, Pt__

$$NiL_4 + CF_3COOH \longrightarrow \qquad P(OR)_3$$
$$[NiL_4H]OOCCF_3$$
$$NiL_2(CO)_2 + NOBr \longrightarrow \qquad P(OR)_3$$
$$NiL_2(NO)Br$$
$$PtL_2X_2 + X_2 \longrightarrow PtL_2X_4 \qquad P(OR)_3$$

3. SUBSTITUTION.

__Mo__

$$Mo(CO)_5L + PR_3 \longrightarrow \qquad P(OR)_3$$
$$Mo(CO)_4L(PR_3)$$

__Mn__

$$Mn(CO)_3L_2H + X_2 \longrightarrow$$
$$Mn(CO)_3L_2X$$

__Re__

$$Re(PR_3)_{2-x}L_xH_3 + X_2 \longrightarrow \qquad P(OR)_3$$
$$Re(PR_3)_{2-x}L_xX_3$$
$$x = 0 \text{ or } 1$$

__Fe, Ru, Os__

$$\text{diolefin}Fe(CO)_2L + \qquad\qquad P(OR)_3$$
$$PMe_3 \longrightarrow Fe(CO)_2L(PMe_3)_2$$
$$Ru(CO)_3L_2 + (CF_3)_2C=O \longrightarrow \qquad P(OR)_3$$
$$Ru(CO)_2L_2OC(CF_3)_2$$
$$Ru(CO)_3L_2 + C_2X_4 \longrightarrow \qquad P(OR)_3$$
$$Ru(CO)_2L_2(C_2X_4)$$
$$X = CF_3, \text{ F, or CN}$$

$$Os(CO)_2L_2Cl_2 + AlCl_3 \xrightarrow{\text{CO}}$$
$$[Os(CO)_3L_2Cl]^+ \xrightarrow[\text{AlCl}_3]{\text{CO}}$$
$$[Os(CO)_4L_2]^{2+}$$

__Co, Rh__

$$CoL_4H + CO \longrightarrow \qquad P(OR)_3$$
$$HCo(CO)L_3$$
$$Rh(CO)L_2Cl + NCS^- \longrightarrow \qquad P(OR)_3$$
$$\xrightarrow{\text{NCS}^-}$$
$$[RhL_2(NCS)]_2 \longrightarrow$$
$$[RhL_2(NCS)_2]^-$$

__Ni, Pd, Pt__

$$C_2H_4NiL_2O_2 \longrightarrow NiL_2O_2 \qquad P(NR_2)_3$$

The O_2 molecule apparently remains largely undissociated in this complex.

$$ML_2Cl_2 + I^- \longrightarrow ML_2I_2 \qquad P(OR)_3, P(NR_2)_2$$

M = Pd or Pt
$Cl(R_3P)Pt[P(R)_2O]_2Pt(PR_3)Cl +$
$OH^- \longrightarrow (HO)(R_3P)Pt[P(R)_2O]_2-$
$Pt(PR_3)(OH)$

<u>Au</u> $AuLX + RMgX \longrightarrow RAuL$ $P(OR)_3$

4. ADDITION, REMOVAL, OR HYDROLYSIS.

<u>Mo, W</u> $CpMo(CO)_2L(COR) \longrightarrow$ $P(OR)_3$
$\overline{(=M)}$ $CpMo(CO)_2LR + CO$
         $[CpMo(CO)_2L]_2Hg \longrightarrow$ $P(OR)_3$
         $[CpMo(CO)_2L]_2 + Hg$

The following reaction is an example of addition in
which insertion between two groups takes place:

         $2 [CpM(CO)_2L]^- + Hg^{2+} \longrightarrow$ $P(OR)_3$
         $[CpM(CO)_2L]_2Hg$

<u>Mn</u> $[Mn(CO)_4L]^- +$ $P(OR)_3$
         $[Au(PR_3)]^+ \longrightarrow$
         $(R_3P)AuMn(CO)_4L$
         $[Mn(CO)_{4-x}Lx]^- + Hg^{2+} \longrightarrow$ $P(OR)_3$
         $Hg[Mn(CO)_{4-x}Lx]_2$
x = 1 or 2
         $Mn(CO)_{4-x}Lx(OCR) \longrightarrow$ $P(OR)_3$
         $Mn(CO)_{4-x}LxR$
x = 1 or 2

<u>Fe, Ru</u> $[FeS_4C_4X_4]^- \longrightarrow$ $P(OR)_3$
         $[Fe(S_4C_4X_4)L]^-$

<u>Co</u> $2 [Co(CO)_2L_2]^- + Hg^{2+} \longrightarrow$ $P(OR)_3$
         $[Co(CO)_2L_2]_2Hg$
         $Co_2(CO)_6L_2 + SnCl_2 \longrightarrow$ $P(OR)_3$
         $[(OC)_3LCo]_2SnCl_2$

<u>Pd, Pt</u> $ML_2Cl_2 + NR_3 \longrightarrow$ $P(OR)_3$
$\overline{(=M)}$ $ML_2(NR_3)_2Cl_2$

5. POLYMER CLEAVAGE.

<u>Fe</u> $Hg[Fe(CO)_2(NO)L_2] +$ $P(OR)_3$
         $HgI_2 \longrightarrow 2 IHg(CO)_2(NO)L$
         $[CpFeL_2]_2 + I_2 \longrightarrow$ $P(OR)_3$
         $CpFeL_2I$

The above cleavage reaction takes place oxidatively.

<u>Co, Rh</u> $[Co(CO)_3L]_2 + NO \longrightarrow CoNO(CO)_2L$

Substitution of CO also occurs in the above reaction.

$$[Co(CO)_{4-x}L_x]_2 + Na \longrightarrow \quad P(OR)_3$$
$$[Co(CO)_{4-x}L_x]^- + Na^+$$
x = 1 or 2

$$(RhL_2Cl)_2 + \qquad\qquad P(OR)_3$$
$$2\ (MeCOCHCOMe)^- \longrightarrow$$
$$2\ [RhL_2(MeCOCHCOMe)] + 2\ Cl^-$$

Cleavage is accompanied by substitution of Cl^- in the above reaction.

$$[CuLX]_{2,3,4} + NR_3 \longrightarrow \quad P(OR)_3$$
$$CuL(NR_3)X$$

The degree of polymerization in the reactant depends on the solvent, X, and L. The product may also be polymeric.

Pd, Pt
(=M)

$$[MLX_2]_2 + NR_3 \longrightarrow \qquad\qquad P(OR)_3$$
$$ML(NR_3)X_2 \longrightarrow$$
$$ML(NR_3)_2X_2 \longrightarrow$$
$$ML(NR_3)_3X_2$$

IV.   ALTERATION OF COORDINATED L

Some rather novel routes to complexes of L have been obtained by solvolyzing, dehydrating, dealkylating, or deprotonating L species in intermediates. The last two processes convert L to a coordinated anion. Substitution of L with other ligands has also given new complexes.

1.   SOLVOLYSIS OR DEHYDRATION OF L. The earliest reported synthesis of a complex of L in 1870 contains a description of the solvolysis in water and alcohols of phosphorus trichloride complexes of platinum.[472,473] It has only been during the last two decades, however, that substantial interest in this synthetic method has been revived. It should be noted that several of the starting materials are complexes which do not contain ligands of type L. The solvolyses of these complexes are included here to minimize needless proliferation of synthetic categories. Some of the steps are aided by the presence of OR^-.

Mo

$$Mo(CO)_{6-x}[PF_2CCl_3]_x + R_2NH \longrightarrow$$
$$Mo(CO)_{6-x}[PF_2NR_2]_x + HCCl_3$$
x = 2 or 3

It is interesting to note in the preceding reaction that the CCl_3 group is solvolyzed in preference to the fluoride.

__Mn__ $Mn_2(CO)_9PF_3$ + MeOH $\longrightarrow$
             $Mn_2(CO)_9(PF_2OMe)$

                                              THF
__Fe__ $Fe(PF_3)_5$ + $Ba(OH)_2$ $\xrightarrow{\quad}$
                                     $(Et_3O)BF_4$
             $Fe(PF_3)_4PF_2O^-$ $\xrightarrow{\hspace{2cm}}$
             $Fe(PF_3)_4PF_2OEt$

This reaction is not strictly a solvolysis but the principle is the same.

__Co__ $CoNO(PF_3)_3$ + MeOH $\longrightarrow$
             $CoNO(PF_3)_2(PF_2OMe)$ $\longrightarrow$
             $CoNO(PF_3)(PF_2OMe)_2$ $\longrightarrow$
             $CoNO(PF_2OMe)_3$ $\longrightarrow$
             $CoNO[P(OMe)_3]_4$

__Ni, Pd,__ $Ni(PF_3)_4$ + MeOH $\longrightarrow$
__Pt__ $Ni[P(OMe)_3]_4$

The only example of the conversion of a coordinated L to a coordinated metal L' complex is the dehydration reaction shown below:

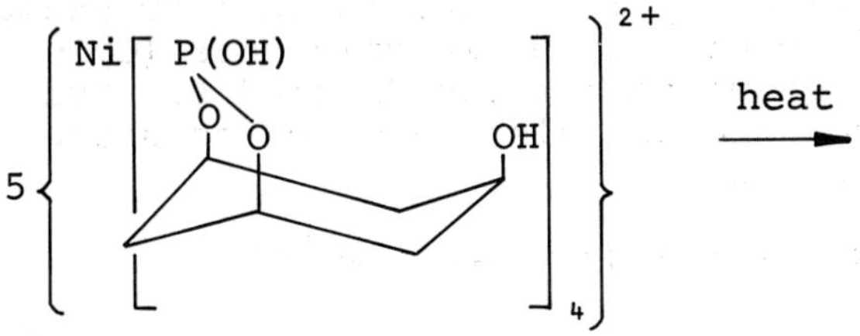

$4\ \{Ni[P(OCH)_3(CH_2)_3]_5\}^{2+}$ + Ni^{2+} + 5 H_2O
    $[M(PCl_3)_2X_2]$ + ROH $\longrightarrow$
    $M[P(OR)_3]_2X_2$
M = Pd, Pt
    $M(PCl_3)_2Cl_2$ + R_2NH $\longrightarrow$
    $M[P(NR_2)_3]_2Cl_2$
M = Pd, Pt
    $Pt(PR_2Cl)ZX_2$ + ROH $\longrightarrow$
    $Pt(PR_2OR)ZX_2$
Z = PR_3 or AsR_3

__Au__ $Au(PCl_3)Cl$ + ROH $\longrightarrow$
             $Au[P(OR)_3]Cl$

__B__ H_3BPF_3 + RNH_2 $\longrightarrow$

$$H_3BPF_2NHR \longrightarrow$$
$$H_3BPF(NHR)_2 \longrightarrow$$
$$H_3BP(NHR)_3$$

2. CONVERSION OF L TO A COORDINATED ION. Such a conversion can occur by hydrolysis of coordinated $P(OR)_3$, deprotonation of a coordinated $P(OH)(OR)_2$, or by a novel form of the Michaelis-Arbuzov reaction to form anionic ligands:

$$M[P(OR)_3] + H_2O \longrightarrow$$
$$M[P(O)(OR)_2]^- + ROH + H^+$$
$$M[P(OH)(OR)_2] + OH^- \longrightarrow$$
$$M[P(O)(OR)_2]^- + H_2O$$
$$M[P(OR)_3]Y \longrightarrow$$
$$M[P(O)(OR)_2] + RY$$

Y = nucleophile

Mild conditions generally suffice for the hydrolysis and deprotonation while heat aids the progress of the Michaelis-Arbuzov reaction.

Mo $[CpMo(CO)_2L_2]-$ $P(OR)_3$
           $[CpMo(CO)_3] \longrightarrow$
           $CpMo(CO)_2L[P(O)(OR)_2]$ +
           $CpMo(CO)_3R + CpMo(CO)_2LR$

The above reaction is reminiscent of that of $R_3SiMn(CO)_5$ and $P(OR)_3$ (see category I.1.) and is another example wherein a metalic nucleophile (in this case $[CpMo(CO)_3]^-$) can attack the R group on the coordinated L. Evidently some form of rearrangement also takes place since $CpMo(CO)_2LR$ is formed.

Ru $HRu(PR_3)_3X + P(OR)_3 \longrightarrow$ $P(OR)_3$

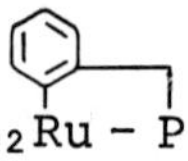

           $L_2Ru - P(OR)_2X + H_2$

R = aryl

Although this reaction does not strictly fall into this category, the conversion of L to a coordinated anion may be thought of as taking place after initial coordination of the triaryl phosphite. The anionic coordinated L is bidentate.

                                      heat
Ir $Ir[P(OPh)_3]_3HCl_2 \longrightarrow$
           $Ir[P(OPh)_3][P(OPh)_2(OC_6H_4)]_2Cl$

Pt $PtL_2(NH_3)_2Cl_2 \longrightarrow$ $P(OR)_3$
           $PtL[P(O)(OR)_2](NH_3)Cl$ +
           $RCl + NH_3$

$$PtL(NH_3)_3Cl_2 \longrightarrow \qquad P(OR)_3$$
$$Pt[P(O)(OR)_2](NH_3)_2Cl +$$
$$RCl + NH_3$$
$$PtL_2Cl_2 \longrightarrow \qquad P(OR)_3$$
$$PtL[P(O)(OR)_2]Cl$$

Only one case is known where a coordinated L is converted to a cationic ligand. The complex $(OC)_5WP(OCH_2)_3P$
reacts with Me_3OBF_4 to give $[(OC)_5WP(OCH_2)_3PMe]BF_4$.

3. SUBSTITUTION OF L. The ligands (L) we are considering are susceptible to replacement by neutral groups
such as CO or NR_3 and by electronegative groups such as
X^- in which case oxidation can be considered to occur.

Ru

$$RuL_4X_2 + CO \longrightarrow \qquad P(OR)_3, \; P(NR_2)_3,$$
$$RuL_3(CO)X_2 \qquad\qquad\qquad P(OR)_xX_{3-x}$$
$$\qquad\qquad\qquad oxid.$$
$$RuL_4X_2 + NOCl \longrightarrow \qquad P(OR)_3$$
$$RuL_2(NO)Cl_3$$

Rh

$$[RhL_4]BPh_4 \longrightarrow \qquad P(OR)_4$$
$$[RhL_2BPh_4]$$

The preceding reaction is most unusual in that a phenyl
group on the BPh_4 anion becomes coordinated by displacing
L.

Ni, Pd,
Pt

$$NiL_4 + RNC \longrightarrow \qquad P(OR)_3$$
$$NiL_2(CNR)_2$$
$$PdL_2X_2 + NR_3 \longrightarrow \qquad P(OR)_3$$
$$PdL(NR_3)Cl_2$$
$$\qquad\qquad\qquad oxid.$$
$$PtL_4 + HF \longrightarrow \qquad P(OR)_3$$
$$PtL_2F_2$$
$$\qquad\qquad\qquad chel.$$
$$PtL_2[P(O)(OR)_2]_2 \longrightarrow \qquad P(OR)_2OH$$
$$Pt(chel)[P(O)(OR)_2]_2$$
$$chel = Ph_2PCH_2CH_2PPh_2$$

B. STRUCTURE AND PROPERTIES

The reasons cited most often for the syntheses of complexes
of L have been the study of their bonding, structural, and
kinetic properties. A review in depth of the results of
metal-phosphorus bonding studies is not within the scope
of this chapter since such a discussion would necessarily
include all the spectroscopic investigations done with PR_3
ligands as well. A critical treatment of all the kinetic
results would be similarly impossible here.

We shall therefore be concerned in this section with
the overall structural, chemical, and physical properties
of the complexes. Because these characteristics depend
highly on the acceptor atom involved, they will be discussed by chemical family where possible and by element
where necessary. Much of the chemistry of the complexes
appeared in the section on synthetic routes (Section A).
Thus chemical reactions of complexes of L to form new coordination compounds of L will not be repeated here. Only
general characteristics of the complexes will be given for
the most part while more specific details can be found in
the compilation and in the accompanying references. Occasionally tables will be used to summarize properties of
the complexes, but it should be kept in mind that the information in these tables is only a rough indicator to the
properties inasmuch as several data are not available for
most complexes.

Every compound appearing in the compilation is not
necessarily accounted for in the discussion below. The
criteria for inclusion used in this section are that the
compound must have been structurally characterized to a
reasonable degree or that it must be analogous to a compound which has met this requirement. Admittedly, a certain amount of subjectivity is associated with these criteria. In the relative absence of diffraction studies,
much emphasis has been placed upon infrared and NMR structural studies of phosphorus complexes. In order to avoid
a great deal of repetition, some generalizations concerning certain typical spectroscopic and physical characteristics will now be given.

Because most of the complexes of L also contain carbonyl groups, much information on the symmetry of the complex can be gained from the number and intensities of the
CO stretching frequencies in the 2000 cm^{-1} region of the
infrared spectrum. The assignment of these bands for the
various ligand-substituted carbonyl geometries has been
reviewed elsewhere.[160] In the discussion below, the number of expected and observed bands when indicated will
not be further justified. A reasonably general
rule for the influence of the phosphorus ligand L on
the CO stretching frequency is that it rises as the electronegativity of the substituents on phosphorus increases.
The reasons for this are still held to be controversial
and will not be reviewed here. Other group frequency patterns (M-H, M-X, M-OH, etc.) can also yield important information on the overall geometry as well as an indication
of the particular isomer in certain instances and these
will be discussed where appropriate. The presence of some
groups which are often encountered in complexes of L is
revealed by a characteristic band in the near infrared.
Thus MCOR($\sim$C-O, 1650 cm^{-1}), M-H($\sim$2000 cm^{-1}), CN($\sim$2000 cm^{-1})

groups can generally be detected in the absence of inter-
fering bands.
    The nuclear magnetic resonance method has been par-
ticularly useful in arriving at proper formulations and is
becoming increasingly effective in revealing the isomerism
present in a complex. The degree of substitution, for
instance, can often be quickly determined by integrating
proton spectra if the metal substrate retains a group containing hydrogen. Moreover the chemical shifts of protons
on the free ligand generally move downfield to varying
degrees (depending on the distance from the donor site)
when coordinated to a neutral or positively charged acceptor and upfield when attached to anions. Except for hydrogen itself, proton NMR absorptions in most organic groups
attached to the metal do not change markedly on coordination. Coordinated hydrogen absorptions are found up to
20 ppm above TMS and the splittings due to coordinated
phosphorus (^2JPMH) can be up to 185 Hz. There seems to
be no general trend concerning the movement of the ^{31}P
chemical shift upon coordination of L although consistent
upfield shifts with increasing positive charge have been
noted in complexes which are both isoelectronic and isostructural.[142]
    Three-bond splittings of the type ^3JPYCH (where Y =
N, O, or S) appear to be insensitive to coordination except in polycyclic ligands of the type shown in (1) and
(2) where they increase to more than twice the free-ligand
value in a positive manner upon coordination of L while
^2JPCH interactions very probably approach zero and can become negative.[52] It will be noted in the tables that
^3JPYCH seems to vary rather widely, particularly in the
case where Y = O (2 to 15 Hz). The reason for this lies
in the relationship between this coupling and the POCH
dihedral angle. In the two phosphites below, for example,
^3JPOCH increases from ∿2 Hz in the free ligand (1) (POCH
angle = 60°) to ∿5 Hz in a complex while in (2) (POCH
angle = 180°), the corresponding values are ∿6 to ∿15 Hz.
There appears to be a linear correlation between ^3JPOCH
and the chemical shift of the affected hydrogen in a variety of carbonyl complexes and adducts of (1)[405] and
(2).[520b]

```
         O — CH₂
        /        \
   P — O — CH₂ — CR
        \        /
         O - CH₂

          (1)
```

(2)

A phenomenon based on the virtual spin-spin coupling
of chemically equivalent phosphorus nuclei has been utilized quite advantageously in assigning cis and trans

positions to two chemically equivalent coordinated L
molecules in complexes where other methods do not resolve
the question.   It has been known for some time that such
ligands in trans positions yield ^{1}H spectra which usually
show a 1:2:1 or nearly 1:2:1 "triplet," and when they are
cis only the expected doublet or a poorly resolved "triplet" is observed.[280a]   The theory associated with such
spectra has been reviewed recently.[397c]   The method is
not without pitfalls[405] but it has been remarkably suc-
cessful in spite of the lack of an adequately detailed
understanding of its theoretical basis.[54,405]   An exten-
sive study of the signs and magnitudes of virtually coupled phosphorus nuclei in cis and trans $M(CO)_4L_2$ complexes
(where M = Cr, Mo, or W) showed that the sign is always
negative in the cis complexes ranging from -10 to -80 Hz
while in the trans systems the vast majority of the signs
were positive with coupling values from -40 to +320 Hz.[54,405]
In both cases a rough correlation of $^2J_{PMP}$ with the
electronegativity of L was observed and interpretations
of the signs and trends were attempted.

Other empirical NMR structural criteria which seem to
be emerging with more or less reliability are the larger
$^2J_{PMH}$ couplings in trans octahedral hydride complexes
compared to cis[116b] and the larger $^3J_{PMCH}$ couplings involving the Cp protons in complexes having structure (3)
than in structure (4)[146,304,355,356] (see later discussion).   When Y = H, larger couplings to the phosphorus in
(3) ($^2J_{HMP}$) are observed when H and L are trans[370] than
when they are cis.[33,304,356]   It should be noted that an
exception to the general rule that $^2J_{PMH}$ is larger in
trans octahedral hydrides than in the cis analogs has recently appeared.[497b]

(3)   (4)

There are a few generalizations which can be made concerning the physical properties of the complexes insofar
as they depend on L.   For a given series of complexes
varying only in L, the color often becomes more intense as
the electronegativity of the ligand decreases.   Thus $P(OR)_3$
and $P(OR)_xX_{3-x}$ complexes tend to be colorless or light
yellow.   This is perhaps in large measure due to the high
ligand field properties of these ligands.[524,525]   Second,
halogenated phosphorus ligands and those with R groups

larger than Me tend to give complexes which are low-melting solids or liquids while $P(OMe)_3$ or $P(NMe_2)_3$, for example, generally yield crystalline products. In contrast, polycyclic phosphorus ligands such as (1) and (2) tend to form high-melting complexes which is probably a result of their unusually high dipole moments and their highly ordered structures.[235,520a] There are also strong indications that the nature of the phosphorus-metal interactions in polycyclic ligands is different from those operative in analogous complexes of open-chain phosphorus systems.[520b]

The discussion in this section again follows the periodic chart by subgroup beginning with vanadium. In general the pattern within each subgroup will be to begin with complexes in which the element treated is in the zero-valent and negative one oxidation state and then to progress to more positive oxidation states. The oxidation state of the metal is obvious in most cases, bearing in mind our previous definitions. It remains, however, to point out our criterion for defining the oxidation states where NO groups are involved. Although this fragment is often considered to possess a formal monopositive charge, there is evidence that it can be negative.[269a] We shall consider the NO group to be electron withdrawing, and therefore formally it causes the oxidation state of the metal to increase positively. Thus we consider $Co(NO)_2$-$[P(OR)_3]Cl$ as a Co(III) complex. This is done to maintain consistency with the synthetic routes in the previous section. Thus the reaction of NOCl with RuL_4X_2 to give $RuL_2(NO)Cl_3$ is viewed as an oxidation in the same way as is the reaction of PtL_2X_2 with X_2 to give PtL_2X_4.

It will be helpful to recall some generalizations concerning coordination numbers in metal carbonyl complexes. Carbonyl metal compounds, much more often than not, adopt the electronic configuration of the next inert gas. Chromium and nickel carbonyl, for example, must have the formulas $Cr(CO)_6$ and $Ni(CO)_4$, respectively, since zerovalent chromium has six electrons and nickel eight. Six and four CO groups must coordinate to the respective metals for then the sum of the electrons is 18 assuming each CO contributes two electrons. Coordinated groups such as PR_3, NR_3, OR_2, olefin, etc. can be considered to contribute two electrons each while halogens and NO contribute two and four, respectively. Using these simple rules it is easy to see why $Co(NO)_2[P(OR)_3]Cl$ is a reasonable formula. The Co(III) has six electrons and the sum of the electrons from all the ligands is 12, giving a total of 18 electrons.

The coordination numbers most often enountered are two, four, five, and six. Corresponding structures which are well-established are linear, tetrahedral (or square planar), trigonal bipyramidal, and octahedral, respectively.

The choice between square planar and tetrahedral is generally not difficult in the systems we will cover although
our treatment here is somewhat formalistic. When the complex is diamagnetic, the geometry can be predicted by considering all coordinated R groups and halogens as negatively charged and therefore as two-electron donors. An
NO group can also be considered here as negative (in keep-
ing with our oxidation state discussion) but then formally
it becomes a four-electron donor. A few examples will
illustrate the procedure. If the manganese atom in
$Mn(NO)_3L$ were truly trivalent, it would have four electrons
in its 3d orbitals. If we allow two of the four available
electrons on each NO group to formally occupy the remaining 3d orbital vacancy, the four ligand groups together
would donate four pairs of electrons to the 4s and 4p
orbitals yielding an sp^3 tetrahedral hybrid. Similar arguments apply to divalent iron $(3d^6)$ in $Fe(NO)_2(CO)L$, trivalent cobalt $(3d^6)$ in $Co(NO)_2LX$, and monovalent cobalt
$(3d^8)$ in $Co(NO)(CO)_2L$. In cases such as $Ni(CO)_2L_2$ and
CuL_4^+, the tetrahedral geometries are obviously attained
via the available 4s and 4p orbitals since the metal electrons fill the 3d subset. Diamagnetic complexes in which
the metal contains eight d electrons [e.g., rhodium(I) and
platinum(II)] and in which only two-electron donors are
present are square planar since the metal hybrid set involving the ligands then includes one d, an s, and two p
orbitals leading to dsp^2 square planar geometry.

The paramagnetic four-coordinate complexes we shall
encounter are those of cobalt(II) (d^7) and nickel(II) (d^8).
The hybridization approach leads to sp^3 involvement for
the ligand electrons donated since the metal electrons reside in the 3d orbitals. Thus cobalt(II) complexes will
possess three unpaired d electrons and nickel(II) two. In
the simple crystal field model, the d orbital splittings
of the divalent metals in a tetrahedral field are those
shown below wherein the energy difference Δ is a measure

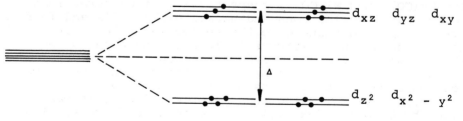

M^{2+} surrounded by
spherical ligand
field at large distance

CoL_4^{2+} NiL_4^{2+}

of the crystal field strength offered by the ligands. The
crystal field stabilization energy for complexes is aug-

mented by filling orbitals below the degenerate level of the spherical ligand field. Thus if the field is sufficiently strong, a square-planar array as shown below for a group of nickel(II) complexes allows six of the electrons to reside in stabilizing levels. While dsp^2 hybrid orbitals on the metal accommodating the ligand lone pairs in covalent bonds would lead to the same conclusion, the concept of a ligand field leads to a more satisfying interpretation of the visible spectra. Although the hybridization and crystal field formalisms represent extremes in a bonding picture which in any case is far more accurately represented by molecular orbital approaches, they are sufficient for our purpose.

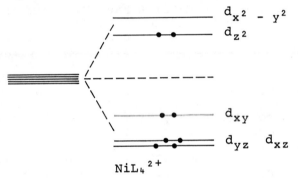

$$NiL_4{}^{2+}$$

(1) V

The only complexes of vanadium are of the type $CpV(CO)_3L$. The orange complex where $L = P(NMe_2)_3$[297] exhibits four CO bands in the IR while the red oils or yellow solids (m. 114-164°C) where $L = P(OR)_3$[174,493b] show one to three CO bands from 1839-1983 cm^{-1}. The IR and NMR data are consistent with the existence of a pyramidal arrangement of the ligands although diffraction work has not been carried out.

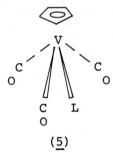

(5)

(2) Cr, Mo, W

Zerovalent carbonyl derivatives of the type $M(CO)_{6-x}L_x$ (M = Cr, Mo, or W, and x = 1-4) are octahedral. When x = 2-4, the isomers shown in (6) to (11) are possible and have been characterized.

```
    L             L             L             L             L             L
OC──┬──L     OC──┬──CO    OC──┬──L     OC──┬──L     OC──┬──L     L──┬──CO
OC──┴──CO    OC──┴──CO    OC──┴──L     OC──┴──CO    OC──┴──L     OC──┴──L
    C             L             C             L             L             L
    O                           O
   (6)           (7)           (8)           (9)          (10)          (11)
   cis          trans       fac(facial)      mer           cis          trans
                            (meridional)
```

Complexes of the type $M(CO)_5L$ have C_{4v} symmetry and characteristically exhibit three CO bands unless the E mode degeneracy is split in which case four are observed. The ranges for these bands along with other pertinent properties are summarized in Table 1. The structure of $Cr(CO)_5P(OPh)_3$[427] determined by x-ray crystallography reveals that the CrP bond (2.309 Å) is appreciably shorter than in the analogous PPh_3 complex (2.422 Å). ^{1}JWP values in the tungsten complexes vary from 276 to 416 Hz and appear to be a linear function of the electronegativity of the Y substituent.[293]

The difunctional ligand $P(OCH_2)_3P$ possesses two coordination sites and therefore its complexes can exist as linkage isomers.[52,520b] The compounds $M(CO)_5P(OCH_2)_3P$ (M = Cr, Mo, or W) are white and show POCH coupling constants and ^{31}P chemical shifts for the coordinated phosphorus similar to those found for $M(CO)_5P(OCH_2)_3CMe$.[516,522] The ligand-bridged complexes, $(OC)_5MP(OCH_2)_3PM(CO)_5$ (M = Cr, Mo, or W), were similarly characterized as well as the novel charged ligand complex cation $[W(CO)_5P(OCH_2)_3-PMe]^+$.[52]

Disubstituted $M(CO)_4L_2$ complexes occur as cis and trans isomers whose properties are given in Table 2. Although three bands are expected for the trans D_{4h} symmetry, often only one band is observed as the two higher frequency modes are weak. Mixed trans complexes of the type $M(CO)_4(PR)_3-[P(OR)_3]$[209,405] where M = Cr, Mo, or W as well as those of the type $M(CO)_4[P(NR_2)_3][P(OR)_3]$[405,406] have also been made which are colorless solids. The solid cis complexes $Mo(CO)_4LP(OCH_2)_3CMe$ where L = $P(NMe_2)_3$[406] or $P(NMe_2)_2CN$[144b] as well as cis-$Mo(CO)_4[P(NMe_2)_2CN][P(OMe)_3]$[144b] have also been characterized.

The difunctional ligands $EtN(PF_2)_2$ with chromium, molybdenum, and tungsten, and $RN(PPh_2)_2$ with molybdenum yield cis complexes as expected from the configurations of the ligand. The former complexes are colorless liquids or

Table 1. Properties of M(CO)$_5$L Complexes

M	L	Color	m. or dec. (°C)	ν_{CO} (cm^{-1})	δ^{31}P[a]	^{3}JPYCH[b]
Cr	P(OR)$_3$	White	58–218	1955–2082	−92 to −180	4.1–12.6
	P(NR$_2$)$_3$	White	145	1932–2055		10.1
	P(OMe)$_2$Me	Yellow	6	1947–2071		11.5
	P(NMe$_2$)$_2$Me	Yellow	74	1938–2061		10.0
Mo	P(OR)$_3$	White	Glass to 210	1952–2085	−93 to −162	4.2–12.7
	P(NR$_2$)$_3$			1938–2066		10.2
	P(OR)$_2$R	White, Yellow	Liquid to 54	1946–2078		12.0
	P(OR)X$_2$			1975–2091		
	P(NMe$_2$)$_2$Me		45	1946–2070		10.5
W	P(OR)$_3$	Yellow	69–228	1944–2088	−93 to −137	4.2–12.8
	P(NR$_2$)$_3$	White	83–84	1903–2072	−112	4.8–15.6
	P(SR)$_3$	White	158–160	1931–2081	−31.1	4.0–12.95
	P(OR)xR$_{3-x}$	White or yellow		1945–2077		12.0
	P(NMe$_2$)$_2$Me	Yellow		1928–2069		11.0

[a] Values relative to external 85% H$_3$PO$_4$. Coupling values are in hertz.
[b] Y = O, S, N as appropriate to the ligand L.

31

Table 2. Properties of $M(CO)_4L_2$ Complexes

M	L	Isomer	Color	m. or dec. (°C)	ν_{CO} (cm^{-1})	$\delta^{31}P$[a]	$^3J_{PYCH}$[b]
Cr	P(OR)₃	Cis	White	278-332	1909-2043	-164	4.2-11.6
		Trans	White or green	Liquid, 148	1872-2030	-172	4.1-11.3
	P(NR₂)₃	Cis			1900-2015		
		Trans			1868	-178	9.7-9.8
	P(OMe)₂R	Cis-trans[c]	Yellow	99-101	1902-2014		10.9, 11.1[c]
	P(OMe)₂Me	Cis	Yellow	75	1909-2022		11.5
	P(NMe₂)₂Me	Trans	White	88	1883		10.0
	P(NMe₂)₂CN	Trans	Yellow	126	1910		11.8
Mo	P(OR)₃	Cis	White to yellow	Glass or solid	1920-2050	-164	4.3-11.0
		Trans	White	Glass, 267	1921-2055	-174	4.2-11.6
	P(NMe₂)₃	Cis	Yellow	96	1882-2125	-154	9.2-10.9
		Trans	Yellow		1885		10.0-10.2
	P(OEt)Cl₂	Cis			1967-2063		
	P(OMe)₂Ph	Trans	White	90	1908-2028		11.6

	Cis/Trans	Color				
P(OMe)₂Me	Cis	White	67	1921–2031		12.0
P(OEt)₂CN	Cis	White		1970–2065		7.0
P(NMe₂)₂Me	Trans	Yellow	88	1894		10.5
P(NMe₂)₂CN	Cis	White	87	1930–2035		11.6
W P(OR)₃	Cisᵈ	White	117–272	1917–2053		4.4
	Transᵈ	White	312	1924–1931	−133	4.3
P(NR₂)₃	Cis			1866–2004		
	Trans	Yellow		1870	−134	10.4
P(OR)₂R	Trans	White	96–98	1901–228		12.4
P(OMe)₂Me	Cis	White	49	1912–2028		12.0
P(NMe₂)₂Me	Trans	Yellow	91	1886		10.5
P(NR)F₂	Cis	Colorless	Liquid, 38	1942–2057	−180 to −181	11.2

ᵃValues in ppm relative to external 85% H₃PO₄.

ᵇy = O or N, as appropriate. This J value (Hz) also contains an additional contribution from ⁵JPMPYCH – see previous discussion.

ᶜMixture.

ᵈSome of these compounds exist as cis-trans oily mixtures which were not separated.

low-melting solids while the latter compounds are solids melting from 202 to 210°C. Most of them exhibit the four CO bands (1887-2068 cm^{-1}) required for their C_{2v} symmetry. The cis structure in the case of Mo(CO)$_4$(PPh$_2$)$_2$NEt was confirmed by x-ray diffraction work.[416] The difunctional ligand P(OCH$_2$)$_3$P is unable to coordinate in a chelating fashion and gives rise to three possible cis linkage isomers of which Cr and Mo(CO)$_4$[P(OCH$_2$)$_3$P]$_2$ and Mo(CO)$_4$-[P(OCH$_2$)$_3$P][P(CH$_2$O)$_3$P] have been isolated.[52,520b] All are white and show only three of the expected four bands from 1938-2048 cm^{-1}. Only one proton chemical shift for the first two compounds is seen while two are seen for the last one. In the first two linkage isomers the ^3J(POCH) couplings are similar to those observed in the analogous P(OCH$_2$)$_3$CMe complexes while in the last a nearly composite spectrum of a coordinated P(CH$_2$O)$_3$CMe and P(OCH$_2$)$_3$CMe ligand is observed.[52] The ^{31}P chemical shifts for these compounds also support the existence of the linkage isomers. The mixed-ligand cis complexes Mo(CO)$_4$LP(OMe)$_3$ and Mo(CO)$_4$LP(NMe$_2$)$_3$ where L = P(OCH$_2$)$_3$CEt have also been identified.[406]

Some rather subtle factors, only part of which may be steric, seem to govern the formation of cis and trans isomers. Whereas trans isomers of P(OCH)$_3$(CH$_2$)$_3$ (2) are made with all three metals,[237,51b] only the trans chromium complex could be made with P(OCH$_2$)$_3$CMe (1)[516,522] which is only slightly less bulky. Furthermore, cis isomers can be made for all three metals with both ligands. Merely substituting a propyl group for the methyl at the end of ligand (1), however, allows formation of cis-M(CO)$_4$-[P(OCH$_2$)$_3$CPr]$_2$ [488] where M = Mo and W but only the trans chromium analog has been isolated. A more bizzare illustration of the questionability of steric arguments is the ease of the isomerization of cis-Mo(CO)$_4$[P(NMe$_2$)$_3$]$_2$ to the trans form compared to the remarkable stability of the cis P(NMePh)$_3$ analog.[405]

Preliminary evidence has been obtained for the novel dimers having the formula (OC)$_4$M[Q$_2$P(CN)]$_2$M(CO)$_4$ where M = Cr or Mo, and Q = EtO or Me$_2$N. The ligands apparently bridge the metals in cis positions by means of the nitrogen on the cyano group.[144b]

Trisubstituted complexes of the type M(CO)$_3$L$_3$ occur as fac or mer isomers (Table 3). The mer (C_{2v}) isomers generally show three CO bands and the fac (C_{3v}) forms two unless the E mode of the latter is resolved. Support for the mer configuration has been given to the P(OMe)$_3$ and P(OEt)$_3$ complexes for which two ^{31}P resonances have been distinguished.[322] The fac isomers on the other hand reveal only one ^{31}P absorption[322] but this could conceivably arise from rapid ligand exchange. In several cases fac-mer mixtures are suspected because of more than the

Table 3. Properties of $M(CO)_3L_3$ Complexes

M	L	Isomer	Color	m. or dec. (°C)	ν_{CO} (cm^{-1})	$\delta^{31}P$ [a]	^3JPYCH [b]
Cr	$P(OR)_3$	Fac	Yellow	Liquid, 126	1795-2008		4.4-4.7
	$P(OR)_2R$	Mer	Yellow	86-88	1866-1969		11.1, 11.4
	$P(OMe)_2Me$	Fac	Yellow	152	1871-1960		11.0
		Mer	Yellow	152	1877-1972		11.0
Mo	$P(OR)_3$	Fac	White	Glass, 142	1790-1998		4.3-10.9
		Mer	White	Glass, 142	1877-1993	-166	10.8, 11.1
	$P(OR)_2R$	Mer	White	121	1880-1980	-164 to -177	10.8 [c]
	$P(OMe)_2Me$	Fac	White	171	1891-1970		11.0
	$P(OR)_xX_{3-x}$	Fac	White	Liquid, 47-90	1930-2041	-150 to -171	
		Mer			1903-2042		
W	$P(NR_2)_xX_{3-x}$	Fac	White	Oil, 108	1860-2000	-177 to -182	
	$P(NR_2)RF$		White	127-139	1858-1970	-186	
	$P(OR)_3$	Fac	White	130-180	1845-1996	-123	4.5-4.7
	$P(OR)X_2$						
	$P(OMe)_2Me$	Fac	Yellow		1885-1966		11.0
		Mer	Yellow		1899-1984		11.0

[a] Values in ppm relative to external 85% H_3PO_4.
[b] See discussion of virtually coupled spectra in text.
[c] Only one doublet resolved due to accidental overlap.

expected number of CO stretching bands.

The tetrasubstituted compounds $M(CO)_2L_4$ have been found only for molybdenum and here L can be $P(OR)_3$,[429],[430] $P(OEt)Cl_2$,[430] or $P(NR_2)F_2$.[463] Examples of cis (C_{2v}) forms exist for all three types of ligands and a trans (D_4h) form (ν_{CO} = 1970 cm^{-1}) has been isolated in the case of $P(OPh)_3$.[200] All are white solids (m. or dec. 142-180°C) and the cis isomers show the expected two CO bands (1856-1998 cm^{-1}).

The vast majority of complexes of the type chelM(CO)$_3$L where M = Cr, Mo, or W, and L = $P(OR)_3$ have the fac configuration (12). The chelates can be of the o-phenanthroline, bipyridyl, or α,ω-alkane diamine or dithiane types.[17,151,198,274a,482] Most of these compounds have not been isolated and have only been detected in solution by their characteristic three-band spectra in the carbonyl region (1740-1692 cm^{-1}). The o-phenanthroline and bipyridyl complexes are purple-black or red solids having m. or dec. points from 180 to >325°C. The mer isomers show only two CO stretching bands (1862-1988 cm^{-1}) and must have structure (13) since the chelate is unable to span trans positions on the octahedron. The ligand HP(O)(OPh)$_2$ apparently rearranges to form fac-chelMo(CO)$_3$[P(OH)(OPh)$_2$][274a]

(12)   (13)

which possesses the three expected CO bands (1807-1944 cm^{-1}).

Except for one mer isomer which was isolated, the remaining few were detected by IR spectroscopy in solution, usually as mixtures. The solid yellow mer-Mo(CO)$_3$-(Ph$_2$PCH$_2$CH$_2$PPh$_2$)[P(OEt)$_3$][158,209] showed three ^{31}P chemical shifts which is consistent only with the unsymmetrical mer structure shown in (13).

Although complexes of the type chelM(CO)$_2L_2$ can exist as three possible isomeric structures (14-16), only (14) is believed to have been observed where the chelate is o-phenanthroline or bipyridyl, L = $P(OR)_3$, and M = Cr, Mo, W. The trans arrangement of the L groups arises from kinetic and solubility considerations. All of the complexes are purple-black solids having two characteristic CO stretching bands from 1739-1874 cm^{-1}. A red Mo(CO)$_2$-(bipyridyl)[P(OH)(OPh)$_2$]$_2$[274a] of structure (14) is also known. The yellow Mo(CO)$_2$[P(OEt)$_3$]$_2$(Ph$_2$PCH$_2$CH$_2$PPh$_2$)[158] may have either structure (14) or (15).

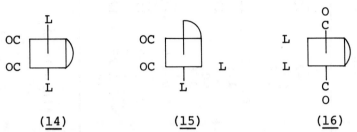

(<u>14</u>)   (<u>15</u>)   (<u>16</u>)

It is not possible to distinguish from the four bands in the carbonyl stretching region (1820-1973 cm^{-1}) the isomeric form of [CpM(CO)$_2$L]$_2$Hg (M = Mo or W)[304,370] in which each half of the molecule could have either a cis or trans relationship between the Hg and L [P(OR)$_3$] substituent. One of the yellow compounds does exhibit a 1-Hz PMCH coupling which may be indicative of a trans arrangement[370] (see below).

The anion [CpM(CO)$_2$L]$^-$ (M = Mo or W)[304,356,370] exhibits two carbonyl bands (1721-1834 cm^{-1}) indicative of · a structure like its manganese analog (see later).

The formally monovalent Cp(OC)$_3$MoMo(CO)$_2$[P(OPh)$_3$]Cp[221] and CpL(OC)$_2$MoMo(CO)$_2$LCp dimers [where L = P(OR)$_3$] are red or red-purple solids (m. or dec. 96-176°C),[219a,b,370] the former displaying five CO bands (1837-1976 cm^{-1}) while the latter show two such bands (1850-1883 cm^{-1}) and two equivalent Cp protons in the ^{1}H NMR spectra. The data for the latter complexes are not inconsistent with the C$_2$h structure shown (<u>17</u>) or the isomeric form in which Cp is cis to

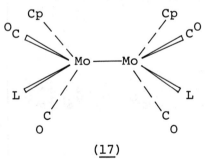

(<u>17</u>)

L on each metal unit. The monosubstituted P(OPh)$_3$ complex probably has a similar overall geometry.

The formally divalent yellow Cp(OC)$_2$Mo[P(OPh)$_2$]$_2$Mo-(CO)$_2$Cp[221] probably contains Mo-P(OPh)$_2$-Mo bridges as well as a metal-metal bond. This compound can be formally considered as a dimer of the trivalent phosphorus compound P(OPh)$_2$Mo(CO)$_2$Cp.

Tungsten(II) and molybdenum(II) complexes of the type CpM(CO)$_2$P(OR)$_3$Y, where Y = the groups shown in Table 4, occur in both possible isomeric forms, (<u>3</u>) and (<u>4</u>), and

Table 4. Properties of CpM(CO)$_2$[P(OR)$_3$]Y Complexes

M	Y	Isomer	Color	m. or dec. (°C)	ν_{CO} (cm^{-1})	$^3J_{PMCH}$[a]	$^3J_{POCH}$
W	Halogen	Mix.	Red or orange		1884-1972	2	12
	Me	Mix.	Yellow		1855-1942	1-2	12
	H	Cis		110-226	1842-1971	~0	
Mo	Halogen	Trans	Red	50-81	1908-1983	1.8	
		Cis	Red	98-154	1891-1996	~0	
	R	Mix.	Yellow	97-113	1867-1968	~0, 1-2[b]	12
		Trans				1.5	10
	H	Mix.	Colorless	Oil	1877-1957	~2	12
		Cis	Buff	42-188	1878-1974	~0	
	SnR$_3$	Trans	White	181-211	1855-1931	~0.8	
	SnX$_3$	Trans	Yellow-red	169-215	1904-1984	1	12
	RCO	Trans	Yellow	Oil, 171	1860-1983	1.0-1.5	5-10
	HgX	Trans	Yellow		1872-1940	1-2	12
W	Halogen	Mix.	Red to orange		1884-1972	~0, 1-2[b]	12
	Me	Mix.	Yellow		1855-1942	~0, 1-2[b]	12
	H	Mix.	Yellow		1866-1954	~0, 1-2[b]	12
		Cis		110-226	1842-1971	~0	

[a]Couplings in Hz of phosphorus to the protons on Cp.
[b]The near-zero value applies to the cis while the 1-2 Hz range refers to the trans isomer.

most of them are mixtures. The assignment of the isomerism in compounds of this type stems from the finding that cis complexes give two strong CO bands of nearly equal intensity while the high-frequency band in the trans will generally be of lower intensity than the low-frequency band. In the great majority of compounds it is also found that the coupling ^{3}JPMCH involving the Cp protons is small but detectable ($\sim$1-2 Hz) in the trans isomer and unresolved in the cis. The properties of this wide range of compounds are summarized in Table 4. Several complexes found in the literature with unassigned stereochemistries were included in Table 4 when infrared and NMR data warranted. The complex CpMo(CO)$_2$[P(NMe$_2$)$_3$]I[297] also occurs as a cis-trans mixture of isomers. Although the compound CpMo(CO)$_2$-[P(OPh)$_3$]O$_2$CCF$_3$ seems to be cis on the basis of the IR and NMR criteria, no evidence is presently available for elucidating the isomerism in the recently obtained CpMo(CO)$_2$-[P(OR)$_3$][P(O)(OR)$_2$] systems.[219a,b] The anion formulation is substantiated by the presence of a P=O stretching band at about 1160 cm^{-1}. Insufficient IR and NMR data for a soluble and insoluble isomer of CpMo(CO)$_2$[P(OPh)$_3$]SnMe$_3$[304] prevent the assignment of structures at present.

Relative intensities of the two CO bands indicates a trans structure for areneMo(CO)$_2$[P(OR)$_3$](COR) where arene = π-indenyl anion.[225]

It is possible to obtain the molybdenum(II) species {CpMo(CO)$_2$[P(OR)$_3$]$_2$}$^+$ as BPh$_4$$^-$ or [CpMo(CO)$_3$]$^-$ salts and the relationship of the intensities of the two CO bands is suggestive of a trans configuration.[219a,b]

The complex CpMo(CO)[P(OR)$_3$]$_2$Y where Y = I[220] or SnCl$_3$[371] gives no JPMoCH coupling and its lone CO band (1873-1895 cm^{-1}) gives no information on the arrangement of the CO, L, or Y groups.

Because the arene ion C$_7$H$_7$$^-$ possesses eight π electrons, only three monodentate groups are accommodated in the brown complex C$_7$H$_7$Mo(CO)[P(NMe$_2$)$_3$]I[297] and only one CO band is observed at 1915 cm^{-1}. Its structure is not known.

A singlet and doublet (JPMoCH = 3 Hz) NMR absorption for the Cp protons in CpMo(NO)I$_2$P(OPh)$_3$[299] is consistent with the presence of two isomers. Molecular weight and conductivity data indicate that P(OPh)$_3$ is dissociated in nonpolar solvents while I$^-$ is liberated in polar solvents.

(3) Mn

The dinuclear L(OC)$_4$MnMn(CO)$_4$L where L = P(OR)$_3$[123,332-334] is believed to have the diaxial structure (18) with staggered carbonyl planes on the basis of the CO stretching frequency data and the binary combination data in the 4000 cm^{-1} region. The yellow compounds melting at 80-147°C exhibit two strong characteristic bands in the CO stretch-

(18)

ing region from 1968 to 2000 cm^{-1} but can show as many as eight, most of which are weak. The yellow P(NMe$_2$)$_3$[302] complex probably has the same configuration. The related compound (RO)$_3$P(OC)$_4$MnHgMn(CO)$_4$P(OR)$_3$[250,412] from a simi- lar study of the CO fundamentals in the 1954-2055-cm^{-1} range and the 4000-cm^{-1} overtone region was also concluded to be D$_4$d. The monosubstituted P(OMe)$_x$F$_{3-x}$ complexes prob- ably have a C$_4$v structure.[173] Although {Re(CO)$_3$[P- (OPh)$_3$]$_2$}$_2$[411] is diamagnetic in the solid, its paramagnet- ism in solution may be caused by partial dissociation into monomeric species. Presumably the completely substituted {Re[P(OPh)$_3$]$_5$}$_2$ is also of D$_4$d symmetry.[188]

All that can be said concerning the two strong CO stretching bands observed for {Mn(CO)$_2$[P(OMe)$_3$]$_3$}$^-$ is that they probably reflect a cis relationship between the car- bonyl groups.[502] The yellow P(OPh)$_3$ analog and the yellow [Mn(CO)$_4$L]$^-$[250] and [Mn(CO)$_3$L$_2$]$^-$[412] where L = P(OPh)$_3$ would have related structures.

The octahedral manganese(I) complexes of the formula YMn(CO)$_{5-x}$L$_x$ where x = 1 or 2 are most abundant for L = P(OR)$_3$. When x = 1, these complexes are cis when Y = halogen, p- or m-FC$_6$H$_4$, or AuPPh$_3$,[10,77,292] trans when Y = C(CN)$_3$[47] and cis and/or trans when Y = COR, R, AuP- (OPh)$_3$, or SnR$_3$.[77,250,263,292a,311a] The few trans isomers characterized possess three CO stretching bands (1984-2087 cm^{-1}) and are yellow. The cis compounds tend to show four CO stretching bands (1945-2110 cm^{-1}) although only three are observed in several cases. The cis structures of Ph$_3$PAuMn(CO)$_4$P(OPh)$_3$[354] and PhCOMn(CO)$_4$P(OCH$_2$)$_3$CEt[71] have been confirmed by x-ray diffraction and dipole moment studies, respectively. The cis complexes tend to be white or yellow with m. or dec. points from 70 to 200°C although the only case of a rhenium complex in this category, cis- Re(CO)$_4$[P(OPh)$_3$]H, is a viscous liquid.[185] Those in which Y = halogen have only been detected in solution. Where L = P(NMe$_2$)$_3$, again only three CO bands are observed in the white Mn(CO)$_4$LSnMe$_3$[302] complex but a cis configuration is probable from the similarity of the spectra to those of other cis compounds. The presence of five or six CO bands in the yellow Mn(CO)$_4$[P(NMe$_2$)$_3$]R (m. 73-144°C) and Mn(CO)$_4$- [P(NMe$_2$)$_3$]X systems (m. 106-120°C) were interpreted as splittings arising from conformational isomers.[302]

Besides the facial isomer (19) for YMn(CO)$_3$L$_2$, where

$Y = R$, COR, H, or halogen, and $L = P(OR)_3$[36,359,502] or $P(OR)_2R$,[36] only the symmetrical mer form (20) has been detected and not the unsymmetrical mer geometry (21).

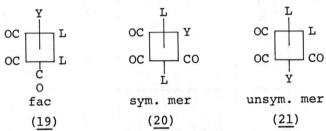

fac	sym. mer	unsym. mer
(19)	(20)	(21)

The majority of fac isomers have been observed only in solution (Y = halogen) while the few that have been isolated (Y = COR, Br, or H) are orange, yellow, or white with melting points from 107 to 290°C. Their symmetry (C_s) results in three strong CO stretching bands in the 1927-2053 cm^{-1} range and in a few cases a doublet or partially resolved "triplet" NMR absorption for the POCH$_2$ protons on the ligand, which was taken to support a cis relationship for the phosphorus ligands. The only rhenium complex of this type is facRe(CO)$_3$[P(OPh)$_3$]$_2$I,[190] a white compound exhibiting the expected three strong carbonyl IR bands.

The properties of the mer isomers are summarized in Table 5. When POCH or PCH NMR absorptions are observed, their "triplet" appearance has been interpreted to support the trans configuration of the ligands L. Like the fac isomers, the CO groups also produce three bands but their relative intensities are not the same. Further support for a mer configuration in two cases stems from dipole moment studies[12,359] in which the moments obtained were similar to that of the parent carbonyl compound as expected.

Interesting examples of manganese(I) stereochemistry are the YMn(CO)$_2$L$_2$ and YMn(CO)$_3$L complexes where Y is the chelating [F$_3$CCOCHCOCF$_3$]$^-$[226] or [S$_2$P(OPh)$_2$]$^-$[316] anion, respectively, and L = P(OR)$_3$. The former orange to purple substances, which are oils or solids (m. 40-90°C), show two strong CO bands (1885-1985 cm^{-1}) which are expected for either possible structure (22 or 23). The 1:2:1 "triplet" NMR spectrum of the protons in the P(OMe)$_3$

(22)	(23)

Table 5. Properties of mer-YMn(CO)$_3$L$_2$ Complexes

Y	L	Color	m. or dec. (°C)	ν_{CO} (cm^{-1})	$^3J_{POCH}$[a]
OCR	P(OR)$_3$	Yellow or white	Liquid, 135-197	1906-2058	4.6
	P(OR)$_2$R		138	1912-2049	
R	P(OR)$_3$	White	37-197	1905-2083	4.7-10.8
	P(OR)$_2$R		138-175	1912-2041	2[b]
Halogen	P(OR)$_3$	Orange or yellow	147-154	1927-2070	
	P(OR)$_2$R			1896-2051	
H	P(OR)$_3$	Yellow or white	84-91	1937-2090	

[a]See footnote b, Table 2.
[b]This value represents $^2J_{PCH}$ augmented by any $^4J_{PMPCH}$ coupling which may be present.

complex was felt to be consistent only with (22). In yellow-orange $S_2P(OPh)_2^-$ complexes (dec. 85-86°C), a fac-mer
mixture was suspected from the IR data. The $P(OMe)_3$
ligands in $Mn(CO)_4L_2S_2P(OMe)_2$[227] were also concluded to
be trans from the observation of the 1:2:1 POCH proton
NMR "triplet."

The compound $PhCOMn(CO)_2[P(OCH_2)_3CEt]_3$ can have the
possible isomeric structures (24-26).

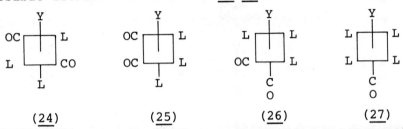

| (24) | (25) | (26) | (27) |

A strong band at 1933 cm^{-1} and a weak one at 2024 cm^{-1} was
felt to be most consistent with (24), as (25) and (26)
might be expected to give two bands of more equal intensity. The orange compounds $Mn(CO)_x[P(OMe)_3]_{5-x}Br$, where
$x = 1$ and 2, show one and two strong CO bands, respectively, which indicates a cis arrangement of the CO groups in
the latter compound. Similar spectra were reported for
the white compounds where Br is replaced by Me.[502] The ^1H
NMR spectra of the Me compounds revealed a quintuplet and
a quartet Me proton absorption for $x = 1$ and 2, respectively, which constitutes good evidence for structures (26)
and (27), respectively, since the ^{31}P atoms must be fixed
in equivalent positions barring rapid exchange. It should
be noted that the observation of two strong bands in the
case of structure (26) does add credence to structure (24)
for $PhCOMn(CO)_2[P(OCH_2)_3CEt]_3$. Although structure (25)
was postulated to account for the rhenium compound $Re(CO)_2$-
$[P(OPh)_3]_3I$[190] which showed two strong CO bands and a dipole moment of 4.3 D, structure (26) is not completely
ruled out. A ^{31}P chemical shift study could distinguish
the two possibilities.

The red compound $MnNO(CO)_3[P(NMe_2)_3]$[302] possesses a
three-band CO stretching spectrum and a single NO band
which is consistent with structures in which the NO and L
ligands are in a cis arrangement in a trigonal bipyramidal
array. The cations $[CpMn(NO)(CO)L]^+$, $[CpMn(NO)L_2]^+$, and
$[CpMn(NO)(\gamma-MeC_5H_4N)L]^+$ [where $L = P(OPh)_3$] characterized
by the conductivity of their hexafluorophosphates are yellow or red solids (m. 134-167°C) which all give one CO
and/or one NO band (2062 and 1762-1821 cm^{-1}, respectively).[278] Small nuclear couplings from phosphorus to the Cp
protons (ca. 2 Hz) were also detected. The structures are
undoubtedly pseudotetrahedral (28).

Pale yellow compounds of the type $CpMn(CO)_2P(OR)_3$[469],

[493b] are oils or solids (m. 113-165°C) for which only one
CO band (1945-1965 cm^{-1}) is recorded although two are ex-
pected if the reasonable structure (28) applies.
   The sextet hydride NMR spectrum for ReP(OPh)$_3$(Ph$_2$PCH$_2$-
CH$_2$PPh$_2$)$_2$H[188] would suggest that rapid polytopal rearrange-
ment of the proton is taking place.

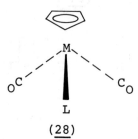

(28)

   Mn(NO)$_3$L complexes [where L = P(OR)$_3$,[241,265] P(NMe$_2$)$_3$,
[302] or P(SR)$_3$][265] are tetrahedral giving two bands indicative of (C$_{3v}$) symmetry between 1690 and 1805 cm^{-1} unless
the E mode is split into two bands. Two bands are also
seen for the green, red, or brown five-coordinate Mn(NO)$_2$-
[P(OR)$_3$]$_2$X[265] series (1688-1756 cm^{-1}) indicating a cis arrangement of the NO groups.
   Preliminary x-ray evidence indicates that the structure
of H$_2$C$_2$S$_2$Mn$_2$(CO)$_4$[P(NMe)$_3$]$_2$[301] is that given schematically
in (29).

(29)

   Formulations (30) and (31) shown for the manganese(II)
complex can not be distinguished from the available evidence. Although the geometry of the complex is not known,
an Mn-P bond is reasonable since the PH and P=O stretching
bands of the ligand from which it is made (32) do not appear in the complex.[279] Interestingly, the isomeric form
of the ligand (33) is incapable of forming this complex.[279]
   Blue crystalline trivalent rhenium complexes of the
formula Re[P(OR)$_3$]$_3$I$_3$[189,352] (m. 107-134°C) which are nonelectrolytes have been characterized. The relatively low
dipole moments (∿1.6 C) indicates that a rather symmetrical

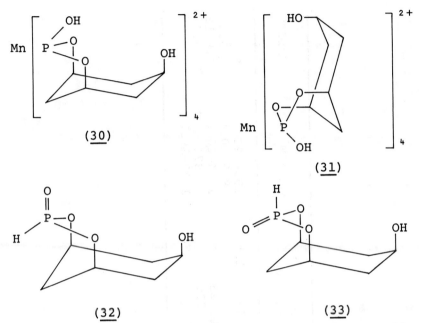

octahedral species is probably stabilized. Mixed yellow
derivatives of the type $Re(PPh_3)P(OPh)_3X_3$ and $Re(PPh_3)_2$-
$[P(OR)_3]_2Y_3$ where $Y = Cl$, Br, or H, are seven-coordinate
but their structures are unknown. The eight-coordinate
systems, $[Re[P(OPh)_3]_{4-x}(PPh_3)_xH_4]Cl$[188] where $x = 1$ or 2,
have been formulated as 1:1 electrolytes but their structures are undefined.

(4) Fe, Ru, Os

Both bridging (1782–1855 cm^{-1}) and terminal (1985–2091
cm^{-1}) CO bands are observed for the dark green $Fe_3(CO)_{12-x}$-
$[P(OMe)_3]_x$ complexes where $x = 1$, 2, or 3.[432] No firm
structural conclusions can be inferred, however, and the
same is true for $Ru_3(CO)_9[P(OPh)_3]_3$.

Properties of the mono-, and di-, and trisubstituted
complexes of $Fe(CO)_5$ are summarized in Table 6. In the
monosubstituted $Fe(CO)_4L$ [$L = P(OR)_3$[145a,436,437] or
$P(NMe_2)_3$],[297] the four carbonyl IR bands are consistent
with an axial position of L on the trigonal bipyramid.
Disubstituted complexes have always been found to be trans
diaxial (34) and show three CO bands when all the modes
are resolved.[145a,297]

Linkage isomerism is again evident in the iron carbonyl
complexes of $P(OCH_2)_3P$. Using NMR methods similar to those
described previously for the Group VI complexes of this
ligand, axial-$Fe(CO)_4P(OCH_2)_3P$[520b] and axial-$Fe(CO)_4P$-
$(CH_2O)_3P$[520b] were characterized. Spectroscopic data indicative of a nonaxial isomer of the former complex were

Table 6. Properties of Fe(CO)$_{5-x}$L$_x$ Complexes

x	L	Color	m. or dec. (°C)	ν_{CO} (cm^{-1})	JPYCH[a]
1	P(OR)$_3$	Yellow	Liquid to 195	1950-2067	5.1-15
	P(NMe$_2$)$_3$	White		1930-2041	9.6
2	P(OR)$_3$	Yellow or white	61-253	1884-2006	4.9-12
	P(NMe$_2$)$_3$	White	220	1871	
3	P(OMe)$_3$	White		1879-1937	9.4

[a]See footnote b, Table 2.

later shown to arise[520b] from a mixture of the former compound and a trans disubstituted complex to be described
shortly. The novel complex $(OC)_4FeP(OCH_2)_3PFe(CO)_4$ in
which iron is bridged in axial positions was also isolated.[520b] A crystal structure of the disubstituted trans-
$P(OCH_2)_3PFe(CO)_3P(CH_2O)_3P$ confirms the conclusion based on
NMR studies that linkage isomerism is present in the complex.[520b] The mixed complex trans-$Fe(CO)_3[P(OCH_2)_3CPr]$-
$[P(NMe_2)_3]$[406] has also been made. Solution IR data
which was reported to be consistent with the presence of
two cis isomers of $Fe(CO)_3[P(OMe)_3]_2$ upon later reinvesti-
gation was found to stem from the presence of (butadiene)
$Fe(CO)_2P(OMe)_3$.[437] The trisubstituted compound $Fe(CO)_2$-
$[P(OMe)_3]_3$[363,436,437] is felt to be of C_{2v} symmetry since
two CO bands are observed. From the ratio of the carbonyl
band intensities, the (CO)Fe(CO) angle is concluded to be
120° and so the CO groups reside in the equatorial plane
of the trigonal bipyramid. The related trisubstituted
complexes $Fe(CO)_2[P(OMe)_3]_2(PEt_3)$ (35) and $Fe(CO)_2[P(OMe)_3]$-
$(PEt_3)_2$ (36) also were synthesized.[437] Their structures
were concluded to be as shown from carbonyl band intensity
arguments. Thus the small increase in intensity of the B_1
mode in (36) compared to $Fe(CO)_2(PEt_3)_3$ was felt to indicate

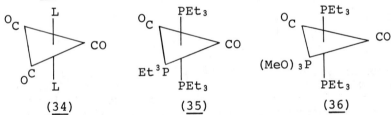

(34)   (35)   (36)

that the $P(OMe)_3$ group replaced a PEt_3 ligand in an equatorial position wherein steric interactions with the CO
groups were minimal. Similarly the substantial reduction
of the B_1 mode intensity in (35) compared to $Fe(CO)_2$-
$[P(OMe)_3]_3$ was taken to mean that the PEt_3 group went into
axial position where steric interactions are maximal. The
observation of only two ^{19}F and ^{31}P chemical shifts in
$Fe(PF_2OEt)(PF_3)_4$[314] is most reasonable if ligand exchange
were occurring.
     Olefin complexes of the straightforward type diene-
$Fe(CO)_2P(OMe)_3$ have been detected in solution and the
three carbonyl IR bands in some cases strongly suggest
the presence of more than one isomer.[363,437] The yellow
oils of the formula $C_4H_6Fe(CO)_2L$ where L = $P(OR)_{3-x}R_x$ are
probably similar.[437] Also detected in solution is the
novel complex (cycloheptatriene)$Fe(CO)_2P(OMe)_3$ in which it
is proposed that only two of the olefin groups are coordinated. The unusual zerovalent complex shown in (37) may
be of a similar nature.[531a]

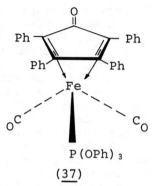

P(OPh)₃

(37)

On the basis of the similarity of the IR spectrum of
$Fe_2(CO)_7[P(OMe)_3](SnR_2)_2$ in the CO stretching region to
$M(CO)_3L'_2L$ compounds, the structure of the former is believed to be (38).[290]

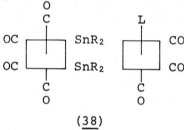

(38)

The six or seven CO bands found in a series of
$Hg\{Fe(CO)_2(NO)[P(OR)_3]\}_2$ complexes were considered to be
consistent with a structure in which two trigonal bipyramids are axially joined by the Hg atom with the L molecules occupying the terminal axes. The trigonal planes
containing the CO and NO ligands would very likely be
staggered.[94b] The $IHgFe(CO)_2(NO)[P(OR)_3]$ series[94b] could
be analogous in that two CO bands are observed.

Several interesting $(CF_2)_4Fe(CO)_{4-x}[P(OR)_3]_x$ complexes,
where x = 1 or 2, have been structurally characterized
from infrared and NMR data.[116b] Where x = 1, two strong
and one weak CO band indicate that two of the CO groups
are trans thus favoring (38a) rather than (38b). The in-
equivalence of all the fluorines also supports structure
(38a) owing to the unsymmetrical position of L.[116b] Where
x = 2, two strong CO bands rule out (38c) and suggest
either (38d) or (38e). The symmetry of the ¹⁹F resonance
favors (38e), however.

A number of examples are now known of the type trans-
$\{M[P(OR)_3](Et_2PCH_2CH_2PEt_2)H\}BPh_4$ where M = Fe, Ru, or
Os.[35b] The quintet of doublets in the ¹H NMR spectra of
these colorless compounds strongly implies that the hydride proton is trans to L while four equivalent phosphorus

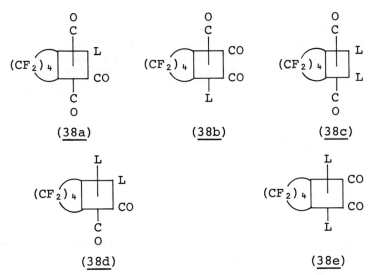

(38a)   (38b)   (38c)

(38d)   (38e)

atoms from the chelates are in the equatorial square plane.
In all cases, $^2J\text{HMP}_{cis}$ is less than $^2J\text{HMP}_{trans}$.

The infrared spectrum of $Fe(CO)(NO)_2P(OPh)_3$ consisting
of two NO bands (1787 and 1739 cm^{-1}) and one CO band
(2030 cm^{-1}) is in accord with a tetrahedral configuration.
Similarly the red or orange $Fe(NO)_2L_2$ complexes, where
$L = P(OPh)_3$[44,46,259,349,517] or $P(NMe_2)_3$,[298] while the orange $\{Fe(CO)_2NO[P(NMe_2)_3]_2\}^+$ [298] ion is probably trigonal
bipyramidal.

Each iron is quite likely to be in a pseudotetrahedral environment in the orange dimer $Cp[(PhO)_3P]_2FeFe[P(OPh)_3]_2Cp$.
[387b,389,390a,b] Interestingly no coupling of the phosphorus
nuclei to Cp protons was in evidence in the ^1H NMR spectrum.

The NMR spectra of the Cp protons in $Cp_2Fe_2(CO)_3-$
$[P(OR)_3]$ complexes consist of two singlets indicating the
possibility of two isomers. Two terminal CO bands (1944-
1971 cm^{-1}) and one bridging CO absorption (1750-1760 cm^{-1})
were taken to suggest cis and trans isomers of the type
shown in (39).

(39)

Although a mixture of isomers was indicated by the
five-band carbonyl IR spectra of $Cp(OC)_2FeSnCl_2Fe(CO)-$
$[P(OR)_3]Cp$, two terminal carbonyl (1939-1968 cm^{-1}) bands

and one bridging carbonyl absorption (1749-1760 cm^{-1}) were felt to be consistent with the presence of more than one isomer in $Cp_2Fe_2(CO)_3P(OR)_3$ systems.[216] This gains support from the two Cp proton NMR resonances and the evidence was interpreted in terms of the isomeric structures (40). Using the chelating ligand $Ph_2PN(Et)PPh_2$, the diamagnetic

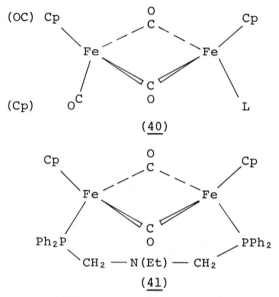

(40)

(41)

bicyclic system (41) can be isolated in which the Cp protons are not split by phosphorus and only bridging CO groups are detected in the IR spectra.[218]

Various divalent CpFe(CO)LY species have been characterized and their properties are condensed into Table 7. Their configurations are undoubtedly pseudotetrahedral [cf. (27)] although no definitive structural work has been carried out on these complexes. Optical isomers are possible but no resolution studies have been reported. Two more highly substituted complexes are also known and probably have the same configuration: the black $CpFe[P-(OPh)_3]_2I$[387b,389,390a] and the yellow hydride analog[389,390a] which interestingly gives no evidence of coupling from the Cp protons to the two phosphorus nuclei even though the hydride-phosphorus coupling is 82 Hz. The related $[CpFe(CO)_2L]^+$ complexes [L = $P(OR)_3$[136] or $P(NMe_2)_2$[247]] are yellow, show the expected two carbonyl stretches (1991-2075 cm^{-1}), and exhibit ~0 to 1.4 Hz coupling between the phosphorus and the Cp protons. This coupling is about the same as that observed in the isoelectronic, isostructural $[CpMn(NO)(CO)L]^+$ ions[278] discussed earlier.

An x-ray diffraction study of what was thought to be

Table 7. Properties of CpFe(CO)[P(OR)$_3$]Y Complexes

Y	Color	m. or dec. (°C)	ν_{CO} (cm^{-1})	$^3J_{PFeCH}$[a]
Halogen	Green or brown	126	1944–1993	∼0
CN		185	2002	1
SeCN	Orange to red	119	1993	1.0
SC$_6$F$_5$	Brown	71	1964	
COMe	Yellow or orange	Liquid, 65	1937–1959	∼0
Ph deriv.	Yellow or orange	91–145	1944	
SiMe$_3$	Orange	95	1953	1.5
SnMe$_3$	Orange	99	1949	∼0

[a]Coupling (Hz) of phosphorus to Cp protons.

an isomer of $CpFe[P(OPh)_3]_2I$ was shown to have the novel structure (42) while $CpFe[P(OPh)_3]_2I$ has a similar configuration except for the C-C bond between the aromatic rings.[6]

(42)

Although earlier workers excluded a cis arrangement in $H_2Fe[P(OEt)_3]_4$ at low temperatures from NMR studies,[315] others reported NMR evidence[497b] which indicates that the cis arrangement predominates at low temperature. Furthermore, peaks for both isomers of $H_2Fe[P(OEt)_2Ph]_4$ can be resolved at -50°C which give a time-averaged quintet spectrum at room temperature suggesting a rapid polytopal rearrangement via a transition state involving the L ligands in a tetrahedral array around the metal with a mobile hydrogen on the tetrahedral faces.[497b] Although the precise configuration of the complex $[FeL_4(H_2O)_2](ClO_4)_2$ [which has a formulation similar to that in structure (29)] is not known, the evidence for rearrangement of the ligand (31) as shown in (29) is analogous to that already given for the manganese complex.[279]

Ligands of the type $P(OR)_3$ are capable of stabilizing the green iron(III)[341] and blue iron(IV)[466] complexes $[Fe(S_2C_4Y_4)_2L]^-$ and its neutral analog, respectively. Their structures very likely resemble that shown schematically in (43).

Although four isomers are possible for the bivalent ruthenuim complex $Ru(CO)_2[P(OMe)_3]_2(O_2CCF_3)_2$, only one has been isolated.[88] The two strong CO bands (2021 and 2091 cm^{-1}) indicate a cis arrangement for the CO groups and the well-resolved "triplet" ^1H spectrum of the OCH_3 group strongly suggests structure (44).

(43)

(44)

Two isomers have been isolated for the white

or yellow zerovalent ruthenium complexes $Ru(CO)_2L_2Z$ where
Z is an unsaturated species spanning cis positions. The
Z groups are olefins such as $(CF_3)_2CC(CN)_2$ or Z can be

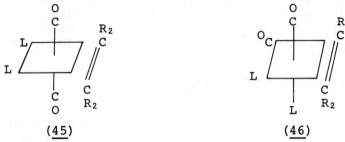

(45)   (46)

$(CF_3)_2C=O$.[137] With the olefin cited, both isomers have
been isolated where $L = P(OCH_2)_3CEt$. Isomer (45) shows
only one CO stretch (2061 cm^{-1}) while (46) shows two as
expected. Isomer (46) also shows two OCH_2 proton doublets
indicative of two chemically different ligands in cis
positions for which virtual coupling is small. In similar
fashion isomer (46) has also been characterized where Z =
trans-$CF_3(CN)CC(\overline{CN})CF_3$ while isomer (45) seems to be preferred for $L = P(OR)_3$ and $Z = (CF_3)_2CO$ or CF_2CF_2.[137] The
yellow or white CF_2CF_2 complexes are liquids or solids
(m. or dec. 59-115°C) which are quite volatile.
     Ruthenium(II) and osmium(II) complexes are most prev-
alent and are of the general formula $M(CO)_{2-x}L_{2+x}X_2$ where
x can be from 0 to 2 and X is most often halogen. A
"triplet" proton resonance and two CO infrared bands observed for the complex where M = Ru, x = 0, $L = P(OMe)_3$,
and $X = O_2CCF_3$ indicated isomer (47).[88] Similar results
obtained for the yellow $Ru(CO)_2[P(\overline{OMe})_3]_2Cl_2$ and white
$Ru(CO)_2[P(OCH_2)_3CEt]_2(I)CF_3$ suggest a corresponding stereochemistry for the yellow or white ruthenium complexes

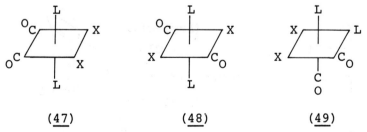

(47)   (48)   (49)

(m. or dec. 125-260°C) showing two CO stretches (2017-
2085 cm^{-1}) but which do not show "triplet" POCH NMR proton
spectra owing to the nature of the R groups on the ligand.[88]
The same is probably true of the colorless osmium analogs
(m. 178°C; two CO bands 2001-2068 cm^{-1}).[223,251,327] The
yellow $Ru(CO)_2[P(OCH_2)_3CEt]_2I_2$ exists as an inseparable
mixture of (47) and (48) since two POCH proton "triplets"

and three CO bands are observed.[88]   On the other hand
$Ru(CO)_2[P(OCH_2)_3CAm]_2X_2$ is probably exclusively (48) since
it exhibits only one CO band and one POCH triplet.[520b]
Structure (49) was postulated for $Os[P(OPh)_3]_2(CO)_2Cl_2$ on
the basis of the two CO stretching bands (1999 and 2065
cm^{-1}) and its dipole moment of 4.43 D.[253]   Cations of the
type $\{Os(CO)_4[P(OPh)_3]_2\}^{2+}$ [251] and $\{Os(CO)_3[P(OPh)_3]_2-$
$Cl\}^+$ [251,253] have been reported and a symmetrical meridional
structure was suggested for the latter on the basis of the
two strong CO bands at 2108 and 2052 cm^{-1}.

Mixed-ligand complexes of the type $Ru(CO)L(PR_3)_2XX'$
where $L = P(OEt)_3$ or $P(OMe)_2Ph$, and $X = X' = Cl$ or $X =$
Cl and $X' = H$ have been concluded to possess structures
(50) and (51).[164]   The hydride NMR absorptions of (50)
(~16 ppm upfield of TMS) reveal a small coupling (~24 Hz)
due to two identical phosphorus ligands (PR_3) and a larger

(50)   (51)

one (161-185 Hz) due to L.   The six-coordinate complex
$Ru[P(OPh)_3]_4ClH$ is also known[307a,415a,b] and its stereochemistry (52) was inferred from the hydride resonance data
obtained.   Thus ^2JHRuP(trans) was represented by a 174-Hz
doublet and the same coupling cis was signalled by a 28-Hz
doublet.   A 24-Hz triplet was deduced to have arisen from
the mutually trans phosphorus ligands.   The Cl group therefore must be cis to the hydride.

(52)   (53)

The stereochemistries of the cream to orange $Ru(CO)-$
L_3X_2[327] complexes where $L = P(OR)_3$, and X is Cl, Br, or I
(m. or dec. 128-148°C; single ν_{CO}(2040-2051 cm^{-1}), the
white to yellow RuL_4X_2[327] compounds (m. or dec. 151-173°C),
and the white OsL_4X_2[327,452a] analogs (m. 159-212°C) are

not known. The interesting white $RuP(OPh)_2[P(OPh)_3]_3X^{307a,}$

^{415a,b} complexes (m. 161-185°C) and the white $RuP(OPh)_2-$
$[P(OPh)_3]_2COX$ compounds^{327} (m. 161-185°C; single ν_{CO} 2024-
2026 cm^{-1}) contain Ru-Ph bonds for which the evidence is
the detection of H_2 evolution in the synthesis when
$P(OPh)_3$ is added to the starting material $Ru(PPh_3)_3ClH$,
the lack of an RuH stretching band, IR bands suggestive
of aryl substitution, and a proton NMR spectrum more complicated than that for RuL_4X_2. The detection of two sets
of overlapping triplets and a pair of overlapping doublets

in the ^{31}P NMR spectrum of $RuP(OPh)_2[P(OPh)_3]_3X$ sys-
tems^{415a,b} was interpreted to be consistent with (53)
since all the $^{31}P-^{31}P$ couplings were similar in magnitude.
A trans stereochemistry for the orange trivalent anion
$[Ru[P(OPh)_3]_2Cl_4]^-$ was suggested on the basis of analogy
with similar PR_3 complexes for which NMR and IR evidence
indicated this configuration.^{491a}
    Only mixed complexes of the type $Os[P(OR)_3](PR_3)_2Cl_3$
are known for trivalent osmium.^{163,165} The stereochemistry of these red substances (m. 178-198°C) is not known.
The same is true for the colorless liquid tetravalent
$OsL(PR_3)_2H_4$ compounds where L can be $P(OR)_3$ or $P(OMe)_2Ph$.^{163}
Coupling of the PR_3 phosphorus nuclei to the high-field
hydrides (~-19 ppm) is about 9 Hz and the same coupling
to the $P(OR)_3$ or $P(OMe)_2Ph$ ligand is about 14 and 9 Hz,
respectively. Although intermediate virtual coupling of
the PR_3 ligands seems indicated, the stereochemistry of
these seven-coordinate complexes has not been elucidated.

(5) Co, Rh, Ir
Molecular weight and infrared spectra (seven to eight CO
frequencies) are consistent with the formulations $M_4(CO)_8-$
$[P(OPh)_3]_4$ where M = Co, Rh, or Ir.^{409b} Structure (54) is
felt to apply to these systems since an analogous structure
was found for a related compound involving PPh_3.^{409a} The
possibility of isomerism arising from different apical
stereochemistries was recognized. Black or brown and red
$M_4(CO)_{12-x}L_x$ (x = 1, 2, or 3) species are also formed
where M = Co and Rh, respectively.^{409b} Trisubstituted
iridium derivatives where L = $P(OEt)_3$ and $P(O-o-tolyl)_3$
were also reported.^{409b}
    The presence of four CO stretching bands observed for
$Co_2(CO)_7P(OR)_3$^{496} complexes (1952-2088 cm^{-1}) has been
interpreted to indicate a structure like that of (55) with

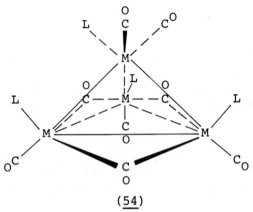

(54)

a CO in place of one of the axial ligands. The two terminal CO stretching bands (1981-1970 cm^{-1}) seen for the red,
orange, or brown zerovalent $Co_2(CO)_6[P(OR)_3]_2$[43,206,357,
440,459] complexes (m. or dec. 103-173°C) is consistent
with either a D_{3h} or D_{3d} configuration. Not unlike the
analogous $Mn_2(CO)_8L_2$ complexes discussed earlier,[123,332-
334] the staggered configuration (55) is believed to be

(55)   (56)

sterically favored. The detection of a terminal as well
as a bridging CO band in $Co_2(CO)_6[P(OPh)_3]_2$ indicates
structure (56) but the broadness of the bands prevented
further speculation on the isomerism.[206]   The yellow
iridium compound $[Ir(CO)_2P(O-p-MeC_6H_4)_3]_2$[539] probably has
a similar structure. By analogy to structural conclusions
from IR studies of PR_3 complexes, the one bridging and three terminal CO stretching bands observed for
the brown compound $Co_2(CO)_5L_3$ and the two bridging and two
terminal CO absorptions observed for the yellow-brown
$Co_2(CO)_4L_4$ complex where L = $P(OCH_2)_3CEt$[70] were suggested
to indicate structures (57) and (58) although the precise
geometry around cobalt or the overall isomerism could not
be determined. Similar complexes where one of the bridging CO groups is replaced by a CF_3CF group and where one
L group [$P(OR)_3$] and two carbonyls are bonded to each
cobalt probably have analogous structures.[73]   By analogy

to $Co_2(CO)_8$, the diamagnetic $\{M[P(OR)_3]_4\}_2$ dimers,[452a] where M = Rh or Ir, or the weakly paramagnetic cobalt analog[528a] may be metal-metal bonded. Each cobalt in $[Co(CO)_3P-(NMe_2)_3]_2Hg$[297] and $[Co(CO)_3[P(OPh)_3]_2SnCl_2$[68,69] is undoubtedly five-coordinate but the IR spectra in the carbonyl region (two bands, 1936-2044 cm^{-1}) do not allow more precise structural conclusions to be made. The five bands observed in the CO infrared region of $[Co(CO)_3P(OPh)_3]_2Hg$, however, are consistent with structure (55) wherein a mercury bridge connects the cobalt atoms.

Cobalt nitrosyl complexes of the formulas $CoNO(CO)_{3-x}$ L_x[46,94a,124,193,248,255,256,349,440,499,500a,517] where x = 1 to 3 are tetrahedral and their properties are summarized in Table 8. Several mixed complexes of the type $CoNO(CO)LL'$ and $CoNOL_2L'$ where L and L' can include PF_3, $P(OMe)_xF_{3-x}$, and $P(OMe)_3$ have been detected in solution by their characteristic IR patterns in the 1760-2060-cm^{-1} region.[124]

Examples of the cationic series $[Co(CO)_{5-x}L_x]^+$ with various anions (e.g., ClO_4^-,[143] BPh_4^-,[30,70] NO_3^-,[143] and $[Co(CO)_4]^-$[30,70]) where x = 2-5 are known and all the cations are yellow. The only case where x = 2 is for L = $P(NMe_2)_3$[297] which displays two carbonyl stretching frequencies and a "triplet" $PNCH_3$ proton NMR spectrum owing to virtual coupling. This is suggestive of a trans configuration for the ligands L although a CO band is missing in the IR. A similar complex cation was reported where L = $P(OPh)_3$[246] but only one CO band was recorded. The presence of two CO bands in the IR spectrum of the $Co-(CO)_2[P(OR)_3]_3^+$[30] cation suggests a cis arrangement of the CO groups on a square pyramid or trigonal bipyramid. In one case only a single NMR ligand proton resonance was detected[30] but the longer time scale in the NMR could be responsible for this observation inasmuch as nonequivalent ligands could not be detected if their exchange were sufficiently rapid. Although two CO stretches in solution IR spectra of $\{Co(CO)[P(OR)_3]_4\}^+$ species indicate isomerism, the precise stereochemistry involved is not known.[70] In the solid state, the $P(OMe)_3$ complex shows only a single CO band suggestive of a single isomer.[30] A single CO band is also observed for $\{Ir(CO)(PMePh_2)_2[P(OMe)_3]_2\}^+$ but the medium was not specified.[115] In the completely substituted $\{Co[P(OR)_3]_5\}^+$ cations, the structure is almost certainly trigonal bipyramidal.[143] A crystal structure study of the isoelectronic $[NiL_5]^{2+}$ species has been carried out[448] (see later) in which nearly perfect trigonal bipyramidal symmetry was found. The great similarity in the electronic spectra of the $[ML_5]^{1+}$ and $^{2+}$ species is good evidence for identity in the structures as well. The $\{Rh[P(OMe)_3]_5\}^+$ cation is also known[402a] and again a trigonal bipyramidal geometry is not unlikely.

Table 8. Properties of CoNO(CO)$_{3-x}$L$_x$ Complexes

x	L	Color	m. or dec. (°C)	ν_{CO} (cm^{-1})	ν_{NO} (cm^{-1}) [a]
1	P(OR)$_3$	Yellow to red	Liquid, 55-150	1988-2099[b]	1754-1806
	P(NMe$_2$)$_3$			1995-2052[b]	1764[b]
	P(OMe)F$_3$			2011-2071[b]	1780-1798
2	P(OR)$_3$	Yellow or orange	Liquid to 86	1980-2008[a]	1727-1787
	P(OMe)F$_3$			2010-2035[a]	1760-1798
3	P(OR)$_3$				1708-1738
	P(OMe)F$_2$				1795

[a]One band.
[b]Two bands.

The predominant coordination number of monovalent cobalt, rhodium, and iridium complexes is five although the
question of stereochemistry still remains unanswered in
most cases. A summary of the properties of the neutral
$Co(CO)_{4-x}[P(OR)_3]_xY$ series appears in Table 9. Although
the coordination number of all these compounds is five,
the stereochemistry could be complicated by the presence
of isomers which is evident in the observation of two to
three bands when x = 3. When x = 2, the two bands observed are in accord with a cis relationship of the CO
groups and when x = 1, three bands have been taken to indicate that the CO groups are equatorial. In some complexes where x = 1, however, only one or two bands are
seen. Two CO bands and the lack of a 1:2:1 "triplet" in
the broad singlet CH_2 NMR resonance of $Co(CO)_2[P(OCH_2)_3-$
$CMe]_2[(NC)_2C_6F_3]$ strongly indicate that both L molecules
occupy the equatorial plane with a CO molecule.[72c] Similarly the broad singlet CH_2 resonance in the trisubstituted
analog is consistent with all three L moieties being in the
equatorial plane of a trigonal bipyramid.[72c]
    The yellow mixed complex $Co(CO)_2[P(OMe)_3](PPh_3)COMe$
has been prepared[231] and the brown $Co(CO)_2[P(OPh)_3](PPh_3)X$
(X = Cl or I) compounds exhibit two CO bands (1941-2013
cm[1]).[246] The isomeric form of the complexes has not been
defined. A monovalent hydride series of type $M(CO)_x-$
$[P(OR)_3]_{4-x}H$ has also been recognized but properties are
available only for the three highest members where x =
0,[326,456] 1,[70] and 2.[247] It is not possible to conclude
from the triplet[247] and quartet[5c] hydride NMR resonances
in $Co(CO)_{4-x}[P(OPh)_3]_xH$ where x = 2 and 3, respectively,
whether the two phosphorus nuclei are geometrically equivalent, are exchanging rapidly, or are inequivalent with
rapid polytopal exchange of hydrogen. The $P(OCH_2)_3CEt$
analog where x = 2 may exist as isomers since three CO
stretching frequencies are observed.[233] The related complex $Ph_3PAuCo(CO)_3P(OPh)_3$ shows only two CO stretching
bands[77] and so little can be concluded concerning the isomeric form of this five-coordinate bimetallic complex.
C_{3v} symmetry (59) has been postulated for $Ir(CO)[P(OPh)_3]_3H$
on the basis of a single CO stretching vibration and a
hydride NMR quartet indicating coupling (11 Hz) to chemically equivalent phosphorus ligands.[5b] Complications like

(57)   (58)

Table 9. Properties of Co(CO)$_{4-x}$[P(OR)$_3$]$_x$Y

x	Y	Color	m. or dec. (°C)	ν_{CO} (cm^{-1})[b]
1	COR	Colorless to yellow	Liquid, solid[a]	1960-2083
	COOEt	Yellow	70	1999-2079
	R	Yellow	10-190	2012-2087
2	I	Yellow-brown	65-261	2020-2045
	H	White	88	1935-2010
	COMe	White or yellow-green	35-185	1935-2000
3	Me	Yellow	50-185	1920-1995
	MeCO	Colorless or yellow	49	1940-1985
	Me	Colorless	185	

[a]Solids decarbonylate.
[b]See text for discussion of number of bands.

those cited for the cobalt analog can not be ruled out.
Unfortunately the stereochemical problem was not solved
by the study of Ir(CO)[P(OCH$_2$)$_3$CPr](PPh$_3$)$_2$H and Ir(CO)-
[P(OCH$_2$)$_3$CPr]$_2$(PPh$_3$)H.[520b] Although each shows only one
CO band, the PIrH couplings for each ligand are sufficiently different in the two complexes that no stereochemical conclusions could be drawn. Low-temperature
studies may be of help in determining the role of ligand
exchange if any. Again the stereochemistry seems to be a
little clearer at least in the solid state for CoL$_4$H where
L = P(OR)$_3$. Thus Co[P(OPh)$_3$]$_4$H is isomorphous with a nickel analog which has been found to possess a tetrahedral
array of phosphorus atoms with the H atom likely to be
close to a face of the tetrahedron. In all cases, however, only a quintet hydride NMR spectrum is observed
(JPCoH = 11-17 Hz)[326] indicative of a square pyramidal structure in solution (60) or rapid ligand exchange. ^{31}P chemical shifts for these complexes range from -116 to -168
ppm.[315,326] A recent crystal structure determination
shows that the L ligands in Co[P(OEt)$_2$Ph]$_4$H are pseudo-
tetrahedrally arranged around the cobalt with the hydrogen on a face of the tetrahedron (61).[409b] A quintet
hydride spectrum (^{2}JPCoH = 21 Hz) observed for the complex is consistent with rapid polytopal rearrangement.[409b]
The white iridium and rhodium analogs exhibit PIrH and
PRhH couplings (25[195] and ∿45 Hz,[520b] respectively) as
well as RhH coupling of about 8 Hz and a triplet for the
POCH proton resonance in one of the rhodium complexes due
to virtual coupling. The latter coupling is not observed
in the cobalt analog. The lemon-yellow five-coordinate
Co[P(OEt)$_3$]$_4$Cl[382a,527] has also been reported.
    Orange or yellow monovalent cobalt complexes of the
type π-C$_4$H$_7$Co(CO)[P(OR)$_3$]$_2$ (m. or dec. 87-155°C),[70] π-
C$_3$H$_5$Co(CO)[P(OCH$_2$)$_3$CEt]$_2$ (m. or dec. 155-230°C),[233] HC-
(CHR)$_2$Co(CO)[P(OCH$_2$)$_3$CEt]$_2$,[233] and C$_8$H$_{13}$Co(CO)[P(OPh)$_3$]$_2$[508]
all display one CO band (1935-1970 cm^{-1}) and probably have
pseudotetrahedral configurations. The same is probably
true for the yellow π-C$_3$H$_5$Co(CO)$_2$P(OR)$_3$[233] compounds. The
fact that the latter complex shows three bands in the carbonyl IR spectrum is perhaps indicative of isomerism inasmuch as the C$_3$H$_5$ group is not symmetrical with respect to
the other three ligand positions.
    The yellow, green, or red CpRh(CO)$_{2-x}$[P(OR)$_3$]$_x$ series[471]
are oils or solids and show a doublet splitting of the Cp
proton spectrum of 1.2 Hz when x = 1 and a triplet splitting of the same magnitude when x = 2. The structures are
probably pseudotrigonal planar.
    The normal proton NMR spectrum of Rh[P(OPh)$_3$]$_2$-
[MeCOCHCOMe] indicates that the yellow compound is diamagnetic and therefore it probably adopts a square planar
configuration since a tetrahedral array of ligands on a

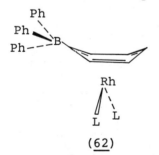

(59)   (60)   (61)

d^8 metal would be paramagnetic. By analogy the yellow {Rh[P(OR)$_3$]$_4$}$^+$[214,215] salts; the yellow neutral compounds Rh[P(OR)$_3$]$_3$X,[214,215,295,296,376,514,515] Co[P(OEt)$_3$]$_3$Cl,[527] and Ir(CO)[P(OPh)$_3$]X;[252,495a,b] and the dimers diolefinRhX$_2$RhL$_2$ and L$_2$RhX$_2$RhL$_2$ [where L = P(OPh)$_3$[214,215] or P(NMe$_2$)F$_2$[126]] are likely to have the same configurations. This is probably also true for the orthosubstituted Rh[P(OPh)$_3$]$_3$-[P(OPh)$_2$(OC$_6$H$_4$)] complex.[5b]

A very surprising reaction takes place upon allowing Rh[P(OMe)$_3$]$_5$BPh$_4$ to stand in air. A crystal structure determination of the product reveals that the product has the configuration (62) in which a phenyl group is pi coordinated.[402a] An NMR spectrum of the nonconducting substance in solution also shows the presence of coordinated

Ph
Ph
Ph---B
Rh
L   L

(62)

and noncoordinated phenyl groups. The yellow solid Rh$_2$-(SCN)$_2$[P(O-p-MeC$_6$H$_4$)$_2$]$_2$[514] has been postulated to have bridging SCN groups from the CN vibration frequency position. It is diamagnetic and apparently three-coordinate.

Anionic complexes of cobalt(I) such as the yellow [Co(CO)$_3$L]$^-$[260,261] and colorless [Co(CO)$_2$L$_2$]$^-$[43,247] ions [where L = P(OPh)$_3$] are four-coordinate. The only example of an anionic cobalt(I) complex in which five donors are probably involved is the [CoL(NO$_3$)$_2$]$^-$ species.[143] That the nitrates are bidentate is supported by infrared evidence and the observation that structural work on nitrate complexes generally reveals bidentate coordination.[143] The specific geometry of the anion is not as yet known. Anionic rhodium(I) complexes include the types {Rh[P(OR)$_3$]$_2$-S$_2$C$_2$(CN)$_2$}$^-$[134] and Rh[P(OPh)$_3$]$_2$(NCS)$_2$$^-$.[283] All are probably cis square planar since two CN stretching frequencies were noted for the latter compound in solution and the former incorporates a bidentate ligand.

Compounds in the series CoL_2X_2 where L = PhP(NEt)$_2$[317] or P[N(CH$_2$CH$_2$)$_2$O]$_3$[162] are tetrahedral as their paramagnetism (4.34-4.83 BM) would indicate. Cobalt-phosphorus and/or cobalt-nitrogen bonds in the P[N(CH$_2$CH$_2$)$_2$O]$_3$ complexes are involved in a weak ligand field (three unpaired electrons). The weak conduction of the PhP(NEt)$_2$ complexes in nitrobenzene may reflect the tendency for one of the ligands to become bidentate (perhaps using two nitrogen coordination sites) upon dissociation of an anion. Several complexes of the type Co[P(OR)$_3$]$_x$Cl$_2$ where x = 4, 5, and 6 have been reported[384,385] to exist in solution but their characterization is incomplete. The novel complex shown below in (63) very probably does not involve a cobalt-phosphorus bond since its paramagnetism (4.26 BM) is best explained on the basis of tetrahedrality around cobalt.[267] The dimethyl glyoxime complexes Co[P(OMe)$_3$]-[HONC(Me)CMeNO]$_2$Me[468a] and [CoP(O-p-MeC$_6$H$_4$)$_3$HONC(Me)MeNO]$_2$[467]

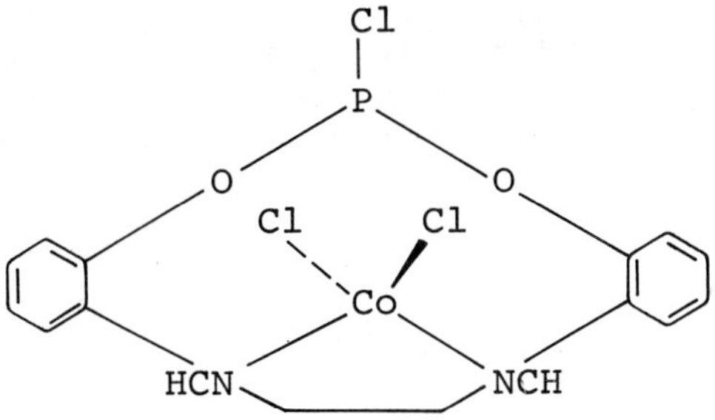

(63)

are known. The first could very well be octahedral with the phosphorus and methyl ligands trans in order to permit electron delocalization in the plane defined by the chelating rings and the metal atom. The stereochemistry of the second is similar except that a cobalt-cobalt bond replaces the methyl groups. The black compound [Co(NO)-[P(OPh)$_3$]SPh]$_2$[249] is likely to be tetrahedral with bridging SPh$^-$ anions.

Nitrosyl cobalt(III) complexes of the formula Co(NO)$_2$-[P(OPh)$_3$]X (where X = halogen,[256] R,[44,46] SPh,[243] or NCS[243]) are black solids (m. 45-57°C) showing two NO stretching frequencies (1785-1850 cm^{-1}) which would be indicative of a tetrahedral configuration. The blue compound Co(NO)L$_2$Cl$_2$[256] is apparently five-coordinate but the stereochemistry is not known. The red [Ir(CO)[P(OR)$_3$]-I$_3$]$_2$[539] dimers are probably iodine bridged while the ortho-substituted {Ir[P(OPh)$_2$(OC$_6$H$_4$)]$_2$Cl}$_2$ dimer is probably chlorine bridged.[5b] The iridium(III) systems Ir[P(OPh)$_3$]$_2$-

$[P(OPh)_2(OC_6H_4)]HCl$, $Ir[P(OPh)_3][P(OPh)_2(OC_6H_4)]_2H$, and $Ir[P(OPh)_3][P(OPh_2)(OC_6H_4)]_2Cl$ are all known[5b] but the exact stereochemistries remain to be determined. The presence of two equivalent hydride protons and PEt_2Ph phosphorus nuclei as shown by NMR studies of $Ir[P(OMe)_3]-[PEt_2Ph]_2H_3$ and its $P(OMe)_2Ph$ analog, limits the stereochemical possibilities to (63a).[339b]

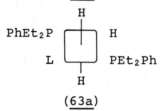

(63a)

Symmetrical mer structures are postulated for a $Rh[PR_3]_2[P(OR)_3]Cl_3$ series from the equivalence of the PR_3 phosphorus atoms in the ^{31}P spectra and a comparison of [1]JRhP values in fac- and mer-$Rh(PR_3)_3Cl_3$ complexes.[5c,d] The 1:1 electrolyte $[Rh[P(OR)_3]_4Cl_2]Cl\cdot2H_2O$ aquates in solution displacing chloride,[523,524] and the neutral $Rh[P(OPh)_3](PR_3)Cl_3$[56,405,419] is believed to have the symmetrical mer configuration (64) since only a single [2]JPMP coupling (30 Hz) is observed in the ^{31}P NMR spectrum. The

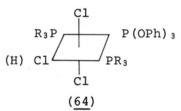

(64)

structure of the iridium hydride derivative (64) was inferred by analogy with the known configuration where the phosphite ligand is replaced by PR_3.[66] The rhodium(III) hydride complex $[Rh[P(OPh)_3]_4H_2][BPh_4]$ is stable only in a hydrogen atmosphere.[215] Although six-coordination is probably involved, the stereochemistry is not known. The same stereochemical question also exists for the six-coordinate complex $Ir[P(OPh)_3]_3Cl_2H$.[5b,452a]

All of the fully substituted $Co[P(OR)_3]_6{}^{3+}$ compounds are diamagnetic and their colorless or off-white appearance is indicative of the strong ligand field surrounding the d^6 cation which is only exceeded by CN^- in $[Co(CN)_6]^{3-}$.[143,275]

Anionic cobalt(III) complexes have only been isolated for $[Co(S_2C_2Y_2)_2P(OR)_3]^-$ (where $Y = CF_3$ or CN), $[Co(o-S_2C_6Cl_4)_2P(OEt)_3]^-$, and $[Co(o-S_2C_6H_3Me)_2P(OEt)_3]^-$ systems.[341] Although these green or orange-brown species appear to be five-coordinate and perhaps square pyramidal

(65), their exact stereochemistry is not known.

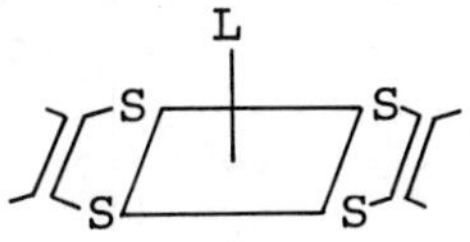

(65)

Dithiolene ligands are capable of stabilizing cobalt(IV) complexes of P(OPh)$_3$ of the formula [Co(S$_2$C$_2$Y$_2$)$_2$P(OPh)$_3$] where Y = Ph,[135,192,341,466] or CF$_3$.[34]   The latter complex is diamagnetic in the solid state and the dimeric structure (66) has been proposed to account for this.   In solution it is monomeric with a magnetic moment consistent with the expected presence of one unpaired electron (d^5 strong field electron configuration).   The structure could well be square pyramidal as in (65).

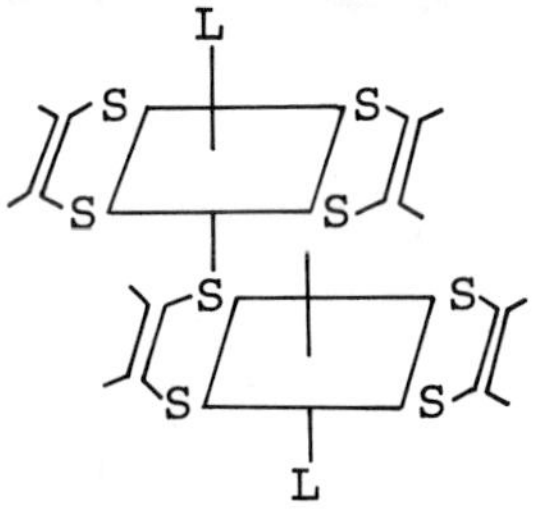

(66)

(6) Ni, Pd, Pt

The most numerous compounds of L are those of the tetrahedral Ni(CO)$_x$L$_{4-x}$ series whose properties are summarized in Table 10.   The ligand P$_4$(NMe)$_6$ (67) is capable of coordinating up to four Ni(CO)$_3$ groups to the phosphorus atoms

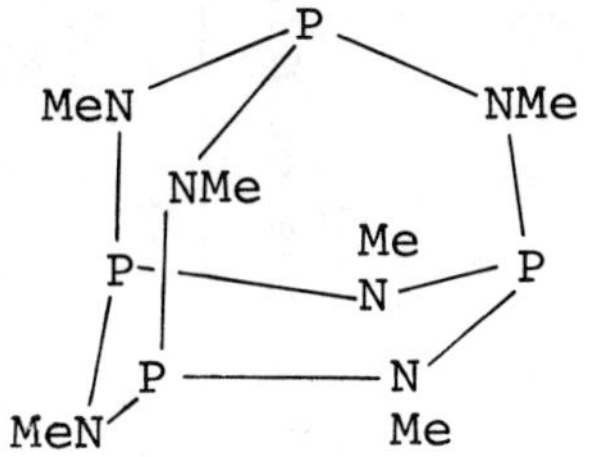

(67)

as shown from the characteristic two-band spectra in the CO IR stretching region (1990-2065 cm^{-1}) and ^{31}P chemical shift studies.[449,450]   Mixed dicarbonyl complexes of the type Ni(CO)$_2$[P(OR)$_3$](PR$_3$) also show two bands in the carbonyl stretching region (1950-2035 cm^{-1}) while monocarbonyl complexes exhibit only one.[57,62,373]   Mixed complexes of

Table 10. Properties of Ni(CO)$_x$L$_{4-x}$ Complexes

x	L	Color	m. or dec. (°C)	ν_{CO} (cm^{-1})	δ^{31}P	JPYCH[a]
0	P(O-alkyl)$_3$	Colorless	Liquid, 108-365		-93 to -163	3.8 to 10.8
	P(O-aryl)$_3$	Colorless to yellow	74-151		-130 to -131	
	P(OR)$_{3-x}$Clx	Colorless to yellow	Gel, 98		-104 to -177	13
	P(OR)$_{3-x}$Fx		129		-140 to -157	
	P(NR$_2$)$_{3-x}$Fx	Colorless	46-164		-168	
1	P(O-alkyl)$_3$	White	0-328	1954-1997	-92 to -165	3.6 - 11.6
	P(O-aryl)$_3$	White	Liquid, 99-145	2004		
	P(OBu)$_{3-x}$Clx			2004-2037		

2	$P(O\text{-alkyl})_3$	Colorless to Yellow	Liquid, 225-278	1964-2048	-93 to -165	4 - 11.8
	$P(O\text{-aryl})_3$	White to yellow	76-135	1996-2050	-146	9.3
	$P(NR_2)_3$	White to yellow	69-86	1929-1996		
	$P(OBu)_{3-x}Cl_x$			1996-2068		
	$P(NR_2)_{3-x}(CN)_x$	Colorless to orange	Liquid, 40	1961-2083	-165 to -181	
		Brown	45	1976-2028		
3	$P(O\text{-alkyl})_3$	Colorless	Liquid, 128-142	2004-2090	-93 to -161	3.9 - 13
	$P(O\text{-aryl})_3$	White	70	1997-2086		
	$P(NR_2)_3$	White	103	1984-2066		
	$P(OBu)_{3-x}Cl_x$			2020-2098		
	$P[N(CF_3)_2](CF_3)_2$			2021-2117		

aSee footnote b in Tables 1 and 2.

the general formula $NiL_3(PR_3)$,[120] $NiL_3(PF_3)$,[312] and NiL_2Z_2 where $Z = NCR$,[372,480b] $SC(CR_3)_2$,[82] $OC(CF_3)_2$,[82] O_2,[533] and C_2H_4[480b] are also known. The crystal structure of $Ni(PF_2-NC_5H_{10})_4$ confirms the expected tetrahedral geometry of these complexes.[204] Some NiL_4 complexes where L is the unusual ligand $P(NCO)_3$[534] or $P(NCS)_3$[534] have been reported and when $L = F_2POCH_2CH_2OPF_2$ or $o-(F_2PO)_2C_6H_4$, polymeric species are encountered having decomposition temperatures above 250°C.[462,465] Tetrasubstituted complexes of the NiL_4 type can dissociate to NiL_3 quite appreciably in solution and in some cases can be isolated.[480b] The complex $Pt[P(OPh)_3]_3$[350,351] has also been isolated and the structure of such complexes is likely to be nearly trigonal planar. The carbonyl series where palladium[347,348,372,383] or platinum[342,350,351,372,383] is the metal is unknown except where x = 0. Where $L = P(OR)_3$, the compounds are colorless and melt between 65 and 170°C. Mixed complexes such as $Pt[P(OR)_3]_3PR_3$[351] and $Pd[P(OR)_3]_3(CNR)$[348] are also known.

Divalent complexes make up the remainder of the nickel group and the largest category is represented by the general formula ML_2X_2. Within this class, most of the compounds are diamagnetic which is consistent with a square planar array of groups for a d^8 metal. The only reported examples of paramagnetic complexes of L in the nickel family are in the $Ni\{P[N(CH_2CH_2)_2O]_3\}_2X_2$ series where X = halogen or NO_3.[162] These compounds are green or blue and have magnetic moments ranging from 3.00-3.72 BM, which are in accord with two unpaired electrons in a tetrahedral complex. When X = NCS, however, a diamagnetic complex is obtained[162] which is probably square planar but its isomeric form (cis or trans) is not known. An unusual paramagnetic nickel complex in which a trivalent phosphorus moiety apparently remains uncoordinated is depicted in the proposed structure (63) in which Co is substituted by Ni.[267]

The general properties of the diamagnetic ML_2X_2 and the $MLZX_2$ complexes are summarized in Tables 11 and 12, respectively. The possibility of a square planar geometry was advanced for the $Ni[P(OR)_3]_2X_2$ complexes although none were crystallized and therefore could not be characterized completely. Using $P(OMe)_3$ or the caged phosphite $P(OCH_2)_3$-CAm, dark violet, crystalline, diamagnetic nonelectrolytes having the formula NiL_3I_2 were isolated.[520b] Thus it may be that the earlier $P(OR)_3$ complexes also contain five-coordinate species. A single CN stretching vibration in the 2100-2165-cm^{-1} range indicates a trans configuration for the $NiL_2(CN)_2$ complexes[102,144a] which in the presence of excess L, however, do become five-coordinate (see later). The infrared technique is also useful in determining the isomeric form of the palladium and platinum complexes in

Table 11. Properties of ML$_2$X$_2$ Compounds

M	L	X	Isomer	Color	m. or dec. (°C)	ν_{MX} (cm^{-1})	^{3}JPYCH[a]
Ni	P(OR)$_3$	Halogen		Black to violet	Oils		
	PPh$_2$(OR)	CN	Trans	Yellow	136		
	PPh(NEt$_2$)$_2$	NCS		Red	99		
Pd	P(O-alkyl)$_3$	Cl	Cis	White to yellow	128	305–332	5.5–12.9
			Trans	White			
	P(NMe$_2$)$_3$	Cl	Cis	Red to yellow	120		9.7
	PPh(NEt$_2$)$_2$	Cl	Trans			357	10.1
	P(OCH$_2$)$_3$CAm	I	Cis	Yellow, orange	109–115		2
			Trans	Red			
			Trans	Red			10.3
Pt	P(NMe$_2$)$_3$	I		White or pink[b]			
	P(OPh)$_3$	F	Cis	White	182	289–334	5.7–12.7
	P(O-alkyl)$_3$	Cl	Trans				10.1
	P(NMe$_2$)$_3$	Cl	Mix.	Yellow and white			10.1,[c] 9.4[d]

Table 11 (Continued)

M	L	X	Isomer	Color	m. or dec. (°C)	ν_{MX} (cm^{-1})	$^3J_{PYCH}$[a]
	$PPh(NEt_2)_2$	Cl		Yellow	100		
	$P(OEt)_3$	Br	Cis				
	$P(OR)_3$	I	Cis	Yellow	201–215	150, 157	5.7–12.9
	$P(NMe_2)_3$	I	Trans	Orange	150		10.0
	$P(OR)_{3-x}(OH)_x$	x $P(O)(OR)_2$		White	Liquid, 36–166		
	$(Ph_2PCH_2)_2$	$P(O)(OR)_2$	Cis[e]		214–276		
	Z[f]	$P(O)(OR)_2$	Trans		138–175		

[a] See footnote b in Tables 1 and 2.
[b] White form turns pink on heating.
[c] Trans.
[d] Cis.
[e] Cis geometry imposed by chelating $Ph_2PCH_2CH_2PPh_2$.
[f] Z = pyridine, PR_3, or AsR_3.

Table 12. Properties of MLZX₂ Compounds

M	L	Z	X	Isomer	Color	ν_{MX} (cm^{-1})	m. or dec. (°C)
Pd	P(OR)₃	NR₃	Cl	Trans	Yellow to orange		90–131
	P(OR)₃	NR₃	Cl	Trans	White		
	P(OPh)₃	NR₃	I	Trans			
Pt	P(O alkyl)₃	NR₃	Cl	Trans	Green to yellow		92–98
	P(OPh)₃	PR₃	Me	Cis			82–84
	P(OPh)₃	PR₃	Me,Cl[b]	Cis			101–103
	P(OPh)₃	PR₃	Cl	Cis	White		152–154
	P(OPh)₃	PR₃	Cl	Trans	Yellow		113–117
	PR₂OH	ER₃[a]	Cl	Cis	Cream to white	264–311	133–227
	PR₂OMe	PR₃	Cl	Cis	White		184–220
	PPh₂OH	PR₃	Br	Cis	Cream to white	178–207	189–204
	PPh₂OH	PR₃	I	Cis	Yellow	149, 162	174

E = P or As.
One of the X groups is Cl and the other Me in this compound.

71

many cases. Thus two metal-halogen bands can often be
identified in cis compounds while only one is expected
for the trans (Tables 11 and 12). Cis complexes generally have large dipole moments ($\sim$10 D) compared to their
trans counterparts but the moments of trans complexes are
not usually zero owing to high apparent atom polariza-
tions.[113] Measuring the change in dielectric constant,
$\Delta\varepsilon$, with respect to mole fraction χ circumvents this problem in trans complexes and a constant value of $\Delta\varepsilon/\chi$ between 1.8 and 2.8 is obtained.[113] This value in cis complexes lies in the 150 to 170 range and thus isomers can
be assigned quite unambiguously. Using this method the
trans configuration of the yellow Pd[P(OR)$_3$]ZCl$_2$ complexes
was deduced.[106] The white analogs may be cis inasmuch as
their method of synthesis was different. Dielectric measurements are not made in many cases, however, and the de-
cision regarding the isomer present is predicted on the
observation that cis compounds are generally lighter in color and
only slightly soluble in nonpolar and soluble in polar organic solvents while the opposite is true for trans compounds. Because of geometrical restrictions of the chel-
ates, Pt[P(OR)$_3$]Cl(OCMeCHCMeO),[274b] Pt(diphos)[OP(OMe)$_2$]$_2$,[426b] and Pd[EtN(PPh$_2$)$_2$]Cl$_2$ are cis and this was confirmed
for the last compound by an x-ray diffraction study.[416]
Although the isomeric form of Pt(PEt$_3$)(p-toluidine)-
(P(O)Ph$_2$)Cl is unknown, the phosphorus-containing anion
was shown to be coordinated through phosphorus by virtue
of the P=O link the stretching frequency of which was observed in the 1050-1100-cm^{-1} region.[108] Similar phos-
phorus-platinum bonds have been postulated in Pt(diphos)-
[OP(OR)$_2$]$_2$,[426b] trans-PtZ$_2$[OP(OR)$_2$]$_2$,[5d,426b] trans-
Z$_2$Pt[OP(OR)$_2$]X,[5d,426b] trans-[P(OH)(OR)$_2$]ZMCl[OP(OR)$_2$],[5d,426b] and M[P(OH)(OR)$_2$]$_2$[OP(OR)$_2$]$_2$[5d,426b] (M = Pd or Pt)
because of the high platinum-phosphorus coupling values
(3400-5800 Hz) which are typical of Pt-L complexes. A trans configuration for PtZ$_2$[OP(OR)$_2$]$_2$ was
assigned from the relatively small ^{2}JPPtP coupling ($\sim$38 Hz)
seen in the ^{31}P A$_2$X$_2$ spectra where Z = PBu$_3$.[426b] Equivalence of the phosphorus atoms in Z$_2$Pt[OP(OR)$_2$]X where Z =
PEt$_3$ indicated that these systems must be trans.[426b]
Strong coupling (547 Hz) of the PEt$_3$ phosphorus atom in
Cl(PEt$_3$)Pt[P(OH)(OMe)$_2$][P(OH)(OMe)$_2$] to the protonated
form of L suggested that these ligands are trans in complexes of this type.[5d,426b] Equivalence of the phosphorus
atoms in M[P(OH)OR)$_2$]$_2$[OP(OR)$_2$]$_2$ complexes showed that proton exchange was occurring.[5d]
    The normal NMR spectra of Ni[P(OMe)$_3$]$_2$(CF$_2$)$_4 \cdot$0.5
C$_6$H$_6$,[149b] Pt[P(OR)$_3$](PR$_3$)X$_2$,[5d] Pt[P(OPh)$_3$](PBu$_3$)Cl$_2$,[5d] and
Pt[P(OPh)$_3$](PBu$_3$)Me$_2$[5d] show that these complexes are diamagnetic and therefore square planar. Fluorine-19 NMR
results are consistent with the presence of a five-membered

ring for the $Ni(CF_2)_4$ moiety and hence a cis relationship
for the L molecules.[149b] Trans and cis forms of Pt-
$[P(OPh)_3](PBu_3)Cl_2$ were assigned on the basis of the larger
^2JPPtP coupling observed for the former (709 Hz) compared
to the latter (20.0 Hz).[5d] The small ^2JPPtP values for
the $Pt[P(OPh)_3][PBu_3]ClMe$ and $Pt[P(OPh_3][PBu_3]Me_2$ complexes
similarly lead to cis geometrical assignments.[5d] The presence of nonequivalent CH_3 groups in the ^1H NMR spectrum
confirmed this conclusion in the latter compound.[5d] The
structures of $[Ni[P(O-o-PhC_6H_4)_3]_2Me_2$[532] and $Ni[P(OPh)_3]_2-$
$NOBr$[242] are probably also square planar.

Cis and trans dimeric nickel(II) complexes of the type
shown in (68) have been identified from the two- and one-
band NO stretching spectra expected, respectively, in the
$1815-cm^1$ region.[45,241,242] Dimeric palladium(II) and
platinum(II) complexes schematically shown in (69) are
known which contain a variety of bridging links (Y =
halide, SR_2, and $OPPh_2$), terminal anions (X = halide, OH,

(68)   (69)

$P(O)(OR)_2$, and $P(O)R_2$), and terminal ligands (Z = L, AsR_3,
Pr_3).[108,330,426a] Most appear to have trans configurations
of the X and Z groups. The yellow compounds formulated in
the 1870s as $PtLCl_2$[472-477] were probably dimers and over
50 years ago Werner suggested the cis and trans forms of
(69) as possibilities in interpreting the early molecular
weight results on these compounds. A summary of the properties of these compounds appears in Table 13 and it can
be seen that the formulations and trans (D_2h) stereochemistries are supported by the infrared data except in the
case of $[R_2(O)P](PEt_3)Pt(SEt)_2Pt(PEt_3)[P(O)R_2]$[108] where
the isomerism can not be deduced from the available data.
The last compound in the table is probably trans by analogy
to the Br, OH, and Cl analogs. The sixth compound in
Table 13 was included since the X groups might have been
coordinated in the fashion $PtOPR_2$ in which case it could
formally have been an uncoordinated $P(OR')R_2$ ligand where
R' is a metal group. The bridging groups in the last four
compounds of Table 13 are precisely such ligands except
that they are dimerized by phosphorus coordination. Bridging apparently is a more powerful driving force here than
P=O bonding and indeed pi bonding in this bond may still
be important by virtue of the delocalization of electrons
which can occur around the ring using d electrons on the
metal. In contrast, NMR equivalence of the phosphorus
nuclei in the third compound in Table 13 mitigates against
$PdOP(OPh)_2Pd$ bonds since bridging and terminal ^{31}P resonances should then have been observed.[426b] Proton exchange

Table 13. Properties of trans-$XZM(Y)_2MZX$ Complexes

M	Y	Z	X	Color	m. or dec. (°C)	ν_X (cm^{-1})	ν_Y (cm^{-1})	ν_{MP} (cm^{-1})
Pd	Cl	P(OR)$_3$	Cl	Yellow to brown Liquid,	114-171	364[b]	266-305[c]	364
	Br	P(OEt)$_3$	Br	Red	Liquid	272[b]	173,[c] 189[c]	354
	Cl	P(OH)(OPh)$_2$	OP(OPh)$_2$	White	170-195			
Pt	Cl	P(OEt)$_3$	Cl	White	170-183	354,[b] 361[b]	264,[c] 325[c]	361
	SEt	PEt$_3$	P(O)R$_2$	White	66-180	1103[d]	969-1010[f]	
	OPR$_2$	EEt$_3$[e]	Cl	White	155	272-290[b]	892-1002[f]	
	OPPh$_2$	PEt$_3$	OH	White	185	3520-3610[g]	967[f]	
	OPPh$_2$	PEt$_3$	Br	Yellow	205	187[b]	991[f]	
	OPPh$_2$	PEt$_3$	I					

[a] The frequency associated with some mode of X as specified by the appropriate footnote.
[b] The absorption(s) associated with the terminal M-halogen stretches of which one strong band is expected, the weak one probably being due to the ^{37}Cl-Pt vibration.
[c] Bridging M-halogen frequencies of which two are expected.
[d] The P=O stretching frequency.
[e] E = P or As.
[f] The P-O stretching frequency.
[g] OH vibration region.

among the phosphorus ligands must also occur in order to account for the single ^{31}P resonance.[426b]
    The method used by bridging OPR_2 systems seems to be $MOP(R)_2M$ but the geometrical arrangement of the two bridges relative to one another is not unambiguously clear [see structures (70) and (71)].   It seems likely that (71) is

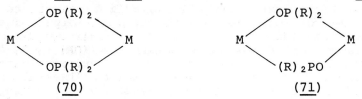

(70)   (71)

the preferred arrangement although (70) is not ruled out. The isomerism problem is further complicated by the possibility of two different "trans" arrangements, (72) and (73), and present evidence allows this distinction to be

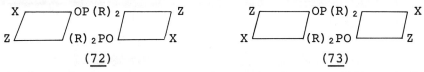

(72)   (73)

made in only one case.   Two isomeric forms, (72) and (73), of $(HO)(Et_3P)Pt[OP(Ph)_2][(Ph)_2PO]Pt(PEt_3)(OH)$ can be isolated both of which are trans since only one OH stretch is observed in both cases.[108]   Because of the lower symmetry, cis complexes can only exist as one isomer (74) if the assumption of the bridging geometry as shown in (71) is correct.   In support of the existence of such cis isomers

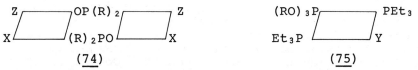

(74)   (75)

is the observation of two chemically inequivalent PMe_3 groups in $Cl(Me_3P)Pt[OP(Ph)_2][(Ph)_2PO]Pt(PMe_3)Cl$ and the bromo analog.[108]   The dimeric platinum dications $\{L_2Pt-[P(O)(OPh_2]_2PtL_2\}^{2+}$ (where $L = PEt_3$) and $ZLPt[P(O)(OPh)_2]_2-PtLZ^{2+}$ [where $L = P(OPh)_2OH$ and $Z = AsEt_3$] have also been reported.[426a]
    The 1:1 electrolyte series $\{Pt[P(OR)_3](PEt_3)_2Y\}^+$ is known where $Y = FC_6H_4$, Cl, Br, or H.[116]   They are assigned the trans configuration (75) on the grounds that the methyl protons in the PEt_3 occur as a quintet owing to virtual coupling which is favored when chemically equivalent phosphorus nuclei are trans.   Furthermore in the hydride derivatives the proton is coupled to two identical and one unique phosphorus thus ruling out structure (76).   Recently, however, cations of the type $\{Pt[P(OR)_3][AsEt_3]_2H\}^+$ were shown to be cis-trans mixtures from the presence of two PtH

stretching frequencies and two hydride proton NMR resonances.[116b] Complexes which very probably contain di-
positive cations are known for nickel and palladium(II)
but the lack of magnetic data for the nickel analog of
(30) or (31)[279] and Pd[P(OEt)$_3$]$_4$[PdCl$_4$][328,330] leaves
their stereochemistry somewhat in doubt.

Nickel(II) complexes very often are five-coordinate.
Two such series of complexes are the trigonal-bipyramidal
NiL$_3$(CN)$_2$ and the [NiL$_5$]$^{2+}$ systems. The former complexes
are red, orange, or yellow solids where L = P(OR)$_{3-x}$Ph$_x$[102,447] but red oils for L = P(OR)$_3$[102,144a,447] except in the
case of P(OMe)$_3$[144a] which is a yellow solid. The solids
melt from 105-137°C and all the compounds exhibit a single
CN stretching frequency indicative of a trans orientation
for these groups. Geometry (77) was confirmed in the case
of Ni[P(OEt)$_2$Ph]$_3$(CN)$_2$ by means of a crystal structure

Et$_3$P ⎯⎯⎯⎯⎯ P(OR)$_3$

Et$_3$P ⎯⎯⎯⎯⎯ H

(76)

N
C
|
L
L
L
|
C
N

(77)

determination.[487] The orange or yellow [NiL$_5$]$^{2+}$ salts are
all diamagnetic and behave as 1:2 electrolytes as salts
of monovalent anions such as ClO$_4^-$ or BF$_4^-$.[143,275,279,520b,525] An x-ray crystal structure investigation showed
nearly perfect trigonal-bipyramidal symmetry for {Ni[P-
(OCH)$_3$(CH$_2$)$_3$]$_5$}(ClO$_4$)$_2$ but with the curious feature that
the axial NiP distances were slightly but measurably decreased over the equatorial distances.[448] Geometrical
constraints due to the tridentate ligand in the complex
shown in (78) are undoubtedly instrumental in forcing the
phosphite to adopt the axial position in a trigonal-
bipyramidal geometry.[167] The red Ni[P(OR)$_3$]$_4$Br$_2$ complexes
are ionic but whether the cation is four or five-coordinate
is not certain.[193,194]

Hydride complexes of nickel of the type [Ni[P(OR)$_3$]$_4$H]$^+$
have been detected in solution and they are five-coordinate
as shown by the quintet hydride proton and the doublet [31]P
NMR spectra.[166,500d,520b] No narrowing of the hydride resonance occurs down to -60°C and the geometry adopted by
these systems may well be similar to that of the pseudotetrahedral CoL$_4$H systems (61) described earlier. A few
organometallic derivatives of nickel(II) such as π-C$_3$H$_5$Ni-
[P(OPh)$_3$]$_2$Br[529a] and CpNiP(OPh)$_3$I[536] are known but their
structures are not.

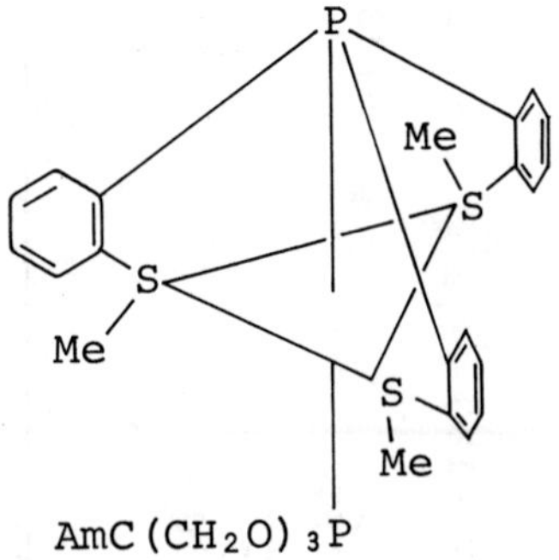

AmC(CH$_2$O)$_3$P

(78)

The only six-coordinate metal complex in this subgroup seems to be the yellow compound Pt(CO)$_2$[P(OPh)$_3$]$_2$F$_2$ whose three-band carbonyl spectrum permitted no deductions for its isomeric form or forms.[340] First reports of {Ni-[P(OR)$_3$]$_6$}$^{2+}$ [275] and {Ni[P(OCH$_2$)$_3$CMe]H$_2$O}$^{2+}$ [525] were later corrected when elemental analyses were discovered to be untrustworthy. These species involve the five-coordinate [NiL$_5$]$^{2+}$ [143] cations discussed earlier.

(7) Cu, Ag, Au

All L complexes of this subgroup contain the metal in the monovalent state. Complexes of the formula CuLX can be monomeric, dimeric or tetrameric in solution depending on the solvent. In the solid state, CuP(OMe)$_3$Cl has structure (79) since it is of the same cubic space group as [CuAsEt$_3$I]$_4$ which is known to have the analogous metal cluster structure.[141] Triphenyl phosphite is a sufficiently strong ligand in these species that it can be used as an extraction agent for copper.[392] The properties of this type of complex are given in Table 14. The novel (MeO)$_3$-P(CuH)$_2$ is insufficiently characterized for guesses to be made concerning its structure although bridging hydrides may be involved.[155] Complexes of the formula CuL$_2$X (Table 15) can be dimeric or tetrameric depending on the solvent used for the measurement.[19,22,28,392,393] The structure of the dimer is likely to be a bridged one as in (80) but

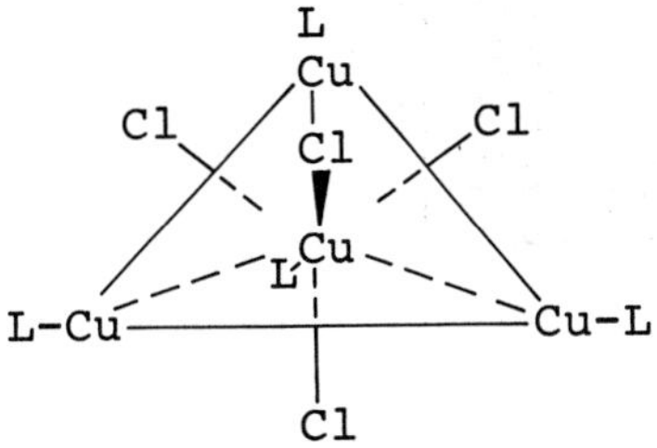

(79)

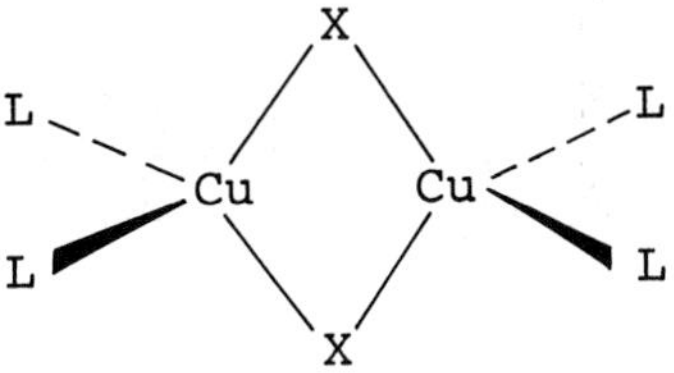

(80)

Table 14. Properties of CuLX Compounds

L	X	Color	m. or dec. (°C)	$\delta^{31}P$
P(OR)₃	Cl	White	Liquid, 88–208	–118 to –132
P(OR)₂Ph	Cl	White	141	
P(NMe₂)F₂	Cl	White		
P(OPh)₂OH	Cl	White	120	
P(OR)₃	Br	White	Liquid, 27–180	–112
P(OR)₂Ph	Br	White	143	
P(OR)₃	I	White	61–199	–103
P(OR)xR₃₋x	I		126–167	
P(NEt₂)xPh₃₋x	I	White	194–225	
P(OPh)₂NEt₂	I		111	
P(OR)₃	CN	Colorless	Liquid	–113
P(O-i-Pr)₃	CNS		95	
P[O₂(CH₂)₃](OCH₂Ph)	CNS		150	

Table 15. Properties of CuL_2X Compounds

L	X	Color	m. or dec. (°C)
$P(OR)_3$	Cl	White	Liquid, 69-144
$P(NMe_2)F_2$	Cl	Colorless	
$P(OR)_3$	Br	White	58-148
$P(OR)_3$	I	White	70-187
$P(OR)_3$	CN	White	Liquid, 60-128

the nature of the tetramer is not known. Mixed complexes
of the formula $Cu[P(OPh)_3]Z$ (halogen) where $Z = P(O-alkyl)_3$[32]
or AsR_3[31,32] are also known. The related series $CuLZ$ (halogen) where Z can be a variety of monodentate organic nitrogen
bases have also been realized[31,32,392] and their general properties are recorded in Table 16. The white compound Cu-
$[P(OEt)_3][\alpha,\alpha\text{-bipyridyl}]Cl$[392,395] very likely is a four-
coordinate tetrahedral monomer since the nitrogen base can
chelate.

Most of the remaining silver and copper complexes involve four L ligands and are of the formula $[ML_4]^+$. The
copper(I) complexes [L = $P(O-alkyl)_3$[142,143,236,294,305,
483,523,524] or $P(OEt)_2Ph$[409a]] and silver(I) compounds [L =
$P(O-alkyl)_3$[142,143,279,294,523,524]] are white solids and
the anions can be BR_4^-, ClO_4^-, or Cl^-. The cations are
isoelectronic and undoubtedly isostructural.[142] X-ray
diffraction work has shown that the $\{Ag[P(OCH_2)_3CMe]_4\}^+$
is a nearly perfect tetrahedron.[511] The salt $Ag[OP(OEt_2]_2$
from infrared[485] and x-ray studies[490] contains an AgOP
link and hence the phosphorus remains trivalent in this
case. From infrared studies, the nitrate anion was indeed found to be coordinated in the nonelectrolyte Ag-
$[P(OMe)_3]_2NO_3$.[143] Since the nitrate ion generally behaves in a
bidentate fashion when coordinated, structure (81) is probable. A few trimeric $\{Ag[P(OR)_3]X\}_3$ (X = Cl, Br, or I)
compounds (m. 29-163°C) have been reported[22,28] and $[AgL_2]^+$
and $[AgL_3]^+$ species have been observed in solution in
potentiometric titration studies involving $P(OBu)_3$.[492]

Most monovalent gold complexes of L are of the formula
$Au[P(OR)_3]Y$ and range in color from white (Y = Cl or
$CCPh$[22,127,336]) to yellow [Y = $Mo(CO)_3Cp$[222]] to brown
[Y = $M(CO)_5$[77,292a]]. Compounds where Y = m- or $p\text{-FC}_6H_4$[391]
have also been characterized. The geometry around gold in
these systems is undoubtedly very nearly linear as the
large dipole moments of some $Au[P(OR)_3]Cl$ systems suggest
(6-7 D).[28] A dimeric compound $\{Au[P(OCH_2)_3CMe]_2Cl\}_2$ may
be a chloro bridged species with tetrahedral gold units.[523,524]

A copper(II) complex similar to the cobalt compound
shown in (63) has been reported but again the phosphorus
moiety is not likely to be coordinated.[267]

(8) Zn, Cd, Hg
It is strange that the only reported complexes of L with
divalent zinc and cadmium where all four coordination sites
are occupied by L are the colorless $[ZnL_4]^{2+}$ and $[CdL_4]^{2+}$
cations whose formulations are presumably similar to either
(30) or (31) for the same reasons as those given for the
manganese(II) analog.[279] The remainder of the complexes
in this subgroup are those of divalent cadmium and mercury.
The structure of the few colorless $[HgCl_2]_2P(OR)_3$[25]

Table 16. Properties of CuL(N-base)X Complexes

L	X	Color	m. or dec. (°C)
P(OEt)$_3$	Cl	White	56–80
P(OPh)$_3$	Cl	White	122–126
P(OEt)$_3$	Br	White	66–108
P(OPh)$_3$	Br	White or yellow	125–135
P(OPh)$_3$ [a]	Br		116
P(OEt)$_3$	I	White	60–97

[a] AsPh$_3$ is present in place of an N base.

complexes (m. 93-114°C) is unknown but the yellow to white
Hg[P(NR$_2$)$_2$Ph]I$_2$[173] (m. 134-192°C) complexes are substantially dimerized (82). The isomeric forms present, however, are not known. Dimers are then probably also present when L = P(OR)$_3$ and P(NMe$_2$)$_3$ in the HgLX$_2$ systems[25]
and in Cd[P(NMe$_2$)$_3$]I$_2$.[404] The colorless HgL$_2$I$_2$ complexes

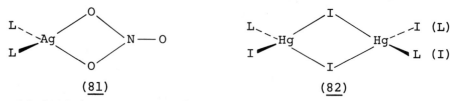

(81)   (82)

(m. 86-213°C) are monomeric and tetrahedral when L =
P(NR$_2$)$_2$Ph[173,404] and this is probably also the case when
L = P(NMe$_2$)$_3$.[404]

The [HgP(O)(OR)$_2$X]$_2$ dimers (X = Br or Cl) seem to re-
tain a great deal of double bond character in the P=O bond
although it is lowered substantially in frequency (1171-
1188 cm^{-1}).[89] This observation along with molecular weight
data led[87b,185] to the postulate of structure (83) involving three-coordinate (probably trigonal planar) mer-
cury(II). A similar P=O bond was postulated in Hg[P(O)-
(OR)$_2$]$_2$.[185,516b] The former compound, however, was also

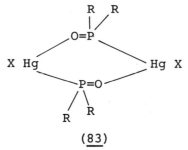

(83)

suggested to involve HgOP bonds from other infrared evidence.[484] A recent x-ray structural determination of
Hg[OP(OEt)$_2$]Cl shows[48b] that half of the mercury coordination spheres are distorted octahedra with approximately
linear P-Hg-Cl groups. The other four "octahedral" sites
are occupied by one oxygen and three chlorine atoms from
neighboring molecules. The remaining mercury coordination
geometries resemble distorted trigonal bipyramids with
linear P-Hg-Cl moieties, with the trigonal plane containing a chlorine and two oxygens from nearby molecules. Although Hg[OP(OEt)$_2$]$_2$ is dimeric in solution, the single
^{31}P resonance[48b] suggests a labile complex.

A series of green HgLI$_3$ compounds where L = [MePR$_2$-
Ph]$^+$[173] (m. 59-132°C) as well as the yellow Hg[Me$_2$NPMe$_3$]I$_3$[84]
have been formulated as [L]$^+$[HgI$_3$]$^-$ salts. In the absence

of conductivity data, the possibility of a monomeric four-coordinate complex of structure (84) or (85) can not be ruled out.

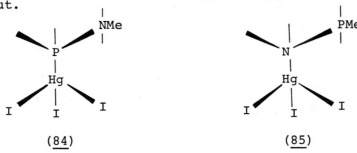

(84)   (85)

(9) B, Al

Boron hydrides,[114,235,236,320,441,442] alkyls,[235,521] and halides[235] as well as aluminum alkyls[129,149a] and halide-alkyl[130] compounds have been used as electron deficient systems with which to make adducts of L. In Table 17 are summarized some general properties of the series Y_3ML where M = B or Al. The BH_3 adducts exhibit two bands characteristic of tetrahedral (C_{3v}) symmetry in the BH IR region (2260-2426 cm^{-1})[235] and where both couplings have been observed, $^1JBP = {}^1JBH$ (96-141 Hz) for reasons as yet not understood.[377,521]

From a comparison of the calculated dipole moments of the possible isomers of $H_3B(MeO)P(OCH_2)_2C(CH_2Cl)Me$ and the measured value of a known mixture of two isomers stereochemically different only at C_4, the geometry at phosphorus was concluded to be as shown in (86a) and (86b).[531b] This configuration at phosphorus in the thermodynamically stable form of six-membered phosphite borane adducts was confirmed

CH₃ structure diagram and CH₂Cl structure diagram

$$BH_3-P \quad CH_2Cl \qquad BH_3-P \quad CH_3$$

(86a)   (86b)

in a recent crystal structure investigation which revealed the result shown in (87).[452b] The adduct (88) isomeric at phosphorus in (87) has also been characterized.[520]

The difunctional ligand $P(OCH_2)_3P$ interestingly adds a BH_3 to the phosphite phosphorus first giving $H_3BP(OCH_2)_3P$ as shown from a comparison of the pertinent NMR parameters of $H_3BP(OCH_2)_3CMe$.[520b] Another H_3B unit then adds to the phosphine phosphorus to give $H_3BP(OCH_2)_3BH_3$ as was shown in a similar manner using $H_3BP(CH_2O)_3CMe$ to generate NMR

Table 17. Properties of Y₃ML Compounds[a]

M	Y	L	m. or dec. (°C)	$\delta^{31}P$	^3JPYCH[b]
B	H	P(OR)₃	Liquid, 46–247	−97 to −118	4.2–12.5
		P(NR₂)₃[c]	9.5–74	−62	5–14
		P(NR₂)₃−xXx[c]	Liquid, 24	−130	
		[P(O)(OR)₂]⁻	50–164[d]	−95	
		PMe(NEt)₂(O-i-Pr)		−117	10.5
	Me	P(OR)₃	Liquid, 8.8	−13	2.2–6.7
	R	P(NMe₂)₃	Liquid, 8.4–9.6		
		P(NMe₂)₃−xMex	40		
	F	P(OR)₃		−93 to −138	
		P(NR₂)₃−xFx		−138 to −143	
Al	Et	P(OR)₃	Liquid	−24 to +41	10.2
		P(NR₂)₃−xRx[a]	Liquid, 80	−59 to −64	10.7
		RN(PPh₂)₂	70–130		3.7
	Et, 2Cl	P(OR)₃	Liquid		6.9

[a] All are colorless or white.
[b] See footnote b, Table 1.
[c] R can be H also.
[d] These temperatures reflect in part the nature of the mono or divalent cation used for isolation.

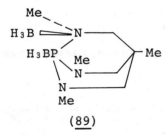

(87)   (88)

parameters for comparison.[520b] The polycyclic amino phosphine P(NMeCH₂)₃CMe has the unique property among amino
phosphines of being able to add a second BH₃ unit after
the phosphorus site is occupied to give (89) as shown from
the more complicated CH₂ proton NMR spectrum and from vapor
tension studies.[319]

(89)

The difunctional ligands P(NMeMeN)₃P, PhP(NMeMeN)₂PPh,
and Ph₂PNMeMeNPPh₂ all add Et₃Al to both phosphorus nuclei.
[486]   The white compounds (m. 79-181°C) exhibit chemically
equivalent methyl groups in the ¹H NMR spectra, and in the
di-adduct of the first ligand, P-P coupling is evident in
the methyl proton NMR absorption.
    The formulation of the adducts where L = [P(O)(OR)₂]⁻[445]
seems reasonable on the grounds that a relatively large
(141 Hz) P-B coupling was observed.[360]
    Boron-phosphorus rather than boron-oxygen or boron-
nitrogen bonds in the compounds listed in Table 17 are
generally postulated[360,441,521] on the grounds that large
BP couplings are recorded (96-141 Hz) and the protons in
ligands such as P(OR)₃ and P(NR₂)₃ remain chemically equivalent even at low temperatures where exchange among the
oxygens and/or nitrogens might be expected to be noticeably
reduced. Boron-phosphorus stretching frequencies have also
been assigned in several of the compounds in the 759-869
cm⁻¹ region.[236] A B-N bond may occur in F₃BP(NMe₂)F₂,
however.[132] All of the compounds in Table 17 are tetrahedral around the boron as was found to be the case for
H₃BP(NH₂)₃ by x-ray diffraction analysis.[402b]   In the
compound Et₃Al(PPh₂)₂NMe, both phosphorus nuclei are coordinated since only one JPNCH coupling is observed.[125] If
the nitrogen were coordinated, a second ligand might be
expected to add as is characteristic for amines. Thus
aluminum in these compounds is five-coordinate because of
the bidentate ligand unless rapid exchange of phosphorus

ligand sites is occurring. The liquid aluminum hydride adduct $HEt_2AlP(NEt_2)Ph_2$ has also been characterized and its ^1H NMR spectrum is consistent with the formulation shown.[125]

The series of BCl_3 adducts $Cl_3BO{=}P(H)(OR)_2$ is only stable at low temperatures.[48] Infrared evidence in the $P{=}O$ stretching region suggests that this ligand does not rearrange to the $P(OH)(OR)_2$ form in the adduct. Attempts to make $Cl_3BP(NMe_2)(CF_3)_2$ resulted in decomposition above 50°C to give the boron-nitrogen ring system (90).[205] The B-N bond was postulated as a result of the lack of observed P-B coupling and a lowered P-N stretching frequency from the free ligand. In contrast, $(CF_3)_2PN(t{-}Bu)BCl_2$ does not dimerize.[205] The evidence for this suggestion is a ^{11}B chemical shift which is typical of three-coordinate boron and a P-N stretching frequency which is characteristic of uncoordinated nitrogen. The steric requirements of the t-Bu group are probably responsible for the monomeric state.[205]

The structure of the $H_7B_3P(OR)_3$ species is unknown but it may resemble the NH_3 analog which consists of a triangular boron framework with the nitrogen bonded to one of the borons.[235] The only H_8B_4L adduct reported characterized is that in which $L = P(NMe_2)F_2$.[101] The crystal structure determination showed the configuration depicted in (91) wherein the configuration of each nitrogen group is planar.[318] The ^{19}F spectrum is consistent with

$$Cl_2B{-\!-\!-}N(Me) \quad P(CH_3)_2$$

$$(CF_3)_2P(Me)N{-\!-\!-}BCl_2$$

(90)

the presence of an additional isomer, namely that in which the ligand and the geminal hydrogen on the boron are interchanged.[101]

(91)

Numerous $H_{12}B_{10}L_2$ adducts are known where $L = P(OR)_3$,[145b,431,489] $P(OR)_2Ph$,[489] $P(SR)_3$,[489] $P(OR)Ph_2$,[468b] $P(OR)_2$-NR_2,[145b] $P(NR_2)_3$,[145b] $P(NR_2)_2Ph$,[145b] $P(NR_2)Ph_2$,[468b] $P(NR_2)_2X$,[145b] $P(NR_2)PhX$,[145b] and $P(NR_2)Ph_2$.[145b]   Where $L = P(OR)_2OH$,[468b] apparently the boron is bonded to phosphorus in contrast to similar BCl_3 adducts.   The structures have not been determined but they are likely to resemble that of the bis-NCMe analog (92).

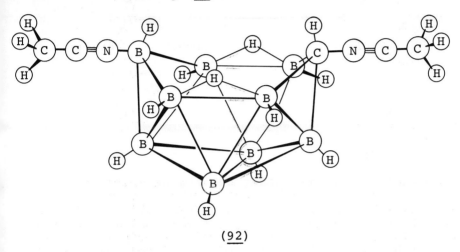

(92)

(10) Ge
The only complex of this element is the yellow liquid $I_2GeP(NMe_2)_3$[297] which although not well characterized is probably similar to the I_2GePR_3 series.   Its structure should be trigonal pyramidal around germanium since this atom still contains an electron lone-pair.

(11) N, P
A black mono adduct of N_2F_4 with $P(OPh)_3$ has been described which is diamagnetic in solution but slightly paramagnetic in the solid state.[292b]
The white solids F_5PL where $L = P(OR)_3$,[81] $P(NMe_2)_3$,[80] and $P(NMe_2)_3-xMe_x$[80,81] decompose easily but P-P bonding is suggested on the basis of IR and NMR evidence.   A CH stretching band (2800 cm^{-1}) which is not observed in coordinated amines remains in the complex.   The ^1H NMR spectrum shows that JPNCH and the chemical shift of the NCH_3 protons remain close to those of the free ligand while the chemical shift of the PCH_3 protons alters appreciably.[80]

(12) O
Phosphite-ozone adducts of the type $O_3P(OR)_3$ are unstable above -15°C and yield the corresponding phosphate and oxygen upon decomposing.[307b,498b]   The compound $O_3P(OCH_2)_3CEt$

is stable up to room temperature, however, and its increased robustness has been ascribed to the rigid and relatively strain-free bicyclic ligand.[78b] All the adducts appear to be sources of singlet oxygen[78b,382b] and this was confirmed from EPR measurements in the case of $O_3P(OPh)_3$.[529b] Possible structures[498b] for these systems are shown in (93) and (94).

(93)   (94)

C. LIST OF COMPOUNDS

The complexes are ordered under the ligands which appear in the order: $P(OR)_3$, $P(OR)_2(OR')$, $P(OR)_2Y$ (Y = R, NR_2, X), $P(OR)Y_2$ ($Y_2 = R_2$, X_2), $P(NR_2)_3$, $P(NR_2)(NR'_2)R$, $P(NR_2)_2Y$ (Y = R, X), $P(NR_2)Y_2$ ($Y_2 = R_2$, X_2), $P(NR_2)(OR)R$, $P(NR_2)RX$, $P(SR)_3$, $P(SR)X_2$, PX_3, cationic ligands, anionic ligands, ligands with multiple bonding sites and miscellaneous. The ligand groups are further broken down into the specific ligands by the number of carbon and hydrogen atoms in the ligand. If the number of carbon atoms is equal to that in the nonidentical ligands, then the lesser number of hydrogens is the criterion. Complexes containing more than one identical ligand L will appear in succeeding lists (in increasing order of the number of identical ligands) under the carbon-hydrogen index of the particular ligand involved. Ligands containing isomeric organic moieties are listed separately in increasing order of complexity. Because of the variable polymeric behavior of the copper(I) halide and pseudohalide complexes in solution, all of them will be listed for L = 1 if the ratio of L to Cu = 1, L = 2 if this ratio is 2, etc. Complexes containing nonidentical L ligands will be found in the ligand group containing the smaller number of carbon atoms. Within the group corresponding to a given ligand, the complexes are arranged by acceptor atom according to the subgroups in the periodic chart. For a given acceptor, the compounds will be listed by increasing formal oxidation state employing the definitions discussed in the preceding sections.
    Many compounds will appear in the compilation for which data are very sparse or even nonexistent. This is justified here on the ground that a wide range of complexes can

be synthesized by analogy to those which are well-charac-
terized. In some cases the isomeric form is not specified
because sufficient data are lacking. On the other hand,
isomerically unassigned compounds from earlier reports
were given predicted assignments in cases where it was
reasonably obvious from later studies which isomeric form
was involved.

Following each compound is a number which identifies
a synthetic method discussed earlier and in some instances
several methods are given. For some compounds more than
one reaction is involved and so the route given may be only
a main step. The ^{31}P chemical shifts are given in ppm with
respect to 85% phosphoric acid. Upfield and downfield
shifts are positive and negative, respectively. The solvent
used appears in parentheses. Where it was necessary to
convert from P_4O_6 as a standard to 85% H_3PO_4, the ^{31}P
chemical shift of P_4O_6 was assumed to be -112 ppm from 85%
H_3PO_4.

Where more than one value for the recorded numerical
data is available for a complex, only one is given if they
do not greatly differ and the choice was necessarily often
arbitrary. References are given, however, to all reports
in which the data appear. When widely differing values
for data were found, they are given and their sources
cited. If a complex is discussed only in a single reference, the source is indicated with the synthesis route.

Several implications of the abbreviations used may not
be obvious and these will now be discussed briefly. If
the known cis or trans relationships of groups in the complex were not clear using the usual format (e.g., cis-
$Cr(CO)_2(bipy)L_2$) then the groups whose geometry is to be
implied immediately follow the Greek prefix (i.e., cis-
$(OC)_2Cr(bipy)L_2$). The melting points (m.) which are quoted
often occur with decomposition. Melting and decomposition
were not differentiated because true melting implies re-
solidification at the same temperature. This seldom occurs with the coordination compounds discussed here because
decomposition generally is not visually apparent.

P(OR)$_3$

TYPE: C_3H_5;L = $\overline{PO(OCH_2)_2}\overset{\frown}{C}H$

H_3BL.   I.1.f.[520b] IR

TYPE: $2C_3H_5$;L_2 = $[\overline{PO(OCH_2)_2}\overset{\frown}{C}H]_2$

PtL_2Cl_2.   IV.1.[475] Colorless. Liquid.
$[PtLCl_2]_2$.   IV.1.[475,477] Yellow. Liquid.

TYPE: C_3H_9; L = $P(OMe)_3$

CpV(CO)$_3$L. I.1.a.[493b] Red. b.$_{0.001}$ ∿122. IR, anal.
Cr(CO)$_5$L. I.1.a.[362] White.[339c] Liquid.[339c] IR,[339c]
    $^{31}P(C_6H_6)$ =179.6,[362] NMR.[339c]
trans-Cr(CO)$_4$L[P(NMe$_2$)$_2$CN]. I.1.f.[144c] Yellow. IR, anal.
Mo(CO)$_5$L. I.1.a,[152,430] I.1.b.[153] m. 8.[430] ^{31}P -162,[322,]
    [362] ^1H,[53,322] IR,[59b,133,150,152,430] Raman,[430] anal.,[152]
    kinetics.[153]
cis-Mo(CO)$_4$LP(OCH$_2$)$_3$CEt. I.1.a.[406] $^{31}P(CDCl_3)$ -164.3(L),
    -139.6[P(OCH$_2$)$_3$CEt], ^{2}JPMoP = 48,[405,406] ^1H,[406] IR,[405]
    mol. wt.[406]
cis-Mo(CO)$_4$L[P(NMe$_2$)$_2$CN]. I.1.f.[144c] White. ^1H, IR,
    anal.
trans-Mo(CO)$_4$LP(OCH$_2$)$_3$CEt. I.1.a.[406] $^{31}P(CDCl_3)$ -176 (L) ^{2}JPP =
    185.[405,406] IR,[406] ^1H,[406] mass. spec.,[406] mol. wt.[406]
fac-Mo(CO)$_3$(bipy)L. I.1.a.[151,274a] Red,[274a] m. 183.[151]
    IR,[151] kinetics.[274a]
fac-Mo(CO)$_3$(o-phen)L. I.1.a.[151,274a] Red,[274a] m. 231.[151]
    IR,[151] anal,[274a] kinetics.[274a]
fac-Mo(CO)$_3$(tmen)L. I.1.c.[158] IR, kinetics.
Na[CpMo(CO)$_2$L]. III.3.[370] ^1H,[356] IR.[356]
trans-CpMo(CO)$_2$LHgCl. I.1.a.[370] Yellow. ^1H, IR, anal.,
    mol. wt.
trans-CpMo(CO)$_2$LHgBr. I.1.a.[370] Yellow. ^1H, IR, anal.,
    mol. wt.
trans-CpMo(CO)$_2$LHgI. I.1.a.[370] Yellow. ^1H, IR, anal.,
    mol. wt.
CpMo(CO)$_2$LSnCl$_3$. III.1.[371] Yellow, m. 180. ^1H, IR,
    anal., mol. wt.
CpMo(CO)$_2$LSnBr$_3$. III.3.[371] Yellow, m. 198. ^1H, IR,
    anal., mol. wt.
CpMo(CO)$_2$LSnI$_3$. III.3.[371] Orange-red, m. 2.5. ^1H, IR,
    anal., mol. wt.
trans-CpMo(CO)$_2$LSnPh$_3$. III.1.[356] White, m. 181-183. ^1H,
    IR, anal.
cis-CpMo(CO)$_2$LCl. III.3.[33] m. 98-100. ^1H, IR, anal.
CpMo(CO)$_2$LCl(cis-trans mixture). III.3.[370] Orange. ^1H,
    IR, anal.
CpMo(CO)$_2$LCl. I.1.[369] kinetics.
CpMo(CO)$_2$LBr(cis-trans mixture). III.2.[370] Orange. ^1H,
    IR, anal.
trans-CpMo(CO)$_2$LI. I.1.a.[355,356] Red,[355] m. 82-84. ^1H,[355,]
    [356] IR,[355,356] anal.[355]
CpMo(CO)$_2$LI(cis-trans mixture). III.2[370] or from isomer-
    ization of trans form in CS$_2$.[355] Red.[370] ^1H, IR,
    anal., mol. wt.
trans-CpMo(CO)$_2$LMe. III.4.[42] ^1H.
CpMo(CO)$_2$LMe(cis-trans mixture). III.2,[370] IV.2.[219b]
    Yellow.[370] ^1H,[219b] IR,[219a,b,370] anal.,[219a,b,370]
    mol. wt.[37,219b]

cis-CpMo(CO)$_2$LH.   I.1.a,[33] III.3.[356] m. 42. ^{1}H,[33,356]
   IR,[33,356] anal.[33]
CpMo(CO)$_2$LH(cis-trans mixture).   III.2.[370] Colorless oil.
   ^{1}H, IR, anal., very sensitive to oxygen.
trans-CpMo(CO)$_2$L(COMe).   I.1.d.[42,92,146,148,225]   Yellow.[92]
   ^{1}H,[42,146] IR,[92,146,225] anal.[92,146]
trans-CpMo(CO)$_2$L(COEt).   I.1.d.[146]   Low melting solid.
   ^{1}H, IR, anal.
(π-Indenyl)Mo(CO)$_2$L(COMe)(cis-trans mixture).   I.1.d.[225]
   IR, kinetics.
Na[CpW(CO)$_2$L].   III.3.[370]
W(CO)$_5$L.   I.1.a.[208,362] White.[339c] Liquid.[339c] ^{31}P(C$_6$H$_6$)
   -137.3,[362] ^{1}JWP = 398,[208] ^{1}H,[208,339c,362] IR.[208]
CpW(CO)$_2$LCl(cis-trans mixture).   III.3.[370] Orange. ^{1}H,
   IR, anal.
CpW(CO)$_2$LBr(cis-trans mixture).   III.2.[370] ^{1}H, IR.
CpW(CO)$_2$LI(cis-trans mixture).   III.2.[370] Red. ^{1}H, IR,
   anal., mol. wt.
CpW(CO)$_2$LMe(cis-trans mixture).   III.2.[370] Yellow. ^{1}H,
   IR, anal., mol. wt.
CpW(CO)$_2$LH(cis-trans mixture).   I.1.a,[291] III.2.[370] Yel-
   low.[370] ^{1}H,[291,370] IR,[291,370] anal.[370]
Mn$_2$(CO)$_9$L.   IV.1.[123] IR.
CpMn(CO)$_2$L.   I.1.a.[493b] Yellow,[493b] b.$_{0\ 01}$ ~110,[493b]
   anal.,[493b] IR,[493b,494a] useful as herbicide.[469]
Mn(CO)$_4$L(COMe)(cis-trans mixture).   I.1.d.[311a] Yellow.
   ^{1}H, IR.
cis-Mn(CO)$_4$LMe.   I.1.a.[311a] Oil. ^{1}H, IR.
cis-Mn(CO)$_4$L(C$_6$H$_4$-p-F).   I.1.a.[491b] Oil. ^{19}F, IR.
cis-Mn(CO)$_4$L(C$_6$H$_4$-m-F).   I.1.a.[491b] Oil. ^{19}F, IR.
(R$_2$Sn)$_2$Fe$_2$(CO)$_7$L.   I.1.a.[290] IR.
IHgFe(CO)$_2$(NO)L.   III.3.[94b] IR, very unstable.
Fe$_3$(CO)$_{11}$L.   I.1.a.[132] Green, m. 67. ^{1}H, IR, anal., mol.
   wt.
Fe(CO)$_4$L.   I.1.f.[145a,436,437] m. 45-46.[145a] ^{1}H,[145a] IR,[59b,
   145a,436,437] Raman,[145a] anal.,[145a] mass spec.[78a,83]
Fe(CO)$_2$L(PMe$_3$)$_2$.   I.1.a.[363] IR.
Fe(CO)$_2$L(PEt$_3$)$_2$.   I.1.c.[437] IR.
C$_4$H$_6$Fe(CO)$_2$L.   IR.[437]
C$_7$H$_8$Fe(CO)$_2$L.   IR.[437]
C$_8$H$_{12}$Fe(CO)$_2$L.   I.1.c.[363] IR.
Cp(OC)Fe(CO)$_2$FeLCp(cis-trans mixture).   I.1.a.[216,217]
   Red.[217] ^{1}H,[216,217] IR,[216,217] conductivity,[217] anal.[217]
CpFe(CO)LSPh.   I.1.b.[136] Brown, m. 71. IR, UV, anal.
trans-[Fe(Et$_2$PCH$_2$CH$_2$PEt$_2$)$_2$LH]BPh$_4$.   I.1.b.[35b] ^{1}H, Moss-
   bauer, anal.
trans-[Ru(Et$_2$PCH$_2$CH$_2$PEt$_2$)$_2$LH]BPh$_4$.   I.1.b.[35b] White. ^{1}H,
   anal.
trans-[Os(Et$_2$PCH$_2$CH$_2$PEt$_2$)$_2$LH]BPh$_4$.   I.1.b.[35b] White. ^{1}H,
   anal.
Co$_2$(CO)$_7$L.   I.1.a.[496] IR.

Co(CO)$_2$L(PPh$_3$)(COMe). I.1.a.[231] Yellow, m. 105-106.5.
(π-C$_3$H$_5$)Co(CO)$_2$L. I.1.a.[230] Kinetics.
Co(CO)$_3$L(COMe). II.[231] Oil. IR.
Co(NO)(CO)$_2$L. I.1.a.[500a] Red liquid.[500a] IR,[124,500a]
   anal.,[500a] mol. wt.,[500a] kinetics.[500a]
[Ph$_4$As][CoL(NO$_3$)$_2$]. II.[143] Blue, diamagnetic. IR, con-
   ductivity, anal.
CoL[ONC(Me)C(Me)NOH]$_2$Me. IR,[468a] anal.
CpRh(CO)L. I.1.a.[471] Green. Oil. ^{1}H, IR, anal.
mer-RhL(PBu$_3$)$_2$Cl$_3$. ^{31}P -92.4 (L), -6.5 dd (PBu$_3$),[94b] ^{1}H.
IrL(PEt$_2$Ph)$_2$H$_3$. II.[339b] ^{1}H.
Ni(CO)$_3$L. I.1.a.[57,62] Liquid.[62] ^{31}P -161.4,[362] dipole
   moment 3.22 D,[61] ^{1}H,[364] IR,[59b,62,76,493b,500b] Raman.[76]
trans-PdL[HN(CH$_2$)$_5$]Cl$_2$. III.3.[105,106,112] Orange-yellow,[112] m. 90-91.[112] IR,[106] anal.[112]
trans-PdL(H$_2$N-p-MeC$_6$H$_4$)Cl$_2$. I.1.f.[106,112] Yellow-
   orange,[112] m. 131-131.5.[112] IR,[106] anal.,[112] mol.
   wt.[112]
PdL(H$_2$N-p-MeC$_6$H$_4$)Cl$_2$. IV.3.[184] White.
trans-PdL(H$_2$N(CH$_2$)$_7$Me)Cl$_2$. I.1.f.[106] IR.
PdL(NH$_3$)Cl$_2$. IV.3.[183] Colorless.
PdL(NC$_5$H$_5$)Cl$_2$. IV.3.[184]
trans-PtL[HN(CH$_2$)$_5$]Cl$_2$. III.5.[110] Green-yellow,[110] m.
   98-99.[110] IR,[104,105] UV-visible,[107] conductivity,[110]
   anal.[110]
PtLCl(OCMeCHMeCO). II.[274b] m. 90-91. ^{1}H, IR, anal.
trans-[PtL(PEt$_3$)$_2$H]ClO$_4$. I.1.b.[116a] White. ^{1}H, IR,
   conductivity, anal.
[PtL(AsEt$_3$)$_2$H]ClO$_4$ (cis-trans mixture). I.1.b.[116b] ^{1}H,
   IR, anal.
trans-[PtL(PEt$_3$)$_2$(p-FC$_6$H$_4$)]ClO$_4$. I.1.b.[116a] ^{1}H, conductivity, anal.
trans-[PtL(PEt$_3$)$_2$(m-FC$_6$H$_4$)]ClO$_4$. I.1.b.[116a] ^{1}H, conductivity, anal.
trans-[PtL(PEt$_3$)$_2$Br]ClO$_4$. I.1.b.[116a] IR, conductivity,
   anal.
trans-[PtL(PEt$_3$)$_2$Cl]ClO$_4$. I.1.b.[116a] IR, conductivity,
   anal.
PtL(P(OEt)$_3$)Cl$_2$. IV.1.[476] Colorless oil.
CuLCN. I.1.f.[393] Colorless. Liquid. Anal.
CuLI. m. 175-177.[19,228]
CuLBr. m. 180-182.[19,228]
CuLCl. IV.1.[51,154] White,[154] m. 190-192,[19] 200-217.[154]
   ^{31}P -132,[483] ^{1}H,[483] conductivity,[483] anal.[13] Crystal-
   lizes in same cubic space group as (Et$_3$AsCuI)$_4$ and
   presumably has a similar type of tetrameric structure[141]
   (see text), useful as preparative reagent for vinyl
   phosphate synthesis[51] and as a paint ingredient.[228]
CuL[P(OPh)$_3$]Br. I.1.f.[32] White, m. 46-50. Anal.
(CuH)$_2$L. I.1.b.[155] Anal.
AgLCl. I.1.f.[483] ^{31}P -135, ^{1}H, conductivity.

AuLCl. I.1.b,[22] I.1.f,[22] IV.1.[22,336] m. 101-102.[22] Dipole moment 7.08 D.[28]

AuLCCPh. I.1.f.[127] Liquid.

(HgCl$_2$)$_2$L. I.1.f.[25] Colorless, m. 114. Anal.

HgLBr$_2$. I.1.f.[25] Colorless, m. 107-109.

H$_3$BL. I.1.f,[236,320] I.1.b.[441,442] b.$_2$86,[441] ^{31}P(MeCN) -118,[521] ^{1}H,[377,521] IR,[236] anal.,[441] mol. wt.,[441] Faraday effect.[320]

D$_3$BL. I.1.f.[236] IR.

Me$_3$BL. I.3.[521] ^{31}P(MeCN) -106, ^{1}H.

Et$_3$AlL. I.3.[129,149a] Liquid.[129] ^{1}H, anal.

Cl$_2$EtAlL. I.3.[130] Anal., rearranges rapidly to (MeO)$_2$-MePOAlEtCl$_2$.

F$_5$PL. I.3.[81] Solid. Decomposes at 0°C to F$_4$POMe, FP(OMe)$_2$, [(MeO)$_3$PMe]PF$_6$, and F$_2$P(OMe).

TYPE: 2 C$_3$H$_9$; L$_2$ = [P(OMe)$_3$]$_2$

cis-Cr(CO)$_4$L$_2$. I.1.a.[339c] White. Liquid. ^{31}P(C$_6$H$_6$) -180.2, IR.

trans-Cr(CO)$_4$L$_2$. I.1.a,[339c] I.1.c.[176,363,405] White,[339c] m. 81.[339c] ^{31}P(C$_6$H$_6$) -193.1,[339c] ^{2}JPCrP = -15.0,[54] IR,[176,405] mass spec.,[78a,405] isomerizes to cis form at -15°C.[363]

cis-Mo(CO)$_4$L$_2$. I.1.c.[176,363,406,429,430] Yellow. Liquid.[363,429] ^{31}P(C$_6$D$_6$) -164.77, ^{2}JPMoP = -40.5,[54] ^{1}H,[53,54,322,364,405] IR,[133,176,363,430] anal.,[363] kinetics.[176]

trans-Mo(CO)$_4$L$_2$. I.1.a.[176] m. 89. ^{31}P(C$_6$D$_6$) -174.2, ^{2}JPMoP = +162,[53,54] ^{1}H,[53,54,322,405] IR,[59b,133,152,176,430] kinetics.[176]

cis-(OC)$_2$Mo(o-phen)L$_2$. I.1.a.[274a] Purple-black. Anal.

cis-(OC)$_2$Mo(bipy)L$_2$. I.1.a.[274a] Purple-black. Anal.

Hg[CpMo(CO)$_2$L]$_2$. I.1.a.[370] Yellow. ^{1}H, IR, anal., mol. wt.

[CpMo(CO)$_2$L]$_2$. I.1.a,[219a,b] III.2.[370] Red.[370] ^{1}H,[370] IR, anal.,[370] mol. wt.[370]

CpMoCOL$_2$SnCl$_3$. III.1.[371] Yellow-orange. ^{1}H, IR, anal., mol. wt.

[CpMo(CO)$_2$L$_2$][CpMo(CO)$_3$]. II.[219a,219b] IR.[219b]

trans-[CpMo(CO)$_2$L$_2$]BPh$_4$. II.[219a,219b] Yellow. ^{1}H,[219b] IR,[219b] conductivity,[219b] anal.[219b]

cis-W(CO)$_4$L$_2$. I.1.b,[339c] I.1.c.[156,157,201,363] White.[339c] Liquid.[339c] ^{31}P(C$_6$H$_6$) -141.1,[339c] IR,[339c] some isomerization to trans occurs,[339c] kinetics.[156,157,201]

trans-W(CO)$_4$L$_2$. I.1.b.[339c] White, m. 94. ^{31}P(C$_6$H$_6$) -147, IR.

[Mn(CO)$_4$L]$_2$. I.1.a,[332] IV.1.[123] IR,[123] anal.,[332] mass spec.[333]

mer-Mn(CO)$_3$L$_2$Me. I.1.a.[359] White, m. 60.0-61.5. ^{1}H, IR, anal., kinetics.

mer-Mn(CO)$_3$L$_2$Ph. I.1.a.[36] m. 80.5-81.[37] IR.

mer-Mn(CO)$_3$L$_2$(COMe). I.1.d.[359] Yellow.[359] Oil.[359]
^{1}H,[359] IR,[359] kinetics.[455]

fac-Mn(CO)$_3$L$_2$Br. III.3.[502] Orange, m. 110. ^{1}H, IR, anal.

cis-(OC)$_2$MnL$_2$[S$_2$P(OMe)$_2$]. I.1.a.[227] ^{1}H, IR.

cis-(OC)$_2$MnL$_2$[CF$_3$COCHCOCF$_3$]. I.1.a.[226] Purple, m. 40.
^{1}H, IR.

[Co(CO)$_3$L]$_2$. I.1.a,[357] II.[244] Red,[357] m. 130.[357] IR,[357]
anal.[357]

Hg[Fe(CO)$_2$(NO)L]$_2$. I.1.a.[94b] IR, anal., unstable.

Fe$_3$(CO)$_{10}$L$_2$. I.1.a.[432] Green, m. 105. ^{1}H, IR, anal.,
mol. wt.

cis-Fe(CO)$_3$L$_2$. The IR data consistent with this formulation were later[437] shown to be due to the presence of
coordinated olefin.

trans-Fe(CO)$_3$L$_2$. I.1.a,[436] I.1.f.[145a] m. 71-72.[145a]
^{1}H,[145a] IR,[59b,145a,437] Raman,[145a] anal.[145a]

Fe(CO)$_2$L$_2$(PEt$_3$). I.1.a.[437] IR.

trans-(OC)$_2$-cis-L$_2$Ru(CF$_2$)$_2$. III.2.[137] Yellow. ^{1}H, IR,
anal., mass spec.

cis-(OC)$_2$-trans-L$_2$Ru(O$_2$CCF$_3$)$_2$. III.2.[88] m. 125-128. ^{1}H,
IR, anal.

cis-(OC)$_2$-trans-L$_2$RuCl$_2$. III.2.[88] White, m. 125-127.
^{1}H, IR, anal., mass spec., mol. wt.

Co(CO)$_2$L$_2$Me. III.2.[231] Yellow, m. ∿50. IR, anal.

Co(CO)$_2$L$_2$(COMe). I.1.d.[231] White, m. 35. IR, anal.

Co(NO)(CO)L$_2$. I.1.a,[93,94a,499] IV.1.[124] Liquid.[94a]
IR,[124] kinetics.[93,499]

CpRhL$_2$. I.1.a.[471] Red. Oil. ^{1}H, IR, anal.

[RhL$_2$][BPh$_2$]. III.4.[402a] Yellow. ^{1}H, conductivity, x-ray
structure shows arene of BPh$_4$ π-bonded to Rh, Rh-P =
2.185Å, P-Rh-P = 90°, P, Rh, P plane makes angle of
91° with mean plane of bonded phenyl ring.

[Ir(CO)L$_2$(PPh$_2$Me)]ClO$_4$. I.1.b.[115] Anal.

Ni(CO)$_2$L$_2$. I.1.a.[57,62,440] Liquid.[440] ^{31}P(C$_6$H$_6$)
-165.1,[362] ^{2}JPNiP = 10,[405] dipole moment 3.11 D,[61]
^{1}H,[362,364,405] IR.[62,76,517]

Raman,[76] useful as catalyst for cyclizing olefins
and for dimerization of conjugated dienes.[74]

NiL$_2$(CNC$_6$H$_{11}$)$_2$. IV.3.[372] IR.

NiL$_2$(CF$_2$)$_4$·½PhH. III.2.[149b] m. 89-91. ^{1}H, IR, mass
spec.

NiL$_2$I$_2$. I.2.[285] Oil.

NiL$_2$Br$_2$. I.2.[285] Oil.

NiL$_2$Cl$_2$. I.2.[285] Oil.[285] ^{1}H.[170]

cis-PdL$_2$I$_2$. III.3.[281] Orange, m. 115-146. Ir, anal.

PdL$_2$I$_2$ (cis-trans mixture). ^{1}H.[405]

cis-PdL$_2$Cl$_2$. I.1.a,[281,282] IV.1.[182,183,184] Colorless,[281]
m. 128-129.[281] ^{31}P(CDCl$_3$) -96.29, ^{2}JPPdP = +79.9,[54]
^{1}H,[54,405,465] IR,[281] anal.[282]

trans-$[PdLCl_2]_2$. I.1.b.[111] Yellow-orange, m. 114-115.
  Anal., mol. wt., conductivity.
cis-PtL_2I_2. ^{1}H.[405]
PtL_2I_2. III.3.[24] m. 145-147.
PtL_2Br_2. IV.1.[454] Anal.
cis-PtL_2Cl_2. I.1.f,[24] IV.1.[24,281,453,475,477] White,[453]
  m. 137-138.[24] ^{1}H,[405] IR,[281] dipole moment 9.23 D,[28]
  anal.,[453] mol. wt.[453]
$PtL_2(H_2NC_6H_5)Cl_2$. III.4.[128]
$PtL_2(H_2N$-p-$MeC_6H_4)Cl_2$. III.4.[128]
$PtL_2(NH_3)_2Cl_2$. III.4.[474,477] Anal.[477]
$[PtLCl_2]_2$. IV.1.[24,475,476] Yellow,[476] orange,[24] m. 131-
  132.[24] Anal.,[476] unstable.[24]
CuL_2CN. I.1.f.[393] m. 60-61. Tetrameric at freezing
  point but dimeric in organic solvents.
CuL_2I. m. 69-70.[19]
$[AgL_2NO_3]$. I.2.[143] Colorless. IR, anal., mol. wt.,
  conductivity.
$H_{12}B_{10}L_2$. I.1.b,[145b] II.[145b] White, m. 58-66. UV, anal.

TYPE: 3 C_3H_9; L_3 = $[P(OMe)_3]_3$

fac-$Cr(CO)_3L_3$. I.1.a.[176,339c,425] White. ^{31}P(C_6H_6)
  -186, IR.[176,425]
mer-$Cr(CO)_3L_3$. I.1.a.[339c] White. Liquid. ^{31}P(C_6H_6)
  -194.7 (two trans L), -188 (unique L), IR.
fac-$Mo(CO)_3L_3$. I.1.c.[158,423,425,429] White,[158,425] m.
  140.[425] ^{31}P(CCl_4) -166s,[322] ^{1}H,[322,425] IR,[133,152]
  anal.,[425] kinetics.[158,423,425]
mer-$Mo(CO)_3L_3$. I.1.c.[429] ^{31}P(CCl_4) -177d(trans L),
  -165t(cis L), ^{2}JP cis MoP trans = 47,[322] ^{1}H,[322,405]
  IR,[133,430] anal.[429]
$Mo(CO)_3L_3$ (fac-mer mixture). I.1.c.[430] Crystalline. IR.
fac-$W(CO)_3L_3$. I.1.b.[339c] White, m. 145. ^{31}P(C_6H_6)
  -145.3, IR.
mer-$W(CO)_3L_3$. I.1.b.[339c,422-425] White. ^{31}P(C_6H_6)[59b]
  -148.6 (two trans L), -142.8 (unique L), IR,
  anal.,[425] kinetics.[422,425]
Na$[Mn(CO)_2L_3]$. III.1.[502] IR.
cis-$(OC)_2MnL_3Me$. I.1.a.[502] White, m. 117. ^{1}H, IR.
cis-$(OC)_2MnL_3Br$. III.3.[502] Orange, m. 161. ^{1}H, IR, anal.
$Fe(CO)_2L_3$. I.1.a,[436] I.1.c.[363] White.[363] IR,[363,436,437]
  anal.[363]
$Fe_3(CO)_9L_3$. I.1.a.[432] Green, m. 107. ^{1}H, IR.
$Co(CO)L_3Me$. I.1.a.[231] Yellow, m. 49-50. IR, anal.
$Co(NO)L_3$. IV.1.[124] IR.
$[Co(CO)_2L_3][BPh_4]$. II.[30] IR, anal.
$[Co(CO)_2L_3][Co(CO)_4]$. II.[30] Yellow. IR, anal.
RhL_3Cl. I.1.c.[376]
$Ni(CO)L_3$. I.1.a.[57,62] White,[62] m. 97-98.[62] ^{31}P(C_6H_6)
  -165.8,[362] dipole moment 2.80 D,[61] ^{1}H,[362,364]

IR,[57,62,76,517] Raman.[76]
trans-$NiL_3(CN)_2$. I.l.f.[144a] Red-orange. [1]H, IR, UV, anal., mol. wt.
NiL_3I_2. I.l.f,[520b] II,[520b] IV.1.[520b] Purple. [1]H, UV, diamagnetic, nonconductor, anal.

TYPE: 4 C_3H_9; L_4 = $[P(OMe)_3]_4$

cis-$Cr(CO)_2L_4$. I.l.a,[339c] I.l.c.[429,430] White,[430] m. 290.[339c] [31]P(C_6H_6) – 187 (two trans L), -198 (two cis L),[339c] [1]H,[322] IR,[339c] anal.
cis-$Mo(CO)_2L_4$. I.l.a,[339c] I.l.c.[429,430] Yellow,[339c] white,[430] m. 200.[339c] IR.
cis-$W(CO)_2L_4$. I.l.a,[339c] I.l.c.[339c] White, m. 100. [31]P(C_6H_6) -148.6, IR.
$Hg[CpW(CO)_2L_2]_2$. I.l.a.[370] Yellow. IR, anal., mol. wt.
trans-$Mn(CO)L_4Me$. I.l.a.[502] White, m. 176. [1]H, IR, anal.
trans-$Mn(CO)L_4Br$. III.3.[502] Orange, m. 131. [1]H, IR, anal.
CoL_4H. II.[382a] Useful as a gasoline additive and hydrogenation catalyst.
$[Co(CO)L_4][BPh_4]$. I.l.a.[30] Yellow. IR, anal.
$[RhL_4][BPh_4]$. I.l.b.[214] Anal., mol. wt.
NiL_4. I.l.a,[57,62] II,[143,285,526] IV.[313] White,[64] m. 132.[313] [31]P(C_6H_6) -163,[142,500c] [1]H,[142,313,362] IR,[57] anal.,[285] useful as a gasoline additive and hydrogenation catalyst.[382a]
$[NiL_4H]Cl$. III.4.[166] [1]H.
$[NiL_2(CN)_2]_2$. III.4.[144a] Yellow. IR, UV, anal., mol. wt., unstable.
PtL_4. [195]Pt.[342]
$[PtL_4]Cl_2$. I.l.b.[506] Yellow, m. 100-125.
$[CuL_4]ClO_4$. II.[143] White.[143] Diamagnetic.[143] [31]P(CH_2Cl_2) -125, ^{1}H,[142,305] anal.,[143] conductivity.[143]
$[CuL_4]NO_3$. II.[143] Colorless. Diamagnetic. IR, anal., conductivity.
$[AgL_4]ClO_4$. I.2.[143][483] White.[143] [31]P(Me_2CO) -132,[142] [1]H,[142,483] anal.,[143] conductivity.[143]
$[AgL_4]NO_3$. I.2.[483] [31]P -132, anal., conductivity.

TYPE: 5 C_3H_9; L_5 = $[P(OMe)_3]_5$

$[CoL_5]ClO_4$. II.[143] Yellow. Diamagnetic. IR, anal., conductivity.
$[CoL_5]NO_3$. II.[143] Yellow.[143] Diamagnetic.[143] [31]P(Me_2CO) -147,[142] [1]H,[142] IR,[143] anal.,[143] conductivity.
$[CoL_5]BPh_4$. I.l.a.[30] Yellow. Anal.
$[CoL_5]_2[Co(NO_3)_4]$. II.[143] Red. IR, conductivity.
$[RhL_5][BPh_4]$. I.l.c.[402a]
$[NiL_5](ClO_4)_2$. I.2.[143] Orange.[143] Diamagnetic.[143] [31]P (Me_2CO) -110,[142] [1]H,[142] anal.,[143] conductivity.[143]

TYPE: 6 C_3H_9; L_6 = P(OMe)$_3$]$_6$

[CoL$_6$](ClO$_4$)$_3$. II.[143] Off white. Diamagnetic. IR, anal.,
   conductivity.
[CoL$_5$][CoL(NO$_3$)$_2$]. II.[143] Violet. Diamagnetic. IR, conductivity.

TYPE: C_5H_9; L = P(OCH$_2$)$_3$CMe

Cr(CO)$_5$L. I.1.a.[516,522] White,[522] m. 212.[522] ^{1}H,[405,522]
   IR,[516,522] anal.,[522] mol. wt.[522]
cis-Cr(CO)$_4$L[P(NMe$_2$)$_2$CN]. I.1.f.[144b] White. ^{1}H, IR.
fac-Cr(CO)$_3$(bipy)L. I.1.a.[13] Black. IR, anal., kinetics.
fac-Cr(CO)$_3$(4,4'-diMebipy)L. I.1.a.[13] Kinetics.
fac-Cr(CO)$_3$(o-phen)L. I.1.a.[159] IR, kinetics.
fac-Cr(CO)$_3$(4,7,diPh-o-phen)L.[15]I.1.a.   IR, kinetics.
fac-Cr(CO)$_3$(3,4,7,8-tetraMe-o-phen)L. I.1.a.[15] IR,
   kinetics.
fac-Cr(CO)$_3$(o-phen deriv.)L. Where o-phen derivatives
   are 5,6-diCl, 5-NO$_2$, 3-Me, 3,5,7-triMe, and 3,4,6,7-
   tetraMe. I.1.a.[15] Kinetics.
Cr(CO)$_3$(diphos)L(fac-mer mixture). I.1.a.[175] Kinetics.[175]
fac-Cr(CO)$_3$(NH$_2$CHMeCH$_2$NH$_2$)L. I.1.a.[15] Kinetics.[15]
Mo(CO)$_5$L. I.1.a.[14,516,522] White,[522] m. 197.[522] IR,[516,]
   [522] ^{1}H,[522] anal.,[522] mol. wt.[522]
fac-Mo(CO)$_3$(bipy)L. I.1.a.[197] Red, black. IR, anal.,
   kinetics.
fac-Mo(CO)$_3$(3,4,7,8-tetraMe-o-phen)L. I.1.a.[15,198] Anal.[15]
fac-Mo(CO)$_3$(o-phen deriv.)L(o-phen deriv. = o-phen, 3,4,
   6,7-o-phen, 3,5,7-triMe-o-phen, 3-Me-o-phen, 4,7-
   diMe-o-phen, 5-NO$_2$-o-phen, and 5,6-diCl-o-phen).
   I.1.a.[198] Kinetics.
fac-Mo(CO)$_3$(NH$_2$CH(Me)CH$_2$NH$_2$)L. I.1.a.[15,198] Brown.[15]
   Anal.,[15] kinetics.[15,198]
Mo(CO)$_3$(diphos)L(fac-mer mixture). I.1.a.[175] IR, kinetics.
trans-CpMo(CO)$_2$L(COMe). I.1.d.[146] m. 170.5. ^{1}H, IR,
   anal.
trans-CpMo(CO)$_2$L(COEt). I.1.d.[146] m. 149. ^{1}H, IR, anal.
cis-CpMo(CO)$_2$LH. I.1.a.[33] m. 188. ^{1}H, IR, anal.
W(CO)$_5$L. I.1.a.[522] White,[522] m. 228.[522] ^{1}H,[522] IR,[516,]
   [522] anal., mol. wt.
fac-W(CO)$_3$(o-phen deriv.) L (where o-phen deriv. = o-phen
   itself, 5-NO$_2$, or 3,4,7,8-tetraMe-o-phen). I.1.a.[198]
   Kinetics.[198]
fac-W(CO)$_3$(bipy)L. I.1.a.[197] Kinetics.
cis-Mn(CO)$_4$LMe. I.1.a.[202,311a] White,[311a] m. 180-200,[311a]
   192-193.[202] ^{1}H,[202,311a] IR,[202,311a] anal.[311a]
cis-Mn(CO)$_4$L(COMe). I.1.d.[202,203,311a] White,[311a] m.
   140-210,[311a] 186-187.[202] ^{1}H,[202,203,311a] IR,[202,203,]
   [311a] anal.[311a] Decarbonylates on heating.[202]
cis-Mn(CO)$_4$LI. I.1.b.[8] IR, kinetics.

cis-Mn(CO)$_4$LBr. I.l.a.[10] IR,[9,10] kinetics.[9,10]
Co(NO)(CO)$_2$L. I.l.a.[500a] Yellow-orange, m. 150. IR,
    anal., mol. wt., kinetics.
Fe(CO)$_4$L. I.l.a.[522] Yellow,[522] m. 179.[522] ^{1}H,[522] IR,[516],
    [522] anal.,[522] mol. wt.[522]
Fe$_3$(CO)$_{11}$L. I.l.a.[18] IR, kinetics
CpRh(CO)L. I.l.a.[471] Yellow, subl. 120. ^{1}H, IR, anal.,
    mol. wt.
Ni(CO)$_3$L. I.l.a.[522] White.[522] ^{1}H,[522] IR,[500b,516,522]
    anal.,[522] mol. wt.[522]
(CuH)$_2$L. I.l.b.[155]
cis-PdLCl[(CH$_2$)$_2$CMe]. ^{1}H.[528c]
H$_3$BL. I.l.f.[235] White,[235] subl. 199.[235] ^{11}B,[521] IR,[236]
    ^{31}P -97,[521] dipole moment 8.60 D,[235] ^{1}H,[521] anal.,[235]
    mol. wt.,[235] conductivity.[235]
D$_3$BL. I.l.f.[235] White.[235] IR.[236]
H$_7$B$_3$L. I.l.f.[235] White,[235] m. 167.[235] IR, anal.[235]
Me$_3$BL. I.3.[235] White.[235] ^{31}P(MeCN) -91,[521] ^{11}B,[521]
    ^{1}H,[521] IR.[236] Method of purification of BMe$_3$ by de-
    composing with heat.[235]
F$_3$BL. I.3.[235] White.[235] m. 40-45.[235] ^{31}P(MeCN) -93,[521]
    ^{11}B,[521] ^{1}H,[521] IR.[236]

TYPE: 2 C$_5$H$_9$; L$_2$ = [P(OCH$_2$)$_3$CMe]$_2$

cis-Cr(CO)$_4$L$_2$. I.l.c.[201,516] ^{1}H,[405] IR,[516] anal.[516]
trans-Cr(CO)$_4$L$_2$. I.l.a.[516] IR, anal.
Cr(CO)$_4$L$_2$(cis-trans mixture).[516] I.l.a.[522] White,[522] m.
    278.[522] IR,[522] ^{1}H,[237] ^{2}JPCrP = 9,[237] ^{2}JPCrP ≃ 1.[522]
cis-Mo(CO)$_4$L$_2$. I.l.c.[15,198,516] IR, anal., kinetics.[15,198]
Mo(CO)$_4$L$_2$(cis-trans mixture). I.l.a,[237,522] I.l.b.[197]
    White.[522] ^{1}H,[237,522] IR,[522] anal.,[197,522] mol. wt.,[522]
    kinetics.[197]
[CpMo(CO)$_2$L]$_2$(cis-trans mixture). I.l.a.[219b] Red. IR,
    anal., mol. wt., conductivity.
[CpMo(CO)$_2$L$_2$][CpMo(CO)$_3$]. II.[219b] IR.
trans-[CpMo(CO)$_2$L$_2$]BPh$_4$. II.[219b] Yellow. IR, anal.
cis-W(CO)$_4$L$_2$. I.l.c.[516] ^{1}H,[522] ^{2}JPWP ≃ 8, IR,[516,522]
    anal.,[522] mol. wt.[522]
trans-W(CO)$_4$L$_2$. I.l.c.[516] IR, prep could only be repeated
    once.
fac-Mn(CO)$_3$L$_2$(COMe). I.l.d.[359] White, m. 290-320. ^{1}H,
    IR, anal.
fac-Mn(CO)$_3$L$_2$Br. I.l.a.[9-11] IR,[9] kinetics.[9-11] Isomer-
    izes to fac-mer mixture at 50°C.
trans-Fe(CO)$_3$L$_2$. I.l.a.[522] Yellow,[522] m. 231.[522] ^{1}H,[237],
    [522] ^{2}JPFeP = 300,[237] IR,[516,522] anal.,[522] mol. wt.[522]
Co(NO)(CO)L$_2$. I.l.a.[499] IR, kinetics.
CpRhL$_2$. I.l.a.[471] Yellow, m. 215. ^{1}H, IR, anal., mol.
    wt.

Rh(CO)L₂Cl. I.l.f.[520b] Yellow. IR, anal.
Ni(CO)₂L₂. I.l.a.[522] Colorless,[522] m. 278.[522] ¹H,[237,]
    [405,522] ²JPNiP ≃ 0,[522] IR,[516,522] anal.,[522] mol. wt.[522]
    Useful as catalyst for polymerization of olefins.[181]
PdL₂Cl₂. I.2.[523,524] White.[524] Anal.,[524] conductivity.[523]
PtL₂Cl₂. I.2.[523,524] White.[524] Anal.,[524] conductivity.[523]

TYPE: 3 C₅H₉; L₃ = [P(OCH₂)₃CMe]₃

fac-Cr(CO)₃L₃. I.l.c.[15,488] ¹H,[488] IR,[488] kinetics.[15]
fac-Mo(CO)₃L₃. I.l.a,[488] I.l.c.[15,198,488] ¹H,[488] IR,[197]
    [488] anal.,[197,488] kinetics.[15,197,198]
fac-W(CO)₃L₃. I.l.a,[488] I.l.c.[424] ¹H,[488] IR,[488] anal.,[488]
    kinetics.[424]
Ni(CO)L₃. I.l.a.[522] White, m. 328. ¹H, ²JPNiP ≃ 1, IR,
    anal.
NiL₃(CN)₂. I.2.[144a] Yellow. IR, UV, anal.

TYPE: 4 C₅H₉; L₄ = [P(OCH₂)₃CMe]₄

[RhL₄Cl₂]Cl·2 H₂O. I.2.[524] Pink.[524] Anal.,[524] conduc-
    tivity.[523]
NiL₄. I.l.a,[522] II.[275,294] White,[522] m. 365.[522] ¹H,
    ²JPNiP∿15,[522] IR,[275,294] anal.[522]
[NiL₄H]BF₄. III.4.[520b] Off white, m. > 360. ¹H, IR,
    anal.
PdL₄. I.l.b.[294] White. IR, anal.
PtL₄. I.l.b.[294] White. IR, anal.
[CuL₄]ClO₄. II.[524] White.[523,524] IR,[294] anal.[524]
[CuL₄]NO₃. II.[524] White.[524] IR,[236] anal.,[524] conduc-
    tivity.[523]
[AgL₄]ClO₄. I.2.[524] White.[524] IR,[294] anal.,[524] conduc-
    tivity.[523] x-Ray structure shows regular tetrahedral
    symmetry around Ag, AgP = 2.436Å,[511] IR.[294]
[AgL₄]NO₃. I.2.[524] White.[524] Anal.,[524] conductivity.[523]
[AuL₂Cl]₂. I.l.b.[524] Colorless.[524] Anal.,[524] mol. wt.,[524]
    conductivity.[523]

TYPE: 5 C₅H₉; L₅ = [P(OCH₂)₃CMe]₅

[CoL₅]ClO₄. II.[525] Yellow.[525] Diamagnetic. Conductiv-
    ity,[525] IR.[525]
[CoL₅]NO₃. II.[525] Yellow. ³¹P(H₂O) -137, IR, ¹H, anal.,
    conductivity.
[NiL₅](BF₄)₂. I.2.[520b] Yellow, m. ∿220. IR, anal.
[NiL₅](ClO₄)₂. I.2.[143] Yellow.
[NiL₅H₂O](ClO₄)₂. I.2.[525] Yellow. Diamagnetic. Anal.,
    later shown to be previous compound.[143]

TYPE: 6 C₅H₉; L₆ = [P(OCH₂)₃CMe]₆

[CoL₆](ClO₄)₃. II.[525] White.[525] IR,[525] anal.,[525]

conductivity.[275] Explodes.
[CoL$_6$](NO$_3$)$_3$. II.[525] White. Anal., diamagnetic, decomposes readily in water.
[NiL$_6$](ClO$_4$)$_2$. I.1.b.[525] Yellow. Diamagnetic. Anal., conductivity, later shown to be [NiL$_5$](ClO$_4$)$_2$.[143]

TYPE: C$_6$H$_6$Cl$_3$; L = P(OCH$_2$CCl$_3$)$_3$

Ni(CO)$_3$L. IR.[500b]

TYPE: 4 C$_6$H$_6$Cl$_3$; L$_4$ = [P(OCH$_2$CCl$_3$)$_3$]$_4$

NiL$_4$. I.1.b.[500c] White, m. 200-203. ^{31}P(CDCl$_3$) -150.2, ^{1}H, anal.

TYPE: C$_6$H$_6$F$_3$; L = P(OCH$_2$CF$_3$)$_3$

H$_3$BL. I.1.b.[442]

TYPE: C$_6$H$_9$; L = P(OCH)$_3$(CH$_2$)$_3$

Cr(CO)$_5$L. I.1.a.[237] White, m. 218. ^{1}H, IR, mol. wt.
fac-Cr(CO)$_3$(bipy)L. I.1.a.[13] Kinetics.
Mo(CO)$_5$L. I.1.a.[237] White, m. 210. ^{1}H, IR, anal., mol. wt.
fac-Mo(CO)$_3$(chel)L$_2$ (chel = o-phen, 4,7-diPh-o-phen, and bipy). I.1.a.[197,198] Kinetics.
W(CO)$_5$L. I.1.a.[237] White, m. 221. ^{1}H, IR, anal., mol. wt.
fac-W(CO)$_3$(bipy)L. Kinetics.[197]
Fe$_3$(CO)$_{11}$L. I.1.a.[18] IR, kinetics.
Fe(CO)$_4$L. I.1.a.[237] White, m. 195. ^{1}H, IR, anal., mol. wt.
Ni(CO)$_3$L. I.1.a.[237] White, m. 142. ^{1}H, IR, anal., mol. wt.
H$_3$BL. I.1.f.[520a] White, m. 247-251. ^{31}P(MeCN) -117,[521] dipole moment 8.82 D,[520a] ^{1}H,[521] ^{11}B,[521] IR,[520a] anal.[520a]
D$_3$BL. I.1.f.[236] IR.
H$_7$B$_3$L. I.1.f.[520a] White.[520a] IR,[236] anal.[520a]
Me$_3$BL. I.3.[520a] ^{1}H,[520a,521] IR.[236] Dissociates easily.
F$_3$BL. I.3.[520a] ^{31}P(MeCN) -138,[521] ^{11}B,[521] ^{1}H,[521] ^{19}F,[521] IR.[236]

TYPE: 2 C$_6$H$_9$; L$_2$ = [P(OCH)$_3$(CH$_2$)$_3$]$_2$

cis-Cr(CO)$_4$L$_2$. I.1.a.[516] IR, anal.
trans-Cr(CO)$_4$L$_2$. I.1.a.[516] IR, anal.
Cr(CO)$_4$L$_2$(cis-trans mixture[516]). I.1.a.[237] White, m. 332. ^{1}H, IR, anal.

cis-Mo(CO)$_4$L$_2$. I.l.c.[516] White. IR, anal.
trans-Mo(CO)$_4$L$_2$. I.l.a.[516] White. IR, anal.
Mo(CO)$_4$L$_2$(cis-trans mixture). I.l.a.[237] White, m. 267.
    ^{1}H, IR, anal., kinetics.[197]
cis-W(CO)$_4$L$_2$. I.l.c.[516] ^{1}H,[14] IR,[237,516] anal.[237,516]
trans-W(CO)$_4$L$_2$. I.l.a.[237,516] White,[237] m. 312.[237] IR,
    anal.[237,516]
trans-Fe(CO)$_3$L$_2$. I.l.a.[237] White, m. 253. ^{1}H, IR, anal.
Ni(CO)$_2$L$_2$. I.l.a.[237] White, m. 225. ^{1}H, IR, anal., mol.
    wt.

TYPE: 3 C$_6$H$_9$; L$_3$ = [P(OCH)$_3$(CH$_2$)$_3$]$_3$

fac-Mo(CO)$_3$L$_3$. I.l.c.[197] IR, kinetics.
Ni(CO)L$_3$. I.l.a.[237] White,[237] m. 305.[237] ^{1}H,[237] IR,[237,]
    [516] anal.[237]

TYPE: 4 C$_6$H$_9$; L$_4$ = [P(OCH)$_3$(CH$_2$)$_3$]$_4$

NiL$_4$. I.l.a.[237] II.[275,294] White,[237] m. > 350.[237]
    ^{31}P(CH$_3$CN) -153,[142] ^{1}H,[142,237] IR,[294,516] anal.[237]
PdL$_4$. I.l.b.[294] White. IR, anal.
PtL$_4$. I.l.b.[294] White. IR, anal.
[AgL$_4$]ClO$_4$. I.2,[279] IV.1.[279] Anal., conductivity.
[AgL$_4$]NO$_3$. I.2,[279] IV.1.[279] ^{31}P(Me$_2$SO) -134,[142] ^{1}H,[142]
    anal., conductivity.

TYPE: 5 C$_6$H$_9$; L$_5$ = [P(OCH)$_3$(CH$_2$)$_3$]$_5$

[CoL$_5$]ClO$_4$. II.[275] Yellow.[275] ^{31}P(CH$_2$Cl$_2$) -157,[142] dia-
    magnetic,[275] ^{1}H,[142] IR,[275] anal.,[275] conductivity.[275]
[NiL$_5$](BF$_4$)$_2$. I.2,[143] IV.1.[279] Yellow.[143] Diamagnetic,[143]
    IR,[143] anal.[143]
[NiL$_5$](ClO$_4$)$_2$. IV.1.[279] Yellow.[142] ^{31}P(dimethylformamide)
    -125,[142] x-ray analysis[448] shows the cation to be tri-
    gonal bipyramidal with a slight axial compression
    (NiP$_{ax}$ = 2.14 Å; NiP$_{eq}$ = 2.19 Å), ^{1}H.[142]
[CuL$_4$]ClO$_4$. II.[279] Conductivity.
[CuL$_4$]NO$_3$. II.[279] Anal.

TYPE: 6 C$_6$H$_9$; L$_6$ = [P(OCH)$_3$(CH$_2$)$_3$]$_6$

[CoL$_6$](ClO$_4$)$_3$. II.[275] Colorless. Diamagnetic, IR, anal.,
    conductivity.
[NiL$_6$](ClO$_4$)$_2$. I.2.[275] Yellow. Diamagnetic, IR, anal.,
    conductivity, subsequently shown to be [NiL$_5$](ClO$_4$)$_2$.[143]

TYPE: C$_6$H$_{11}$; L = P(OCH$_2$)$_3$CEt

Mo(CO)$_5$L. I.l.a.[199] Kinetics.
cis-Mo(CO)$_4$L[P(NMe$_2$)$_3$]. I.l.a.[406] ^{31}P(CDCl$_3$) -138.0[406]

(L), -145.0[406] (PN$_3$), ^{2}JPMoP = 39,[405,406] ^{1}H,[406] IR,[406]
anal.[405,406]

trans-Mo(CO)$_4$L[P(NMe$_2$)$_3$].   I.l.a.[406] ^{31}P(CDCl$_3$) -147.0[406]
(L), -153.7[406] (PN$_3$), ^{2}JPMoP = 141,[405,406] ^{1}H,[406]
IR,[406] anal.[406]

trans-CpMo(CO)$_2$L(COC$_3$H$_5$).   I.l.d.[147]   Yellow, m. 92. ^{1}H,
IR, anal.

trans-CpMo(CO)$_2$L(COCH$_2$Ph).   I.l.d.[147]   Yellow, m. 136-138.
^{1}H, IR, anal.

W(CO)$_5$L.   I.l.b.[16,276]   Kinetics.[16,276]

cis-CpW(CO)$_2$LH.   I.l.a.[33]   m. 226-227. ^{1}H, IR, anal.

Na[Mn(CO)$_4$L].   III.1.[71]

cis-Mn(CO)$_4$LMe.   III.2,[202] III.4.[202]   m. 138-140. ^{1}H,
IR, anal.

Mn(CO)$_4$LPh(cis-trans mixture).   III.4.[71]   White, m. 107-
108.   Dipole moment 4.0D, ^{1}H, IR, anal.

cis-Mn(CO)$_4$L(COMe).   I.l.d.[202]   m. 155-156. ^{1}H, IR, anal.
Decarbonylates on heating.

cis-Mn(CO)$_4$L(COPh).   I.l.d.[71]   Yellow.[71]   m. 105.[71]   Dipole
moment 5.8D,[71] ^{1}H,[71] IR,[71] anal.,[71] stable to isomerization,[71] kinetics.[72b]

trans-Mn(CO)$_4$L(COPh).   III.4.[71]   Yellow, m. 102. ^{1}H, IR,
anal., mol. wt., stable to isomerization.

Co(CO)$_3$LMe.   III.2.[231]   IR.

(π-C$_3$H$_5$)Co(CO)$_2$L.   I.l.a.[233]   Yellow-orange, m. 107-112.
IR, anal., kinetics.

Co(CO)$_3$LH.   I.l.a.[233]   Colorless.   Unstable above 0°C.

Co(CO)$_3$L(COMe).   I.l.a.[231]   Colorless.   IR, anal.

Ni(CO)$_3$L.   IR.[500b]

trans-(OC)$_2$-cis-L$_2$Ru(CF$_2$)$_2$.   III.2.[137]   Yellow. ^{1}H, IR,
anal., mass spec.

trans-(OC)$_2$-cis-L$_2$Ru[C(CF$_3$)$_2$C(CN)$_2$].   III.2.[137]   m. 250.
^{1}H, IR, anal., isomerizes to cis-(OC)$_2$-cis-L$_2$ analog.

cis-(OC)$_2$-cis-L$_2$Ru[C(CF$_3$)(CN)C(CF$_3$)(CN)].   III.2.[137]   m.
250. ^{1}H, IR, anal.

cis-(OC)$_2$-cis-L$_2$Ru[C(CF$_3$)$_2$C(CN)$_2$].   Isomerization of trans-
(OC)$_2$-cis-L$_2$ analog.[137]   m. 250. ^{1}H, IR, anal.

trans-(OC)$_2$-cis-L$_2$Ru[(CF$_3$)$_2$CO].   III.2.[137]   m. 108-112.
^{1}H, IR, anal.

cis-(OC)$_2$-trans-L$_2$RuICF$_3$.   III.2.[88]   Yellow, m. 260. ^{1}H,
IR, anal., mass spec.

Ru(CO)$_2$L$_2$I$_2$[cis-(OC)$_2$-trans-L$_2$ and trans-(OC)$_2$-trans-L$_2$
mixture].   III.2.[88] ^{1}H, IR, anal.

O$_3$L.   I.3.[78b]   Fairly stable, decomposes at room temperature to O=L and O$_2$, source of singlet O$_2$.

TYPE: 2 C$_6$H$_{11}$; L$_2$ = [P(OCH$_2$)$_3$CEt]$_2$

trans-Cr(CO)$_4$L$_2$.   I.l.a.[405] ^{1}H, ^{2}JPCrP $\simeq$ 9.
trans-Mo(CO)$_4$L$_2$.   I.l.a.[406]
[Mn(CO)$_4$L]$_2$.   I.l.a.[71]

mer-Mn(CO)$_3$L$_2$Ph.   III.4.[71]   White, m. 197. ^{1}H, ^{2}JPMnP =
   110, dipole moment 1.69D, IR, anal.
fac-Mn(CO)$_3$L$_2$(COMe).   I.l.d.[202]   m. 214-215. ^{1}H, IR,
   anal.
fac-Mn(CO)$_3$L$_2$(COPh).   I.l.d.[71]   Yellow, m. 107.   Dipole
   moment 6.2D, ^{1}H, IR, anal., mol. wt.
mer-Mn(CO)$_3$L$_2$(COPh).   I.l.d.[71]   Yellow,[71]   m. 197.[71]   Dipole moment 2.5D, ^{1}H, IR, anal., mol. wt.
trans-Fe(CO)$_3$L$_2$.   I.l.a.[405] ^{1}H, ^{2}JPFeP = 300.
(π-C$_3$H$_5$)Co(CO)L$_2$.   I.l.a.[233]   Yellow, m. 230-232.   IR,
   anal., kinetics.
(π-ClC$_3$H$_4$)Co(CO)L$_2$.   I.l.a.[233]   Yellow, m. 155-157.   IR,
   anal., kinetics.
(π-C$_4$H$_7$)Co(CO)L$_2$.   III.4.[70]   Yellow, m. 155d.   IR, anal.
(π-H$_2$CCHCHCH$_2$COMe)Co(CO)L$_2$.   I.l.a.[233]   Orange, m. 172-
   173.   IR, anal., kinetics.
[π-MeO(CH)$_3$CH$_2$COPh]Co(CO)L·OC$_4$H$_8$.   I.l.a.[233]   Yellow, m.
   145-147.   IR, anal., kinetics.
L(OC)$_2$Co(CO)(CF$_3$CF)Co(CO)$_2$L.   I.l.a.[73]   Yellow, m. 130.
   IR, anal.
Co(CO)$_2$L$_2$Me.   III.2.[231]   Yellow, m. 185-200.   IR, anal.
Co(CO)$_2$L$_2$[C$_6$F$_3$(CN)$_2$].   III.2.[726]   Yellow, m. 180. ^{1}H,
   IR, anal.
Co(CO)$_2$L$_2$(COMe).   I.l.a.[231]   Yellow-green, m. 185.   IR,
   anal.
Co(CO)$_2$L$_2$H.   I.l.a.,[70,233] III.3.[231]   Tan,[231] m. 155-160,[231]
   255-260.[233]   IR.
Co(CO)$_2$L$_2$I.   III.3.[232]   Yellow, m. 261-263.   IR, anal.
NiL$_2$[(CF$_3$)$_2$CO].   I.l.h.[82]   Yellow, m. 203. ^{1}H, ^{19}F, IR,
   anal.
NiL$_2$[(CF$_3$)$_2$CS].   I.l.c.[82]   Yellow, m. 259. ^{19}F, anal.,
   mol. wt.
cis-PtL$_2$I$_2$.   III.3.[281]   Yellow.   IR, anal.
cis-PtL$_2$Cl$_2$.   I.l.b.[281]   Colorless.   IR, anal.

TYPE: 3 C$_6$H$_{11}$; L$_3$ = [P(OCH$_2$)$_3$CEt]$_3$

fac-Cr(CO)$_3$L$_3$.   I.l.a.[488] ^{1}H, IR.
fac-Mo(CO)$_3$L$_3$.   I.l.c.[488] ^{1}H, IR, anal.
fac-W(CO)$_3$L$_3$.   I.l.a.[488] ^{1}H, IR, anal.
trans-(OC)$_2$MnL$_3$(COPh).   I.l.d.[71]   Yellow, m. 132. ^{1}H,
   IR, anal., mol. wt.
Co$_2$(CO)$_5$L$_3$.   II.[70]   Brown, m. 95.   IR, anal.
Co(CO)L$_3$[C$_6$F$_3$(CN)$_2$].   III.2.[72c]   Orange-yellow, m. 160.
   ^{1}H, IR, anal.
Co(CO)L$_3$Me.   III.2.[231]   Colorless, m. 185.   IR, anal.
Co(CO)L$_3$(COMe).   I.l.a.[231]   Colorless.   IR, anal.
Co(CO)L$_3$H.   I.l.a.[70]   White, m. 180.   IR, anal.
[Co(CO)$_2$L$_3$][BPh$_4$].   I.l.a,[70] II.[70]   Yellow, m. 160.   IR,
   anal.
[Co(CO)$_2$L$_3$][Co(CO)$_4$].   I.l.a,[70] II.[70]   Yellow, m. 135.

IR, anal., conductivity.

TYPE: 4 C_6H_{11}; L_4 = [P(OCH$_2$)$_3$CEt]$_4$

[Co(CO)L$_4$][BPh$_4$]. I.1.a,[70] II.[70] Yellow, m. 200. IR,
    anal.
[Co(CO)L$_4$][Co(CO)$_4$]. I.1.a,[70] II.[70] Yellow, m. 200.
    IR, anal.
Co$_2$(CO)$_4$L$_4$. II.[70] Yellow-brown, m. 100. IR, anal.
NiL$_4$. III.1.[275] Colorless. Diamagnetic, anal., conduc-
    tivity.
[AgL$_4$]NO$_3$. I.2.[142] ^{31}P(CH$_2$Cl$_2$) -99, ^{1}H.

TYPE: 5 C_6H_{11}; L_5 = [P(OCH$_2$)$_3$CEt]$_5$

[CoL$_5$]ClO$_4$. II.[275] Yellow. Diamagnetic, IR, anal.,
    conductivity.
[NiL$_5$](ClO$_4$)$_2$. I.2.[143] Yellow.[143] ^{31}P(CH$_2$Cl$_2$) -108,[142]
    ^{1}H.[142]

TYPE: 6 C_6H_{11}; L_6 = [P(OCH$_2$)$_3$CEt]$_6$

[CoL$_6$]ClO$_4$. II.[275] Colorless. Diamagnetic, IR, anal.,
    conductivity.
[NiL$_6$](ClO$_4$)$_2$. I.2.[275] Yellow. Diamagnetic, anal.,
    conductivity, later shown to be [NiL$_5$](ClO$_4$)$_2$.[143]

TYPE: C_6H_{12}; L = P(OCH$_2$CH$_2$Cl)$_3$

CpV(CO)$_3$L. I.1.a.[493b] Red-orange, b$_{0.001}$ ∿128. IR, anal.
CpMn(CO)$_2$L. I.1.a.[493b] Yellow, b. 254. IR, anal.
Ni(CO)$_3$L. IR.[493b,500b]
CuLI. ^{31}P -102.5.[360]
CuLBr. ^{31}P -112.1.[360]
CuLCl. ^{31}P -118.7.[360]
H$_3$BL. I.1.b.[441,442] Anal.[441]

TYPE: 3 C_6H_{12}; L_3 = [P(OCH$_2$CH$_2$Cl)$_3$]$_3$

RhL$_3$Cl. I.1.c.[376]

TYPE: 4 C_6H_{12}; L_4 = [P(OCH$_2$CH$_2$Cl)$_3$]$_4$

RuL$_4$Cl$_2$. I.1.b.[327] Cream, m. 173-174. Anal.
CoL$_4$. II.[382a] Useful as a gasoline additive and a hydro-
    genation catalyst.
CoL$_4$H. II.[382a] Useful as a gasoline additive and a hydro-
    genation catalyst.
NiL$_4$. I.1.a,[119] II.[127,382a,407] White,[407] m. 138-140.[407]
    ^{31}P(toluene) -156.7,[500c] anal.[407] Useful as catalyst
    for dimerization of olefins[121] and as gasoline additive

and hydrogenation catalyst.[382a]

TYPE: C_6H_{15}; L = P(OEt)$_3$

CpV(CO)$_3$L. I.1.a.[493b] Red, b$_{0.001}$ $\sim$ 110. IR, anal.
fac-Cr(CO)$_3$(o-phen)L. I.1.a.[274a] Purple-black. IR.
fac-Cr(CO)$_3$(bipy)L. I.1.a.[13,274a] Purple-black.[274a]
    IR,[274a] anal.[274a]
Cr(CO)$_4$(diphos)L(fac-mer mixture). I.1.a.[175] IR.
Mo(CO)$_5$L. I.1.a.[14,152,199,429,430] Liquid,[152,429] b$_{0.2}$
    76-78.[152] IR,[152] kinetics.[14,199]
fac-Mo(CO)$_3$(o-phen)L. I.1.a.[198,273,274a,482] Red,[273,274a]
    m. 178-179. IR,[273] anal.,[482] kinetics.[198,273,274a,482]
fac-Mo(CO)$_3$(4,7-diPh-o-phen)L. I.1.a.[198] Kinetics.
fac-Mo(CO)$_3$(bipy)L. I.1.a.[197,273] Red,[273] m. 194-195.[273]
    IR,[273] kinetics.[197,273]
fac-Mo(CO)$_3$(tmen)L. I.1.a.[158] IR, kinetics.
mer-Mo(CO)$_3$(diphos)L. I.1.a.[158,209] Yellow,[209] m. 117-
    118.[209] ^{31}P -175.4(L), -53.8 (PPh$_2$ cis to L), -64.5
    (PPh$_2$ trans to L), ^{2}J$_{PMoP_{trans}}$ = 104,[209] ^{1}H,[209] IR,[158]
    anal.,[209] kinetics.[158]
trans-CpMo(CO)$_2$LI. I.1.a.[355] Red, m. 50-51. ^{1}H, IR,
    NMR shows two isomers but IR only one, anal.
trans-CpMo(CO)$_2$L(COMe). I.1.d.[146] m. 35. ^{1}H, IR, anal.,
    kinetics.
CpMo(CO)$_2$LEt(cis-trans mixture). IV.2.[219a,b] IR,[219b]
    anal.[219a]
W(CO)$_5$L. I.1.a.[293] ^{31}P,[208] IR.[208,293]
fac-W(CO)$_3$(o-phen)L. I.1.a.[198,274a] Purple-black.[274a]
    IR,[274a] kinetics.[198,274a]
fac-W(CO)$_3$(bipy)L. I.1.a.[197,274a] Purple-black.[274a]
    IR,[274a] kinetics.[197,274a]
Mn(CO)$_3$L[S$_2$P(OPh)$_2$](fac-mer mixture). I.1.a.[316] Yellow-
    orange. IR.
CpMn(CO)$_2$L. I.1.a.[493b,494a] Yellow.[493b] b$_{0.01}$ $\sim$ 113,[493b]
    IR,[493b,494a] anal.[493b]
cis-Mn(CO)$_4$L(C$_6$H$_4$-p-F). I.1.a.[491b] White, m. 39-40.
    ^{19}F, IR, anal.
cis-Mn(CO)$_4$L(C$_6$H$_4$-m-F). I.1.a.[491b] White, m. 53-54.
    ^{19}F, IR, anal.
Mn(NO)$_3$L. IR.[46,517]
Fe(CO)$_4$L. I.1.f.[145a] Yellow,[145a] b$_{0.01}$ 100.[145a] ^{1}H,[145a]
    IR,[145a] Raman,[145a] anal.,[145a] mass spec.[83]
C$_4$H$_6$Fe(CO)$_2$L. I.1.a.[386] Yellow, b$_{4.0}$ 124-125. Anal.
Cp(OC)Fe(CO)$_2$FeLCp(cis-trans mixture). I.1.a.[216,217]
    Red.[217] ^{1}H,[217] IR,[217] anal.,[217] conductivity.[217]
Fe(CO)$_3$L(CF$_2$)$_4$. I.1.a.[116b] m. 62. IR, ^{1}H, anal., mol.
    wt.
CpFe(CO)LI. I.1.a,[79] III.2.[217] CO(CH$_2$Cl$_2$) 1944,[217]
    kinetics.[79]
[Fe(S$_2$C$_6$Cl$_4$)$_2$L]$^{1-}$. Paramagnetic,[341] UV.

trans-(CO)Cl-trans-HLRu[PPh(n-Pr)$_2$]$_2$. III.3.[164] [1]H.
OsL(PPhMe$_2$)$_2$Cl$_3$. III.1.[163,165] Red,[163] m. 198-204.[163]
    Anal.[163]
OsL(PPhMe$_2$)$_2$H$_4$. III.2.[163,165] Colorless.[163] [1]H, IR.
Co$_2$(CO)$_7$L. I.1.a.[496] IR.
Co(NO)(CO)$_2$L. Mass spec.[12]
[Bu$_4$N]{CoL[SC(CN)C(CN)S]$_2$}. I.1.f.[341] Orange-brown,
    m. 120-122. IR, anal., conductivity, volatammetric
    study.
Ni(CO)$_3$L. I.1.a.[57,62,373] Colorless.[373] Oil.[373]
    [31]P(neat) -157,[374] IR,[373,493b,500b] anal.[373]
Ni(CO)$_2$L(PBu$_3$). III.3.[373] IR.
PdL(NH$_3$)Cl$_2$. IV.3.[183] Colorless. Anal.
PdL[HN(CH$_2$)$_5$]Cl$_2$. IV.3.[183,184]
PdL[H$_2$N-p-MeC$_6$H$_4$]Cl$_2$. IV.3.[183,184] White.[183]
Pt(CO)LCl$_2$. IV.3.[476] Yellow. Oil. Anal.
PtL[P(OH)$_3$]Cl$_2$. III.4.[476] Anal.[473]
PtL(PCl$_3$)Cl$_2$. III.5.[476] Anal.[473]
trans-PtL(H$_2$N-p-MeC$_6$H$_4$)Cl$_2$. III.5.[110] Green-yellow,[110]
    m. 92-93.[110] IR,[104] anal.,[110] conductivity.[110]
PtL(H$_2$N-p-MeC$_6$H$_4$)Cl$_2$. III.5.[128,476] Colorless.[476]
    Anal.[476]
PtL(H$_2$N-p-MeC$_6$H$_4$)(OH)$_2$. III.3.[476] Colorless. Anal.
cis-PtL(H$_2$NC$_6$H$_5$)Cl$_2$. IV.3.[454] Yellow. Anal., conduc-
    tivity.
trans-PtL(H$_2$NC$_6$H$_5$)Cl$_2$. IV.3.[454] White. Anal., conduc-
    tivity.
PtL(H$_2$NC$_6$H$_5$)Cl$_2$. III.5.[128]
cis-PtL(H$_2$NC$_6$H$_5$)Br$_2$. Yellow.[454] Anal.
trans-PtL(H$_2$NC$_6$H$_5$)Br$_2$. White.[454]
trans-PtL(NC$_5$H$_5$)Cl$_2$. White.[454]
cis-PtL(NC$_5$H$_5$)Cl$_2$. Yellow.[454] Anal.
[PtL(NH$_3$)$_2$Cl]Cl. III.3,[454] III.5.[239,474,476] White.[239]
    Anal.,[473] conductivity.[454]
PtLCl$_4$. III.2.[433] Yellow. Unstable to H$_2$O.
PtLCl$_2$Br$_2$. III.2.[433] Red. Unstable to H$_2$O.
CuLCN. I.1.f.[64,393] Colorless.[393] Liquid.[393] Anal.[393]
    Useful as insecticide.[64]
CuLI. I.1.f.[392] White,[392] m. 109-110,[19] 110-111.[392]
    Anal.[392] Trimeric to tetrameric.[392]
CuLBr. m. 27-28.[19,228] Useful as paint ingredient.[228]
CuLCl. I.1.f.[51] Liquid.[19] [31]P -128,[483] [1]H,[483] anal.[483]
    Useful as paint ingredient,[228] a catalyst for making
    polyesters,[179] and a reagent for making vinyl phos-
    phate derivatives.[51]
CuL[P(OPh)$_3$]I. III.5.[32] Colorless,[32] m. 69-70.5.[31]
    Anal.,[32] mol. wt.[32]
CuL[P(OPh)$_3$]Br. III.5.[32] Colorless,[32] m. 66-67.5.[31]
    Anal.[32] Approximately monomeric.[32]
CuL[P(OPh)$_3$]Cl. III.5.[392] White, m. 75-76. Anal.
CuL(NC$_5$H$_5$)Br. III.5.[392] White, m. 66-66.5,[395] 105-106.[392]

Anal.[392]   Useful as insecticide.[395]

CuL(amine)I.   III.5.[392]   White,[392] m. 72-73 (NC$_5$H$_5$),
  59-60 (α-picoline), 84-85 (β-picoline), 76-77 (γ-
  picoline), 80-81 (H$_2$NPh), 60-62 (H$_2$N-m-ClC$_6$H$_4$), 72-73
  (H$_2$N-p-ClC$_6$H$_4$), 60-61 (H$_2$N-p-MeC$_6$H$_4$), 66-67 (H$_2$N-p-
  MeOC$_6$H$_4$), 97-99 (2,4-lutidine), 70-71 (2,6-lutidine).[392]
  Anal.[392]   Useful as insecticides.[395]

CuL(amine)Cl.   III.5.   White, m. 56-57 (NC$_5$H$_5$),[392,395]
  105-106 (bipyridyl),[392,395] 75-76.5,[395] 80-81[392]
  (H$_2$NPh).   Anal.[392]   Useful as insecticides.[395]

AuLCl.   IV.1.[336]

AuLCCPh.   I.1.f.[127]   Liquid.

(HgCl$_2$)$_2$L.   I.1.f.[25]   m. 93-96.

HgLBr$_2$.   I.1.f.[25]   m. 69-73.

HgLCl$_2$.   I.1.f.[25]   m. 73-75. ^{31}P -115,[483] ^{1}H,[483] anal.,[483]
  between monomer and dimer,[25] NMR.[483]

H$_3$BL.   I.1.b.[441,442]   b$_{0.4}$39.[441] ^{31}P -114,[97] anal.,[441]
  mol. wt.[441]

(i-Bu)$_3$AlL.   I.3.[460]   Liquid.   Decomposes at 80-100°C to
  give (EtO)$_2$EtPOAl(i-Bu)$_3$.

Cl$_2$EtAlL.   I.3.[130] ^{1}H, anal.   Rearranges on heating to
  (EtO)$_2$EtPOAlEtCl$_2$.

F$_5$PL.   I.3.[81]   Solid.   Decomposes at 0°C to very toxic
  products.   Products include [Et$_3$O]PF$_6$, OP(OEt)F$_2$,
  OPEtF$_2$, OP(OEt)$_2$F, P(OEt)F$_2$, and P(OEt)$_2$F.

O$_3$SL.   I.3.[509]   Useful in sulfonation of octanol and poly-
  styrene.

O$_3$L.   I.3.[382b]   Decomposes to O=L and O$_2$ above -20°C,
  source of singlet oxygen, EPR studies, can react with
  singlet O$_2$ acceptors in gas phase.

TYPE: 2 C$_6$H$_{15}$; L$_2$ = [P(OEt)$_3$]$_2$

Cr(CO)$_4$L$_2$(cis-trans mixture).   I.1.c.[156]   IR.[158]

cis-Cr(CO)$_4$L$_2$.   I.1.c.[176]   IR.

trans-Cr(CO)$_4$L$_2$.   I.1.c.[156,158,176,177]   Colorless.[158]
  IR,[176] anal.,[158] kinetic study.[177]

cis-(OC)$_2$Cr(o-phen)L$_2$.   I.1.a.[274a]   Purple-black.   IR,
  anal.

cis-(OC)$_2$Cr(bipy)L$_2$.   I.1.a.[274a]   Purple-black.   IR, anal.

cis-Mo(CO)$_4$L$_2$.   I.1.c.[158,176,429,430]   Glass.[429]   IR,[176]
  goes to cis-trans mixture on heating,[430] kinetics.[158,176]

trans-Mo(CO)$_4$L$_2$.   I.1.c.[176,429,430]   Crystalline,[430]
  glass.[429]   IR, kinetics.[176,197]

cis-(OC)$_2$Mo(bipy)L$_2$.   I.1.a.[273]   Purple-black, m. 240.
  IR, anal., kinetics.

cis-(OC)$_2$Mo(o-phen)L$_2$.   I.1.a.[273]   Purple-black, m. 190.
  IR, anal., kinetics.

cis-(OC)$_2$Mo(diphos)L$_2$.   I.1.a.[158]   Yellow.   IR, anal.,
  kinetics.

[CpMo(CO)$_2$L]$_2$. I.l.a.[219a,b] Red.[219b] ^{1}H,[219b] IR,[219b]
  anal.,[219b] mol. wt.,[219b] conductivity.[219b]
[CpMo(CO)$_2$L$_2$][CpMo(CO)$_3$]. II.[219a,b] IR, cation is
  trans.[219b]
trans-[CpMo(CO)$_2$L$_2$]BPh$_4$. II.[219a,b] Yellow. ^{1}H,[219b]
  IR,[219b] anal.,[219b] conductivity.[219b]
W(CO)$_4$L$_2$ (cis-trans mixture). I.l.c.[153,157] Oil.[156]
  IR, anal.,[156] kinetics.[156,157]
cis-(OC)$_2$W(o-phen)L$_2$. I.l.a. [274a] Purple-black. IR,
  kinetics.
[Mn(CO)$_4$L]$_2$. I.l.a.[332] Anal.,[332] mass spec.[333]
trans-L$_2$Mn(CO)$_3$(C$_6$H$_4$-p-F). I.l.a.[491b] White, m. 39-41.
  ^{19}F, IR, anal.
trans-L$_2$Mn(CO)$_3$(C$_6$H$_4$-m-F). I.l.a.[491b] White, m. 37-38.
  ^{19}F, IR, anal.
mer-Mn(CO)$_3$L$_2$C$_6$F$_5$. I.l.a.[56] White, m. 75-76. ^{1}H, IR,
  anal.
Mn(NO)$_2$L$_2$I. III.2.[265] Red.[265] IR,[46] anal.[265]
[Co(CO)$_3$L]$_2$. I.l.a.[459] Orange-red,[459] m. 103.[459] Useful
  in hydroformylation of olefins.[172]
Hg[Fe(CO)$_2$(NO)L]$_2$. I.l.a.[94b] Orange,[258] m. 103-104.[94b]
  IR,[94b] anal.,[94b] decomposes easily.[94b]
trans-Fe(CO)$_3$L$_2$. I.l.f.[145a] m. 61-62. ^{1}H, IR, Raman,
  anal.
Fe(CO)$_2$(CF$_2$)$_4$. I.l.a.[116b] m. 180. IR, anal., mol. wt.,
  NMR.
trans-(OC)$_2$-cis-L Ru(CF$_2$)$_2$. III.2.[137] White, m. 59. ^{1}H,
  IR, mass spec.
Co(NO)(CO)L$_2$. I.l.a.[93] Kinetics.
[Bu$_4$N]{RhL$_2$[SC(CN)C(CN)S]}. I.l.a.[134]
[Ir(CO)LI$_3$]$_2$. Red,[537] m. 270.
Ni(CO)$_2$L$_2$. I.l.a.[57,62,373] Yellow.[373] Liquid.[62] ^{31}P
  (neat) -160,[324] ^{1}H,[374] IR,[373] anal.[373]
NiL$_2$(CNC$_6$H$_{11}$)$_2$. IV.3.[372] IR.
NiL$_2$[P(OPh)$_3$]$_2$. Brown.[120] Liquid.
[NiL$_2$(Ph$_2$PCH$_2$CH$_2$PPh$_2$)H]AlCl$_4$. I.l.c.[470a] ^{1}H.
NiL$_2$I$_2$. I.2.[285] Black, violet. UV.
NiL$_2$Br$_2$. I.2.[285] Black.
NiL$_2$Cl$_2$. I.2.[285] Oil.[285] ^{31}P -118,[483] ^{1}H,[170,483]
  anal.,[483] conductivity.[483]
cis-PdL$_2$Cl$_2$. I.l.b.[328] Yellow.[328] Dipole moment
  10.17D,[328] IR,[330] UV.[330]
trans-PdL$_2$Cl$_2$. White.[330] IR, UV.
PdL$_2$Cl$_2$. IV.1,[182-184] I.l.b.[3] Red,[183,184] beige.[3]
trans-[PdLBr$_2$]$_2$. I.l.b.[3] Red. Oil. IR.
trans-[PdLCl$_2$]$_2$. I.l.b.[3] Brown. Oily solid, b. 40-60.
  IR.
(PtLCl$_2$)$_2$. IV.1.[24,239,472,473,475,476] Yellow,[476] m.
  83.[476] IR,[3,476] decomposes to HCl, C$_2$H$_4$, C$_2$H$_4$Cl$_2$,
  and H$_3$PO$_4$ among other products on heating.[474]
[PtL(NO$_3$)$_2$]$_2$. Anal.[473]

cis-PtL$_2$I$_2$. III.3.[24] m. 85-86.[24] [195]Pt.[421]
cis-PtL$_2$Br$_2$. [195]Pt.[421]
cis-PtL$_2$Cl$_2$. I.2,[24] IV.1,[24] III.4.[475] Colorless,[477]
    m. 59-60.[24] Dipole moment 9.12D,[28] [1]H,[420,421,342]
    [195]Pt,[342] anal.[473]
PtL$_2$(H$_2$NC$_6$H$_5$)Cl$_2$. III.4.[128]
PtL$_2$(H$_2$N-p-MeC$_6$H$_4$)Cl$_2$. III.4.[128]
[PtL$_2$(NH$_3$)$_2$]Cl$_2$. III.4.[473,474] White.[454] Anal.[454]
[PtL$_2$(NH$_3$)$_2$]PtCl$_4$.[212]
[PtL$_2$(H$_2$NCH$_2$CH$_2$NH$_2$)]PtCl$_4$.[212]
PtL$_2$Cl$_2$Br$_2$. III.2.[433]
CuL$_2$S$_2$COEt. Useful as fungicide, bactericide, insecti-
    cide, lube and fuel additive, and cleansing agent.
CuL$_2$CN. I.l.f.[393,396] White,[393] m. 121-122.[393] Anal.[393]
    Tetrameric at freezing point in benzene and dimeric
    at room temperature in organic solvents,[393] useful as
    insecticide[64] and bactericide.[396]
CuL$_2$Cl. I.l.f.[65] Oil.[228] Useful as paint ingredient.[228]
HgL$_2$Cl$_2$. I.l.f.[65] Bactericide and defoliant.[63]
H$_2$B$_{10}$L$_2$. II,[431,489] III.3.[431] White,[431] m. 90-91,[431]
    84-85.[489] Anal.[431,489]

TYPE: 3 C$_6$H$_{15}$; L$_3$ = [P(OEt)$_3$]$_3$

fac-Mo(CO)$_3$L$_3$. I.l.a.[158,428,429] Glass.[429] IR,[133,428]
    anal.,[428] kinetics.[158,197]
mer-Mo(CO)$_3$L$_3$. I.l.c.[429,430] Crystalline,[430] glass.[429]
    [31]P(CCl$_4$) -172d,[322] -164t,[322] [2]JP_{cis}MoP_{trans} = 51,[322,]
    [405] [1]H,[322,405] IR,[133,430] anal.[430]
Mo(CO)$_3$L$_3$(fac-mer mixture). I.l.c.[430] Crystals. IR,
    anal.
CoL$_3$Cl. II.[527] Dark green, m. 87-89. [1]H, anal., mol. wt.
CoCl$_2$L$_3$. I.2.[385] UV-visible equilibrium study.[385]
RhL$_3$Cl. I.l.c.[376]
Ni(CO)L$_3$. I.l.a.[57] Colorless,[353] m. 0.[353] [31]P (neat)
    -163,[374] [1]H,[374] IR,[373] anal.[373]
NiL$_3$(CN)$_2$. I.l.f.[102] Red-orange. Oil. IR, UV.

TYPE: 4 C$_6$H$_{15}$; L$_4$ = [P(OEt)$_3$]$_4$

FeL$_4$H$_2$. I.2.[315] m. > 180.[315] [31]P(CS$_2$) -184.7d,[315]
    [1]H,[315,497b] sensitive to O$_2$.[315]
CoL$_4$H. II.[315,382a] m. > 180,[315] [31]P(CS$_2$) -168.3,[315]
    [1]H.[315] Sensitive to O$_2$, useful as gasoline addi-
    tive,[382a] hydroformylation catalyst,[382a] and homogene-
    ous hydrogenation catalyst.[380]
CoL$_4$Cl. II.[527] Yellow,[527] m. 65.[527] [1]H,[527] anal.,[527]
    mol. wt.[527] Useful as gasoline additive and hydro-
    genation catalyst.[382]
CoL$_4$Cl$_2$. I.2.[385] Blue. UV-visible equilibrium studies.
[RhL$_4$][BPh$_4$]. I.l.b.[214] Anal., mol. wt.

$Ir_4(CO)_8L_4$. I.l.a.[461b]

NiL_4. I.l.a,[57,62] II.[91,165,285,372,526] White,[526] m. 108.[526] ^{31}P(cyclohexane) -160,[374,500c] ^1H,[98,374] IR,[294,383] UV,[409a] mol. wt. indicates appreciable dissociation[526] although when measurements are carried out under rigorously anaerobic conditions, the expected value is observed.[520b] Useful as gasoline additive,[91,382a] hydrogenation catalyst,[91,382a] hydroformylation catalyst.[91]

$[NiL_4H]Cl$. III.4.[166] ^{31}P($CHCl_3$) -135d,[166] ^1H,[166] UV,[500d] also detected using HSO_4^- [166,500d] and CF_3COO^- [166] anions, kinetics.[500d]

NiL_4Br_2. I.2.[193] Red.[193] Conducts.[194] UV-visible equilibrium studies.[193,194]

PdL_4. II.[372] White,[372] m. 65.[372] IR,[383] anal.[372]

$[PdL_4]PdCl_4$. I.l.b.[328] IR,[330] UV.[330]

$[PdL_4]Cl_2$. I.l.b.[328]

PtL_4. II.[372] Colorless,[372] m. 114.[372] IR,[383] anal.[372]

$[PtL_4]PtCl_4$. I.l.b.[212,213,504] Reacts with $CHCl_3$ to give PtL_2Cl_2.[213]

$[CuL_4]ClO_4$. II.[50] m. 185.[50] ^{31}P(CH_2Cl_2) -122,[305] ^1H,[305] anal.[50]

$[AgL_4]ClO_4$. I.2.[483] ^{31}P -128, ^1H, anal., conductivity.

TYPE: 6 C_6H_{15}; L_6 = $[P(OEt)_3]_6$

CoL_6Cl_2. UV-visible studies.[384]

TYPE: 8 C_6H_{15}; L_8 = $[P(OEt)_3]_8$

$[CoL_4]_2$. III.1.[528a] White, m. 222-225. Weakly paramagnetic, anal., mol. wt.

TYPE: C_7H_{13}; L = $P(OCH_2)_3CPr$

$CpV(CO)_3L$. I.l.a.[174] IR.

$Cr(CO)_5L$. I.l.a.[488] ^1H, IR, anal.

$Mo(CO)_5L$. I.l.a,[488] I.l.c.[488] ^1H, IR, anal.

$W(CO)_5L$. I.l.a.[488] ^1H, IR, anal.

$Fe(CO)_4L$. I.l.a.[406,488] ^1H, IR, anal.

$Ni(CO)_3L$. IR.[500b]

TYPE: 2 C_7H_{13}; L_2 = $[P(OCH_2)_3CPr]_2$

trans-$Cr(CO)_4L_2$. I.l.a.[488] ^1H, IR, anal.

cis-$Mo(CO)_4L_2$. I.l.a,[488] I.l.c.[488] ^1H,[405,488] ^2JPMoP = 50,[405] IR,[488] anal.[488]

trans-$Mo(CO)_4L_2$. I.l.a,[488] I.l.c.[488] ^1H,[405,488] ^2JPMoP = 210,[405] IR,[516] anal.[488]

cis-$W(CO)_4L_2$. I.l.a.[488] ^1H, ^2JPWP = 35,[405] IR,[488] anal.[488]

cis-W(CO)$_4$L$_2$. I.l.a.[488] ^{1}H, ^{2}JPWP = 35,[405] IR,[488]
    anal.[488]
trans-W(CO)$_4$L$_2$. I.l.a.[488] ^{1}H,[405 488] ^{2}JPWP = 140,[405]
    IR,[516] anal.[488]
trans-Fe(CO)$_3$L$_2$. I.l.a.[488] ^{1}H, IR, anal.
Rh(CO)L$_2$Cl. I.l.a.[520b] Yellow. IR.
Ir(CO)L(PPh$_3$)$_2$H. I.l.a.[520b] White. ^{1}H, ^{2}JPIrP = 175,
    IR.
Ir(CO)L$_2$(PPh$_3$)H. I.l.b.[520b] White. ^{1}H, ^{2}JPIrP = 57, IR.

TYPE: 3 C$_7$H$_{13}$; L$_3$ = [P(OCH$_2$)$_3$CPr]$_3$

fac-Cr(CO)$_3$L$_3$. I.l.a.[488] ^{1}H, IR, anal.
fac-Mo(CO)$_3$L$_3$. I.l.a,[488] I.l.c.[488] ^{1}H, IR, anal.
fac-W(CO)$_3$L$_3$. I.l.a.[488] ^{1}H, IR, anal.
NiL$_3$I$_2$. I.l.f,[520b] IV.3.[520b] Blue-black. Diamagnetic,
    anal., nonconductor, UV-visible.

TYPE: 4 C$_7$H$_{13}$; L$_4$ = [P(OCH$_2$)$_3$CPr]$_4$

CoL$_4$H. II.[520b] White. ^{1}H, anal.
RhL$_4$H. I.l.b.[520b] White. ^{1}H.
NiL$_4$. ^{31}P(CHCl$_3$) -128, NMR.[142]

TYPE: 5 C$_7$H$_{13}$; L$_5$ = [P(OCH$_2$)$_3$CPr]$_5$

[CoL$_5$]ClO$_4$. II.[142] ^{31}P(CH$_2$Cl$_2$) -138, NMR.

TYPE: C$_9$H$_{12}$; L = P(OCH$_2$CH$_2$CN)$_3$

Ni(CO)$_3$L. IR.[500b]

TYPE: C$_9$H$_{15}$; L = P(OCH$_2$CH = CH$_2$)$_3$

Ni(CO)$_3$L. IR.[500b]

TYPE: C$_9$H$_{17}$; L = P(OCH$_2$)$_3$CAm

W(CO)$_5$L. I.l.b.[293] White, m. 147-148. ^{31}P(CH$_2$Cl$_2$) -115,
    ^{1}H, IR, anal., mol. wt.
[NiLP(o-MeSC$_6$H$_4$)$_3$](ClO$_4$)$_2$. I.2.[167] Purple, m. 173.
    Explodes, diamagnetic, UV, anal., conductivity.

TYPE: 2 C$_9$H$_{17}$; L$_2$ = [P(OCH$_2$)$_3$CAm]$_2$

Rh(CO)L$_2$Cl. I.l.a.[520b] Yellow. ^{1}H, IR, anal.
trans-PdL$_2$I$_2$. III.3.[281] Red.[281] ^{1}H,[405] IR,[281]
    anal.[281]
cis-PdL$_2$Cl$_2$. I.l.b.[281] Yellow. IR, anal.

trans-PdL$_2$Cl$_2$. ^{1}H.[405]
cis-PtL$_2$I$_2$. III.3.[281] Yellow.[281] ^{1}H, ^{2}JPPtP = 14,[405]
    anal.[281]
cis-PtL$_2$Cl$_2$. I.1.b.[281] Colorless.[281] ^{1}H, ^{2}JPPtP =
    35,[405] IR,[281] anal.[281]

TYPE: 3 C$_9$H$_{17}$; L$_3$ = [P(OCH$_2$)$_3$CAm]$_3$

NiL$_3$I$_2$. I.1.f,[520b] IV.3.[520b] Blue-black. Diamagnetic,
    anal., nonconductor.

TYPE: 5 C$_9$H$_{17}$; L$_5$ = [P(OCH$_2$)$_3$CAm]$_5$

[NiL$_5$](ClO$_4$)$_2$. I.2.[143] Yellow. Diamagnetic, UV, anal.

TYPE: C$_9$H$_{21}$O$_3$; L = P(OPr)$_3$

C$_4$H$_6$Fe(CO)$_2$L. I.1.a.[386] Yellow, b$_2$ 170-171. Anal.
Ni(CO)$_3$L. I.1.a.[57]
CuLCN. I.1.f.[393] Colorless. Liquid. Anal.
CuLI. I.1.f.[392] White,[392] m. 64-65,[19] m. 61-63.[392]
    Anal.[392] Trimeric to tetrameric.[392]
CuLBr. Liquid.[19]
CuLCl. I.1.f.[51] Liquid.[19] Useful as reagent in synthesis
    of vinyl phosphate derivatives.[51]

TYPE: 2 C$_9$H$_{21}$O$_3$; L$_2$ = [P(OPr)$_3$]$_2$

Ni(CO)$_2$L$_2$. I.1.a.[57]
(PtLCl$_2$)$_2$. IV.1.[434a] Yellow.
CuLCN. I.1.f.[393] Colorless.[393] Liquid.[393] Anal.[393]
    Useful as a bactericide.[396]

TYPE: 3 C$_9$H$_{21}$O$_3$; L$_3$ = [P(OPr)$_3$]$_3$

CoL$_3$Cl$_2$. I.1.f.[385] UV-visible equilibrium studies.
Ni(CO)L$_3$. I.1.a.[57] Subl$_{0.001}$10.$^{-3}$

TYPE: 4 C$_9$H$_{21}$O$_3$; L$_4$ = [P(OPr)$_3$]$_4$

CoL$_4$Cl$_2$. I.1.f.[385] Blue. UV-visible equilibrium studies.
NiL$_4$. I.1.a.[98] White. ^{31}P -162, anal., Faraday effect
    study.
NiL$_4$Br$_2$. Conducts.[194] UV-visible equilibrium studies.
[PtL$_4$]PtCl$_4$. I.1.b.[504]

TYPE: C$_9$H$_{21}$O$_3$; L = P(O-i-Pr)$_3$

fac-Cr(CO)$_3$(o-phen)L. I.1.a.[159] IR, kinetics.
fac-Mo(CO)$_3$(o-phen)L. I.1.a.[274a] Purple black. IR, anal.,
    kinetics.

fac-Mo(CO)$_3$(bipy)L.   I.1.a.[274a]   Red.   IR, anal., kinetics.
fac-Mo(CO)$_3$(tmen)L.   I.1.a.[158]   IR, kinetics.
CpMo(CO)$_2$L(i-Pr)(cis-trans mixture).   IV.2.[219a,b]   Anal.
W(CO)$_5$L.   I.1.a.[208]   [31]P, IR.
CpMn(CO)$_2$L.   III.3.[379]   Mass spec.
IHgFe(CO)$_2$(NO)L.   III.3.[94b]   IR.   Very unstable.
Cp(OC)$_2$FeSnCl$_2$Fe(CO)LCp.   III.1.[216]
Cp(OC)Fe(CO)$_2$FeLCp(cis-trans mixture).   I.1.a.[216,217]   Red.
   [1]H,[217] IR,[217] anal.[217]
CpFe(CO)LI.   III.2.[216,217]   IR.
[CpFe(CO)$_2$L]I.   III.2.[216]
[CpFe(CO)$_2$L][BPh$_4$].   III.2.[216,217]   Yellow.[217]   [1]H,[217]
   IR,[217] anal.[217]
Co$_2$(CO)$_7$L.   I.1.a.[496]   IR.
Ni(CO)$_3$L.   IR.[500b]
CuLCNS.   m. 95-96.[22]   Between dimer and trimer.
CuLCN.   I.1.f.[393]   Colorless.   Liquid.   IR, anal.
CuLI.   I.1.f.[392]   White,[392] m. 199-200,[392] 184-185.[19]
   Dipole moment 2.57D,[28] anal.,[392] trimer,[22] mol. wt.[392]
CuLBr.   I.1.f.[392]   White,[392] m. 149-150.[22] Dipole moment
   2.78D,[28] anal.,[392] trimer,[22,28] trimeric to tetra-
   meric.[392]
CuLCl.   I.1.f.[392]   White,[392] m. 117,[392] 112-114.[22]
   Anal.,[392] trimer,[22] between trimer and tetramer.[392]
   Useful as a paint ingredient.[228]
CuL[P(OPh)$_3$]Br.   III.5.[32]   White,[32] m. 115-117.[31]   Anal.,[32]
   approximately monomeric.[32]
CuL(NC$_5$H$_5$)Br.   III.5.[32]   White,[32]   m. 108-109.[31]   Anal.,[32]
   approximately monomeric.[32]   Decomposes on heating to
   [CuLBr]$_3$.
HgLI$_2$.   I.1.f.[25]   Yellow, m. 120.
HgLBr$_2$.   I.1.f.[25]   m. 128-130.
HgLCl$_2$.   I.1.f.[25]   m. 117.
H$_3$BL.   I.1.b.[441-443]   b$_{0.1}$42-43.[441]   [31]P (neat) -115,[377]
   [11]B,[377] [1]H,[360,377] anal.,[441] mol. wt.[441]   Useful as
   antioxidant, polymerization catalyst, or blowing
   agent.[444]

TYPE: 2 C$_9$H$_{21}$O$_3$; L$_2$ = [P(O-i-Pr)$_3$]$_2$

trans-Cr(CO)$_4$L$_2$.   I.1.c.[176]   IR, kinetics.
cis-Mo(CO)$_4$L$_2$.   I.1.c.[158,176]   IR,[176] kinetics.[176,158]
[CpMo(CO)$_2$L]$_2$.   I.1.a.[219a,b]   Red.   [1]H,[219b] IR,[219b]
   anal.,[219b] mol. wt.,[219b] conductivity.[219b]
trans-Mo(CO)$_4$L$_2$.   I.1.c.[156,176]   IR,[176] kinetics.[156,176]
cis-(OC)$_2$Mo(o-phen)L$_2$.   I.1.a.[274a]   Purple-black.   IR,
   anal., kinetics.
cis-(OC)$_2$Mo(bipy)L$_2$.   I.1.a.[274a]   Purple-black.   IR, anal.,
   kinetics.
trans-[CpMo(CO)$_2$L$_2$][CpMo(CO)$_3$].   II.[219a,b]   IR.[219b]
trans-[CpMo(CO)$_2$L$_2$]BPh$_4$.   II.[219a,b]   Yellow.   [1]H,[219b]

IR,[219b] anal.,[219b] conductivity.[219b]
$W(CO)_4L_2$ (cis-trans mixture). I.l.c.[156] Kinetics.[156,157]
mer-$Mn(CO)_3L_2$(CO-i-Pr). I.l.e.[455] Colorless. ^{1}H, IR,
   anal., mass spec.
$Hg[Fe(CO)_2(NO)L]_2$. I.l.a.[94b] m. 127-129. IR, anal.,
   decomposes easily.
$Co(NO)(CO)L_2$. I.l.a.[93] Kinetics.
RhL_2BPh_4. III.4.[402a]
NiL_2I_2. I.2.[285] Oil.
NiL_2Br_2. I.2.[285] Black, violet.
NiL_2Cl_2. I.2.[285] Oil.
PtL_2I_2. III.3.[24] Yellow, m. 137.
cis-PtL_2Cl_2. I.l.b,[306] I.2.[24] m. 132.5.[306] Dipole
   moment 3.92D,[306] 9.05D.[28]
trans-PtL_2Cl_2. I.l.b.[306] m. 44-46. Dipole moment 1.51D.
$[PtL_2(H_2NSNH_2)_2]Cl_2$. III.3.[306]
CuL_2CN. I.l.f.[393][396] White.[393] Liquid.[396] m. 85-86.[393]
   Anal.,[393] monomer in solution.[393] Useful as bacteri-
   cide.[396]
CuL_2I. I.l.f.[392] White, m. 94-95. Anal., monomer in
   solution.
CuL_2Br. I.l.f.[392] White, m. 58-59. Anal., monomer in
   solution.
CuL_2Cl. I.l.f.[392] White, m. 69-70. Anal., monomer in
   solution.

TYPE: 3 $C_9H_{21}O_3$; L_3 = $[PO-i-Pr)_3]_3$

fac-$Mo(CO)_3L_3$. I.l.c.[158]
CoL_3Cl_2. I.l.f.[385]
$[AgLI]_3$. m. 163-164.[22] Dipole moment 3.43D,[28] mol. wt.[22]
$[AgLBr]_3$. m. 73-75.[22] Dipole moment 3.53D,[28] mol. wt.[22]
$[AgLCl]_3$. m. 29-30.[22] Mol. wt.

TYPE: 4 $C_9H_{21}O_3$; L_4 = $[P(O-i-Pr)_3]_4$

CoL_4Cl_2. I.l.f.[385] Blue. UV-visible studies.
NiL_4. II.[285] Colorless. Diamagnetic, anal.
NiL_4Br_2. I.2.[194] Conductor, UV-visible equilibrium
   studies.
$[PtL_4]PtCl_4$. III.3.[306,504] m. 110-112.[306] Anal.,[306]
   decomposes to cis and trans PtL_2Cl_2.

TYPE: $C_9H_{21}O_6$; L = $P(OCH_2CH_2OCH_3)_3$

$Ni(CO)_3L$.   IR.[500b]

TYPE: $C_{12}H_{27}$; L = $P(OBu)_3$

$CpV(CO)_3L$. I.l.a.[174,493b] Red,[493b] b. 237.[493b] IR,[174,493b] anal.

$Cr(CO)_5L$. Mass spec.[83]
$Mo(CO)_5L$. Mass spec.[83]
fac-$Mo(CO)_3$(o-phen)L. I.l.a.[274a] Purple-black. IR,
    anal., kinetics.
fac-$Mo(CO)_3$(bipy)L. I.l.a.[274a] Purple. IR, kinetics.
[$CpMo(CO)_2LBu$] (cis-trans mixture). IV.2.[219a,b] IR,[219b]
    anal.[219a]
$CpMo(CO)_2L(COMe)$. I.l.d.[90] Yellow. Oil. IR.
$W(CO)_5L$. I.l.b.[16,276] ^{31}P,[208] ^1H,[208] IR,[208] kinetics,[16]
    [276] mass spec.[83]
cis-$Mn(CO)_4LBr$. I.l.a.[10] IR,[9] kinetics.[9,10]
fac-$Mn(CO)_3LP(OPh)_3Br$. I.l.a.[10] Kinetics.
fac-$Mn(CO)_3LZBr$ (Z = PPh_3, PPh_2Cl, $AsPh_3$, $SbPh_3$, PBu_3).
    I.l.a.[10] Kinetics.
$CpMn(CO)_2L$. I.l.a.[493b] Yellow.[493b] b. 250.[493b] IR,[493b,]
    [494a] anal.[493b]
$Cp(OC)Fe(CO)_2FeLCp$(cis-trans mixture). I.l.a.[216,217]
    Red.[217] ^1H,[217] IR,[217] anal.[217]
$CpFe(CO)L(COMe)$. I.l.d.[56,90] Yellow-orange.[56] Liquid.[56]
    ^1H,[56] IR,[56,90] kinetics.[90]
$Co_2(CO)_7L$. I.l.a.[496] IR.
$Co(NO)(CO)_2L$. IR, kinetics.
$Ni(CO)_3L$. I.l.a.[57,62] Liquid.[62] IR.[50,493b,517]
$CuLCN$. I.l.a.[393] Colorless.[393] Liquid.[393] Anal. Use-
    ful as catalyst for making polyesters from anhydrides
    and dialcohols.[179]
H_3BL. I.l.b.[441,442] $b_{0.15}$83-85. Anal.,[441] mol. wt.[441]
Et_3AlL. I.3.[460] Decomposes at 80°C to give $(BuO)_2BuPO$-
    $AlEt_3$.

TYPE: 2 $C_{12}H_{27}$; $L_2 = [P(OBu)_3]_2$

trans-$Cr(CO)_4L_2$. I.l.a.[343,365-367] Pale green. Liquid.
    [343,366,367] $b_{0.1}$ 230.[343,367] IR,[343] anal.,[343,367]
    mol. wt.[343,367] Useful as motor fuel additive, oxo
    catalyst, and pesticide.[365]
cis-$(OC)_2Mo$(o-phen)L_2. I.l.a.[274a] Purple-black. IR,
    anal., kinetics.
cis-$(OC)_2Mo$(bipy)L_2. I.l.a.[274a] Purple-black. IR, anal.,
    kinetics.
[$CpMo(CO)_2L$]$_2$. I.l.a.[219a,b,221] Red,[219b,221] m. 96-98.[221]
    ^1H,[219b,221] IR,[219b,221] anal.,[219b] mol. wt.,[219b] con-
    ductivity.[219b,221]
[$CpMo(CO)_2L_2$][$CpMo(CO)_3$]. II.[219a,b] IR.[219b]
trans-[$CpMo(CO)_2L_2$]BPh_4. II.[219a,b] Yellow. ^1H,[219b]
    IR,[219b] anal.,[219b] conductivity.[219b]
fac-$Mn(CO)_3L_2Br$. I.l.a,[12] IV.3.[11] IR,[9] kinetics,[9-12]
    isomerizes to mer form at 50°C,[11] IR.[59b]
mer-$Mn(CO)_3L_2Br$. I.l.a.[12] IR,[12] kinetics.[11,12]
cis-$(OC)_2MnL_2(CF_3COCHCOCF_3)$. I.l.a.[226] Orange. Oil. IR.
$Co(NO)(CO)L_2$. I.l.a.[93] Kinetics.

RhL_2BPh_4. III.4.[402a]
$Ni(CO)_2L_2$. I.1.a.[57,62,478]   Liquid.[62]   IR.[58,517]
NiL_2I_2. I.2.[285]   Oil.
NiL_2Br_2. I.2.[285]   Oil.
NiL_2Cl_2. I.2.[285]
CuL_2CN. I.1.f.[393]   Colorless.[393]   Liquid.[393]   Anal.,[393]
    useful as insecticide.[64]
$[AgL_2]ClO_4$. I.2.[492]

TYPE: 3 $C_{12}H_{27}$; L = $[P(OBu)_3]_3$

fac-$Cr(CO)_3L_3$. I.1.c.[343,365]   Green.   Liquid.[343,365]
    $b_{0.1}230$.[343]   IR,[343]   anal.[343]   Useful as motor fuel
    additive, oxo catalyst, and pesticide.
CoL_3Cl_2. I.1.f.[385]
$Ni(CO)L_3$. I.1.a.[57,62] $Subl_{0.001}100$.[57]   IR.[58,517]
$NiL_3(CN)_2$. I.2.[144a] Red-orange.   Oil.   IR, UV.
$[AgL_3]ClO_4$. I.2.[492]

TYPE: 4 $C_{12}H_{27}$; L_4 = $[P(OBu)_3]_4$

CoL_4Cl_2. I.1.f.[385]   Blue.   UV-visible equilibrium studies.
$[RhL_4][BPh_4]$. I.1.b.[214]   Anal., mol. wt.
NiL_4. I.1.a,[98] II.[285,526]   Colorless.[285]   Oil.[526] ^{31}P
    -162,[98]   anal.[98]
NiL_4Br_2. I.2.[194]   Conductor, UV-visible equilibrium studies.
$[PtL_4]PtCl_4$. I.1.b.[504]
$CuLCl$. Liquid.[19]
$CuLBr$. Liquid.[19]

TYPE: 2 $C_{12}H_{27}$; L = $[P(O\text{-}i\text{-}Bu)_3]_2$

RhL_2BPh_4. III.4.[402a]

TYPE: 3 $C_{12}H_{27}$; L_3 = $[P(O\text{-}i\text{-}Bu)_3]_3$

CoL_3Cl_2. I.1.f.[385]

TYPE: 4 $C_{12}H_{27}$; L_4 = $[P(O\text{-}i\text{-}Bu)_3]_4$

CoL_4Cl_2. I.1.f.[385]   Blue.   UV-visible equilibrium studies.
NiL_4Br_2. I.2.[194]   Conductor.   UV-visible equilibrium
    studies.

TYPE: 2 $C_{15}H_{33}$; L_2 = $[P(OC_5H_{11})_3]_2$

$(PtLCl_2)_2$. IV.1.[472,477]

TYPE: 2 $C_{18}H_{12}F_3$; L_2 = $[P(O\text{-}p\text{-}FC_6H_4)_3]_2$

$[Co(CO)_3L]_2$. I.1.a.[459]   Orange-red, m. 130.

$Ni(CO)_2L_2$. m. 76-77.[353]

TYPE: 3 $C_{18}H_{12}F_3$; $L_3 = [P(O-p-FC_6H_4)_3]_3$

$Ni(CO)L_3$. m. 102.[353]

TYPE: $C_{18}H_{12}Cl_3$; $L = P(O-p-ClC_6H_4)_3$

$Mn(NO)_3L$. IR.[46,517]
$Ni(CO)_3L$. White,[353] m. 70-71. IR.[500b]

TYPE: 2 $C_{18}H_{12}Cl_3$; $L_2 = [P(O-p-ClC_6H_4)_3]_2$

$Mn(NO)_2L_2I$. IR.[46]
$Rh(CO)L_2Cl$. I.l.a.[514] Yellow, m. 180-184. Diamagnetic,
    anal., mol. wt., conductivity.
$[Rh(CN-p-MeOC_6H_4)_2L_2]ClO_4$. Yellow, green,[512] m. 157-160.
$[Co(CO)_3L]_2$. I.l.a.[459] Red.
$Ni(CO)_2L_2$.[353] m. 132.
PtL_2I_2. I.l.f.[351] Yellow, m. 215-220. Anal.

TYPE: 3 $C_{18}H_{12}Cl_3$; $L_3 = [P(O-p-ClC_6H_4)_3]_3$

$Ru(CO)L_3Br_2$. IV.3.[327] Cream, m. 140-142. IR, anal.
$Ru(CO)L_3Cl_2$. IV.3.[327] Yellow, m. 133-134. IR, anal.
RhL_3SCN. III.3.[514] Yellow, m. 158-160. Diamagnetic,
    anal., conductivity.
RhL_3I. III.3.[514] Orange, m. 135-136. Diamagnetic, anal.,
    mol. wt., conductivity.
RhL_3Cl. I.l.b,[513] I.l.a.[514] Yellow,[514] m. 183-186.[514]
    Diamagnetic,[514] anal.,[514] mol. wt.,[514] conductivity.[514]
$Ni(CO)L_3$.[353] m. 142.
NiL_3. II.[326] White, m. 145-155. ^{31}P (C_6H_6) -129.6,
    anal., mol. wt.
$PdL_3(CN-p-MeC_6H_4)$. I.l.e.[346-348] White,[346] m. 105.[348]
    Anal.[346]
$PtL_3(PPh_3)$. I.l.b.[351] Colorless, m. 141. Anal.

TYPE: 4 $C_{18}H_{12}Cl_3$; $L_4 = [P(O-p-ClC_6H_4)_3]_4$

RuL_4Br_2. I.l.b.[327] Cream, m. 164-165. Anal.
RuL_4Cl_2. I.l.b.[327] Cream, m. 158. Anal.
OsL_4Br_2. I.l.b.[327] White, m. 212-215. Anal.
CoL_4H. II.[326] Yellow, m. 144-146. $^{31}P(C_6H_6)$ -140.5,
    ^1H, anal., mol. wt.
PtL_4. I.l.b,[351] II.[351] Colorless, m. 163. Anal.

TYPE: 2 $C_{18}H_{12}(NO_2)_3$; $L_2 = [P(O-p-NO_2C_6H_4)_3]_2$

$Ni(CO)_2L_2$.[353] Yellow, m. 135.

TYPE: 3 $C_{18}H_{12}(NO_2)_3$; $L_3 = [P(O-p-NO_2C_6H_4)_3]_3$

$Ni(CO)L_3$.[353] m. 145.

TYPE: 4 $C_{18}H_9D_6$; $L_4 = [P(O-2,6-D_2C_6H_3)_3]_4$

CoL_4D.   IV.1.[415a,b] Yellow. IR, anal., NMR.
RhL_4D.   III.3.[415a,b] Tan.

TYPE: $C_{18}H_{15}$; $L = P(OPh)_3$

$CpV(CO)_3L$. I.1.a.[493b] Yellow, m. 114.
    IR, anal.
$Cr(CO)_5L$. I.1.a.[343,365,367] White,[343,365,367] m. 59.5.[343,
    367] IR,[343] anal.,[343] mol. wt.,[343] x-ray structure
    Cr-P length = 2.309 Å.[427] Useful as motor fuel addi-
    tive,[365] oxo catalyst,[365] and pesticide.[365]
trans-$Cr(CO)_4L(PBu_3)$. I.1.a.[209] Colorless,[209] m. 49-50.[209]
    [31]P -185(L), $^2JPCrP = 30$,[209,405] IR,[209] anal.[209]
fac-$Cr(CO)_3$(o-phen)L. I.1.a.[17,159] IR,[159] kinetic stu-
    dies.[17,159]
fac-$Cr(CO)_3$(o-phen deriv.)L where o-phen derivatives are
    5-nitro, 4,7-diphenyl, and 3,4,7,8-tetramethyl.
    I.1.a.[17] IR.
$Mo(CO)_5L$. I.1.a,[14,430] I.1.b.[153,199,429] m. 50.[430]
    IR,[133,150,152,430] kinetics.[14,153,199]
$Na[CpMo(CO)_2L]$. III.1.[304,356] Colorless.[304] IR,[356]
    [1]H.[356]
trans-$Mo(CO)_4L(PBu_2Ph)$. I.1.a.[209] Colorless, m. 70-72.
    [31]P -165.2(L), -23.1(PC_3), $^1JPMoP = 112$, IR, anal.
trans-$Mo(CO)_4LPBu_3$. I.1.a.[209] Colorless,[209] m. 49-50.
    [31]P -165(L),[209] -17.2(PBu_3),[209] $^2JPMoP = 112$,[209,405]
    IR,[209] anal.[209]
fac-$Mo(CO)_3$(o-phen)L. I.1.a.[17,198,274a,482] Red.[274a]
    IR,[274a] UV,[17] anal.,[482] kinetics.[17,198,274a]
fac-$Mo(CO)_3$(o-phen deriv.)L where o-phen derivatives are
    5-NO_2, 4,7-diPh, and 3,4,7,8 tetraMe. I.1.a.[17]
    Kinetics.
trans-$CpMo(CO)_2LSnCl_3$. III.1.[371] Yellow, m. 169. [1]H,
    IR, anal., mol. wt.
trans-$CpMo(CO)_2LSnPh_3$. III.1.[356] White, m. 211-212. [1]H,
    IR, anal.
$CpMo(CO)_2LSnMe_3$ (isomer A). I.1.a.[304] White, m. 137-138.
    [1]H, IR, anal.
$CpMo(CO)_2LSnMe_3$ (isomer B). III.3.[304] Yellow, m. 209-210.
    IR, anal.
$Cp_2Mo_2(CO)_5L$. I.1.a.[221] Red, m. 152. IR, anal., mol.
    wt.
cis-$CpMo(CO)_2LCl$. III.3.[33,370] Red,[370] m. 131-132.[33]
    [1]H,[33,370] IR,[33,370] anal.[370]
cis-$CpMo(CO)_2LI$. I.1.a,[220,355,356] III.2.[33,304,370]

Red,[220,304,355] m. 146-148,[33,355] 130-140.[304] ^{1}H,[33,220,304,355,356,370] IR,[33,220,304,355,356,370] anal.[220,304,355]

cis-CpMo(CO)$_2$L(O$_2$CCF$_3$). III.3.[304] Red, m. 115-117. IR.

trans-CpMo(CO)$_2$LMe. III.4.[42] ^{1}H.

CpMo(CO)$_2$LMe(cis-trans mixture). III.2.[304] m. 99-100. ^{1}H, IR, anal.

CpMo(CO)$_2$LEt(cis-trans mixture). III.2.[304] m. 92. ^{1}H, IR, anal.

CpMo(CO)$_2$LCH$_2$CHCH$_2$(cis-trans mixture). III.2.[304] m. 97-98. ^{1}H, IR, anal.

CpMo(CO)$_2$LCH$_2$Ph(cis-trans mixture). III.2.[304] m. 113-115. ^{1}H, IR, anal.

trans-CpMo(CO)$_2$L(COMe). I.1.d.[42,90,146,225] Yellow. Oil,[90] m. 85.[146] ^{1}H,[42,146] IR,[90,146,225] anal.,[146] kinetics.[90,225]

trans-CpMo(CO)$_2$L(COEt). I.1.d.[146] m. 93. ^{1}H, IR, anal.

trans-CpMo(CO)$_2$L(COCH$_2$Ph). I.1.d.[147] Yellow, m. 42. ^{1}H, IR, anal.

(π-indenyl)Mo(CO)$_2$L(COMe)(trans-cis mixture). I.1.d.[225] IR, kinetics.

cis-CpMo(CO)$_2$LH. I.1.a,[33] III.2,[304,370] III.3.[356] Pink,[304] Buff,[370] m. 106-107,[304] 100.[33] ^{1}H,[33,304,356,370] IR,[33,304,356,370] anal.,[33,304,370] mol. wt.[370]

CpMo(NO)LI$_2$. I.1.a.[299] Red, m. 138-141. ^{1}H, IR, anal., mol. wt. suggests L dissociates, conductivity suggests I$^-$ dissociates in polar solvents, NMR suggests two isomers.

W(CO)$_5$L. I.1.b.[16,276] Colorless,[361] m. 69-71. ^{31}P(CH$_2$Cl$_2$) -130.3,[361] ^{31}P,[208] IR,[208,361] kinetics.[16,276,293]

trans-W(CO)$_4$L(PBu$_3$). I.1.a.[209] Colorless, m. 67-68.[209] ^{31}P -137.0(I), 5.3 (PBu$_3$), ^{2}JPWP = 120,[209,405] ^{1}H,[405] IR,[209] anal.[209]

trans-W(CO)$_4$L(PBu$_2$Ph). ^{31}P, ^{2}JPWP = 112.[405]

fac-W(CO)$_3$(o-phen)L. I.1.a.[361] Purple-black,[361] m. > 325.[361] ^{31}P(CH$_2$Cl$_2$) -121.9,[361] ^{1}H,[361] IR,[17,316] kinetics.[17]

fac-W(CO)$_3$(o-phen deriv.)L (o-phen deriv. = 5-NO$_2$, 4,7-diPh, 4,7-diNO$_2$, and 3,4,7,8 tetraMe-o-phen). I.1.a.[17] Kinetics.[17]

cis-CpW(CO)$_2$LH. I.1.a.[33] m. 110-112. ^{1}H, IR, anal.

Na[Mn(CO)$_4$L]. III.1.[250] Yellow. Anal.

Mn$_2$(CO)$_9$L. I.1.a.[530]

trans-Ph$_3$SnMn(CO)$_4$L. IR,[77] anal.

cis-Ph$_3$PAuMn(CO)$_4$L. III.3.[292a] Brown,[292a] m. 90.[292a] IR,[77,292a] anal.,[77,292a] conductivity,[292a] mol. wt.[292a] IR indicates it could be a cis-trans mixture[292a] although solid is cis as shown by x-ray diffraction studies[354] (Mn-P(OPh)$_3$ = 2.27 Å, Mn-Au-PPh$_3$ = 166.5°, P-O = 1.46, 1.62, 1.57 Å).

Mn(CO)$_4$LH. III.2.[250] Yellow-orange. Oil. Anal.

cis-Mn(CO)$_4$LMe. III.2,[250] I.1.d.[311a] Yellow, m. 79.0-

80.5,[311a] 71-72.[250] ^{1}H,[311a] IR,[311a] anal.,[250] mol.
  wt.[250]
trans-Mn(CO)$_4$LC(CN)$_3$. I.l.a.[47] IR.
cis-Mn(CO)$_4$LC$_2$F$_5$. III.4.[263] Yellow. Oil. IR, anal.
cis-Mn(CO)$_4$LPh. III.4.[71] White, m. 85-86. IR, anal.
cis-Mn(CO)$_4$L(C$_6$H$_4$-p-F). I.l.a.[491b] White, m. 69-71.
  ^{19}F, IR, anal.
cis-Mn(CO)$_4$L(C$_6$H$_4$-m-F). I.l.a.[491b] White, m. 65-67.
  ^{19}F, IR, anal.
cis-Mn(CO)$_4$L(COMe). I.l.d.[311a] White,[311a] m. 67.5-
  70.0.[311a] ^{1}H,[311a] IR,[311a] anal.,[311a] kinetics.[368]
cis-Mn(CO)$_4$L(COCF$_3$). III.2.[263] Colorless, m. 109. IR,
  anal.
cis-Mn(CO)$_4$L(COC$_2$F$_5$). III.2.[263] Colorless. IR, anal.,
  decarbonylates on heating.
Mn(CO)$_4$L(COPh) (cis-trans mixture). I.l.d.[71] Yellow,
  m. 72-74. IR, anal., mol. wt.
Mn(CO)$_4$LI. I.l.a.[241]
cis-Mn(CO)$_4$LBr. I.l.a.[10] IR,[9,10] kinetics.[9,10]
Mn(CO)$_3$L[S$_2$P(OPh)$_2$](fac-mer mixture). I.l.a.[316] Yellow-
  orange, m. 85-86. IR, anal.
CpMn(CO)$_2$L. I.l.a.[493b,469] Yellow,[493b] m. 113.[493b]
  IR,[493b,494a] anal.,[493b] mass spec.[379] Useful as
  herbicide.[469]
π-MeC$_5$H$_4$Mn(CO)$_2$L. Useful as herbicide.[469]
Mn(NO)$_3$L. III.2.[241,265] Green.[241] IR.[46,241,517]
[CpMn(NO)(CO)L]PF$_6$. I.l.a.[278] Yellow, m. 167. ^{1}H, IR,
  UV, anal., conductivity.
[(MeCp)Mn(NO)(CO)L]PF$_6$. I.l.a.[278] Red, m. 145-147. ^{1}H,
  IR, UV, anal., conductivity.
[CpMn(NO)L(3-Me pyridine)]PF$_6$. I.l.a.[278] Red, m. 145-
  147. ^{1}H, IR, UV, anal.
cis-Re(CO)$_4$LH. I.l.a.[185] Colorless, b$_{0.001}$ 135. ^{1}H, IR.
Re(diphos)LH. II.[188] ^{1}H, IR, anal.
IHgFe(CO)$_2$(NO)L. III.3.[94b] IR. Very unstable.
Fe(CO)$_4$L. I.l.f.[145a] m. 68-69.[145] IR,[145a,493b] Raman,[145a]
  anal.[145a]
(Ph$_4$C$_5$O)Fe(CO)$_2$L. I.l.a.[531a] m. 216-217. IR.
CpFe(CO)LSnMe$_3$. I.l.a.[303] Orange, m. 99-100. ^{1}H, IR,
  anal.
CpFe(CO)LSiMe$_3$. I.l.a.[303] Orange, m. 95-96. ^{1}H, IR,
  anal.
Cp(OC)$_2$FeSnCl$_2$Fe(CO)LCp. III.1.[216] IR.
Cp(OC)Fe(CO)$_2$FeLCp(cis-trans mixture). I.l.a.[216,217]
  Red.[217] ^{1}H,[217] IR,[217] anal.,[217] conductivity.[217]
(OC)$_3$Fe(SPh)$_2$Fe(CO)$_2$L. I.l.a.[266] m. 120-121. Anal.
CpFe(CO)LPh. I.l.a.[387a,390a,b] Yellow,[387a] m. 144.5-
  145.5.[390a] IR,[387a,390a,b] anal.[390]
CpFe(CO)L-p-ClC$_6$H$_4$. I.l.a.[390a] Yellow, m. 90.5-91.5.
  IR, anal.
CpFe(CO)L-p-FC$_6$H$_4$. I.l.a.[390b] Yellow, m. 130-132. IR, anal.

CpFe(CO)LC$_6$F$_5$. I.l.a.[390a] Orange, m. 128.5-130. IR,
   anal.
CpFe(CO)L(COMe). I.l.d.[56] Yellow-orange, m. 65. [1]H,
   IR, anal., mol. wt.
CpFe(CO)LI. III.2,[216,217] I.l.a.[79,284] Green.[79] IR,[79,]
   [216,217,284] [1]H,[216,284] anal.[79,284]
CpFe(CO)LBr. I.l.a.[79] IR, anal.
CpFe(CO)LSeCN. I.l.a.[284] Orange-red, m. 119-121. [1]H,
   IR, anal., mol. wt.
CpFe(CO)LCN. I.l.a.[284] m. 185-189. [1]H, IR, anal., mol.
   wt.
[CpFe(CO)$_2$L]I. III.2.[216] [1]H, IR.
[CpFe(CO)$_2$L][BPh$_4$]. III.2.[216,217] Yellow.[217] [1]H,[217]
   IR,[217] conductivity.[217]
trans-[Fe(Et$_2$PCH$_2$CH$_2$PEt$_2$)$_2$LH]BPh$_4$. I.l.b.[35b] [1]H, anal.,
   Mossbauer.
π-MeC$_5$H$_4$Fe(CO)LI. I.l.a.[79] IR, anal.
π-MeC$_5$H$_4$Fe(CO)LBr. I.l.a.[79] IR, anal.
π-MeC$_5$H$_4$Fe(CO)LCl. I.l.a.[79] IR.
Fe(NO)$_2$(CO)L. I.l.a.[46] IR.[46,517]
Fe(CO)$_3$LI$_2$. I.l.a.[410] Purple, m. 115. Anal.
Fe(CO)$_3$LBr$_2$. I.l.a.[131] Kinetics.[131]
Fe(CO)$_3$L(CF$_2$)$_4$. I.l.a.[116b] m. 88. [1]H, IR, anal., mol.
   wt.
[Ph$_4$P][FeS$_4$C$_4$(CF$_3$)$_4$L]. I.2.[341] Green, m. 100. Paramag-
   netic, IR, anal.
[Ph$_4$P][FeS$_4$C$_4$(CN)$_4$L]. I.2.[341] Green, m. 85. Paramagnetic,
   IR, anal.
FeS$_4$C$_4$Ph$_4$L. I.l.f.[466] m. 208. Anal.
FeS$_4$C$_4$(CF$_3$)$_4$L. I.l.f.[466] Blue, m. 112-114. Diamagnetic,
   IR, anal., mol. wt.
trans-[Ru(Et$_2$PCH$_2$CH$_2$PEt$_2$)$_2$LH]BPh$_4$. I.l.b.[35b] [1]H, anal.
trans-[Os(Et$_2$PCH$_2$CH$_2$PEt$_2$)$_2$LH]BPh$_4$. I.l.b.[35b] [1]H, anal.
Na[Co(CO)$_3$L]. III.1.[260,261] Yellow.[260] Anal.[260]
Ph$_3$PAuCo(CO)$_3$L. IR,[77] anal.
Co$_2$(CO)$_7$L. I.l.a.[496] IR.
trans-Co(CO)$_3$LMe. III.2.[260,261] Yellow,[260] m. $\sim$ 10,[260]
   anal.[261]
trans-Co(CO)$_3$L(C$_8$H$_{13}$). I.l.a.[456,508] Orange.[508] [1]H,[508]
   anal.[508]
trans-Co(CO)$_3$LCF$_3$. III.4.[263] Yellow, m. 150. IR, anal.
trans-Co(CO)$_3$LC$_2$F$_5$. III.4.[263] Yellow, m. 170. IR, anal.
trans-Co(CO)$_3$LC$_3$F$_7$. III.4.[263] Yellow, m. 190. IR, anal.
trans-Co(CO)$_3$L(COEt). III.2.[245] Yellow, m. 70. IR, anal.
trans-Co(CO)$_3$L(COCF$_3$). III.2.[263] Yellow. IR, anal.,
   decarbonylates.
trans-Co(CO)$_3$L(COC$_2$F$_5$). III.2.[263] Yellow. IR, anal.,
   decarbonylates.
trans-Co(CO)$_3$L(COC$_3$F$_7$). III.2.[263] Yellow. IR, anal.,
   decarbonylates.
Co(CO)$_3$LH. III.2.[260] Off white.[261] Anal.[260]

trans-Co(CO)$_3$LI. III.2.[246] IR.
Co(CO)$_2$L(PPh$_3$)I. I.1.a.[246] Brown, m. 135. IR, anal.
trans-Co(CO)$_3$LCl·HCl. III.3.[246] IR.
Co(CO)$_2$L(PPh$_3$)Cl. I.1.a.[246] Brown, m. 80-90. IR, anal.
Co(NO)(CO)$_2$L. I.1.a.[248,440,500a] Orange,[440] m. 60-62,[248]
   55.[440] Dipole moment 2.27D,[256] IR,[248] anal.[440,500a]
Co(NO)$_2$LCF$_3$. IR.[44,46]
Co(NO)$_2$LC$_2$F$_5$. IR.[46]
Co(NO)$_2$LC$_3$F$_7$. IR.[44,46]
Co(NO)$_2$LNCS. I.1.f.[243] Black, m. 45-47. Anal.
Co(NO)$_2$LSPh. I.1.f.[243] Black, m. 57-60. Anal.
Co(NO)$_2$LCl. I.1.f.[256] m. 101. IR.[44,46]
CoL[SC(Ph)C(Ph)S]$_2$. I.1.f.[466] m. 175.[466] Paramagnetic,[341]
   voltammetric study,[135] ESR study.[192]
CoL[SC(CF$_3$)C(CF$_3$)S]$_2$. I.1.f.[34] Green, m. 103-104. Anal.,
   mol. wt. Diamagnetic in solid, paramagnetic(1.70 BM)
   in solution.
[Bu$_4$N]{CoL[SC(CN)C(CN)S]$_2$}. I.1.f.[341] Orange, brown,
   m. 140-141. IR, anal.
[Et$_4$N]{CoL[SC(CF$_3$)C(CF$_3$)S]$_2$}. I.1.f.[341] Green, m. 175.
   IR.
Rh$_4$(CO)$_{11}$L. I.1.a.[461b]
mer-RhL(PBu$_3$)$_2$Cl$_3$. I.1.f.[5d] Yellow,[5d] m. 135-136.[5d]
   ^{31}P(CH$_2$Cl$_2$) -78.4 (L), ^{2}JPRhP = 30.3,[5d] ^{1}H,[405,419]
   anal.[5d]
IrL(PPh$_3$)$_2$Cl$_2$H. I.2.[66] IR.
Ni(CO)$_3$L. I.1.a.[62] IR.[62,493b,500b,517] Useful as cata-
   lyst for converting olefins to cyclic olefins.[324]
Ni(CO)$_2$L(PPh$_3$). I.1.a.[373] IR.
Ni(CO)$_2$L(PBu$_3$). I.1.a.[373] IR.
Ni(CO)$_2$L[P(CH$_2$CH$_2$CN)$_3$]. I.1.a.[373] IR.
Ni(CO)L(PPh$_3$)$_2$. Useful as polymerization catalyst.[118]
(π-C$_3$H$_5$)NiBrL. I.1.f.[529a] ^{1}H.
CpNiLI. III.2.[536] Bright red. m. 132-133. Anal.
trans-PdL[HN(CH$_2$)$_5$]Cl$_2$. III.5.[105,106,112] Orange-yellow,[112] m. 131-132.[112] IR,[106] anal.,[112] mol. wt.[112]
trans-PdL[H$_2$N-p-MeC$_6$H$_4$]Cl$_2$. III.5.[106,112] Yellow-
   orange,[112] m. 115.5-116.5.[112] IR,[106] anal.,[112] mol.
   wt.[112]
trans-PdL[H$_2$N(CH$_2$)$_7$CH$_3$]Cl$_2$. III.5.[106] IR.[106]
trans-PdL(PBu$_3$)I$_2$. ^{31}P(neat) -102.3 (L), -9.3 (PBu$_3$),
   ^{2}JPPdP = 758,[419] ^{1}H,[405] NMR.[405,419]
trans-PdL(PMe$_2$Ph)I$_2$. ^{31}P(neat) -97 (L), -15.2 (PMe$_2$Ph)
   ^{2}JPPdP = 829,[419] ^{1}H.[405,419]
cis-PtL(PBu$_3$)Me$_2$. III.3.[5d] m. 82-84. ^{31}P(CH$_2$Cl$_2$)
   -104.2 (L), anal.
cis-PtL(PBu$_3$)MeCl. III.3.[5d] m. 101-103. ^{31}P(CH$_2$Cl$_2$)
   -108.5 (L), ^{2}JPPtP = 23.3, anal.
PtLCl(MeCOCHCOMe). II.[274b] m. 170-171. ^{1}H, IR, anal.
cis-PtL(PBu$_3$)Cl$_2$. I.1.f.[5d] White, m. 152-154. ^{31}P(CH$_2$Cl$_2$)
   -59.3 (L), ^{2}JPPtP = 20.0, anal.

trans-PtL(PBu$_3$)Cl$_2$. Yellow,[5d] m. 113-117. ^{31}P(CH$_2$Cl$_2$)
    -88.8 (L), ^{2}JPPtP = 709, anal.
[PtL(PBu$_3$)Cl]Cl. I.l.b.[5d] ^{31}P(CH$_2$Cl$_2$) -65.8 (L),
    ^{2}JPPtP = 24.0
trans-[PtL(PEt$_3$)$_2$H]ClO$_4$. I.l.b.[116a] White. ^{1}H, IR,
    anal., conductivity.
[PtL(AsEt$_3$)H]ClO$_4$ (cis-trans mixture). I.l.b.[116b] ^{1}H,
    IR, anal.
trans-[PtL(PEt$_3$)$_2$(p-FC$_6$H$_4$)]ClO$_4$. I.l.b.[116a] ^{1}H, anal.,
    conductivity.
trans-[PtL(PEt$_3$)$_2$(m-FC$_6$H$_4$)]ClO$_4$. I.l.b.[116a] ^{1}H, anal.,
    conductivity.
trans-[PtL(PEt$_3$)$_2$Br]ClO$_4$. I.l.b.[116a] IR, anal., conduc-
    tivity.
trans-[PtL(PEt$_3$)$_2$Cl]ClO$_4$. I.l.b.[116a] IR, anal., conduc-
    tivity.
CuLCN. ^{31}P -112.9.[360]
CuLBr. I.l.f.[392] White,[392] m. 90.5-91.5.[19] Dipole
    moment 3.39D,[28] anal.,[392] dimer in solution,[392]
    trimer.[28]
CuLCl. I.l.f.[392] White,[392] m. 88-89,[392] 95-96.[19,22,228]
    Dipole moment 3.21D,[28] ^{31}P -112.4,[360] anal.,[392] dimer
    in solution,[392] trimer.[22] Useful as paint ingredient
    for ships' bottoms.[228]
CuL(AsPh$_3$)Br. III.5.[32] Colorless,[32] m. 116-117.[31]
    Anal.,[32] monomer.[32]
CuL(NC$_5$H$_5$)Br. III.5.[32] Colorless,[32] m. 126-7.[31] Anal.,[32]
    monomer.[32]
CuL(NC$_5$H$_5$)Cl. III.5.[32,392] m. 122-123,[31] 126-127.[392]
    Anal.,[392] monomer.[32] Useful as insecticide.[395]
CuL(quinoline)Br. III.5.[32] m. 135-136.[31] Anal.,[32]
    monomer.[32]
AuLMn(CO)$_5$. III.3.[77,292a] Brown,[292a] m. 68.[292a] IR,[77,]
    [292a] anal.,[77,292a] mol. wt.,[292a] conductivity.[292a]
AuLMn(CO)$_4$PPh$_3$. I.l.a.[292a] IR,[77] anal.[77]
AuLMo(CO)$_3$Cp. III.3.[222] Yellow, m. 75-78. ^{1}H, IR,
    anal., mol. wt., conductivity.
AuLCCPh. I.l.f.[127] Colorless, m. 113-114. Dipole mo-
    ment 4.4D, anal., mol. wt.
AuL(m-FC$_6$H$_4$). III.3.[391] IR, NMR.
AuL(p-FC$_6$H$_4$). III.3.[391] IR, NMR.
AuLCl. I.l.b.[292a,310] White,[292a] m. 104-105.[310] Dipole
    moment 6.27D,[28] anal.,[292a] mol. wt.[22]
H$_3$BL. I.l.f,[114] I.l.b.[442] m. 54,[441] 46-48.[114] Anal.[114]
F$_2$N$_2$L. I.3.[292b] Black, m. 58. Diamagnetic in solution,
    anal.
O$_3$L. I.3.[307b,498b] ^{31}P(CH$_2$Cl$_2$) +63, liberates O$_2$ and
    O=L at -15°C.

TYPE: 2 C$_{18}$H$_{15}$; L$_2$ = [P(OPh)$_3$]$_2$

trans-Cr(CO)$_4$L$_2$. I.l.a,[343,365-367] I.l.c.[176] White,[343,365,367] m. 148-149.[343,365-367] IR,[176,343] anal.[343,367] Useful as inhibitor in olefin isomerizations,[169] motor fuel additive,[365] oxo catalyst,[365] pesticide.[365]

[CpMo(CO)$_2$L]$_2$. I.l.a.[221,304] Red,[221,304] m. 176.[221] IR,[221,304] anal.[304]

cis-Mo(CO)$_4$L$_2$. I.l.c.[176,429,430] IR,[430] kinetics.[176]

trans-Mo(CO)$_4$L$_2$. I.l.a,[209,343] I.l.c.[429,430] White,[343] m. 110-114.[343] IR,[133,152,343,430,538] anal.[343]

cis-(OC)$_2$Mo(o-phen)L$_2$. I.l.a.[274a] Purple. IR, anal., kinetics.

cis-(OC)$_2$Mo(bipy)L$_2$. I.l.a.[274a] Purple. IR, anal., kinetics.

Hg[CpMo(CO)$_2$L]$_2$. I.l.a,[370] III.3.[304] Yellow.[304,370] IR,[304,370] anal.[304,370] mol. wt.[370]

CpMoCOL$_2$I. I.l.a.[220] Orange, m. 114-116. ^{1}H, IR, anal., mol. wt., conductivity.

cis-W(CO)$_4$L$_2$. I.l.c.[157,361] Colorless,[361] m. 117-118.[361] ^{31}P (CH$_2$Cl$_2$) -125.5, IR,[361] anal.,[36] kinetics.[157]

trans-W(CO)$_4$L$_2$. I.l.a.[361] ^{31}P(CH$_2$Cl$_2$) -132.3, anal.

Na[Mn(CO)$_3$L$_2$]. III.1.[257] Anal.

K[Mn(CO)$_3$L$_2$]. III.1.[257] Anal.

[Mn(CO)$_4$L]$_2$. I.l.a.[240,254,332,410] Yellow,[410] m. 140.[332] Diamagnetic,[410] Dipole moment = 2.67D,[254] IR,[254,334,410,413] anal.,[334] mol. wt.,[334] mass spec.[333]

fac-Mn(CO)$_3$L$_2$H. I.l.a.[72a] Yellow, Dec. without m. IR, anal.

mer-Mn(CO)$_3$L$_2$H. I.l.a,[72a] III.2.[510] Yellow,[72a] white,[510] m. 84.[72a] ^{1}H,[510] IR,[72a,510] anal.,[72a,510] mol. wt.[510]

Mn(CO)$_3$L$_2$H. III.2.[257] Yellow. Dipole moment 2.09D, anal.

mer-Mn(CO)$_3$L$_2$Me. I.l.d.[359] White, m. 148-150. Dipole moment 2.60D, ^{1}H, IR, anal.

Mn(CO)$_3$L$_2$Me. III.2.[257] Yellow, m. 157. IR, anal.

mer-Mn(CO)$_3$L$_2$Ph. I.l.d.[36,37,71] Cream,[71] m. 137-138,[71] 146.5-147.5.[37] IR,[36,37,71] anal.,[71] mol. wt.[71]

Mn(CO)$_3$L$_2$CF$_3$. III.2.[257] Colorless, m. 126. Anal.

Mn(CO)$_3$L$_2$C$_2$F$_5$. III.2.[257] Yellow, m. 155. Anal.

mer-Mn(CO)$_3$L$_2$(COMe). I.l.d.[36,37,359] White,[359] m. 147-148.[359] Dipole moment 3.01D,[359] ^{1}H,[359] IR,[36,37,359] anal.[359]

mer-Mn(CO)$_3$L$_2$(COPh). I.l.d.[71,455] Yellow,[71] m. 135-136.[71] IR,[71] anal.,[71] mol. wt.,[71] kinetics.[455]

fac-Mn(CO)$_3$L$_2$Br. I.l.a.[9,11,12] Yellow.[12] Dipole moment 4.78D,[12] IR,[12] anal.,[12] kinetics.[9,10] Isomerizes to mer isomer at 50°C.[11]

mer-Mn(CO)$_3$L$_2$Br. I.l.b,[11,12] III.3.[72a,510] Yellow,[510] m. 154-157.[510] Dipole moment 3.53D,[12] IR,[11,12,510] UV,[12] anal.,[12] mol. wt.[12]

mer-Mn(CO)$_3$L$_2$I. III.3,[72a,510] III.2.[257] m. 152.[257] IR,[72a,510] anal.[257]

mer-Mn(CO)$_3$L$_2$Cl.   III.3.[72a]   IR.

cis-(OC)$_2$MnL$_2$(CF$_3$COCHCOCF$_3$).   Orange,[226] m. 90. ^{1}H, IR, anal.

CpMn(CO)L$_2$.   I.1.a.[469] m. 120-121. Useful as herbicide.

π-MeC$_6$H$_4$Mn(CO)L$_2$.   I.1.a.[469] Useful as herbicide.

Mn(NO)$_2$L$_2$I. III.2.[241,265] Red.[265] Dipole moment 3.39D,[265] IR,[46,241] anal.,[241] mol. wt.[241]

Mn(NO)$_2$L$_2$Br. III.2.[265] Ochre. Dipole moment 3.24D,[265] IR,[46] anal.[265]

Mn(NO)$_2$L$_2$Cl. III.2.[265] Yellow-green. IR,[46,216] anal.[265]

[CpMnNOL$_2$]PF$_6$. I.1.a.[278] Yellow, m. 134-137. ^{1}H, IR, UV, anal., conductivity.

Tc(CO)$_3$L$_2$Cl. I.1.a.[262] Colorless. Anal.

fac-Re(CO)$_3$L$_2$I. I.1.a.[190] White, m. 133. Dipole moment 5.7D, IR, anal., mol. wt.

ReL$_2$(PPh$_3$)$_2$H$_3$. III.3.[188] ReH 1865. ^{1}H, anal.

ReL$_2$(PPh$_3$)$_2$Cl$_3$. III.3.[188]

ReL$_2$(PPh$_3$)$_2$I$_3$. III.3.[188]

[ReHL$_2$(PPh$_3$)$_2$]Cl. III.4.[188] ^{1}H.

Hg[Fe(CO)$_2$(NO)L]$_2$. I.1.a.[94b] IR, anal.

trans-Fe(CO)$_3$L$_2$. I.1.a.[145a,478] m. 116-117.[145a] μ 2.31D,[257] diamagnetic,[4] IR,[4,145a] Raman,[145a] anal.[145a]

[CpFeL]$_2$. III.4.[303] m. 133-134.

Fe(NO)$_2$L$_2$. III.1,[259] I.1.a.[349] Orange,[349] m. 62-72.[349] IR,[44,46,517] anal.[349]

CpFeL$_2$H. III.3.[390a] Yellow,[390a] m. 122.[389] ^{1}H,[390a] IR,[390a] anal.[389]

CpFeL$_2$I (mixture of two compounds thought to be isomers). I.1.a,[389] III.2.[390a] Black,[390a] m. 130.[390a] Anal.[389] One of the compounds is π-C$_5$H$_4$FeL P(O-⟨⟩) (OPh)$_2$I (see separate listing under Miscellaneous). The other compound is CpFeL$_2$I. Red,[387b] m. 155-162.[387b] ^{1}H, x-ray crystal structure,[6] Fe-P distance is 2.15 Å.

[Fe(CO)$_3$L$_2$Cl][AuCl$_4$]. III.2.[4] Diamagnetic, IR.

Fe(CO)$_2$L$_2$Br$_2$. I.1.a.[131] Kinetics.

[Fe(CO)$_3$L$_2$HgCl][HgCl$_3$]. III.2.[4] Diamagnetic, IR, conductivity.

cis-(CO)$_2$RuL$_2$Br$_2$. III.5.[327] White, m. 158-160. IR, anal.

cis-(CO)$_2$RuL$_2$Cl$_2$. III.5.[327] White, m. 158-160. IR, anal.

[Me$_4$N][trans-L$_2$RuCl$_4$]. I.1.b.[491a] Orange, m. 158-160. IR, anal.

Ru(NO)L$_2$Cl$_3$. III.2.[327] Green, m. 154-161. IR, anal.

cis-(CO)$_2$-cis-L$_2$-cis-I$_2$Os. I.1.a.[223] Colorless. IR, anal.

cis-(CO)$_2$-cis-L$_2$-cis-Br$_2$Os. I.1.a.[223] Colorless. IR, anal.

cis-(CO)$_2$OsL$_2$Cl$_2$. I.1.b.[327] White, m. 178-179. IR, anal.

cis-(CO)$_2$-cis-L$_2$-cis-Cl$_2$Os. I.1.a,[223] I.1.f.[253] Colorless,[223] m. 174.[253] Dipole moment, 4.43,[253] IR,[223] anal.,[223] mol. wt.[253]

cis-(CO)$_2$-trans-L$_2$OsCl$_2$. I.1.f.[251] Colorless.

[trans-L_2-trans-(CO)ClOs(CO)$_2$][AlCl$_4$].   III.3.[251,253]
  IR.[253]
[OsL$_2$(CO)$_4$][AlCl$_4$]$_2$.   III.3.[251]
Na[Co(CO)$_2$L$_2$].   III.1.[247] Colorless, m. 120. Anal.
Hg[Co(CO)$_3$L]$_2$. IR,[43] anal.
SnCl$_2$[Co(CO)$_3$L]$_2$.   III.1,[68] I.1.a.[69] Yellow,[69] m. 133.[68]
  Diamagnetic,[69] IR,[68,69] anal.,[69] mol. wt.[69]
Co$_2$(CO)$_6$L$_2$.   I.1.a.[357,440,459] Red,[459] m. 173.[440]   IR,[43,
  206,357] Raman,[206] anal.[43]
(π-C$_4$H$_7$)Co(CO)L$_2$.   III.4.[70] Orange, m. 87-88. IR, anal.
(π-C$_8$H$_{13}$)Co(CO)L$_2$. I.1.b.[508] Orange, m. 90-95. IR, anal.
L(OC)$_2$Co(CO)(CF$_3$CF)Co(CO)$_2$L. I.1.a.[73] Oil.
Co(CO)$_2$L$_2$[C$_6$F$_3$(CN)$_2$].   III.2.[72c] Yellow, m. 129-131. IR,
  ^{1}H, anal.
Co(CO)$_2$L$_2$H.   III.2.[247] Colorless, m. 88. ^{1}H, anal.
Co(NO)(CO)L$_2$.   I.1.a.[248,349] Orange,[248] yellow,[349] m.
  86.[349] Diamagnetic,[349] IR,[44,46,248,331,349,517]
  anal.,[349] kinetics.[499]
Co(CO)$_2$L$_2$I.   III.2.[246] Brown, m. 100. IR, anal.
Co(CO)$_2$L$_2$Br.   III.2.[246] Yellow. IR, anal.
Co(CO)$_2$L$_2$Cl.   III.2.[246] Orange, m. $\sim$ 65. IR, anal.
[Co(CO)$_3$L$_2$]AlCl$_4$.   III.3.[246] IR.
[Co(NO)LSPh]$_2$.   III.5.[249] Black, m. 131. IR, anal.
Co(NO)L$_2$Cl$_2$.   II.[256] Blue, m. 117. Anal.
{CoL[SC(CF$_3$)C(CF$_3$)S]$_2$}$_2$. Diamagnetic,[341] IR.
Rh$_4$(CO)$_{10}$L$_2$.   I.1.a.[461b]
RhL$_2$(MeCOCHCOMe). I.1.f.[215] Yellow. ^{1}H, anal., conduc-
  tivity.
Rh(CO)L$_2$Cl. I.1.a.[514] Yellow, m. 160-170. Diamagnetic,
  anal., mol. wt., conductivity.
C$_7$H$_8$Rh(Cl)$_2$RhL$_2$. I.1.c.[215] Yellow. ^{1}H, anal., conduc-
  tivity.
(C$_8$H$_{12}$)Rh(Br)$_2$RhL$_2$. I.1.c.[215] Orange.[215] ^{1}H,[215] anal.,[214]
  mol. wt.,[214] conductivity.
(C$_8$H$_{12}$)Rh(Cl)$_2$RhL$_2$. I.1.c.[215] Orange,[215] ^{1}H, anal.,[214]
  mol. wt.,[214] conductivity.
[Bu$_4$N][cis-RhL$_2$(NCS)$_2$].   III.5,[283] III.4.[283] IR.
[Bu$_4$N]{RhL$_2$[SC(CN)C(CN)S]}. I.1.a.[134] Yellow. Anal.,
  conductivity.
Ir(CO)$_2$L$_2$.   I.1.f.[539] Yellow, m. 164.
Ir(CO)L$_2$I.   III.3.[495a]   IR,[495a,495b] anal.,[495a] kinetics.
  [493a,494b]
Ir(CO)L$_2$Br.   II.[495a]   IR,[495a,495b] anal.,[495a] kinetics.[493a,
  494b]
Ir(CO)L$_2$Cl.   II,[495a] I.1.a.[252] Yellow,[252] m. 154.[252]
  IR,[252,495a,b] anal.,[252] kinetics.[493a,494b]
[Ir(CO)LI$_3$]$_2$.   III.2.[539] Red, m. 214.
Ni(CO)$_2$L$_2$.   I.1.a.[62,75,440,478,501] White,[75] m. 95.[440]
  ^{31}P(C$_6$H$_6$) 146,[374] IR,[62,373,517] anal.[440] Useful as
  Diels-Alder reaction catalyst,[49] olefin cyclization
  catalyst,[87,323,324,439,479,501] and octatriene poly-

merization cocatalyst.[323,439,478]

Ni(CO)L$_2$PPh$_3$. Useful as polymerization catalyst.[118]

NiL$_2$[(CF$_3$)$_2$CO]. I.l.c.[82] Yellow, m. 130-131. ^{1}H, IR, anal., mol. wt.

NiL$_2$[(CF$_3$)$_2$CS]. I.l.c.[82] Yellow, m. 132-134. ^{1}H, IR, anal., mol. wt.

(π-C$_3$H$_5$)NiL$_2$Br. I.l.f.[529a] ^{1}H.

Ni(NO)L$_2$NCS. I.l.f.[243] Black, violet,[243] m. 90-91.[243] IR,[45] anal.[243]

Ni(NO)L$_2$I. I.l.f.[242] m. 134-137. Anal.

Ni(NO)L$_2$Br. I.l.f,[242] III.2.[335] m. 152-155.[242] IR,[335] anal.

trans-L(ON)NiI$_2$Ni(NO)L. I.l.f.[241] IR.[45]

trans-L(ON)NiI$_2$Ni(NO)L. I.l.f.[242] m. 168-173. Anal., IR.[45]

trans-L(ON)NiBr$_2$Ni(NO)L. I.l.f.[242] m. 178-187. Anal., IR.[45]

trans-[PdLCl$_2$]$_2$. I.l.b.[111] Yellow-orange, m. 168-171. Anal., mol. wt., conductivity.

PtL$_2$(C$_2$Cl$_4$). III.1.[67] White, m. 126. Anal., mol. wt.

PtL$_2$I$_2$. I.l.f.[351] Yellow, m. 201. Anal., mol. wt.

cis-PtL$_2$Cl$_2$. I.l.b,[5d] I.l.f.[24] m. 190-191,[24] 155.[454] ^{31}P(CH$_2$Cl$_2$) -58.9,[5d] anal.,[5d,454] mol. wt.[14]

PtL$_2$F$_2$. I.l.b,[340] III.2.[340] White. Anal., turns to pink form on heating.

PtL$_2$(CO)$_2$F$_2$. III.4.[340] Yellow. IR, anal.

CuL$_2$CN. I.l.f.[64,396] m. 128-129.[396] Useful as insecticide[64] and bactericide.[346]

CuL$_2$I. I.l.f.[392] White,[392] m. 70-72,[192] 73-75.[19] Dipole moment 1.60D,[26] anal.,[392] monomer.[22,392]

CuL$_2$Br. I.l.f.[392] White,[392] m. 67-68,[392] 73-74,[22] 73-77.[19] Dipole moment 1.65D,[28] monomer.[22,392]

CuL$_2$Cl. I.l.f.[392] m. 98-99,[392] 70.[22] Dipole moment 1.70D,[28] IR,[392] UV,[392] anal.,[392] monomer.[22,392] Useful as paint ingredient for ships' bottoms.[228]

H$_{12}$B$_{10}$L$_2$. II,[489] III.3.[431] m.[489] 209-210.[431] Anal.,[489] UV.[145b]

TYPE: 3 C$_{18}$H$_{15}$; L$_3$ = [P(OPh)$_3$]$_3$

fac-Cr(CO)$_3$L$_3$. I.l.c.[343,365-367] White,[343,315,367] m. 129.[366] IR,[133,343] anal.[343,367] Useful as motor fuel additive,[365] pesticide,[365] oxo catalyst.[365]

fac-Mo(CO)$_3$L$_3$. I.l.c.[343,429,430] White,[343] m. 142-143.[343,365] IR,[133,343] anal.[343] Useful as motor fuel additive, oxo catalyst, and pesticide.[365]

mer-Mo(CO)$_3$L$_3$. I.l.c.[429,430] Crystalline. IR.[133,430]

Mo(CO)$_3$L$_3$ (fac-mer mixture). I.l.c.[430] Crystalline. IR.

fac-W(CO)$_3$L$_3$. I.l.b.[497a] Colorless,[361] m. 178-180.[361] ^{31}P(CH$_2$Cl$_2$) -121.6,[361] IR,[361] anal.[361]

cis-(OC)$_2$ReL$_3$I. I.l.a.[190] White, m. 147. Dipole moment

4.3D, IR, anal., mol. wt.
ReL$_3$(PPh$_3$)H$_3$.   III.3.[188] ^{1}H, IR, anal.
ReL$_3$I$_3$.   II.[189,352]   Blue,[352] m. 134.[352] Paramagnetic
   1.65 BM,[352] anal.,[352] conductivity.[352]
ReL$_3$(PPh$_3$)I$_3$.   III.3.[188] Yellow.
ReL$_3$(PPh$_3$)Cl$_3$.   III.3.[188] Yellow.
[ReL$_3$(PPh$_3$)H$_4$]Cl.   III.4.[188] ^{1}H.
[Ru(CO)$_3$L]$_3$.   I.1.a.[418]   Orange, m. 77-78. IR, anal.,
   mol. wt.
Ru(CO)L$_3$I$_2$.   III.3.[327]   Orange, m. 148-151.   IR, anal.
Ru(CO)L$_3$Br$_2$.   IV.3.[327]   Cream, m. 141-145.   IR, anal.
Ru(CO)L$_3$Cl$_2$.   IV.3.[327]   Yellow, m. 128-129.   IR, anal.
Co(CO)L$_3$H.   III.2,[260] IV.3.[5c]   Off white,[261] m. 126-128,[5c]
   ^{1}H,[5c] IR,[5c] anal.[260]
Co(NO)L$_3$.   II.[255]   Dipole moment 2.38D,[256] IR,[46,517] anal.[256]
Rh$_4$(CO)$_9$L$_3$.   I.1.a.[461b]
RhL$_3$NCS.   I.1.b,[283] I.1.f.[283]   Yellow, m. 144-146.   IR,
   anal., mol. wt. Dissociation of L indicated by pres-
   ence of two CN stretches.
RhL$_3$I.[515]
RhL$_3$Br.   I.1.c,[215] I.1.f.[214]   Yellow. Anal.,[215] mol.
   wt.,[215] conductivity.[215]
RhL$_3$Cl.   I.1.a,[215,514] I.1.b,[295,296] I.1.c,[376] I.1.f.[214]
   Yellow,[514] m. 170-180.[514] Diamagnetic,[514] anal.,[214]
   conductivity,[514] mol. wt.[214]
RhL$_3$I$_3$.   I.1.b.[515]
IrL$_3$(CO)H.   I.1.b.[5b]
IrL$_3$Cl$_2$H.   I.1.f,[5b] II.[452a]   White.[5b,452a] IR,[452a]
   anal.[5b,452a]
Ni(CO)L$_3$.   I.1.a.[62,122]   White,[353] m. 98.5.[353] IR.[62,517]
   Useful as catalyst for cyclization of olefins,[122,323]
   dimerization of conjugated dienes,[74,117] and polymer-
   ization.[118]
NiL$_3$(PPh$_3$).   I.1.a.[120]   m. 124-126. Useful as catalyst
   for the dimerization of olefins.
NiL$_3$PF$_3$.   IV.3.[312]   Colorless, m. 89. IR, anal.
trans-NiL$_3$(CN)$_2$.   I.2.[144a]   Red-orange. Oil. UV.
NiL$_3$Cl$_2$.   II.[234]   Violet.
PdL$_3$(CN-p-MeOC$_6$H$_4$).   I.1.f.[348]   Colorless, m. 105-110.
   Anal.
PtL$_3$.   III.1.[350,351]   White,[350] Yellow,[351] m. 120-124.[351]
   Anal.[351]
PtL$_3$P(p-ClC$_6$H$_4$)$_3$.   I.1.b.[351]   Colorless, m. 157.

TYPE: 4 C$_{18}$H$_{15}$; L$_4$ = [P(OPh)$_3$]$_4$

cis-(OC)$_2$MoL$_4$.   I.1.c.[200]   White, m. 180. IR, anal.
trans-(OC)$_2$MoL$_4$.   I.1.a,[200] I.1.c.[200]   White, m. 136-138.
   IR.
[CpFeL$_2$]$_2$.   III.1.[387b,389,390a,b]   Orange,[390a] yellow,[389]
   m. 132-133.[390a] ^{1}H, anal.,[389] mol. wt.[390a]

Hg[Mn(CO)$_3$L$_2$]$_2$. III.2,[257,412] III.3.[250] Colorless,[257]
   m. 145, 107.[250] IR,[412] anal.[412]
[Re(CO)$_3$L$_2$]$_2$. I.1.a.[411] Diamagnetic in solid, paramagnetic in C$_6$H$_6$ 1.1 BM.
cis-HClRuL$_4$. Reverse of IV.2.[307a] White,[307a] m. 166-
   169.[415a,b] ^{1}H,[307a,415a] IR.[415a] Converts to
   L$_3$RuP(OPh)$_2$Cl refluxing in solvents.[307a]

RuL$_4$I$_2$. III.3.[327] Yellow, m. 173-175. Anal.
RuL$_4$Br$_2$. I.1.b.[327] Cream, m. 163-166. Anal.
RuL$_4$Cl$_2$. I.1.b.[327,375] Yellow,[375] cream,[327] m. 151-
   152.[327] Anal.[327]
OsL$_4$Br$_2$. I.1.b.[327] White, m. 159-161. Anal.
Hg[Co(CO)$_2$L$_2$]$_2$. III.3.[247] Yellow, m. 171. Anal.
Co$_4$(CO)$_8$L$_4$. I.1.a.[461b] Brown. IR.
[Co(CO)$_2$L$_2$]$_2$. III.1.[70] Yellow-brown, oil. IR, anal.
CoL$_4$H. II,[326] I.1.b.[456] Yellow,[326] m. 165-168.[326]
   ^{31}P(C$_6$H$_6$) -140.2,[326] ^{1}H,[326] anal.,[326] mol. wt.[326]
Rh$_4$(CO)$_8$L$_4$. I.1.a.[461b] Red, IR.
[RhL$_2$NCS]$_2$. III.4.[283] Yellow, m. 175-178. IR, anal.,
   mol. wt.
[RhL$_2$Br]$_2$. I.1.c.[214,215] Yellow,[215] anal.,[214] mol.
   wt.,[214] conductivity.[215]
[RhL$_2$Cl]$_2$. I.1.c.[214,215] Yellow,[215] anal.,[214] mol.
   wt..[214] conductivity.[215]
[RhL$_4$][BPh$_4$]. I.1.c.[215] Yellow. Anal., conductivity.
[RhL$_4$]ClO$_4$. I.1.c.[215] Yellow. Anal., conductivity.
[RhL$_4$H$_2$][BPh$_4$]. III.4.[215] Releases H$_2$ readily.
Ir$_4$(CO)$_8$L$_4$. I.1.a.[461b] IR.
IrL$_4$H. II.[195] White, m. 127. ^{1}H, IR, anal., mol. wt.
NiL$_4$. I.1.a,[98,120] II.[119,121,234,326,407,408a,526]
   White,[408a] m. 146-148.[408a] ^{31}P(C$_6$H$_6$) -129.3,[326,500c]
   anal.,[408a] mol. wt.[408a] Useful as polymerization
   initiator.[35]
PdL$_4$. I.1.f.[347,348] Colorless,[348] m. 120-130.[348]
   Anal.,[348] mol. wt.[348]
PtL$_4$. I.1.b,[350,351] II.[350,351] White,[350] m. 145.[351]
   Anal.[351]
[PtL$_4$]PtCl$_4$. I.1.b.[504]

TYPE: 4 C$_{18}$H$_9$D$_6$; L$_4$ = [P(O-2,6-D$_2$C$_6$H$_3$)$_3$]$_4$

cis-DClRuL$_4$. Reverse of IV.2.[415a] White, m. 168-172. Anal.

TYPE: 8 C$_{18}$H$_{15}$; L$_8$ = [P(OPh)$_3$]$_8$

Co$_2$L$_8$. III.1.[528a] White, m. 222-225. IR, ^{1}H, paramagnetic.
Rh$_2$L$_8$. II.[452a] White. Diamagnetic, ^{1}H, mol. wt.
Ir$_2$L$_8$. II.[452a] White. Diamagnetic, ^{1}H, mol. wt.

TYPE: 10 C$_{18}$H$_{15}$; L$_{10}$ = [P(OPh)$_3$]$_{10}$

[ReL$_5$]$_2$.   III.1.[188]   White.   Anal., mol. wt.

TYPE: C$_{18}$H$_{33}$; L = P(OC$_6$H$_{11}$)$_3$

H$_3$BL.   I.1.b.[441,442]   m. 69.[441]   Anal.,[441] mol. wt.[441]

TYPE: C$_{18}$H$_{39}$; L = P(OC$_6$H$_{13}$)$_3$

CpFe(CO)LI.   IR.[79]
Co$_2$(CO)$_7$L.   IR.[496]

TYPE: C$_{21}$H$_{12}$; L = P(O-p-CNC$_6$H$_4$)$_3$

Ni(CO)$_3$L.   IR.[500b]

TYPE: C$_{21}$H$_{18}$; L = P(O-p-Cl-o-MeC$_6$H$_3$)$_3$

Ni(CO)$_3$L.   IR.[500b]

TYPE: C$_{21}$H$_{21}$O$_3$; L = P(O-o-MeC$_6$H$_4$)$_3$

CpV(CO)$_3$L.   I.1.a.[493b]   Orange, m. 162.   IR, anal.
CpMn(CO)$_2$L.   I.1.a.[493b]   Yellow, m. 144.   IR,[493b,494a]
    anal.[493b]
Ni(CO)$_3$L.   IR.[493b]

TYPE: C$_{21}$H$_{21}$O$_3$; L = P(O-p-MeC$_6$H$_4$)$_3$

Ni(CO)$_3$L.   IR.[493b,500b]
O$_3$L.   I.3.[498b]   Unstable above -20°C.

TYPE: 2 C$_{21}$H$_{21}$O$_3$; L$_2$ = [P(O-o-MeC$_6$H$_4$)$_3$]$_2$

cis-(CO)$_2$RuL$_2$Cl$_2$.   I.1.f.[327]   White, m. 161-163.   IR,
    anal.
NiL$_2$(C$_2$H$_4$).   II.[480b]   Yellow, m. 118-120. ^{31}P(MePh)
    -139.7, IR, UV, anal., mol. wt.
NiL$_2$(C$_2$H$_3$CN).   III.3.[480b] ^{1}H.
NiL$_2$(CO)$_2$.   IV.[196b]

TYPE: 2 C$_{21}$H$_{21}$O$_3$; L$_2$ = [P(O-p-MeC$_6$H$_4$)$_3$]$_2$

cis-(CO)$_2$RuL$_2$Br$_2$.   I.1.f.[327]   White, m. 162-165.   IR, anal.
cis-(CO)$_2$RuL$_2$Cl$_2$.   I.1.f.[327]   White, m. 164-167.   IR, anal.
Ru(NO)L$_2$Cl$_3$.   III.2.[327]   Green, m. 152-154.   IR, anal.
{CoL[HONC(Me)C(Me)NO]}$_2$.   I.2.[467]   IR, anal.
Rh(CO)L$_2$Cl.   I.1.a.[514]   Yellow, m. 170-176.   Diamagnetic,
    anal., mol. wt., conductivity.
RhL$_2$(SCN)$_2$.   I.1.b.[514]   Yellow, m. 150-175.   Diamagnetic,
    anal., mol. wt., conductivity.
[Ir(CO)$_2$L]$_2$.   Yellow.[539]
[Ir(CO)LI$_3$]$_2$.   Red,[539] m. 215.

TYPE: 3 $C_{21}H_{21}O_3$; L_3 = $[P(O-p-MeC_6H_4)_3]_3$

ReL_3I_3.   II.[189]   Blue, m. 107.   Paramagnetic 1.65 BM,
    anal., conductivity.
$Ru(CO)L_3Br_2$.   IV.3.[327]   Yellow, m. 148-150.   IR, anal.
$Ru(CO)L_3Cl_2$.   IV.3.[327]   Yellow, m. 143-146.   IR, anal.
RhL_3SCN.   I.1.b.[514]   Yellow, m. 150-170.   Diamagnetic,
    anal., mol. wt., conductivity.
RhL_3Cl.   I.1.a.[514]   Yellow, m. 166-167.   Diamagnetic,
    anal., mol. wt., conductivity.
NiL_3.   II.[196b]   Red, m. 125-150. $^{31}P(MePh)$ -128.3, IR,
    UV, anal., mol. wt., equilibrium constant study.

TYPE: 4 $C_{21}H_{21}O_3$; L_4 = $[P(O-p-MeC_6H_4)_3]_4$

RuL_4Br_2.   I.1.b.[327]   Cream, m. 155-157.   Anal.
RuL_4Cl_2.   I.1.b.[327]   Cream, m. 146-148.   Anal.
OsL_4Br_2.   I.1.b.[327]   White, m. 158-160.   Anal.

TYPE: 4 $C_{21}H_{21}O_3$; L_4 = $[P(O-o-MeC_6H_4)_3]_4$

$Ir_4(CO)_8L_4$.   I.1.a.[461b]
NiL_4.   I.2.[196b]   Off white, m. 90-105. $^{31}P(MePh)$ -129.0,
    UV, anal., equilibrium constant study.

TYPE: 4 $C_{21}H_{21}O_3$; L_4 = $[P(O-m-MeC_6H_4)_3]_4$

CoL_4H.   II.[326]   Yellow, m. 56-59. $^{31}P(C_6H_6)$ -139.5, ^1H,
    anal.
NiL_4.   II.[326]   White, m. 74-79. $^{31}P(C_6H_6)$ -128.7,[326,500c]
    anal.[326]

TYPE: 4 $C_{21}H_{21}O_3$; L_4 = $[P(O-p-MeC_6H_4)_3]_4$

CoL_4.   II.[382a]   Useful as gasoline additive and hydro-
    genation catalyst.
CoL_4H.   II.[326]   Yellow,[326] m. 140-142.[326] $^{31}P(C_6H_6)$
    -141.7,[326] ^1H,[326] anal.,[326] mol. wt.[326]   Useful as
    gasoline additive and hydrogenation catalyst.[382a]
NiL_4.   II,[326,382a,407] I.1.a.[120]   White,[407] m. 98-100,[407]
    151-152.[326] $^{31}P(C_6H_6)$ -130.6, anal.[326,407]   Useful
    as a catalyst in the dimerization of olefins,[119,120]
    and as a gasoline additive and hydrogenation cata-
    lyst.[382a]

TYPE: 3 $C_{21}H_{21}O_6$; L_3 = $[P(O-p-MeOC_6H_4)_3]_3$

$Ni(CO)L_3$.[353]   White.   Liquid.

TYPE: 4 $C_{21}H_{21}O_6$; L_4 = $[P(O-p-MeOC_6H_4)]_4$

NiL_4.[120,121,407] Yellow,[407] m. 98-100,[407] 134-138.[121]
 [31]P(toluene) -131.6,[500c] anal.[407] Useful as catalyst
for dimerization of olefins.[120,121]

TYPE: $C_{21}H_{21}O_6$; L = $P(OC_6H_4-p-OMe)_3$

π-EtPrC$_6$H$_3$Mn(CO)$_2$L. I.1.a. Useful as herbicide.[469]

TYPE: $C_{21}H_{45}$; L = $P(OC_7H_{15})_3$

H_3BL. I.1.b.[442]

TYPE: 2 $C_{24}H_{27}$; L_2 = $[P(O-2,4-Me_2C_6H_3)_3]_2$

cis-$(CO)_2RuL_2Cl_2$. I.1.f.[327] White, m. 158-160. IR,
   anal.
$Ni(CO)_3$L. I.1.a.[500b] IR.

TYPE: 2 $C_{24}H_{27}$; L_2 = $[P(OC_6H_4Et)_3]_2$

π-Me$_2$C$_6$H$_3$Mn(CO)L$_2$. I.1.a. Useful as herbicide.[469]

TYPE: $C_{24}H_{51}$; L_2 = $[P(OC_8H_{17})_3]_2$

$Ni(CO)_2L_2$. Useful for catalysis of polymerization of conjugated dienes.

TYPE: 4 $C_{24}H_{51}$; L_4 = $[P(OCH_2CHEt(CH_2)_3Me)_3]_4$

CoL_4. II.[382a] Useful as gasoline additive and hydrogenation catalyst.
CoL_4H. II.[382a] Useful as gasoline additive and hydrogenation catalyst.
NiL_4. I.1.a,[120] II.[382a,407] White,[407] m. 112-115.[119]
   Anal.[407] Useful as catalyst in dimerization of
   dienes,[119,120,481] and as a gasoline additive and
   hydrogenation catalyst.[382a]

TYPE: $C_{30}H_{13}$; L = $P(OC_{10}H_{21})_3$

Mn(NO)$_3$L. IR.[46,517]
H_3BL. I.1.b.[442]

TYPE: $C_{30}H_{39}$; L = $P(O-p-t-BuC_6H_4)_3$

O_3L. I.3.[498b] Unstable above -20°C.

TYPE: 2 $C_{30}H_{63}$; L_2 = $[P(OC_{10}H_{21})]_2$

Mn(NO)$_2$L$_2$I. IR.[46]

TYPE: $C_{36}H_{27}$; L = P(O-o-PhC$_6$H$_4$)$_3$

CpV(CO)$_3$L.   I.l.a.[493b]   Yellow, m. 164.   IR, anal.
CpMn(CO)$_2$.   I.l.a.[493b]   Yellow, m. 165.   IR,[493b,494a]
    anal.
Ni(CO)$_3$L.   IR.[493b]

TYPE: 2 $C_{36}H_{27}$; L$_2$ = [P(O-o-PhC$_6$H$_4$)$_3$]$_2$

NiL$_2$(PPh$_3$)$_2$.   III.1.[532]
NiL$_2$Me$_2$.   III.3.[532]   Colorless.

TYPE: $C_{36}H_{75}$; L = P(O-n-C$_{12}$H$_{25}$)$_3$

πMe$_2$C$_5$H$_3$Mn(CO)$_2$L.   I.l.a.   Useful as herbicide.[469]

TYPE: 4 $C_{36}H_{75}$; L$_4$ = [P(O-n-C$_{12}$H$_{25}$)$_3$]$_4$

NiL$_4$.   II.[121]   m. 196-198.   Useful as catalyst for dimer-
    ization of olefins.

P(OR)$_2$(OR')

TYPE: C_3H_7; L = P(OMe)(OCH$_2$)$_2$

CuLI.   m. 132-133.[26]

TYPE: C_6H_{12}; L = P(OCH$_2$)$_2$CCH$_2$ClMe(OMe)

H$_3$BL (mixture of two isomers).   I.l.f.[531b]   Dipole moment
    of mixture 5.9D, ^{1}H, IR.

TYPE: C_6H_{13}; L = P(OCHMe)$_2$CH$_2$(OMe)

H$_3$BL(axial OMe).   I.l.f.[520b]   x-Ray structure confirms
    axial OMe at P and equatorial Me groups on ring car-
    bons,[452b] IR.[520b]
H$_3$BL (equatorial-axial OMe isomer mixture).   I.l.f.[520b]
    IR.

TYPE: C_6H_{13}; L = P(OMe)[(OCH$_2$CH$_2$)$_2$CH$_2$]

CuLI.   m. 142-144.[26]

TYPE: C_7H_7; L = P(OMe)(O-o-OC$_6$H$_4$)

CuLBr.   m. 130-135.[21]

TYPE: C_8H_9; L = P(OEt)(O-o-OC$_6$H$_4$)

CuLBr. m. 142-145.[21]

TYPE: C_8H_9; L = P(OPh)(OCH$_2$)$_2$

Ni(CO)$_3$L.   IR.[500b]

TYPE: C_9H_{11}; L = P(OPr)(O-o-OC$_6$H$_4$)

CuLI. m. 138.[21]

TYPE: C_9H_{11}; L = P(O-i-Pr)(O-o-OC$_6$H$_4$)

CuLI.   m. 178-179.[21]
CuLCl. m. 143.[21]

TYPE: C_9H_{15}; L = P(OMe)(OCMe$_2$CCl$_3$)$_2$

CuLCl. m. 182-183.[2]

TYPE: $C_{10}H_{13}$; L = P(OBu)(O-o-OC$_6$H$_4$)

CuLCl. m. 202.[21]

TYPE: $C_{10}H_{13}$; L = P(O-i-Bu)(O-o-OC$_6$H$_4$)

CuLCl. m. 208-210.[21]

TYPE: $C_{10}H_{13}$; L = P(OCH$_2$Ph)[(OCH$_2$)$_2$CH$_2$]

CuLCNS. m. 150-152.[22]

TYPE: $C_{10}H_{17}$; L = P(OEt)(OCMe$_2$CCl$_3$)$_2$

CuLCl. m. 178.[2]

TYPE: $C_{11}H_{19}$; L = P(OPr)(OCMe$_2$CCl$_3$)$_2$

CuLCl. m. 138.[2]

TYPE: $C_{11}H_{19}$; L = P(O-i-Pr)(OCMe$_2$CCl$_3$)$_2$

CuLCl. m. 184.[2]

TYPE: $C_{12}H_{15}$; L = P(OC$_6$H$_{11}$)(O-o-OC$_6$H$_4$)$_2$

CuLCl. m. 187.[2]

TYPE: $C_{12}H_{21}$; L = P(OBu)(OCMe$_2$CCl$_3$)$_2$

CuLCl. m. 162.[2]

TYPE: $C_{12}H_{21}$; L = P(O-i-Bu)(OCMe$_2$CCl$_3$)$_2$

CuLCl.   m. 191.[2]

TYPE: $C_{12}H_{21}$; L = P(O-t-Bu)(OCMe$_2$CCl$_3$)$_2$

CuLCl.   m. 172.

TYPE: $C_{13}H_{13}$; L = P(OMe)(OPh)$_2$

CuLCl.   I.l.f.[287]   Oil.   Anal.

TYPE: $C_{13}H_{18}$; L = P(O-furfuryl)(OCMe$_2$CCl$_3$)$_2$

CuLCl.   m. 136.[2]

TYPE: $C_{14}H_{17}$; L = P(OPh)(OCMe$_2$CCl$_3$)$_2$

CuLCl.   m. 156.[2]

TYPE: $C_{14}H_{21}$; L = P(OC$_8$H$_{17}$)(O-o-OC$_6$H$_4$)

CuLCl.   Oil.[2]

TYPE: $C_{22}H_{31}$; L = P(OPh)$_2$(OC$_{10}$H$_{21}$)

NiL$_4$.   II.[121]   Liquid.   Useful as a catalyst for dimeriza-
    tion of olefins.

TYPE: $C_{26}H_{47}$; L = P(OPh)(OC$_{10}$H$_{21}$)$_2$

Mn(NO)$_3$L.   I.l.a.[46]   IR.[46,517]

TYPE: 2 $C_{26}H_{47}$; L$_2$ = [P(OPh)(OC$_{10}$H$_{21}$)$_2$]$_2$

Mn(NO)$_2$L$_2$I.   IR.[46]

P(OR)$_2$R

TYPE: C_3H_9; L = P(OMe)$_2$Me

Cr(CO)$_5$L.   I.l.a.[144b]   Yellow, m. 6. ^{1}H, IR, anal.
Mo(CO)$_5$L.   I.l.a.[144b]   Yellow.   Liquid. ^{1}H, IR, anal.
W(CO)$_5$L.   I.l.a.[144b]   Yellow. ^{1}H, IR, anal.
mer-Mn(CO)$_3$L$_2$Ph.   I.l.a.[36] ^{1}H, IR.

TYPE: 2 C_3H_9; L$_2$ = [P(OMe)$_2$Me]$_2$

cis-Cr(CO)$_4$L$_2$.   I.l.a.[144b]   White, m. 75-76. ^{1}H, IR,
    anal.

cis-Mo(CO)$_4$L$_2$. I.1.c.[144b] White, m. 67-68. ^{1}H, IR,
    anal.
cis-W(CO)$_4$L$_2$. I.1.b.[144b] White, m. 49-51. ^{1}H, IR, anal.

TYPE: 2 C$_3$H$_9$; L$_3$ = [P(OMe)$_2$Me]$_3$

fac-Cr(CO)$_3$L$_3$. I.1.c.[144b] Yellow, m. 152-155. ^{1}H, IR,
    anal.
mer-Cr(CO)$_3$L$_3$. I.1.c.[144b] Yellow, m. 152-155. ^{1}H, IR,
    anal.
fac-Mo(CO)$_3$L$_3$. I.1.c.[144b] White, m. 171-173. ^{1}H, IR,
    anal.
fac-W(CO)$_3$L$_3$. I.1.c.[144b] Yellow. ^{1}H, IR.
mer-W(CO)$_3$L$_3$. I.1.c.[144b] Yellow. ^{1}H, IR.

TYPE: C$_4$H$_{11}$; L = P(OMe)$_2$Et

CuLI. m. 160-161.[27]

TYPE: C$_6$H$_{15}$; L = P(OEt)$_2$Et

CuLI. m. 167-168.[27]
H$_3$BL. I.1.b.[442]

TYPE: C$_8$H$_9$; L = P(OCH$_2$)$_2$Ph

Ni(CO)$_3$L. IR.[500b]

TYPE: C$_8$H$_{11}$; L = P(OMe)$_2$Ph

Mo(CO)$_5$L. I.1.a.[280b] White, m. 54-55. ^{1}H, IR, anal.
W(CO)$_5$L. ^{31}P, IR.[208]
C$_4$H$_6$Fe(CO)$_2$L. I.1.a.[386] Yellow, b$_{0.5}$ 179-180. Anal.
trans-(CO)Cl-trans-HLRu[PPh(n-Pr)$_2$]$_2$. III.3.[164] White,
    m. 172-175. IR, anal.
Os(PPhMe$_2$]$_2$LH$_4$. III.2.[163,165] Colorless.[163] Oil.[163]
    ^{1}H, IR.
Os(PPhMe$_2$)$_2$LCl$_3$. III.1.[163,165] Red.[163] m. 178-186.[163]
    IR, anal.
mer-IrL(PEt$_2$Ph)$_2$H$_3$. II.[339b] ^{1}H.
H$_3$BL. I.1.b.[442] b$_{0.2}$ 79.5-80.0.
Et$_3$AlL. I.1.f.[129] Liquid. Anal., decomposes to (MeO)-
    MePhP=OAlEt$_3$.

TYPE: 2 C$_8$H$_{11}$; L$_2$ = [P(OMe)$_2$Ph]$_2$

Cr(CO)$_4$L$_2$(cis-trans mixture). I.1.c.[280b] Yellow, m.
    99-101. (trans), dipole moment 2.4D, ^{1}H, IR.
trans-Mo(CO)$_4$L$_2$. I.1.c.[280b] White, m. 90-91. ^{1}H, IR,
    anal.

trans-W(CO)$_4$L$_2$. I.1.c.[280b] White, m. 96-98. ^{1}H, IR,
  anal.

TYPE: 3 C$_8$H$_{11}$; L$_3$ = [P(OMe)$_2$Ph]$_3$

mer-Cr(CO)$_3$L$_3$. I.1.c.[280b] Yellow, m. 86-88. ^{1}H, IR,
  anal.
mer-Mo(CO)$_3$L$_3$. I.1.c.[280b] White, m. 121-23. ^{1}H, IR,
  anal.
trans-NiL$_3$(CN)$_2$. I.2.[102,447] Red-orange,[102] m. 137-
  138.[102] IR, UV, anal.

TYPE: 4 C$_8$H$_{11}$; L$_4$ = [P(OMe)$_2$Ph]$_4$

NiL$_4$. II.[409a] Yellow. Diamagnetic, anal.
PdL$_4$. II.[409a] Yellow. Diamagnetic, anal. Unstable to
  air and moisture.
CuL$_4$Cl. II.[409a] Colorless. Diamagnetic, anal.

TYPE: C$_8$H$_{17}$; L = P(OEt)$_2$CHCHOEt

C$_4$H$_6$Fe(CO)$_2$L. I.1.a.[386] b$_{2.0}$ 156-160. Anal.

TYPE: C$_{10}$H$_{15}$; L = P(OEt)$_2$Ph

C$_4$H$_6$Fe(CO)$_2$L. I.1.a.[386] Yellow, b$_{4.0}$ 178. Anal.
Ni(CO)$_3$L. IR.[500b]

TYPE: 2 C$_{10}$H$_{15}$; L$_2$ = [P(OEt)$_2$Ph]$_3$

Co(CO)$_2$L$_2$H. I.1.a,[409b] IV.3.[409b] Anal.
H$_{12}$B$_{10}$L$_2$. IV.[468b] m. 215-216. Anal., mol. wt.
H$_{12}$B$_{10}$L$_2$·2 HNMe$_2$. IV.[468b] m. 224-226. Anal.
CuLI. m. 122.5-123.5.[288] Anal.

TYPE: 3 C$_{10}$H$_{15}$; L$_3$ = [P(OEt)$_2$Ph]$_3$

trans-NiL$_3$(CN)$_2$. I.2.[102,447] Red-orange,[102] m. 105.[102]
  Diamagnetic,[447] IR,[447] UV,[102] anal.,[102] mol. wt.,[102]
  conductivity. x-Ray structure shows[487] nearly perfect
  trigonal bypyramid with trans axial CN groups, Ni-P
  (avg.) = 2.228 Å, P-O (avg.) = 1.58 Å, P-C (avg.) =
  1.45 Å, (N)C-Ni-C(N) = 170.8°, dissociates slightly
  in MeCN.[102]

TYPE: 4 C$_{10}$H$_{15}$; L$_4$ = [P(OEt)$_2$Ph]$_4$

FeL$_4$H$_2$(cis and trans mixture). II.[409b] ^{31}P(liquid -50°C)
  -198.6 P$_{ax}$(cis form), -188 P$_{eq}$(cis form), ^{2}J$_{PaxFePeq}$
  (cis form)= 45.0,[497b] ^{1}H,[497b] IR.[409b]
CoL$_4$H. II.[409b] x-Ray structure indicates pseudotetra-

hedrality (see text), ^{1}H, IR.
[CoL$_4$Cl]Cl.   III.2.[409b]
[CoL$_4$Cl]ClO$_4$.   III.2.[409b]
NiL$_4$.   II.[409a] Yellow.[409a] Diamagnetic,[409a] ^{31}P(PhMe)
  -162.1,[500c] IR,[409a] anal.[409a]
PdL$_4$.   II.[409a] Yellow. Diamagnetic, IR, anal.
CuL$_4$Cl.   II.[409a] Colorless. Diamagnetic, UV, anal.

TYPE: C$_{12}$H$_{19}$; L = P(OPr)$_2$Ph

CuLI.   m. 129.5-130.5.[288] Anal.

TYPE: C$_{12}$H$_{19}$; L = P(O-i-Pr)$_2$Ph

Ni(CO)$_3$L.   IR.[500b]
CuLBr.   I.l.f.[392] White, m. 143-144. Anal., tetrameric
   in solution.
CuLCl.   I.l.f.[392] White, m. 141-142. Anal., tetrameric.

TYPE: 2 C$_{12}$H$_{19}$; L$_2$ = [P(O-i-Pr)$_2$Ph]$_2$

CuL$_2$Br.   I.l.f.[392] White, m. 148-149. Anal., tetrameric.
CuL$_2$Cl.   I.l.f.[392] White, m. 144-145. Anal., tetrameric.

TYPE: 2 C$_{13}$H$_{13}$; L$_2$ = [P(OPh)$_2$Me]$_2$

mer-Mn(CO)$_3$L$_2$H.   I.l.a.[72a] Yellow, m. 89.5-90.5. ^{1}H,
   IR, anal.
unsym. mer-Mo(CO)$_3$L$_2$Me.   I.l.a.[36] m. 138-139. IR, anal.
mer-Mn(CO)$_3$L$_2$Ph.   I.l.a.[37] m. 173-174.5. ^{1}H, IR.
mer-Mn(CO)$_3$L$_2$(COMe).   I.l.d.[37] m. 138-139. IR.
mer-Mn(CO)$_3$L$_2$Br.   III.3.[71] IR.

TYPE: C$_{13}$H$_{17}$; L = P(OEt)$_2$(-1-indenyl)

CuLI.   m. 126-127.[289]

TYPE: C$_{14}$H$_{23}$; L = P(OBu)$_2$Ph

Ni(CO)$_3$L.   IR.[500b]
CuLI.   m. 88-89.[288] Anal.
H$_{12}$B$_{10}$L$_2$.   II.[489] White, m. 126-127. Anal.

TYPE: C$_{18}$H$_{15}$; L = P(OPh)$_2$Ph

Ni(CO)$_3$L.   IR.[500b]
CuLI.   I.l.c.[392] White, m. 183-184. Anal., tetrameric.

TYPE: 2 C$_{18}$H$_{15}$; L$_2$ = [P(OPh)$_2$Ph]$_2$

(π-C$_8$H$_{13}$)CoL$_2$.   I.l.f.[456] ^{1}H.

CuL$_2$I.   I.l.f.[392]   White, m. 187-189.   Anal., dimeric.

TYPE: 4 C$_{18}$H$_{15}$; L$_4$ = [P(OPh)$_2$Ph]$_4$

CoL$_4$H.   I.l.c.[456]

P(OR)$_2$NR$_2$

TYPE: 2 C$_4$H$_{10}$; L$_2$ = [P(OCH$_2$)$_2$NMe$_2$]$_2$

H$_{12}$B$_{10}$L$_2$.   I.l.b.[145b]   White, m. 200.   Anal.

TYPE: C$_{14}$; L = P(OPh)$_2$NEt$_2$

CuLI.   m. 111.[20]   Trimer.

P(OR)$_2$X

TYPE: C$_2$H$_6$F; L = P(OMe)$_2$F

Mn$_2$(CO)$_9$L.   IV.1.[123]   IR.
Co(NO)(CO)$_2$L.   IV.1.[124]   IR.
Co(NO)(CO)L[P(OMe)$_3$].   IV.1.[124]   IR.

TYPE: 2 C$_2$H$_6$F; L$_2$ = [P(OMe)$_2$F]$_2$

Co(NO)(CO)L$_2$.   IV.1.[124]   IR.

TYPE: 4 C$_2$H$_6$Cl; L$_4$ = [P(OMe)$_2$Cl]$_4$

NiL$_4$. ^{31}P(C$_6$H$_6$) -172.5,[362] ^{1}H.[364]

TYPE: 4 C$_4$H$_{10}$; L$_4$ = [P(OEt)$_2$Cl]$_4$

NiL$_4$.   I.l.a.[96]   Yellow.   Gel.   Anal., Faraday effect.

TYPE: C$_4$H$_{11}$; L = P(OEt)$_2$OH

H$_{12}$B$_{10}$L$_2$.   IV.[468b]   m. 216.   Anal., mol. wt.

O=P(OR)$_2$H*

TYPE: C$_4$H$_{11}$; L = O=P(OEt)$_2$H

*Here X is formally OH but the ligand apparently does not rearrange to give P(OR)$_2$OH in the next five complexes which are grouped together out of the normal order.

Cl_3BL. I.3.[48a]

TYPE: C_6H_{15}; L = O$=$P(OPr)$_2$H

Cl_3BL. I.3.[48a] IR.

TYPE: C_6H_{15}; L = O$=$P(O-i-Pr)$_2$H

Cl_3BL. I.3.[48a] IR.

TYPE: C_8H_{19}; L = O$=$P(OBu)$_2$H

Cl_3BL. I.3.[48a] IR.

TYPE: C_8H_{19}; L = O$=$P(O-i-Bu)$_2$H

Cl_3BL. I.3.[48a] IR.

TYPE: 2 C_4H_{11}; L_2 = [P(OEt)$_2$OH]$_2$

$\{PtL_2[(H_2N_2CS]_2\}Cl_2$.[212]

TYPE: C_5H_{10}; L_2 = [P(OEt)$_2$CN]$_2$

[Cr(CO)$_4$L]$_2$. I.1.c.[144c] Yellow. IR.
cis-Mo(CO)$_4L_2$. I.1.f.[144c] White. ^{1}H, IR.
[Mo(CO)$_4$L]$_2$. I.1.c.[144c] Yellow. ^{1}H, IR, anal.

TYPE: 3 C_6H_4F; L_3 = [P(O-o-OC$_6$H$_4$)F]$_3$

Cr(CO)$_3L_3$. I.1.a.[465] Useful as polymerization catalyst.[465]
fac-Mo(CO)$_3L_3$. I.1.a.[462,465] Colorless,[462] m. 89.5-91.[462]
    ^{31}P(C$_6$H$_6$) -17.6,[438] ^{19}F,[438] IR,[462] anal.,[462] mol.
    wt.[462] Useful as polymerization catalyst.[465]

TYPE: 4 C_6H_4F; L_3 = [P(O-o-OC$_6$H$_4$)F]$_3$

NiL$_4$. I.1.a.[465] m. 129-130.[465] ^{19}F, ^{2}JPNiP = 21.[339a,438]
    Useful as polymerization catalyst.[465]

TYPE: 4 C_6H_4Cl; L_4 = [P(O-o-OC$_6$H$_4$)F]$_4$

NiL$_4$. I.1.a.[438] ^{31}P(PhNO$_2$) -177.4, IR, anal., mol. wt.

TYPE: C_6H_{11}; L = P[O$_2$C$_6$H$_9$(OH)]OH [structure (30) or (31)]

[MnL$_4$](ClO$_4$)$_2$. I.2.[279] Anal., IR.
[FeL$_4$](ClO$_4$)$_2$. I.2.[279] Anal., IR.
[NiL$_4$](BF$_4$)$_2$. I.2.[279] Green. Anal.
[NiL$_4$](ClO$_4$)$_2$. I.2.[279] Green. Anal.
[ZnL$_4$](ClO$_4$)$_2$. I.2.[279] Colorless. Anal.

$[ZnL_4](BF_4)_2$. I.2.[279] Colorless. Anal.
$[CdL_4](ClO_4)_2$. I.2.[279] Colorless. Anal.

TYPE: 4 C_6H_{14}; $L_4 = [P(OPr)_2Cl]_4$

NiL_4. I.1.a.[96] Yellow. Gel. Anal., Faraday effect.

TYPE: C_6H_{15}; $L = P(O-i-Pr)_2OH$

$\{PtL[H_2N)_2CS]_3\}Cl_2$. IV.3.[306] m. 167-168.

TYPE: 2 C_6H_{15}; $L_2 = [P(OPr)_2OH]_2$

PtL_2Cl_2. I.1.b.[503,505]

TYPE: C_8H_{18}; $L = P(OBu)_2Cl$

$Ni(CO)_3L$. IR.[58,59b,517]

TYPE: 2 C_8H_{18}; $L_2 = [P(OBu)_2Cl]_2$

$Ni(CO)_2L_2$. IR.[58,493b,517]

TYPE: 3 C_8H_{18}; $L_3 = [P(OBu)_2Cl]_2$

$Ni(CO)L_3$. IR.[58,517]

TYPE: 4 C_8H_{18}; $L_4 = [P(OBu)_2Cl]_4$

NiL_4. I.1.a.[96] Yellow. Gel. Anal., Faraday effect.

TYPE: 2 C_8H_{19}; $L_2 = [P(OBu)_2OH]_2$

PtL_2Cl_2. III.3.[503,505]

TYPE: $C_{12}H_{11}$; $L = P(OPh)_2OH$

fac-$Mo(CO)_3$(bipy)L. I.1.a.[274a] Red. IR, kinetics.
fac-$Mo(CO)_3$(o-phen)L. I.1.a.[274a] Red. IR, kinetics.
CuLCl. I.1.f.[287] White, m. 120. Anal.

TYPE: 2 $C_{12}H_{11}$; $L_2 = [P(OPh)_2OH]_2$

cis-$(OC)_2Mo$(bipy)L_2. I.1.a.[274a] Red. IR, anal., kinetics.

$P(OR)R_2$

TYPE: C_5H_{13}; $L = P(OMe)Et_2$

cis-[PtL(PEt)$_3$Cl$_2$].   IV.1.[108,109] Colorless,[108] m. 184–189.[108] Anal.[108]

TYPE: C$_{13}$H$_{13}$; L = P(OMe)Ph$_2$

W(CO)$_5$L.   [31]P,[208] IR.
cis-[PtL(PEt$_3$)Cl$_2$].   IV.1.[108,109] Colorless,[108] m. 220–222.[108] Anal.[108]
Ni(CO)$_3$L.   IR.[500b]

TYPE: 2 C$_{13}$H$_{13}$; L$_2$ = [P(OMe)Ph$_2$]$_2$

H$_{12}$B$_{10}$L$_2$.   IV.[468b] m. 227. Anal., mol. wt.

TYPE: C$_{14}$H$_{15}$O; L = P(OEt)Ph$_2$

Ni(CO)$_3$L.   IR.[500b]
H$_3$BL.   I.1.b.[442]

TYPE: 2 C$_{14}$H$_{15}$O; L$_2$ = [P(OEt)Ph$_2$]$_2$

trans-NiL$_2$(CN)$_2$.   I.2.[102] Yellow.[102] IR,[102] UV,[102] anal.[102] Goes to NiL$_3$(CN)$_2$ in excess L.[447]
H$_{12}$B$_{10}$L$_2$.   III.3.[431] White, m. 213–215. Anal.

TYPE: 3 C$_{14}$H$_{15}$O; L$_3$ = [P(OEt)Ph$_2$]$_3$

trans-NiL$_3$(CN)$_2$.   I.3.[102,447] Orange.[102] UV.[102,447]

TYPE: 2 C$_{14}$H$_{15}$O$_2$; L$_2$ = [PPh$_2$OCH$_2$CH$_2$OH]$_2$

H$_{12}$B$_{10}$L$_2$.   IV.[468b] m. 188. Anal., mol. wt.

TYPE: 2 C$_{15}$H$_{17}$; L$_2$ = [P(OPr)Ph$_2$]$_2$

trans-NiL$_2$(CN)$_2$.   I.2.[102] Yellow.[102] m. 136–137.[102] mol. wt.,[102] IR,[102] UV-visible.[102] Goes to NiL$_3$(CN)$_2$ in excess L.

TYPE: 3 C$_{15}$H$_{17}$; L$_3$ = [P(OPr)Ph$_2$]$_3$

trans-NiL$_3$(CN)$_2$.   I.3,[447] III.2.[29] m. 135–137.[29] CN 2110,[29] UV.[447]

TYPE: C$_{18}$H$_{14}$; L = P(O-o-ClC$_6$H$_4$)Ph$_2$

Ni(CO)$_3$L.   IR.[500b]

TYPE: C$_{18}$H$_{15}$; L = P(OPh)Ph$_2$

Ni(CO)$_3$L.   IR.[500b]

$P(OR)X_2$

TYPE: $C_1H_3F_2$; $L = P(OMe)F_2$

$Mn_2(CO)_9L$. IV.1.[123] IR.
$CoNO(CO)_2L$. IV.1.[124] IR.
$CoNO(CO)(PF_3)L$. IV.1.[124] IR.
$CoNO(PF_3)_2L$. IV.1.[124] IR, anal.
$CoNO(CO)L[P(OMe)_2F]$. IV.1.[124] IR.

TYPE: 2 $C_1H_3F_2$; $L_2 = [P(OMe)F_2]_2$

$CoNO(CO)L_2$. IV.1.[124] IR.
$CoNO(PF_3)L_2$. IV.1.[124] IR.

TYPE: 3 $C_1H_3F_2$; $L_3 = [P(OMe)F_2]_3$

$Co(NO)L_3$. IV.1.[124] IR, anal.

TYPE: 4 $C_1H_3Cl_2$; $L_4 = [P(OMe)Cl_2]_4$

NiL_4. $^{31}P(C_6H_6)$ -164,[362] ^1H.[364]

TYPE: $C_2H_5F_2$; $L = P(OEt)F_2$

$FeL(PF_3)_4$. IV.1.[314] Yellow, m. -4.5, b. 121. ^{31}P(neat)
    $-170.9(L)$, $-161.5(PF_3)$, ^{19}F, IR, anal., mass spec.
$NiL(PF_3)_3$. IV.1.[314] Colorless, m. -67. Gas. ^{31}P
    $-142.5(PF_2)$, ^{19}F, IR, anal., mol. wt.

TYPE: $C_2H_5Cl_2$; $L = P(OEt)Cl_2$

$Mo(CO)_5L$. I.1.a.[429,430] IR,[133,150,430] Raman.[430]

TYPE: 2 $C_2H_5Cl_2$; $L_2 = [P(OEt)Cl_2]_2$

cis-$Mo(CO)_4L_2$. I.1.a.[429,430] IR.[133,430]

TYPE: 3 $C_2H_5Cl_2$; $L_3 = [P(OEt)Cl_2]_3$

fac-$Mo(CO)_3L_3$. I.1.c.[429] IR.[133]
$Mo(CO)_3L_3$ (fac-mer mixture). I.1.c.[430] IR.

TYPE: 4 $C_2H_5Cl_2$; $L_4 = [P(OEt)Cl_2]_4$

cis-$(OC)_2MoL_4$. I.1.c.[430] IR.
NiL_4. I.1.b.[95] Yellow. Gel. Anal., Faraday effect.

TYPE: 2 $C_3H_7F_2$; $L_2 = [P(OPr)F_2]_2$

$Fe(CO)_3L_2$. $b_{1.5}$ 91-115.[465] Useful as polymerization catalyst.[465]

TYPE: 3 $C_3H_7F_2$; L_3 = $[P(OPr)F_2]_3$

fac-$Mo(CO)_3L_3$. I.1.c.[462,465] Colorless,[462] $b_{0.05}$ 125.[462]
    [31]P(neat) -157.5,[438] [19]F,[438] IR,[462] anal.,[462] mol.
    wt.[462] Useful as polymerization catalyst.[465]
fac-$W(CO)_3L_3$. Useful as polymerization catalyst.[465]

TYPE: 4 $C_3H_7F_2$; L_4 = $[P(OPr)F_2]_4$

NiL_4. I.1.a.[462,465] $b_{0.5}$ 140.5-143.[465] [31]P(neat)
    -146.0,[438] [19]F,[438] anal.,[462] mol. wt.[462] Useful as
    polymerization catalyst.[465]

TYPE: 4 $C_3H_7Cl_2$; L_4 = $P(OPr)Cl_2$

NiL_4. I.1.a.[95] Yellow. Gel. Anal., Faraday effect.

TYPE: C_3H_8; L_2 = $[P(OPr)(OH)_2]_2$

$[PtLCl_2]_2$. IV.1.[477]

TYPE: C_4H_9; L = $P(OBu)Cl_2$

$Ni(CO)_3L$. IR.[58,493b,517]

TYPE: 2 C_4H_9; L_2 = $[P(OBu)Cl_2]_2$

$Ni(CO)_2L_2$. IR.[58,517]

TYPE: 3 C_4H_9; L_3 = $P(OBu)Cl_2$

$Ni(CO)L_3$. IR.[58,517]

TYPE: 4 C_4H_9; L_4 = $[P(OBu)Cl_2]_4$

NiL_4. I.1.a.[95] Yellow. Gel. Anal., Faraday effect.

TYPE: 3 $C_6H_5F_2$; L_3 = $P(OPh)F_2$

fac-$Mo(CO)_3L_3$. I.1.c.[462,465] Colorless,[462] m. 47.[462]
    [31]P(neat) -150.1,[438] [19]F,[438] IR.[462] Useful as poly-
    merization catalyst.[465]

TYPE: 4 $C_6H_5F_2$; L_4 = $[P(OPh)F_2]_4$

NiL_4. I.1.a.[462,465] IV.1.[462] [31]P(neat) -139.6,[438] [19]F,[438]
    IR,[462] anal.,[462] mol. wt.[462] Useful as polymerization
    catalyst.[465]

TYPE: $C_6H_5Cl_2$; L_4 = $P(OPh)Cl_2$

NiL_4.[353] Colorless, m. 98. Diamagnetic.

$P(NR_2)_3$

TYPE: H_6; L = $P(NH_2)_3$

H_3BL. IV.1.[308] White.[308] Crystal structure[402b] shows
   tetrahedral B and P in a staggered conformation with
   P-B = 1.887 Å, P-N (avg.) = 1.653 Å, N-P-N = 99.6°,
   100.1°, 116.8°, N-P-B = 123.5°, 108.4°, 109.0°,
   anal.,[308] mol. wt.[308]

TYPE: C_3H_{12}; L = $P(NMeH)_3$

H_3BL. IV.1.[309] White, m. 30.5-31.0. Anal.

TYPE: C_5H_{18}; L = $P(NMeCH_2)_3CMe$

H_3BL. I.1.f.[319] White, m. 74-76. ^1H, anal., mol. wt.
$(H_3B)_2L$. I.1.f.[319] ^1H.

TYPE: C_6H_{18}; L = $P(NMe_2)_3$

$CpV(CO)_3L$. I.1.a.[297] Orange,[297] subl. 87.[297] ^1H,[297]
   IR,[297] anal.,[297] mass spec.[300]
$Cr(CO)_5L$. I.1.a.[297] White, subl. 145. ^1H, IR, anal.,
   mass spec.[78a,300]
$Mo(CO)_5L$. I.1.a.[297] $^{31}P(C_6H_6)$ -145.6,[339c] ^1H,[297] IR,[297]
   mass spec.[300]
$CpMo(CO)_2LI$(cis-trans mixture). I.1.a.[297] Red, m. 156-
   165 or 180-182 depending on isomer ratio. ^1H, IR,
   anal., conductivity.
$CpMo(CO)_2L(COMe)$. I.1.d.[92] Yellow. ^1H, IR, anal.
$C_7H_7Mo(CO)LI$. I.1.a.[297] Brown, m. 113. ^1H, IR, anal.
$W(CO)_5L$. I.1.b.[293] ^{31}P,[208,293] IR,[293] mol. wt.[293]
cis-$Mn(CO)_3LCH_2Ph$. III.2.[302] Yellow-brown, m. 73-74.
   IR, anal.
cis-$Mn(CO)_4LI$. III.2.[302] Yellow, m. 120-122. IR, anal.
cis-$Mn(CO)_4LBr$. III.2.[362] Yellow, m. 106-108. IR, anal.
$Mn(NO)_3L$. III.2.[302] Green. IR, anal.
cis-$Me_3SnMn(CO)_4L$. III.3.[302] White, m. 98-101. IR,
   anal., mol. wt.
$Mn(CO)_3(NO)L$. III.2.[302] Red, m. 107-108. IR, anal.,
   mol. wt.
cis-$Mn(CO)_4LMe$. III.2.[302] Yellow, m. 144-145. ^1H, IR,
   anal.
$Fe(CO)_4L$. I.1.f.[297] White,[297] subl. 46-85.[297] ^1H,[297]
   IR,[297] anal.,[297] mass spec.[78a,300]

trans-Fe(CO)$_3$L[P(OCH$_2$)$_3$CPr]. I.1.a.[406] ^{31}P(CDCl$_3$) -164.3 (L), -168.4 (PO$_3$), ^{2}JPFeP = 183,[406] ^{1}H,[405] IR,[406] mol. wt.[406]

[CpFe(CO)$_2$L]I. I.1.a.[297] Yellow, m. 234. ^{1}H, IR, anal.

Co(NO)(CO)$_2$L. I.1.a.[500a] IR, kinetics.

Ni(CO)$_3$L. I.1.a.[404] Rose,[404] m. 103-113.[404] ^{31}P(C$_6$H$_6$) -144.7,[339c] IR, anal.,[404] mol. wt.[404]

HgLI$_2$. I.1.f.[404] Greenish, m. 235-238. Anal.

H$_3$BL. I.1.b,[444,446] I.1.f.[319] m. 32.5,[446] b$_{0.1}$ 49-50.[446] Anal.[446] Useful as antioxidant, polymerization catalyst, or blowing agent.[444]

Me$_3$BL. I.3.[269b]

Et$_3$BL. I.3.[269b,270] ^{1}H,[270] dissociates readily.[269b]

F$_5$PL. I.3.[80] White. Anal., decomposes on heating to F$_4$PNMe$_2$ and other products.

I$_2$GeL. I.1.f.[297] Yellow. Liquid.

TYPE: 2 C$_6$H$_{18}$; L$_2$ = [P(NMe$_2$)$_3$]$_2$

cis-Cr(CO)$_4$L$_2$. I.1.c.[363] IR. Isomerizes to trans at -15°C.

trans-Cr(CO)$_4$L$_2$. I.1.c.[297,363] Yellow,[297] subl. 143-147.[297] ^{1}H,[297,405] ^{2}JPCrP = -17,[405] ^{31}P(CS$_2$) -178.19,[54] IR,[297] mass spec.[78a,300]

cis-Mo(CO)$_4$L$_2$. I.1.c.[158,363,405] Yellow,[363] m. 96-98. ^{1}H,[339c,405] ^{2}JPMoP = 12.4,[405] ^{31}P(C$_6$H$_6$) -152.4,[338c] IR,[158,363,405] anal.,[363] kinetics.[158]

trans-Mo(CO)$_4$L$_2$. I.1.c.[297] Yellow,[297] subl. 147-148.[297] ^{1}H,[54,297,339c,405] ^{2}JPMoP = +101,[405] ^{31}P(C$_6$D$_6$) -154.39,[54] anal.,[297] mass spec.[300]

cis-W(CO)$_4$L$_2$. I.1.c.[363] IR,[363] mass spec.[300] Small amount of isomerization.[363]

trans-W(CO)$_4$L$_2$. I.1.a.[297] Yellow,[297] subl. 153.[297] ^{1}H, ^{2}JPWP = +81,[405] ^{31}P(C$_6$D$_6$) -134.17,[54] IR.[297]

[Mn(CO)$_4$L]$_2$. I.1.a.[302] m. 198-199. IR, anal.

Mn$_2$(CO)$_4$L$_2$(H$_2$C$_2$S$_2$). I.1.a.[301] Red, brown, m. 138-141. IR, UV, anal.

trans-Fe(CO)$_3$L$_2$. I.1.a,[297] I.1.c,[297] I.1.f.[297] White,[297] m. 200-203.[297] ^{1}H,[54,297,405] ^{2}JPFeP = +65,[54] ^{31}P(C$_6$D$_6$) -170.18, IR,[297] anal.,[297] mass spec.[78a,83,300]

Fe(NO)$_2$L$_2$. III.2.[298] Red, m. 99-101. ^{1}H, IR, anal., mol. wt.

[Fe(CO)$_2$NOL$_2$][Fe(CO)$_3$NO]. III.2.[298] Orange, m. ∿135. ^{1}H, IR, anal., conductivity.

Hg[Co(CO)$_3$L]$_2$. I.1.a.[297] Yellow, m. ∿250. IR, anal.

[Co(CO)$_3$L]$_2$. I.1.a.[297] Red. IR.

[Co(CO)$_3$L$_2$][Co(CO)$_4$]. II.[297] Yellow, m. ∿176. ^{1}H, IR, anal., conductivity.

NiL$_2$(CO)$_2$. I.1.a,[297,404] II.[297] White,[297] m. 86-87.[297] Yellow,[404] m. 69-74.[404] ^{1}H,[297,405] ^{2}JPNiP ≃ 0,[405] ^{31}P(C$_6$H$_6$) -147.1,[339c] IR,[297,517] anal.,[297] mol. wt.[404]

trans-PdL$_2$I$_2$. III.3.[281] Red.[218] ^{1}H,[405] anal.[281]
cis-PdL$_2$Cl$_2$. III.3.[281] ^{1}H.[405]
trans-PdL$_2$Cl$_2$. I.1.a.[281,282] Orange-yellow,[282] m. $\sim$
    120,[281] 357.[281] ^{1}H,[405] IR,[281] anal.[282]
trans-PtL$_2$I$_2$. III.3.[281] Orange,[281] m. $\sim$150.[281] ^{1}H.[405]
PtL$_2$Cl$_2$(cis-trans mixture). IV.1.[281] Colorless and
    yellow microprisms. Anal.
cis-PtL$_2$Cl$_2$. IV.1.[281] Colorless.[281] ^{1}H, ^{2}JPPtP $\simeq$ 0.[405]
trans-PtL$_2$Cl$_2$. IV.1.[281] Yellow.[281] ^{1}H,[405] IR.[405]
CdL$_2$I$_2$. I.1.f.[404] Colorless, m. 164-167. Anal., mol. wt.
[CdLI$_2$]$_2$. I.1.f.[404] Colorless, m. 255-257. Anal., mol.
    wt.
HgL$_2$I$_2$. I.1.f.[404] Colorless, m. 213-219. ^{1}H.[405]
H$_{12}$B$_{10}$L$_2$. I.1.b,[145b]II.[145b] White, m. 288-290. UV, anal.

TYPE: C$_{12}$H$_{24}$; L = P[N(CH$_2$)$_4$]$_3$

H$_3$BL. I.1.b.[444,446] White,[446] m. 58.[446] Anal.[446] Use-
    ful as antioxidant, polymerization catalyst, or blow-
    ing agent.[444]

TYPE: 2 C$_{12}$H$_{24}$O$_3$; L$_2$ = {P[N(CH$_2$CH$_2$)$_2$O]$_3$}$_2$

CoL$_2$(NO$_3$)$_2$. I.2.[162] Magnetic moment 4.41 BM, UV, anal.
CoL$_2$(NCS)$_2$. I.2.[162] Magnetic moment 4.62 BM, UV, anal.
CoL$_2$I$_2$. I.2.[162] Magnetic moment 4.83 BM, UV, anal.
CoL$_2$Br$_2$. I.2.[162] Magnetic moment 4.63 BM, UV, anal.
CoL$_2$Cl$_2$. I.2.[162] Magnetic moment 4.59 BM, UV, anal.
NiL$_2$(NCS)$_2$. I.2.[162] Diamagnetic, UV, anal.
NiL$_2$I$_2$. I.2.[162] Magnetic moment 3.71 BM, UV, anal.
NiL$_2$Br$_2$. I.2.[162] Green. Magnetic moment 3.00 BM, UV,
    anal. Reverts to blue form in solution.
NiL$_2$Br$_2$. I.2.[162] Blue. Magnetic moment 3.72 BM, UV,
    anal.
NiL$_2$Cl$_2$. I.2.[162] Magnetic moment 3.64 BM, UV, anal.
NiL$_2$(NO$_3$)$_2$. I.2.[162] Magnetic moment 3.46 BM, UV, anal.

TYPE: C$_{12}$H$_{26}$; L = P(NMeCH$_2$)$_3$CAm

W(CO)$_5$L. I.1.b.[293] White, m. 83-84. ^{31}P(CH$_2$Cl$_2$) -112.0,
    ^{1}H, IR, anal., mol. wt.

TYPE: 2 C$_{12}$H$_{26}$; L$_2$ = [P(NMeCH$_2$)$_3$CAm]$_2$

cis-PdL$_2$Cl$_2$. I.1.a.[282]

TYPE: C$_{12}$H$_{30}$; L = P(NEt$_2$)$_3$

W(CO)$_5$L.[208] ^{1}JWP = 296, IR.
H$_3$BL. I.1.b.[444] Useful as antioxidant, polymerization
    catalyst, or blowing agent.

TYPE: $C_{15}H_{30}$; L = $[P(NC_5H_{10})_3]$

$Ni(CO)_3L$.   I.1.a.[58]   White.   IR.

TYPE: 2 $C_{15}H_{30}$; L_2 = $[P(NC_5H_{10})_3]_2$

$Ni(CO)_2L_2$.   I.1.a.[58]   White.   IR.
NiL_2O_2.   III.3.[533]   Unstable above 5°C.

TYPE: $C_{18}H_{21}$; L = $P(NPr_2)_3$

H_3BL. ^{31}P -61.6.[360]

TYPE: $C_{21}H_{24}$; L_2 = $[P(NMePh)_3]_2$

cis-$Mo(CO)_4L_2$.   I.1.a.[405] ^1H, ^2JPMoP = 18.2, IR.

TYPE: $C_{24}H_{54}$; L = $P(NBu_2)_3$

H_3BL.   I.1.f.[446]   IR,[446] anal.[446]   Useful as antioxidant,
    polymerization catalyst, or blowing agent.[444]

$P(NR_2)(NR'_2)R$

TYPE: 2 $C_{12}H_{27}$; L_2 = $[P(NMe_2)(NEt_2)Ph]_2$

HgL_2I_2.   I.1.f.[173]   m. 159-161.   Anal.

TYPE: 2 $C_{16}H_{29}$; L_2 = $[P(NEt_2)(NPr_2)Ph]_2$

HgL_2I_2.   I.1.f.[173]   Colorless, m. 102-108.   Anal.

$P(NR_2)_2R$

TYPE: C_4H_{13}; L = $P(NMe_2)_2H$

H_3BL.   I.1.b.[403]   m. 24-25, $b_{0.1}$ 46-47.   Anal., mol. wt.

TYPE: C_5H_{15}; L = $P(NMe_2)_2Me$

$Cr(CO)_5L$.   I.1.a.[144b]   Yellow, m. 74-76. ^1H, IR, anal.
$Mo(CO)_5L$.   I.1.a.[144b]   Yellow, m. 45-47. ^1H, IR, anal.
$W(CO)_5L$.   I.1.b.[144b]   Yellow. ^1H, IR.
Me_3BL.   I.3.[269b]   White, m. -46.
Et_3BL.   I.3.[269b]   Liquid.[270]   Vapor pressure 0.2 mm at
    0.0°C.[270] ^1H.[270]
F_5PL.   I.3.[80]   White.   Decomposes at room temperature to
    give F_4PNMe_2 and other products.

TYPE: 2 C_5H_{15}; L_2 = [P(NMe$_2$)$_2$Me]$_2$

trans-Cr(CO)$_4$L$_2$. I.1.a.[144b] Yellow, m. 88-90. ^{1}H, IR,
   anal.
trans-Mo(CO)$_4$L$_2$. I.1.[144b] Yellow, m. 88-90. ^{1}H, IR,
   anal.
trans-W(CO)$_4$L$_2$. I.1.b.[144b] Yellow, m. 91-93. ^{1}H, IR,
   anal.

TYPE: C_8H_{21}; L = P(NMe$_2$)$_2$Bu

H$_3$BL. I.1.f.[403] b$_{1.5}$ 93. Anal., mol. wt.

TYPE: 2 $C_{10}H_{17}$; L_2 = [P(NMe$_2$)$_2$Ph]$_2$

HgI$_2$L$_2$. I.1.f.[173,404] White,[173] m. 186-190.[404] Anal.,[173]
   mol. wt.[404]
[HgLI$_2$]$_2$. I.1.f.[173,404] Yellow, m. 204-209,[404] 192-
   196.[173] Anal.
H$_{12}$B$_{10}$L$_2$. I.1.b.[145b] White, m. 220-222, UV, anal.

TYPE: $C_{14}H_{15}$; L = P(NEt$_2$)$_2$Ph

CuLI. I.1.f.[317] White, m. 194-198. Anal., tetrameric,
   ˙conductivity.

TYPE: 2 $C_{14}H_{25}$; L_2 = [P(NEt$_2$)$_2$Ph]$_2$

CoL$_2$(NCS)$_2$. I.2.[317] Green, m. 98-102. Magnetic moment
   4.34 BM, anal., conductivity.
CoL$_2$Br$_2$. I.2.[317] Green, m. 112-114. Magnetic moment
   4.43 BM, anal., conductivity.
CoL$_2$Cl$_2$. I.2.[317] Green, m. ∿ 70. Magnetic moment
   4.53 BM, anal., conductivity.
NiL$_2$(NCS)$_2$. I.2.[317] Red, m. 99-105. Diamagnetic, anal.,
   mol. wt., conductivity.
PdL$_2$Cl$_2$. I.1.b.[317] Yellow, m. 109-111. Diamagnetic,
   anal., conductivity.
PtL$_2$Cl$_2$. I.1.b.[317] Yellow, m. 100-104. Diamagnetic,
   anal., conductivity.
CuL$_2$I. I.1.b.[317] White, m. 103. Anal., tetramer, conductivity.
HgL$_2$I$_2$. I.1.f.[173] Colorless, m. 142-143. Anal., mol. wt.
[HgLI$_2$]$_2$. I.1.f.[173] Colorless, m. 176-178. Anal., mol.
   wt.

TYPE: 2 $C_{18}H_{32}$; L_2 = [P(NPr$_2$)$_2$Ph]$_2$

HgL$_2$I$_2$. I.1.f.[173] Colorless, m. 126-127. Anal.
[HgLI$_2$]$_2$. I.1.f.[173] Colorless, m. 134-136. Anal., mol.
   wt.

$P(NR_2)_2X$

TYPE: C_2H_8; $L = P(NMeH)_2F$

H_3BL.   IV.1.[309] Clear. IR, anal.
F_3BL. ^{31}P -138 to -148.[360]

TYPE: $C_4H_{12}F$; $L = P(NMe_2)_2F$

H_3BL.   IV.1.[309] ^{31}P -134.[360]

TYPE: 2 $C_4H_{12}F$; $L_2 = [P(NMe_2)_2F]_2$

$Ni(CO)_2L_2$. I.l.a.[464] Orange.[464] Liquid.[464] ^{31}P(neat)
    -181.0,[438] ^{19}F,[339a,438] IR,[464] anal.[464]

TYPE: 3 $C_4H_{12}F$; $L_3 = [P(NMe_2)_3F]_3$

fac-$Mo(CO)_3L_3$. I.l.c.[464] Colorless,[464] m. 108-110.[464]
    ^{31}P(CDCl$_3$) -182.3,[438] ^{19}F,[438] IR,[464] anal.,[464] mol.
    wt.[464]
$Ni(CO)L_3$. ^{19}F.[339a]

TYPE: $C_4H_{12}Cl$; $L = P(NMe_2)_2Cl$

H_3BL.   I.l.f.[403] $b_{0.4}$ 52. Anal., mol. wt.

TYPE: 2 $C_4H_{12}Cl$; $L_2 = [P(NMe_2)_2Cl]_2$

$Ni(CO)_2L_2$. I.l.a.[404] Oil. IR, anal., mol. wt.
$H_{12}B_{10}L_2$. I.l.b.[145b] White, m. 200-204. UV, anal.

TYPE: C_5H_{12}; $L = P(NMe_2)_2CN$

cis-$Mo(CO)_4LP(OCH_2)_3CCH_3$. I.l.f.[144c] White. ^1H, IR,
    anal.

TYPE: 2 C_5H_{12}; $L_2 = [P(NMe_2)_2CN]_2$

trans-$Cr(CO)_4L_2$. I.l.f.[144c] Yellow, m. 126-128. ^1H, IR,
    anal.
$[Cr(CO)_4L]_2$. I.l.c.[144c] Yellow. ^1H, IR, anal., mass
    spec.
cis-$Mo(CO)_4L_2$. I.l.f.[144c] White, m. 87-89. ^1H, IR,
    anal.
$[Mo(CO)_4L]_2$. I.l.c.[144c] Yellow. ^1H, IR, anal., mass
    spec.
$Ni(CO)_2L_2$. I.l.a.[404] Brown, m. 45-53. IR, anal., mol.
    wt.

P(NR$_2$)R$_2$

TYPE: C$_4$; L = P[N(CF$_3$)$_2$](CF$_3$)$_2$

Ni(CO)$_3$L. I.1.a.[7] IR, anal.

TYPE: C$_4$H$_{12}$; L = P(NMe$_2$)Me$_2$

H$_6$B$_2$L. I.3.[85] White. Decomposes to give H$_2$, (Me$_2$PBH$_2$)$_3$, and (Me$_2$N)$_2$BH.
Et$_3$AlL. I.1.f.[125] m. 80.5. ^{31}P +41, ^{1}H, IR, anal., mol. wt.
H$_3$BL. I.1.f.[85] White, m. 9.5-12.
Me$_3$BL. I.3.[269b] White, m. -46.
Et$_3$BL. I.3.[269b] Liquid.[269b] ^{1}H.[270]
F$_5$PL. I.3.[80] White. ^{1}H, IR. Decomposes to F$_4$PNMe$_2$ and other products.

TYPE: C$_{10}$H$_{24}$; L = P(NMe$_2$)Bu$_2$

H$_3$BL. I.1.f.[403] b$_{1.5}$ 101. Anal., mol. wt.

TYPE: 2 C$_{12}$H$_{12}$; L$_2$ = [PPh$_2$NH$_2$]$_2$

H$_{12}$B$_{10}$L$_2$. IV.[468b] m. 222. Anal., mol. wt.

TYPE: 2 C$_{12}$H$_{15}$; L$_2$ = [PNHNH$_2$Ph$_2$]$_2$

H$_{12}$B$_{10}$L$_2$. IV.[468b] m. 214-215. Anal., mol. wt.

TYPE: 2 C$_{13}$H$_{13}$; L$_2$ = [PNHMePh$_2$]$_2$

H$_{12}$B$_{10}$L$_2$. IV.[468b] m. 226. Anal., mol. wt.

TYPE: 2 C$_{14}$H$_{14}$; L$_2$ = [PPh$_2$N(CH$_2$CH$_2$)]$_2$

H$_{12}$B$_{10}$L$_2$. IV.[468b] m. 202. Anal., mol. wt.

TYPE: 2 C$_{14}$H$_{16}$; L$_2$ = [P(NMe$_2$)Ph$_2$]$_2$

H$_{12}$B$_{10}$L$_2$. I.1.b,[145b] II.[145b] White, m. 228-230. Anal.

TYPE: C$_{15}$H$_{17}$; L = P(NH-i-Pr)Ph$_2$

Et$_3$AlL. I.1.f.[125] Liquid. ^{31}P -24, ^{1}H, IR, anal., mol. wt.

TYPE: 2 C$_{16}$H$_{19}$; L$_2$ = [PNH-n-BuPh$_2$]$_2$

H$_{12}$B$_{10}$L$_2$. IV.[468b] m. 165. Anal., mol. wt.

TYPE: $C_{16}H_{20}$; L = $P(NEt_2)Ph_2$

CuLI.   I.1.b.[173]   m. 225-227. Anal., tetramer.
HEt_2AlL.   II.[125]   m. 163. Anal., ^1H, IR.

TYPE: 2 $C_{16}H_{20}$; L_2 = $[P(NEt_2)Ph_2]_2$

$[HgLI_2]_2$. I.1.f.[173] Yellow, m. 188-190. Anal.

$P(NR_2)X_2$

TYPE: C_1H_4; L = $P(MeNH)F_2$

H_3BL.   IV.1.[309]   b. 129. IR, anal.

TYPE: C_2H_6; L = $P(NMe_2)F_2$

$Ni(CO)_3L$. ^{19}F.[339a,397b]
CuLCl.   I.1.f.[132]   Colorless. IR, ^{19}F, tetrameric.
H_3BL.   IV.1.[309]   m. -56.7,[309] b. 119.4.[309] ^{31}P -129.5,
    -130,[360] anal.,[309] mol. wt.[309]
F_3BL.   I.3.[100] White, subl. 61. Dissociates at room
    temperature.
H_8B_4L.   I.1.a,[318,498a] I.1.f.[101]   Clear liquid,[498a] m.
    -18.[498a] ^{31}P (liquid) -120, ^1JPF = 1240,[360] ^1H,[101]
    ^{19}F,[101] ^{11}B.[101] Crystal structure[318] shows P-B = 1.856 Å
    and planar nitrogens (see text), P-F (avg.) = 1.583 Å,
    P-N = 1.593 Å, N-P-B = 117.6°, F-P-F = 96.4°, P-N-C
    (avg.) = 122.4°, anal.,[498a] mol. wt.[498a]

TYPE: 2 C_2H_6; L_2 = $[P(NMe_2)F_2]_2$

cis-$Cr(CO)_4L_2$. ^{19}F,[397b,405] ^2JPCrP = 62.
cis-$Mo(CO)_4L_2$.   I.1.i.[40] Colorless.[40] ^{31}P(neat) -180.5,[40]
    ^{19}F,[40,405] ^2JPMoP = 38.0, ^1H.[40] IR,[40] anal.[40]
cis-$W(CO)_4L_2$.[397] ^{19}F,[405] ^2JPWP = 21.[405]
trans-$Rh(CO)L_2Cl$. I.1.a.[126] Yellow-brown. ^{19}F, ^2JPRhP =
    49.
$Ni(CO)_2L_2$. I.1.a.[463,465] Colorless.[463] Liquid.[463]
    ^{31}P(neat) -168,[438] ^{19}F,[339a,405] IR,[463] anal.,[463] mol.
    wt.[463] Useful as polymerization catalyst.[465]
CuL_2Cl. I.1.f.[132] Colorless. Diamagnetic.

TYPE: 3 C_2H_6; L_3 = $[P(NMe_2)F_2]_3$

mer-$Mo(CO)_3L_3$.   IV.1.[38] Colorless. IR, anal.
$Mo(CO)_3L_3$.   I.1.c.[463] m. 56-57.[463] IR,[463] anal.,[463] mol.
    wt.[463] Useful as polymerization catalyst.[465]
RhL_3Cl. I.1.a.[126] Yellow-brown.
$Ni(CO)L_3$. ^{19}F.[339a]

TYPE: 4 C_2H_6; L_4 = $[P(NMe_2)F_2]_4$

$[RhL_2Cl]_2$. I.l.c.[126] Yellow.
NiL_4. I.l.a,[463,465] I.l.f.[398,399] m. 110.5-111.5.[463]
   [19]F,[339a] IR,[463] anal.[463]

TYPE: C_4H_6; L_2 = $[P(NMe_2)(CN)_2]_2$

$Ni(CO)_2L_2$. I.l.a.[404] Brown.

TYPE: C_4H_{10}; L = $P(NEt_2)F_2$

$Ni(CO)_3L$. [19]F.[39,339a,397b]

TYPE: 2 C_4H_{10}; L_2 = $[P(NEt_2)F_2]_2$

cis-$Mo(CO)_4L_2$. I.l.f.[40,41] Colorless.[41] [19]F,[39,40,405]
   $^2J_{PMoP}$ = 38.0, IR,[41] anal.[41]
$Ni(CO)_2L_2$. I.l.a.[438] [31]P(neat) -168.5,[438] [19]F,[339a,405,]
   [438] $^2J_{PP}$ ≃ 0,[405] IR,[438] anal.,[438] mol. wt.[438]

TYPE: 3 C_4H_{10}; L_3 = $[P(NEt_2)F_2]_3$

$Mo(CO)_3L_3$. I.l.a.[465] Colorless, $b_{0.05}$ 40-50.[465]
   [31]P(CCl_4) -182.4,[438] [19]F,[438] IR,[463] anal.[463] Useful
   as polymerization catalyst.[465]
$Ni(CO)L_3$. [19]F.[339,337b]

TYPE: 4 C_4H_{10}; L = $[P(NEt_2)F_2]_4$

NiL_4. I.l.a,[438] II.[397a] White,[397a] m. 45.5-46.5.[397a]
   [31]P(neat) -167.9,[438] [19]F,[39,397a] [1]H,[397a] IR,[397a]
   anal.[438]

TYPE: C_5H_{10}; L = $P[N(CH_2)_5]F_2$

$Ni(CO)_3L$. [19]F.[339a,397b]

TYPE: 2 C_5H_{10}; L_2 = $\{P[N(CH_2)_5]F_2\}_2$

cis-$Mo(CO)_4L_2$. I.l.c,[41] IV.[38] Colorless. Liquid,[38] m.
   38-38.5.[41] IR,[38,41] anal.[38,41]
$Ni(CO)_2L_2$. I.l.a.[463,465] m. 39.5.[463] [31]P(PhMe) -164.6,[438]
   [19]F,[339a,438] IR,[463] anal.,[463] mol. wt.[463] Useful as
   polymerization catalyst.[465]

TYPE: 3 C_5H_{10}; L_3 = $\{P[N(CH_2)_5]F_2\}_3$

$Mo(CO)_3L_3$. I.l.c.[463] m. 100.[463] [31]P(neat) -177.4,[438]
   [19]F,[438] IR,[463] anal.,[463] mol. wt.[463]
$Mo(CO)_3L_3$(mer-fac mixture). IV.1.[38] Milky. IR.

$Ni(CO)L_3$. ^{19}F.[339,397b]

TYPE: 4 C_5H_{10}; L_4 = $\{P[N(CH_2)_5]F_2\}_4$

cis-$(OC)_2MoL_4$. I.l.a.[463]   White,[463] m. 142.0-142.8.[463]
^{31}P(neat) -183.1,[438] ^{19}F,[438] IR,[463] anal.,[463] mol.
wt.[463]
NiL_4. I.l.a,[463,465] I.l.f,[398,399] II.[397a] Colorless,[397a]
m. 163.5-165,[397a] 56-57.[46] ^1H,[397] ^{19}F,[39,397a] IR,[397a]
anal.,[397a] mol. wt.[46] x-Ray structure shows[204] tetrahedral configuration (Ni-P=2.2Å). Useful as polymerization catalyst.[465]

$P(NR_2)(OR)R$

TYPE: C_8H_{20}; L = $P(NEt_2)(O-i-Pr)Me$

H_3BL. $^{31}P(CCl_4)$ -115.5, ^{11}B, ^1H.[321]

$P(NR_2)RX$

TYPE: 2 C_3H_9; L_2 = $[P(NMe_2)MeF]_2$

$Ni(CO)_2L_2$. I.l.a.[464]   Yellow.[464]   Liquid.[464] ^{31}P(neat)
-184.4,[438] ^{19}F,[438] IR,[464] anal.,[464] mol. wt.[464]

TYPE: 3 C_3H_9; L_3 = $[P(NMe_2)MeF]_3$

fac-$Mo(CO)_3L_3$. I.l.c.[464]   Colorless, m. 127-128.
^{31}P(neat) -194.6,[438] ^{19}F,[438] IR,[464] anal.,[464] mol.
wt.[414]

TYPE: 2 C_8H_{11}; L_2 = $[P(NMe_2)PhCl]_2$

$H_{12}B_{10}L_2$. I.l.b,[145b] II.[145b] White, m. 174-176. UV,
anal.

TYPE: 3 $C_{10}H_{15}$; L_3 = $[P(NEt_2)PhF]_3$

fac-$Mo(CO)_3L_3$. I.l.c.[464]   Colorless,[464] m. 138-139.[464]
^{31}P(neat) -185.5,[438] ^{19}F,[438] IR,[464] anal.,[464] mol.
wt.[464]
$Ni(CO)L_3$. I.l.a.[464]   Colorless, m. 97. IR, anal.

$P(SR)_3$

TYPE: C_3H_9; L = $P(SMe)_3$

Mo(CO)$_5$L.   I.1.a.[339c] ^{31}P(C$_6$H$_6$) -130.2, IR.
W(CO)$_5$L.   I.1.a.[293]   Yellow. ^{1}H, IR, mol. wt.
Ni(CO)$_3$L.   IR.[59b]

TYPE: 2 C$_3$H$_9$; L$_2$ = [P(SMe)$_3$]$_2$

cis-Mo(CO)$_4$L$_2$. ^{31}P(C$_6$H$_6$) -132.6,[339c] IR.
trans-Mo(CO)$_4$L$_2$. ^{31}P(C$_6$H$_6$) -137.0,[339c] IR.

TYPE: C$_6$H$_{15}$; L = P(SEt)$_3$

Ni(CO)$_3$L.   IR.[59b]

TYPE: 2 C$_6$H$_{15}$; L$_2$ = [P(SEt)$_3$]$_2$

H$_{12}$B$_{10}$L$_2$.   II.[489]   White, m. 92-93.   Anal.

TYPE: C$_9$H$_{17}$; L = P(SCH$_2$)$_3$CAm

W(CO)$_5$L.   I.1.b.[293]   White, m. 158-160. ^{31}P(CS$_2$)-31.1,
     ^{1}H, IR, anal., mol. wt.

TYPE: C$_{18}$H$_{15}$; L = P(SPh)$_3$

Mn(NO)$_3$L.   III.2.[265]   IR.[46,517]
Co(NO)$_2$LCl.   I.1.f.[256]   m. 119-120.   Anal.

P(SR)X$_2$

TYPE: C$_6$H$_5$; L$_4$ = [P(SPh)F$_2$]$_4$

NiL$_4$.   I.1.a.[465]   Useful as polymerization catalyst.

PX$_3$

TYPE: 4 C$_3$N$_3$S$_3$; L$_4$ = [P(NCS)$_3$]$_4$

NiL$_4$.   I.1.a.[534]   Yellow.   Anal.

TYPE: 4 C$_3$N$_3$O$_3$; L$_4$ = [P(NCO)$_3$]$_4$

NiL$_4$.   I.1.a.[534]   Colorless.   Anal., mol. wt.

Cationic Ligands

TYPE: C$_4$H$_9$; L = [P(OCH$_2$)$_3$PMe]$^+$

[W(CO)$_5$]BF$_4$.   IV.2.[52] ^{31}P(Me$_2$CO) -113.00 (PO$_3$), -2.98(PC$_3$),

^{3}JPOCP = +143.2, ^{1}H, IR.

[P(NR$_2$)R$_3$]$^+$ or [P(NR$_2$)(NR$_2'$)R''R''']$^+$

TYPE: C$_5$H$_{15}$; L = [P(NMe$_2$)Me$_3$]$^+$

LHgI$_3$. I.l.f.[84] Yellow, m. > 300. Anal.

TYPE: C$_{11}$H$_{20}$; L = [P(NMe$_2$)$_2$PhMe]$^+$

LHgI$_3$. I.l.f.[173] Green, m. 132-134.

TYPE: C$_{13}$H$_{24}$; L = [P(NMe$_2$)(NEt$_2$)PhMe]$^+$

LHgI$_3$. I.l.f.[173] Green, m. 90-92. Anal.

TYPE: C$_{17}$H$_{32}$; L = [P(NEt$_2$)(NPr$_2$)PhMe]$^+$

LHgI$_3$. I.l.f.[173] Green, m. 82-84. Anal.

TYPE: C$_{17}$H$_{33}$; L = [P(NEt$_2$)$_2$Ph$_2$Me]$^+$

LHgI$_3$. I.l.f.[173] Green, m. 116-118. Anal.

TYPE: C$_{19}$H$_{36}$; L = [P(NPr$_2$)$_2$PhMe]$^+$

LHgI$_3$. I.l.f.[173] Green, m. 59-61. Anal.

Anionic Ligands

  [P(OR)HO]$^-$

TYPE: C$_2$H$_6$; L = [P(OEt)HO]$^-$

Pt(NH$_3$)$_2$LCl. IV.2.[477] Anal.

  [OP(OR)$_2$]$^-$

TYPE: C$_2$H$_6$; L = [OP(OMe)$_2$]$^-$

trans-CpMo(CO)$_2$P(OMe)$_3$L. III.2.[219a,b] Yellow. ^{1}H,[219b]
   IR,[219b] anal.,[219b] mol. wt.,[219b] conductivity.[219b]
cis-[P(OMe)$_2$OH](PEt$_3$)PtLCl. IV.2.[426b] m. 102-104.[426b]
   ^{31}P(C$_6$H$_6$) -54.6 (L), -92.6 [P(OMe)$_2$OH],[5] anal.[426b]
cis-[P(OMe)$_2$OH](AsEt$_3$)PtLCl. IV.2.[426b] m. 80-82. ^{31}P
   (C$_6$H$_6$) 56.2 (L), -81.2 [P(OMe)$_2$OH],[5d] anal.[426b]
trans-Pt(PBu$_3$)$_2$LCl. I.l.b.[426b] Oil.[426b] ^{31}P(CH$_2$Cl$_2$)
   -34.7 (L), ^{2}JPPtP = 18.0.[5d]
[t-BuNH$_3$][H$_3$BL]. IV.2.[445] m. 137-140.[445] ^{31}P -95.0.[360]
   Useful as blowing agent for epoxy resins.[445]

TYPE: 2 C_2H_6; L_2 = [OP(OMe)$_2$]$_2{}^{2-}$

Pt[P(OMe)$_2$OH]$_2$L$_2$.   IV.2.[426b]   White,[426b] m. 175-178.[426b]
     ^{31}P(CH$_2$Cl$_2$) -88.2 [L and P(OMe)$_2$OH],[5d] anal.[426b]
Pt(diphos.)L$_2$.   III.3,[507] IV.3.[426b]   m. 276-277.[507]
     anal.[426b]
Pt[P(OMe)$_3$]$_2$L$_2$.   IV.2.[507]   m. 50.
Pt[P(OMe)$_2$OH]$_2$L$_2$.   IV.2.[211,426a,503,505,506]   Colorless,[506]
     m. 163.[506] 164-167.[426a]   Anal.[426a]

TYPE: C_4H_8; L = [OP(OCH$_2$CH$_2$Cl)$_2$]$^-$

[t-BuNH$_3$][H$_3$BL].   IV.2.[445]   Useful as blowing agent for
     epoxy resins.

TYPE: C_4H_{10}; L = [OP(OEt)$_2$]$^-$

trans-CpMo(CO)$_2$P(OEt)$_3$L.   III.2.[219a,b]   Yellow. ^{1}H,[219b]
     IR,[219b] anal.[219b]
Pd[P(OEt)$_2$OH]LCl.   I.1.e.[329]
Pd[P(OEt)$_3$](NH$_3$)LCl.   IV.1.[183]   Colorless.
Pt[P(OEt)$_3$](NH$_3$)LCl.   IV.2.[447]   Anal.
HgLCl.   White,[48b] m. 81-82.5. ^{31}P(C$_6$H$_6$) -64.6, anal.
     x-Ray structure shows HgP distance ranges from 2.31 to
     2.4 Å (see text).
HgLBr.   White,[48b] m. 86.5-88. ^{31}P(C$_6$H$_6$) -68.5.
HgLI.   White,[48b] m. 98.5-100. ^{31}P(C$_6$H$_6$) -76.0.
HgLAc.   IV.2,[18b] III.3.[48b]   White,[48b] m. 104-106.[48b]
     ^{31}P(C$_6$H$_6$) -58.2, ^{1}JHgP = 12970.[48b]
HgLCNO.   III.3.[48b]   White, m. 111-112.   Anal.
HgLSCN.   White.[48b]   Anal.
[t-BuNH$_3$][H$_3$BL].   IV.2.[445]   m. 101-103.   Useful as blow-
     ing agent for epoxy resins.

TYPE: 2 C_4H_{10}; L_2 = [OP(OEt)$_2$]$_2^-$

PdL$_2$.   I.1.e.[329]   IR,[330] UV.[330]
{Pd[P(OH)(OEt)$_2$]LCl}$_2$.   IR,[330] UV.
trans-Pt(PEt$_3$)$_2$L$_2$.   I.1.b.[426b]   Oily crystals.[426b] ^{31}P
     (CH$_2$Cl$_2$) -73.4 (L), 38.1,[5d] anal.[426b]
trans-Pt(PBu$_3$)$_2$L$_2$.   I.1.b.[426b]   White,[426b] m. 190-192.[426b]
     ^{31}P(CH$_2$Cl$_2$) -72.2 (L), ^{2}JPPtP = 37.2,[5d] anal.,[426b]
     conductivity,[426b] mol. wt.[426b]
Pt[P(OEt)$_2$OH]$_2$L$_2$.   IV.2.[212,213,503,505]   Colorless,[505]
     m. 85 (form a),[505] 95.4-95 (form b).[505] ^{31}P(CH$_2$Cl$_2$)
     -85.9 [L and P(OEt)$_2$OH].[5d]
AgL$_2$.   IV.2.[485]   IR, anal.
[HgLCl]$_2$.   IV.2.[89]   m. 103.5.[89] ^{31}P -65,[89] IR,[89,485]
     anal.[485]
HgL$_2$. ^{31}P(C$_6$H$_6$) -106.8,[43]   mol. wt.
Pb[H$_3$BL]$_2$.   IV.2.[445]   Useful as blowing agent for epoxy

resins.

TYPE: C_6H_{14}; L = [OP(OPr)$_2$]$^-$

HgL(O$_2$CMe).   IV.2.[186] m. 86.3-87.2.   Anal.

TYPE: 2 C_6H_{14}; L$_2$ = [OP(OPr)$_2$]$_2{}^{2-}$

Pt[P(OPr)$_2$OH]$_2$L$_2$.   IV.2.[503,505] Colorless,[505] m. 36-37.[503]

TYPE: C_6H_{14}; L = [OP(O-i-Pr)$_2$]$^-$

trans-CpMo(CO)$_2$P(O-i-Pr)$_3$L.   III.2.[219a,b] Yellow. ^{1}H,[219b]
   IR,[219b] anal.,[219b] mol. wt.,[219b] conductivity.[219b]
HgL(O$_2$CMe).   IV.2,[89] IV.2.[186] m. 146.2-146.8,[186] 114-
   115.[89] ^{31}P -61.5,[89] IR,[89] anal.,[89] mol. wt.[89]
[t-BuNH$_3$][H$_3$BL].   IV.2.[445] m. 144-146. Useful as blowing agent for epoxy resins.
H[H$_3$BL].   IV.2.[445] Useful as blowing agent for epoxy
   resins.

TYPE: 2 C_6H_{14}; L$_2$ = [OP(O-i-Pr)$_2$]$_2{}^{2-}$

[HgLBr]$_2$.   IV.2.[89] m. 104-106. IR, anal.
Cd[H$_3$BL]$_2$.   IV.2.[445] m. 50-53. Useful as blowing agent
   for epoxy resins.
Mg[H$_3$BL]$_2$.   IV.2.[445] m. 110-115. Useful as blowing agent
   for epoxy resins.
Co[H$_3$BL]$_2$.   IV.2.[445] Useful as blowing agent for epoxy
   resins.
Zn[H$_3$BL]$_2$.   IV.2.[445] Useful as blowing agent for epoxy
   resins.

TYPE: C_8H_{18}; L = [OP(OBu)$_2$]$^-$
trans-CpMo(CO)$_2$P(O-n-Bu)$_3$L.   III.2.[219a,b] Yellow. ^{1}H,[219b]
   IR,[219b] anal.[219b]
HgL(O$_2$CMe).   IV.2.[186] m. 80.5-81.0.   Anal.
[t-BuNH$_3$][H$_3$BL].   IV.2.[445] Useful as blowing agent for
   epoxy resins.

TYPE: 2 C_8H_{18}; L$_2$ = [OP(OBu)$_2$]$_2{}^{2-}$

trans-Pt(PBu$_3$)$_2$L$_2$.   I.1.b.[426b] m. 173-175. ^{31}P(CH$_2$Cl$_2$)
   -71.3 (L), ^{2}JPPtP = 3308, 37.8.[5d]
Pt[P(OBu)$_2$OH]$_2$L$_2$.   IV.2.[503,505] Colorless.[505] Syrup.[503]
[HgLBr]$_2$.   IV.2.[89] m. 76-77. ^{31}P -69.5, IR, anal., mol.
   wt.
[HgLCl]$_2$.   IV.2.[89] m. 89-90. ^{31}P -66, IR, anal., mol. wt.

TYPE: C_8H_{18}; L = [OP(O-i-Bu)$_2$]$^-$

HgL(O$_2$CMe).   IV.2.[186]   m. 116.0-116.6.   Anal.

TYPE: C$_{10}$H$_{22}$; L = [OP(OAm)$_2$]$^-$

HgL(O$_2$CMe).   IV.2.[186]   m. 72.5-73.5.   Anal.

TYPE: C$_{12}$H$_{10}$; L = [OP(OPh)$_2$]$^-$

Pd[P(OPh)$_2$OH](PPh$_3$)LCl.   III.5.[426b]   m. 106-108.[426b]
   ^{31}P(C$_6$H$_6$) -73.4 (L), -91.0 [P(OPh)$_2$OH],[5d] anal.[426b]
Pd[P(OPh)$_2$OH](PMe$_2$Ph)LCl.   IV.2.[426b]   m. 107-109.[426b]
   ^{31}P(C$_6$H$_6$) -74.2 (L), -94.7 [P(OPh)$_2$OH].[5d]
cis-[P(OPh)$_2$OH](PEt$_3$)PtLCl.   IV.2.[426b]   m. 126-128.[426b]
   ^{31}P(C$_6$H$_6$) -47.7 (L), -86.7 [P(OPh)$_2$OH],[5d] anal.[426b]
Pt[P(OPh)$_2$OH](PBu$_3$)LCl. ^{31}P(C$_6$H$_6$) -47.8 [P(OPh)$_2$OH].[5d]
cis-[P(OPh)$_2$OH](AsEt$_3$)PtLCl.   IV.2.[426b]   m. 120-121.5.[426b]
   ^{31}P(C$_6$H$_6$) -49.2 (L), -74.4 [P(OPh)$_2$OH],[5d] anal.[426b]
trans-Pt(PEt$_3$)$_2$LCl.   I.l.b.[426b]   Grey.[426b]   m. 129-131.[426b]
   ^{31}P(CH$_2$Cl$_2$) -29.6 (L), ^{2}JPPtP = 23.8,[5d] anal.,[426b]
   mol. wt.,[426b] conductivity.[426b]
trans-Pt(PBu$_3$)$_2$LCl.   I.l.b.[426b]   White,[426b] m. 94-96.[426b]
   ^{31}P(CH$_2$Cl$_2$) -29.7 (L), ^{2}JPPtP = 23.8,[5d] anal.[426b]
trans-Pt(AsEt$_3$)$_2$LCl.   I.l.b.[426b]   m. 111-113.[426b] ^{31}P
   (CH$_2$Cl$_2$) -28.1 (L),[5d] anal.[426b]
trans-Pt(PBu$_3$)$_2$LBr.   III.3.[426b]   m. 99.5-101.5.[426b] ^{31}P
   (CH$_2$Cl$_2$) -29.3 (L), ^{2}JPPtP = 22.1,[5d] anal.[426b]
trans-Pt(PBu$_3$)$_2$LI.   III.3.[426b]   m. 115-116.5.[426b] ^{31}P
   (CH$_2$Cl$_2$) -28.9 (L), ^{2}JPPtP = 19.7,[5d] anal.[426b]
trans-Pt(PBu$_3$)$_2$L(ONO$_2$).   III.3.[426b]   m. 148-151.[426b]
   ^{31}P(CH$_2$Cl$_2$) -7.7 (L), ^{2}JPPtP = 25.5,[5d] anal.[426b]
trans-Pt(PBu$_3$)LN$_3$.   III.3.[426b]   m. 129-130.[426b] ^{31}P(CH$_2$Cl$_2$)
   -30.0 (L), ^{2}JPPtP = 26.5,[5d] anal.[426b]
trans-Pt(PBu$_3$)$_2$L(NCO).   III.3.[426b]   m. 150.5-152. ^{31}P
   (CH$_2$Cl$_2$) -28.5 (L), ^{2}JPPtP = 19.4,[5d] IR,[426b] anal.[426b]
trans-Pt(PBu$_3$)$_2$L(NCS).   III.3.[426b]   m. 166-166.5.[426b]
   ^{31}P(CH$_2$Cl$_2$) -25.2 (L), ^{2}JPPtP = 20.4,[5d] IR,[426b] anal.[426b]
trans-Pt(PBu$_3$)$_2$L(NO$_2$).   III.3.[426b]   m. 123-125.[426b]
   ^{31}P(CH$_2$Cl$_2$) -21.4 (L), ^{2}JPPtP = 28.9,[5d] IR,[426b]
   anal.[426b]
trans-Pt(PBu$_3$)$_2$L(CN).   III.3.[426b]   Oil.[426b] ^{31}P(CH$_2$Cl$_2$)
   -64.7 (L), ^{2}JPPtP = 27.2,[5d] IR.[426b]
trans-Pt(PBu$_3$)$_2$L(OCOMe).   III.3.[426b] ^{31}P(CH$_2$Cl$_2$) -21.7
   (L), ^{2}JPPtP = 26.5,[5d] IR.[426b]
Na[H$_3$BL].   IV.2.[445]   Useful as blowing agent for epoxy
   resins.

TYPE: 2 C$_{12}$H$_{10}$; L$_2$ = [OP(OPh)$_2$]$_2^{2-}$

Pd$_2$L$_2$Cl$_2$[P(OPh)$_2$OH]$_2$.   I.l.b.[426b]   White,[426b] m. 170-
   195.[426b] ^{31}P(CH$_2$Cl$_2$) -67.7 [L and P(OPh$_2$)OH],[5d]
   anal.[426b]

Pd[P(OPh)$_2$OH]$_2$L$_2$.   IV.2.[426b] m. 141-143.[426b] ^{31}P(CH$_2$Cl$_2$)
    -87.3 [L and P(OPh)$_2$OH].[5d] anal.[426b]
Pt[P(OPh)$_2$OH]$_2$L$_2$.   IV.2.[426b] White,[426b] m. 166-166.5.[426b]
    ^{31}P(CH$_2$Cl$_2$) -79.7 [L and P(OPh)$_2$OH].[5d]
trans-PtPy$_2$L$_2$.   I.1.b.[426b] Brown,[426b] m. 175-176.[426a]
    ^{31}P(CH$_2$Cl$_2$) -66.4,[5d] anal.[426b]
trans-Pt(PEt$_3$)$_2$L$_2$.   I.1.b.[426b] White,[426b] m. 166-168.[426b]
    ^{31}P(CH$_2$Cl$_2$) -69.0 (L), ^{2}JPPtP = 37.9,[5d] anal.[426b]
trans-Pt(AsEt$_3$)$_2$L$_2$.   I.1.b.[426b] m. 143-145.[426b] ^{31}P
    (CH$_2$Cl$_2$) -70.4,[5d] anal.[426b]
Pt(diphos.)L$_2$.   IV.3.[426b] White,[426b] m. 214-216.[426b]
    Anal.,[426b] NMR.[426a,b]
{[Pt[P(OPh)$_2$OH](AsEt$_3$)L]$_2$}Cl$_2$.   IV.2.[426a] m. 118-120
{[Pt(PEt$_3$)$_2$L]$_2$}Cl$_2$.   I.1.b.[426a] m. 129-131.
AgL$_2$.   IV.2.[426a]

TYPE: C$_{12}$H$_{26}$; L = [OP(OC$_6$H$_{13}$)$_2$]$^-$

HgL(O$_2$CMe).   IV.2.[186] m. 73.9-74.6.   Anal.

TYPE: C$_{14}$H$_{30}$; L = [OP(OC$_7$H$_{15}$)$_2$]$^-$

HgL(O$_2$CMe).   IV.2.[186] m. 83.2-84.

TYPE: C$_{16}$H$_{34}$; L = [OP(OCH$_2$CHEt(CH$_2$)$_3$CH$_3$)$_2$]$^-$

[t-BuNH$_3$][H$_3$BL].   IV.2.[445] m. 164-166.   Useful as blowing
    agent for epoxy resins.

[P(OR)$_2$]$^-$

TYPE: C$_{12}$H$_{10}$; L = [P(OPh)$_2$]$^-$

Cp(OC)$_2$MoL$_2$Mo(CO)$_2$Cp.   I.1.e.[221]   Yellow, m. 169.   IR,
    anal., conductivity.

[OP(OR)R]$^-$

TYPE: C$_9$H$_{12}$; L = [OP(OPr)Ph]$^-$

CpMo(CO)$_2$]P(OPr)$_2$Ph]L.   II.[219a]

[OP(OR)(OH)]$^-$

TYPE: C$_2$H$_6$; L = [OP(OEt)(OH)]$^-$

Rh$_3$[P(OH)$_3$]$_2$[P(OH)$_2$]$_3$LCl.   IV.1.[461a]
Rh$_3$[P(OEt)$_2$O][P(OEt)O$_2$]LCl$_3$.   IV.1.[461a]

[OP(OR)O]$^{2-}$

TYPE: C$_2$H$_5$; L = [OP(OEt)O]$^{2-}$

$Rh_3[P(OEt)_2O]_4LCl_3$.   IV.1.[461a]

$[OPR_2]^{2-}$

TYPE: 2 C_2H_{10}; L = $[OPEt_2]^{2-}$

$[Pt(PEt_3)L(SEt)]_2$. III.3.[108] Colorless, m. 170-171. IR, anal., mol. wt.
$[Pt(PEt_3)LCl]_2$. IV.2.[108] Colorless, m. 66-68. IR, anal., mol. wt.

TYPE: $C_{12}H_{10}$; L = $[OPPh_2]^-$

$CpMo(CO)_2[P(OPr)Ph_2]L$. II.[219a]
$Pt(PEt_3)(H_2N-p-MeC_6H_4)LCl$. III.5.[108] Colorless, m. 103-106. IR, anal.

TYPE: 2 $C_{12}H_{10}$; L_2 = $[OPPh_2]_2^{2-}$

$[Pt(PEt_3)L(SEt)]_2$. III.3.[108] Colorless, m. 183-187. IR, anal., mol. wt.
$[Pt(PMe_3)L(OH)]_2$. III.3.[108] Colorless, m. 213-215.
$\alpha-[Pt(PEt_3)L(OH)]_2$. IV.2.[108] White, m. 171-175. IR, anal., mol. wt., conductivity. Goes to β isomer in excess base.
$\beta-[Pt(PEt_3)L(OH)]_2$. See α isomer.[108] White, m. 155. IR, anal., mol. wt., conductivity.
$[Pt(PEt_3)_2LI]_2$. IV.2.[108] Yellow, m. 205-207. IR, anal., mol. wt.
$[Pt(PMe_3)LBr]_2$. IV.2.[108] Off white, m. 147-149. ^1H, IR.
$[Pt(PEt_3)LBr]_2$. IV.2.[108] Off white, m. 185-186. IR, anal., mol. wt.
$[Pt(PMe_3)LCl]_2$. IV.2.[108] Colorless, m. 217-222. ^1H, IR, anal., mol. wt., conductivity.
$[Pt(PEt_3)LCl]_2$. IV.2.[108] Colorless, m. 180-181. IR, anal., mol. wt., conductivity.
$[Pt(AsEt_3)LCl]_2$. IV.2.[108] Off white, m. 175. ^1H, IR, anal., mol. wt., conductivity.

TYPE: $C_{18}H_9D_5$; L = $P(O-2,6-D_2C_6H_3)_2(O-⊖\text{phenyl})$

trans-ClP_L-trans-$[P(O-2,6-D_2C_6H_3)_3]_2Ru$. IV.2.[415a] m. 177-179. ^1H, anal.

TYPE: $C_{18}H_{14}$; L = $P(OPh)_2(O-⊖\text{phenyl})$

$Ru[P(OPh)_3]_2(CO)LCl$. III.3.[327] White, m. 161-163. IR, anal., mol. wt.

cis-ClP$_L$-trans-[P(OPh$_3$)$_2$Ru.   IV.2.[415a]   m. 169-173.
^{31}P(CH$_2$Cl$_2$) -152, ^{2}JPRuPcis = 62 and 43 -121,
^{2}JPRuP$_{cis}$ = 50 and 43 -144, ^{2}JPRuP$_{cis}$ = 62 and 50,
anal. Reacts with H$_2$ to give Ru[P(OPh)$_3$]$_4$Cl.
Ru[P(OPh)$_3$]$_3$LBr.   IV.2.[327]   White, m. 182-185. Anal.,
mass spec.
trans-ClP$_L$-trans-[P(OPh)$_3$]$_2$Ru.   IV.2.[307a,327]   White,[307a]
m. 176-179.[327]   Anal.,[327] mol. wt.[327] Reacts with H$_2$
to give Ru[P(OPh)$_3$]$_4$HCl.[307a]
Rh[P(OPh)$_3$]$_3$L.   IV.2.[5b]   Yellow. Anal.
[IrLCl]$_2$.   IV.2.[5b]   White. Anal.
IrP(OPh)$_3$L.   IV.2.[5b]   White. Anal.
IrL$_2$HCl.   IV.2.[5b]   White. Anal.
IrP(OPh)$_3$L$_2$H.   IV.2.[5b]

TYPE: C$_{21}$H$_{20}$; L = P(O-p-MeC$_6$H$_4$)$_2$(O–⊖⟨⟩ Me)

Ru(CO)[P(O-p-MeC$_6$H$_4$)$_3$]$_2$LCl.   III.3.[327]   White, m. 182-195.
CO(CH$_2$Cl$_2$) 2024, anal., mol. wt.
Ru[P(O-p-MeC$_6$H$_4$)$_3$]$_3$LBr.   IV.2.[327]   White, m. 162-166.
Anal., mol. wt.
Ru[P(O-p-MeC$_6$H$_4$)$_3$]$_3$LCl.   IV.2.[327]   White, m. 166-169. Anal.,
mol. wt.

Ligands with Multiple Bonding Sites

TYPE: C$_2$H$_4$; L = F$_2$PO(C$_2$H$_4$)$_2$OPF$_2$

NiL$_2$.   I.1.a.[462,465]   Colorless,[462] m. 280.[465] Anal.,[462]
polymer.[465] Useful as polymerization catalyst.[465]

TYPE: C$_2$H$_5$; L = EtN(PF$_2$)$_2$

Cr(CO)$_4$L.   I.1.c.[286]   Yellow. Liquid. ^{31}P(neat) -170.7,
^{19}F, ^{2}JPNP = 78.5, IR, anal.
Mo(CO)$_4$L.   I.1.c.[286]   Colorless, b$_{0.001}$ 60. ^{31}P(neat)
-144.8, ^{19}F, ^{2}JPMoP (^{2}JPNP) = 121.0, IR, anal.
W(CO)$_4$L.   I.1.c.[286]   Colorless, m. 14-15. ^{19}F, ^{2}JPWP =
155.3, IR, anal.

TYPE: C$_6$H$_4$; L = F$_2$PO-o-(F$_2$PO)C$_6$H$_4$

Ni(CO)$_2$L.   I.1.a.[462,465]   Colorless.[462] Anal.,[462] polymer.[465] Useful as polymerization catalyst.[465]

TYPE: 2 C$_6$H$_4$; L$_2$ = [F$_2$PO-o-(F$_2$PO)C$_6$H$_4$]$_2$

NiL$_2$.   I.1.a.[462]   Colorless, m. > 300. Polymer anal.

TYPE: C_3H_6; L = $P(OCH_2)_3P$

$Cr(CO)_5P(OCH_2)_3P$.   I.l.a.[52] ^{31}P(MeCN) -154.47 (PO_3),
   +68.57 (PC_3), ^1H, IR, mol. wt.
$Cr(CO)_5P(OCH_2)_3PCr(CO)_5$.   I.l.a.[52] ^1H, ^{3}JPOCP = +66.1,
   ^{31}P(MeCN) -155.52 (PO_3), -8.21 (PC_3), IR, mol. wt.
$Mo(CO)_5P(OCH_2)_3P$.   I.l.a.[52] ^{31}P(MeCN) -130.82 (PO_3),
   +68.60 (PC_3), ^1H, ^{3}JPOCP = -4.6, IR, anal., mass spec.
$Mo(CO)_5P(OCH_2)_3PMo(CO)_5$.   I.l.a.[52] ^{31}P(MeCN) -130.97
   (PO_3), +18.94 (PC_3), ^1H, ^{3}JPOCP = 63.7, IR, mass spec.
$Mo(CO)_5P(CH_2O)_3P$.   I.l.b.[520b]   White. ^1H.
$W(CO)_5P(OCH_2)_3P$.   I.l.b.[52] ^1H, ^{3}JPOCP = -0.4, ^{31}P(MeCN)
   -109.09 (PO_3), +68.75 (PC_3), IR, mol. wt.
$(OC)_5WP(OCH_2)_3PW(CO)_5$.   I.l.c.[52] ^1H, ^{3}JPOCP = 1.5
   ^{3}JPOCP = 73.8, ^{31}P(MeCN) -108.37 (PO_3), +36.34 (PC_3),
   IR, mol. wt.
$Fe(CO)_4P(OCH_2)_3P$.   I.l.a.[520b] ^{31}P(MeCN) -157.44 (PO_3),
   +71.38 (PC_3), ^1H, ^{3}JPOCP = +8.6,[52] IR.
"non-ax"$Fe(CO)_4P(OCH_2)_3P$.   The NMR and IR parameters con-
   sistent with this formulation[52] were later all found
   to be due to a fortuitous mixture of products of the
   reaction.[520b]
$Fe(CO)_4P(CH_2O)_3P$.   I.l.a.[520b] ^{31}P(MeCN) -87.37 (PO_3),
   -22.42 (PC_3), ^1H, ^{3}JPCOP = +47.1,[52] IR.
$(OC)_4FeP(OCH_2)_3PFe(CO)_4$.   I.l.a.[520b] ^{31}P(MeCN) -160.40
   (PO_3), -22.06 (PC_3), ^1H, ^{3}JPOCP = +95.3,[52] IR.
$H_3BP(OCH_2)_3P$.   I.l.f.[520b]   White. ^1H.
$H_3BP(OCH_2)_3PBH_3$.   I.l.f.[520b]   White. ^1H, water-sensitive.

TYPE: 2 C_3H_6; L_2 = $[P(OCH_2)_3P]_2$

cis-$Cr(CO)_4[P(OCH_2)_3P]_2$.   I.l.c.[52] ^1H, ^{3}JPOCP = 2.4,
   ^{31}P -157.21 (PO_3), +68.52. (PC_3), IR.
cis-$Mo(CO)_4[P(OCH_2)_3P]_2$.   I.l.c.[520b]   White. ^1H, IR,
   mass spec., mol. wt.
cis-$Mo(CO)_4[P(OCH_2)P][P(CH_2O)_3P]$.   I.l.c.[520b]   White.
   ^1H, IR, mass spec., mol. wt.
trans-$Fe(CO)_3[P(OCH_2)_3P][P(CH_2O)_3P]$.   I.l.a.[520]   White.
   Preliminary x-ray studies confirm the proposed stereochemistry, ^1H, IR, mol. wt., mass spec.

TYPE: C_5H_{18}; L = $P(NMeCH_2)_3CMe$

$(H_3B)_2L$.   I.l.f.[319]   White. ^1H, anal.

TYPE: $C_6H_{18}N_6P_2$; L = $P[N(Me)(Me)N]_3P$

$(Et_3Al)_2L$.   I.l.f.[486]   White, m. 153-155. ^{31}P -11.8s,
   ^1H, IR, anal.

TYPE: $C_6H_{18}N_6P_4$; L = $P_4(NMe)_6$

$Ni(CO)_3L$.   I.l.a.[449] ^{31}P($CHCl_3$) -119.3 (NiP), -93.9 (P
   uncoordinated).
$Ni(CO)_3]_2L$.   I.l.a.[449] ^{31}P($CHCl_3$) -128.5 (NiP), -103.1

(P uncoordinated).
[Ni(CO)$_3$]$_3$L. I.l.a.[449] ^{31}P(CHCl$_3$) -135.7 (NiP), -108.9
   (P uncoordinated).
[Ni(CO)$_3$]$_4$L. I.l.a.[449,450] IR,[450] anal.[450]
H$_3$BL. I.l.f.[451] ^{31}P(CHCl$_3$) -80.5 (PB), -103.2 (P un-
   coordinated), NMR, equilibrium study.
(H$_3$B)$_2$L. I.l.f.[451] ^{31}P(CHCl$_3$) -96.7 (PB), -118.1 (P
   uncoordinated), ^{11}B, ^{1}H, equilibrium study.
(H$_3$B)$_3$L. I.l.f.[451] ^{31}P(CHCl$_3$) -110 (PB), -127.6 (P un-
   coordinated), equilibrium study.
(H$_3$B)$_4$L. I.l.f.[451] White. IR, anal. Reacts with Me$_3$N
   to yield L.

TYPE: C$_{16}$H$_{22}$; PhP[N(Me)(Me)N]$_2$PPh

(AlEt$_3$)$_2$L. I.l.f.[486] m. 181-184. ^{1}H, IR, anal.

TYPE: C$_{24}$H$_{21}$; L = HN(PPh$_2$)$_2$

Mo(CO)$_4$L. I.l.a.[417] m. 204-205. IR, anal., mol. wt.

TYPE: C$_{25}$H$_{23}$; L = MeN(PPh$_2$)$_2$

Mo(CO)$_4$L. I.l.a.[417] m. 189-191. IR, anal.
Et$_3$AlL. I.l.f.[125] m. 130-132. ^{31}P -64, ^{1}H, IR, anal.,
   mol. wt.

TYPE: C$_{26}$H$_{25}$; L = EtN(PPh$_2$)$_2$

Mo(CO)$_4$L. I.l.a.[417] m. 210-212.[417] IR,[417] anal.[417]
   x-Ray structure shows both P atoms coordinated in cis
   positions (Mo-P = 2.505 Å, PNP angle = 103°, PMoP
   angle = 64.8°).[416]
CpFe(CO)$_2$FeCpL. I.2.[218] Green. Diamagnetic, ^{1}H, IR,
   anal., mol. wt.
π-MeC$_6$H$_4$Fe(CO)$_2$Fe(π-MeC$_6$H$_4$)L. I.2.[218] Green. Diamagnetic, ^{1}H, IR, anal., mol. wt.
PdLCl$_2$. x-Ray structure[416] shows both P atoms coordinated
   in cis positions (P-Pd = 2.224 Å, P-N = 1.72 Å, PNP
   angle = 97.7°, PPdP angle - 71.4°).
HEt$_2$AlL. I.l.f.[125] b$_{2.4}$ 163. ^{1}H, IR, anal., mol. wt.
Et$_3$Al. I.l.f.[125] m. 78-82. ^{31}P -59, ^{1}H, IR, anal.,
   mol. wt.

TYPE: C$_{26}$H$_{26}$; Ph$_2$PN(Me)(Me)NPPh$_2$

(Et$_3$Al)$_2$L. I.l.f.[486] m. 79-80. ^{31}P -61.2, ^{1}H, IR, anal.

TYPE: C$_{27}$H$_{27}$; L = PrN(PPh$_2$)$_2$

Mo(CO)$_4$L. I.l.a.[417] m. 202-204. IR, anal.

Miscellaneous

$CoClP(OC_6H_4CHNCH_2)_2Cl_2$. I.l.f.[267] Blue-green. Magnetic
  moment 4.26 BM, IR, UV, anal.
$NiClP(OC_6H_4CHNCH_2)_2Cl_2$. I.l.f.[267] Green. Paramagnetic
  3.08 BM, IR, UV, anal.
$CuClP(OC_6H_4CHNCH_2)_2Cl_2$. I.l.e.[267] Yellow. Magnetic
  moment 1.33 BM, IR, UV, anal.
$[Cl\ BNMeP(CF_3)_2]_2$. Formed by decomposition of unstable
  $Cl_3BP(NMeH)(CF_3)_2$.[205] Colorless, b. ∿ 90. ^{31}P (liq.)
  -50.0, IR, anal., mol. wt.

$\pi\text{-}C_5H_4Fe[P(OPh)_3]P(O$ ⟨◯⟩ $)(OPh)_2I$. III.2.[6] Violet,[387b]
  m. 180-190.[387b] x-Ray structure shows pseudotetrahedral
  arrangement around Fe^6 (see text), Fe-P = 2.14 Å,
  Fe-I = 2.65 Å, P-O = 1.63 Å, ^1H.[387b]

                (received December 27, 1970)

REFERENCES

1. Abel, E. W., M. A. Bennett, and G. Wilkinson, J. Chem.
    Soc., 1959, 2323.
2. Abramov, V. S., and V. K. Khairullin, Zh. Obshch.
    Khim., 27, 444 (1957).
3. Adams, D. M., and P. J. Chandler, J. Chem. Soc. (A),
    1969, 588.
4. Adams, D. M., D. J. Cook, and R. D. W. Kemmitt, J.
    Chem. Soc. (D), 1966, 103.
5a. Adelson, D. W., and R. E. Thorpe, U.S. Pat. 2,645,613;
    C.A., 47, 10837b.
5b. Ainscough, E. W., and S. D. Robinson, J. Chem. Soc.
    (D), 1970, 863.
5c. Allen, F. H., G. Chang, K. K. Cheung, T. F. Lai, and
    A. Pidcock, J. Chem. Soc. (D), 1970, 1297.
5d. Allen, F. H., A. Pidcock, and C. R. Waterhouse, J.
    Chem. Soc. (A), 1970, 2087.
6. Andrianov, V. G., Yu. A. Chapovskii, V. A. Semion,
    and Yu. T. Struch, J. Chem. Soc. (D), 1968, 282.
7. Ang, H. G., J. Organometal. Chem., 19, 245 (1969).
8. Angelici, R. J., Inorg. Chem., 3, 1099 (1964).
9. Angelici, R. J., and F. Basolo, J. Am. Chem. Soc.,
    84, 2495 (1962).
10. Angelici, R. J., and F. Basolo, Inorg. Chem., 2, 728
    (1963).
11. Angelici, R. J., F. Basolo, and A. J. Poe, Nature,
    195, 993 (1962).
12. Angelici, R. J., F. Basolo, and A. J. Poe, J. Am.
    Chem. Soc., 85, 2215 (1963).
13. Angelici, R. J., and J. R. Graham, J. Am. Chem. Soc.,
    87, 5586 (1965).
14. Angelici, R. J., and J. R. Graham, J. Am. Chem. Soc.,

88, 3658 (1966).
15. Angelici, R. J., and J. R. Graham, Inorg. Chem., 6, 988 (1967).
16. Angelici, R. J., and C. M. Ingemanson, Inorg. Chem., 8, 83 (1969).
17. Angelici, R. J., R. E. Jacobson, and C. M. Ingemanson, Inorg. Chem., 7, 2466 (1968).
18. Angelici, R. J., and E. E. Siefert, Inorg. Chem., 5, 1457 (1966).
19. Arbusoff, A., Chem. Ber., 38, 1171 (1905).
20. Arbuzov, A. E., and N. N. Sazonova, Dokl. Akad. Nauk SSSR, 115, 1119 (1947); C.A., 52, 6239i.
21. Arbuzov, A. E., and F. G. Valitova, Bull. Acad. Sci. USSR., Classe Sci. Chim., 1940, 529; C.A., 35, 3990[1].
22. Arbuzov, A. E., and V. M. Zoroastrova, Dokl. Akad. Nauk SSSR, 84, 503 (1952); C.A., 46, 10038f.
23. Arbuzov, A. E., and V. M. Zoroastrova, Izvest. Akad. Nauk SSSR, Otdel. Khim. Nauk, 1952, 809; C.A., 47, 9898h.
24. Arbuzov, A. E., and V. M. Zoroastrova, Izvest. Akad. Nauk SSSR, Otdel. Khim. Nauk, 1952, 818; C.A., 47, 9899f.
25. Arbuzov, A. E., and V. M. Zoroastrova, Izvest. Akad. Nauk SSSR, Otdel. Khim. Nauk, 1952, 826; C.A., 47, 9900a.
26. Arbuzov, A. E., and V. M. Zoroastrova, Izvest. Akad. Nauk SSSR, Otdel. Khim. Nauk, 1952, 770; C.A., 47, 9900e.
27. Arbuzov, B. A., and N. I. Rizpolozhenskii, Izvest. Akad. Nauk SSSR, Otdel. Khim. Nauk, 1952, 854; C.A., 47, 99036.
28. Arbuzov, B. A., and T. G. Shavsha, Dokl. Akad. Nauk SSSR, 84, 507 (1952); C.A., 46, 10729g.
29. Argento, B. J., P. Fitton, E. McKeon, and E. A. Rick, J. Chem. Soc. (D), 1969, 1427.
30. Attali, S., and R. Poilblanc, C.R. Acad. Sci., Paris Ser. C, 267, 718 (1968).
31. Azanovskaya, M. M., Zh. Obschch. Khim., 27, 1363 (1957).
32. Azanovskaya, M. M., and L. A. Davidovskaya, Russ. J. Inorg. Chem., 4, 418 (1960).
33. Bainbridge, A., P. J. Craig, and M. Green, J. Chem. Soc. (A), 1968, 2715.
34. Balch, A. L., Inorg. Chem., 6, 2158 (1967).
35a. Bamford, C. H., and K. Hargreaves, Nature, 209, 292 (1966).
35b. Bancroft, G. M., M. J. Mays, B. E. Prater, and F. B. Stefanini, J. Chem. Soc. (A), 1970, 2146.
36. Bannister, W. D., B. L. Booth, M. Green, and R. N. Hazeldine, J. Chem. Soc. (A), 1969, 698.
37. Bannister, W. D., M. Green, and R. N. Hazeldine, J.

Chem. Soc. (D), 1965, 54.
38. Barlow, C. G., R. Jefferson, and J. F. Nixon, J. Chem. Soc. (A), 1968, 2692.
39. Barlow, C. G., and J. F. Nixon, Inorg. Nucl. Chem. Lett., 2, 323 (1966).
40. Barlow, C. G., J. F. Nixon, and J. R. Swain, J. Chem. Soc. (A), 1969, 1082.
41. Barlow, C. G., J. F. Nixon, and M. Webster, J. Chem. Soc. (A), 1968, 2216.
42. Barnett, K. W., Inorg. Chem., 8, 2009 (1969).
43. Bawer, L. M., and M. H. B. Stiddard, J. Organometal. Chem., 13, 235 (1968).
44. Beck, W., and K. Lottes, Z. Naturforsch., 19b, 987 (1964).
45. Beck, W., and K. Lottes, Z. Anorg. Allgem. Chem., 335, 258 (1965).
46. Beck, W., and K. Lottes, Chem. Ber., 98, 2657 (1965).
47. Beck, M., R. E. Nitsschmann, and G. Neumair, Angew. Chem., Int. Ed. Engl., 3, 380 (1964).
48a. Bedell, R., M. J. Frazer, and W. Gerard, J. Chem. Soc., 1960, 4037.
48b. Bennett, J., A. Pidcock, C. R. Waterhouse, P. Coggon, and A. T. McPhail, J. Chem. Soc. (A), 1970, 2094.
49. Benson, R. E., U.S. Pat. 2,804,936; C.A., 54, 16012a.
50. Bergerhoff, G., Z. Anorg. Allgem. Chem., 327, 139 (1964).
51. Beriger, E., U.S. Pat. 2,931,826; C.A., 54, 15244i.
52. Bertrand, R. D., D. A. Allison, and J. G. Verkade, J. Am. Chem. Soc., 92, 71 (1970).
53. Bertrand, R. D., F. B. Ogilvie, and J. G. Verkade, J. Chem. Soc. (D), 1969, 756.
54. Bertrand, R. D., F. B. Ogilvie, and J. G. Verkade, J. Am. Chem. Soc., 92, 1908 (1970).
55. Besnainou, S., and P. Labarbe, J. Mol. Struct., 2, 499 (1968).
56. Bibler, J. P., and A. Wojcicki, Inorg. Chem., 5, 889 (1966).
57. Bigorgne, M., C. R. Acad. Sci., Paris, Sec. C, 250, 3484 (1960).
58. Bigorgne, M., Proc. Int. Conf. Coord. Chem., 6th, Detroit, 1961, 199.
59a. Bigorgne, M., Rev. Chim. Minerale, 1966, 831.
59b. Bigorgne, M., Spectrochim. Acta, 26A, 1217 (1970).
60. Bigorgne, M., and G. Bouquet, C. R. Acad. Sci., Paris, Ser. C, 264, 1485 (1967).
61. Bigorgne, M., and C. Messier, J. Organometal. Chem., 2, 79 (1964).
62. Bigorgne, M., and H. Zehner, Bull. Soc. Chim. France, 1960, 1986.
63. Birum, G. H., U.S. Pat. 2,844,454; C.A., 52, 20862d.
64. Birum, G. H., U.S. Pat. 2,909,544; C.A., 54, 3843i.

65. Eirum, G. H., U.S. Pat. 2,945,872; C.A., _55_, 393f.
66. Blake, D. M., and M. Kubota, J. Am. Chem. Soc., _92_, 2578 (1970).
67. Bland, W. J., and R. D. W. Kemmitt, J. Chem. Soc. (A), _1969_, 2062.
68. Bonati, F., S. Cenini, D. Morelli, and R. Ugo, Inorg. Nucl. Chem. Lett., _1_, 107 (1965).
69. Bonati, F., S. Cenini, D. Morelli, and R. Ugo, J. Chem. Soc. (A), _1966_, 1052.
70. Booth, B. L., M. Gardner, and R. N. Hazeldine, J. Chem. Soc. (D), _1969_, 1388.
71. Booth, B. L., M. Green, R. N. Hazeldine, and N. P. Woffenden, J. Chem. Soc. (A), _1969_, 920.
72a. Booth. B. L., and R. N. Hazeldine, J. Chem. Soc. (A), _1966_, 157.
72b. Booth, B. L., R. N. Hazeldine, and M. B. Taylor, J. Chem. Soc. (A), _1970_, 1974.
72c. Booth, B. L., R. N. Hazeldine, and N. P. Woffenden, J. Chem. Soc. (A), _1970_, 1979.
73. Booth, B. L., R. N. Hazeldine, P. R. Mitchell, and J. J. Cox, J. Chem. Soc. (A), _1969_, 691.
74. Bosmajian, G., U.S. Pat. 3,004,081; C.A., _56_, 4640g.
75. Bosmajian, G., R. E. Burks, C. E. Feazel, and J. Newcombe, Ind. Eng. Chem., Prod. Res. Develop., _3_, 117 (1964).
76. Bouquet, G., A. Loutellier, and M. Bigorgne, J. Mol. Struct., _1_, 211 (1967).
77. Bower, L. M., and M. H. B. Stiddard, J. Chem. Soc. (A), _1968_, 706.
78a. Braterman, P. S., J. Organometal. Chem., _11_, 198 (1968).
78b. Brennan, M. E., J. Chem. Soc. (D), _1970_, 956.
79. Brown, D. A., H. J. Lyons, A. R. Manning, and J. M. Rowley, Inorg. Chim. Acta, _3_, 346 (1969).
80. Brown, D. H., K. D. Crosbie, G. W. Fraser, and D. W. A. Sharp, J. Chem. Soc. (A), _1969_, 551.
81. Brown, D. H., K. D. Crosbie, G. W. Fraser, and D. W. A. Sharp, J. Chem. Soc. (A), _1969_, 872.
82. Browing, J., C. S. Cundy, M. Green, and G. A. Stone, J. Chem. Soc. (A), _1969_, 20.
83. Bruce, M. I., Advan. Organometal. Chem., _6_, 273 (1968).
84. Burg, A., and P. F. Slota, Jr., J. Am. Chem. Soc., _80_, 1107 (1958).
85. Burg, A. B., and P. J. Slota, Jr., J. Am. Chem. Soc., _82_, 2145 (1960).
86. Burg, A. B., and G. B. Street, Inorg. Chem., _5_, 1532 (1966).
87. Burke, R. E., and A. A. Sekul, U.S. Pat. 2,972,640; C.A., _56_, 4640i.
88. Burt, R., M. Cooke, and M. Green, J. Chem. Soc. (A), _1969_, 2645.

89. Butcher, F. K., B. E. Deuters, W. Gerrard, E. F. Mooney, R. A. Rothenburg, and H. A. Willis, Spectrochim. Acta, 20, 759 (1964).

90. Butler, I. S., F. Basolo, and R. G. Pearson, Inorg. Chem., 6, 2074 (1967).

91. Cannell, L. G., U.S. Pat. 3,102,899; C.A., 60, 1362e.

92. Capron-Cotigny, G., and R. Poilblanc, C. R. Acad. Sci., Paris, Ser. C, 263, 885 (1966).

93. Cardaci, G., S. M. Margia, and G. Reichebach, Inorg. Chim. Acta, 4, 118 (1970).

94a. Cardaci, G., and A. Zoffani, Inorg. Chim. Acta, 2, 252 (1968).

94b. Casey, M., and A. R. Manning, J. Chem. Soc. (A), 1970, 2258.

95. Cassoux, P., and J. Labarre, C. R. Acad. Sci., Paris, Sec. C, 264, 763 (1967).

96. Cassoux, P., and J. Labarre, C. R. Acad. Sci., Paris, Sec. C, 265, 773 (1967).

97. Cassoux, P., J. F. Labarre, G. Commenges, and A. J. P. Laurent, J. Chim. Phys. Physicochim. Biol., 64, 1813 (1967).

98. Cassoux, P., J. M. Savariault, and J. F. Labarre, Bull. Soc. Chim. France, 1969, 741.

99. Cassoux, P., J. M. Savarianult, J. F. Labarre, and F. Gallais, J. Chim. Phys. Physicochim. Biol., 65, 68 (1185).

100. Cavell, R. G., J. Chem. Soc. 1964, 1992.

101. Centofanti, L. F., G. Kodama, and R. W. Parry, Inorg. Chem., 8, 2072 (1969).

102. Chastain, B. B., E. A. Rick, R. L. Pruett, and H. B. Gray, J. Am. Chem. Soc., 90, 3994 (1968).

103. Chatt, J., J. R. Dilworth, and G. J. Leigh, J. Chem. Soc. (D), 1969, 687.

104. Chatt, J., L. A. Duncanson, and L. M. Venanzi, J. Chem. Soc., 1955, 4461.

105. Chatt, J., L. A. Duncanson, and L. M. Venanzi, J. Inorg. Nucl. Chem., 8, 67 (1958).

106. Chatt, J., L. A. Duncanson, and L. M. Venanzi, J. Chem. Soc., 1958, 3203.

107. Chatt, J., G. A. Gamlen, and L. E. Orgel, J. Chem. Soc., 1959, 1047.

108. Chatt, J., and B. T. Heaton, J. Chem. Soc. (A), 1968, 2745.

109. Chatt, J., and B. T. Heaton, Spectrochim. Acta, Part A, 23, 2220 (1969).

110. Chatt, J., and L. M. Venanzi, J. Chem. Soc., 1955, 3858.

111. Chatt, J., and L. M. Venanzi, J. Chem. Soc., 1957, 2351.

112. Chatt, J., and L. M. Venanzi, J. Chem. Soc., 1957, 2445.

113. Chatt, J., and R. G. Wilkins, J. Chem. Soc., 1953, 70.
114. Chopard, P. A., and R. F. Hudson, J. Inorg. Nucl. Chem., 25, 801 (1963).
115. Church, M. J., and M. J. Mays, J. Chem. Soc. (D), 1968, 435.
116a. Church, M. J., and M. J. Mays, J. Chem. Soc. (A), 1968, 3074.
116b. Church, M. J., and M. J. Mays, J. Chem. Soc. (A), 1970, 1938.
117. Cities Service Research and Development Co., Brit. Pat. 944,574; C.A., 60, 10570a.
118. Cities Service Research and Development Co., Brit. Pat. 979,553; C.A., 63, 632h.
119. Clark, R. F., U.S. Pat. 3,152,158; C.A., 61, P1423f.
120. Clark, R. F., and C. D. Storrs, Belg. Pat. 610,876; C.A., 57, 16662c.
121. Clark, R. F., C. D. Storrs, and C. G. McAllister, Belg. Pat. 621,207; C.A., 59, 11342b.
122. Clark, R. F., and C. D. Storrs, U.S. Pat. 3,244,738; C.A., 64, 19492e.
123. Clark, R. J., J. P. Hargaden, H. Haas, and R. K. Sheline, Inorg. Chem., 7, 673 (1968).
124. Clark, R. J., and K. A. Morgan, Inorg. Chim. Acta, 2, 93 (1968).
125. Clemens, D. F., H. H. Sisler, and W. S. Brey, Inorg. Chem., 5, 527 (1966).
126. Clement, D. A., J. F. Nixon, and M. D. Sexton, J. Chem. Soc. (D), 1969, 1509.
127. Coates, G. E., and C. Parkin, J. Chem. Soc., 1962, 3220.
128. Cochin, D., C. R. Acad. Sci., Paris, Ser. C., 86, 1402 (1878).
129. Cohen, B. M., A. R. Cullingworth, and J. D. Smith, J. Chem. Soc. (A), 1969, 2193.
130. Cohen, B. M., and J. D. Smith, J. Chem. Soc. (A), 1969, 2087.
131. Cohen, I. A., and F. Basolo, J. Inorg. Nucl. Chem., 28, 511 (1966).
132. Cohn, K., and R. W. Parry, Inorg. Chem., 7, 46 (1968).
133. Colton, F. A., Inorg. Chem., 3, 702 (1964).
134. Connelly, N. G., and J. A. McCleverty, J. Chem. Soc. (A), 1970, 1621.
135. Connelly, N. G., J. A. McCleverty, and C. J. Winscom, Nature, 216, 999 (1967).
136. Cooke, J., M. Green, and F. G. A. Stone, J. Chem. Soc. (A), 1968, 170.
137. Cooke, M., and M. Green, J. Chem. Soc. (A), 1969, 651.
138. Coover, H. W., U.S. Pat. 3,194,799; C.A., 63, 13440c.
139. Coover, H. W., and F. B. Joyner, U.S. Pat. 3,186,977;

C.A., 63, 5773b.
140. Coover, H. W., and F. B. Joyner, U.S. Pat. 2,967,856;
     C.A., 55, 8941e.
141. Corbridge, D. E. C., Top. Phosphorus Chem., 3, 110
     (1966).
142. Coskran, K. J., R. D. Bertrand, and J. G. Verkade,
     J. Am. Chem. Soc., 89, 4535 (1967).
143. Coskran, K. J., T. J. Huttemann, and J. G. Verkade,
     Advan. Chem., 62, 590 (1967).
144a. Coskran, K. J., J. M. Jenkins, and J. G. Verkade,
      J. Am. Chem. Soc., 90, 5437 (1968).
144b. Coskran, K. J., and C. E. Jones, Inorg. Chem., 10,
      56 (1971).
144c. Coskran, K. J., and C. E. Jones, Inorg. Chem., in
      press.
145a. Cotton, J. D., and R. L. Heazlewood, Aust. J. Chem.,
      22, 2673 (1969).
145b. Cragg, R. H., M. S. Fortuin, and N. N. Greenwood,
      J. Chem. Soc. (A), 1970, 1817.
146. Craig, P. J., and M. Green, J. Chem. Soc. (A), 1968,
     1978.
147. Craig, P. J., and M. Green, J. Chem. Soc. (A), 1969,
     157.
148. Craig, P. J., and M. Green, J. Chem. Soc. (D), 1967,
     1246.
149a. Cullingsworth, A. R., A. Pidcock, and J. D. Smith,
      J. Chem. Soc. (D), 1966, 89.
149b. Cundy, C. S., M. Green, and F. G. A. Stone, J. Chem.
      Soc. (A), 1970, 1647.
150. Dalton, J., I. Paul, J. G. Smith, and F. G. A. Stone,
     J. Chem. Soc. (A), 1968, 1195.
151. Dalton, J., I. Paul, J. G. Smith, and F. G. A. Stone,
     J. Chem. Soc. (A), 1968, 1208.
152. Darensbourg, D. J., and T. L. Brown, Inorg. Chem.,
     7, 959 (1968).
153. Darensbourg, D. J., and T. L. Brown, Inorg. Chem.,
     7, 1679 (1968).
154. Davis, T. L. and P. Ehrlich, J. Am. Chem. Soc., 58,
     2151 (1958).
155. Dilts, J. A., and D. F. Shriver, J. Am. Chem. Soc.,
     91, 4088 (1969).
156. Dobson, G. R., Inorg. Chem., 8, 90 (1969).
157. Dobson, G. R., and G. C. Faber, Inorg. Chim. Acta,
     4, 87 (1970).
158. Dobson, G. R., and L. W. Houk, Inorg. Chim. Acta, 1,
     287 (1967).
159. Dobson, G. R., and L. A. H. Smith, Inorg. Chem., 9,
     1001 (1970).
160. Dobson, G. R., J. W. Stolz, and R. K. Sheline, Advan.
     Inorg. Chem. Radiochem., 8, 1 (1966).
161. Donner, R., and K. L. Lohs, J. Chromatog., 17, 349
     (1965).

162. Donoghue, J. T., J. A. McMillan, and D. A. Peters, J. Inorg. Nucl. Chem., 31, 3661 (1969).
163. Douglas, P. G., and B. L. Shaw, J. Chem. Soc. (A), 1970, 334.
164. Douglas, P. G., and B. L. Shaw, J. Chem. Soc. (A), 1970, 1556.
165. Douglas, P. G., and B. L. Shaw, J. Chem. Soc. (D), 1969, 624.
166. Drinkard, W. C., D. R. Eaton, J. P. Jesson, and R. V. Lindsey, Inorg. Chem., 9, 392 (1970).
167. Dyer, G., and D. W. Meek, Inorg. Chem., 6, 149 (1967).
168. E. I. DuPont de Nemours and Co., Brit. Pat. 832,173; C.A., 54, 18434a.
169. Eisenmann, J. L., U.S. Pat. 3,139,460; C.A., 61, 6918c.
170. Engel, R., J. Chem. Soc. (D), 1970, 133.
171. Engel, R., T. Santo, D. Liotta, and D. Freed, J. Chem. Soc. (D), 1970, 646.
172. Esso Research and Engineering Co., Neth. Appl. 6,400,701; C.A., 62, 5194g.
173. Ewart, G., D. W. Payne, A. L. Porte, and A. P. Lane, J. Chem. Soc., 1962, 3984.
174. Faber, G. C., and R. J. Angelici, Inorg. Chem., 9, 1586 (1970).
175. Faber, G. C., and G. R. Dobson, Inorg. Chim. Acta, 2, 479 (1968).
176. Faber, G. C., and G. R. Dobson, Inorg. Chem., 7, 584 (1968).
177. Faber, G. C., T. D. Walsh, and G. R. Dobson, J. Am. Chem. Soc., 90, 4178 (1968).
178. Faller, J. W., A. S. Anderson, and C. Chen, J. Organometal. Chem., 17, P7 (1969).
179. Farbenfabriken Bayer A-G, Neth. Appl. 6,404,853; C.A., 62, 11979b.
180. Feldman, J., B. A. Saffer, and O. D. Frampton, U.S. Pat. 3,284,529; C.A., 66, 28373k.
181. Feldman, J., B. A. Saffer, and M. Thomas, U.S. Pat. 3.194,848; C.A., 63, 9989.
182. Finck, E., C. R. Acad. Sci, Paris, Ser. C, 115, 176 (1892).
183. Finck, E., C. R. Acad. Sci., Paris, Ser. C, 123, 603 (1896).
184. Finck, E., Chem. Zent., 67 (part 2), 968 (1896).
185. Flitcroft, N., J. M. Leach, and F. J. Hopton, J. Inorg. Nucl. Chem., 32, 137 (1970).
186. Fox, R. B., and D. L. Venezky, J. Am. Chem. Soc., 75, 3967 (1953).
187. Frazer, M. J., W. Gerrard, and J. K. Patel, Chem. Ind. (London), 1959, 728.
188. Freni, M., and P. Romiti, Inorg. Nucl. Chem. Lett., 6, 167 (1970).

189. Freni, M., and V. Valenti, Gazz. Chim. Ital., <u>90</u>, 1436 (1960).
190. Freni, M., V. Valenti, and D. Giusto, J. Inorg. Nucl. Chem., <u>27</u>, 2635 (1965).
191. Gallais, F., and H. Haraldsen, C. R. Acad. Sci., Paris, Ser. C, <u>264</u>, 1 (1967).
192. Genser, E. E., Inorg. Chem., <u>7</u>, 13 (1968).
193. Ginzberg, G. D., and A. D. Troitskaya, Tr. Kazan. Khim.-Tekhnol. Inst., <u>34</u>, 38 (1965); C.A., <u>67</u>, 68166x.
194. Ginzberg, G. D., and A. D. Troitskaya, Tr. Kazan, Khim.-Tekhnol. Inst., <u>36</u>, 124 (1967); C.A., <u>70</u>, 33903v.
195. Giusto, D., Inorg. Nucl. Chem. Lett., <u>5</u>, 767 (1969).
196a. Gmelins Handbuch der Anorganischen Chemie, 8th Ed., Pt. D, No. 68, Verlag Chemie, GMBH., Weinheim/ Bergstrass, 1957, pp. 171, 351, 465.
196b. Gosser, L. W., and C. A. Tolman, Inorg. Chem., <u>9</u>, 2350 (1970).
197. Graham, J. R., and R. J. Angelici, J. Am. Chem. Soc., <u>87</u>, 5590 (1965).
198. Graham, J. R., and R. J. Angelici, Inorg. Chem., <u>6</u>, 992 (1967).
199. Graham, J. R., and R. J. Angelici, Inorg. Chem., <u>6</u>, 2082 (1967).
200. Grasselli, J. G., W. M. Ritchey, and F. J. Knoll, Inorg. Chem., <u>4</u>, 1323 (1965).
201. Graziani, M., F. Zingales, and U. Belluco, Inorg. Chem., <u>6</u>, 1582 (1967).
202. Green, M., R. I. Hancock, and D. C. Wood, J. Chem. Soc. (A), <u>1968</u>, 2718.
203. Green, M., and D. C. Wood, J. Am. Chem. Soc., <u>88</u>, 4106 (1966).
204. Greenberg, B., A. Amendola, and R. Schmutzler, Naturwissenschaften, <u>50</u>, 593 (1963).
205. Greenwood, N. N., and B. H. Robinson, J. Chem. Soc. (A), <u>1968</u>, 226.
206. Griffith, W. P., and A. J. Wickham, J. Chem. Soc. (A), <u>1969</u>, 834.
207. Griffiths, W. I., The Chemistry of the Rarer Pt Metals, Interscience, New York, 1966.
208. Grim., S. O., P. R. McAllister, and R. M. Singer, J. Chem. Soc. (D), <u>1969</u>, 38.
209. Grim, S. O., D. A. Wheatland, and P. R. McAllister, Inorg. Chem., <u>7</u>, 161 (1968).
210. Grinberg, A. A., Dokl. Akad. Nauk SSSR, <u>105</u>, 1256 (1955); C.A., <u>50</u>, 7641c.
211. Grinberg, A. A., T. B. Itskovich, and A. D. Troitskaya, J. Inorg. Chem., <u>3</u>, 30 (1959).
212. Grinberg, A. A., Z. A. Razumova, and A. D. Troitskaya, Bull. Acad. Sci. URSS, Classe Sci. Chim., <u>3</u>, 253

(1946); C.A., 43, 4172g.
213. Grinberg, A. A., and A. D. Troitskaya, Bull. Acad. Sci. URSS, Classe Sci. Chim., 178 (1944); C.A., 39, 1604.[4]
214. Haines, L. M., Inorg. Nucl. Chem. Lett., 5, 399 (1969).
215. Haines, L. M., Inorg. Chem., 9, 1517 (1970).
216. Haines, R. J., and A. L. DuPreez, J. Chem. Soc. (D), 1968, 1513.
217. Haines, R. J., and A. L. DuPreeze, Inorg. Chem., 8, 1459 (1969).
218. Haines, R. J., and A. L. DuPreez, J. Organometal. Chem., 21, 181 (1970).
219a. Haines, R. J., I. L. Mararis, and C. R. Nolte, J. Chem. Soc. (D), 1970, 547.
219b. Haines, R. J., and C. R. Nolte, J. Organometal. Chem., 24, 725 (1970).
220. Haines, R. H., R. S. Nyholm, and M. H. B. Stiddard, J. Chem. Soc. (A), 1967, 94.
221. Haines, R. J., R. S. Nyholm, and M. H. B. Stiddard, J. Chem. Soc. (A), 1968, 43.
222. Haines, R. J., R. S. Nyholm, and M. H. B. Stiddard, J. Chem. Soc. (A), 1968, 46.
223. Hales, L. A., and R. J. Irving, J. Chem. Soc. (A), 1967, 1932.
224. Handley, T. H., and J. A. Dean, Anal. Chem., 33, 1087 (1961).
225. Hart-Davis, A. J., and R. J. Mawby, J. Chem. Soc. (A), 1969, 2403.
226. Hartman, F. A., M. Kilner, and A. Wojcicki, Inorg. Chem., 6, 34 (1967).
227. Hartman, F. A., and A. Wojcicki, Inorg. Nucl. Chem. Lett., 2, 303 (1966).
228. Hattori, J., and Y. Nishizawa, Japan Pat. 1263; C.A., 56, 11745a.
229. Heck, A. F., Adv. Organometallic Chem., 4, 243 (1966).
230a. Heck, R. F., J. Am. Chem. Soc., 85, 655 (1963).
230b. Heck, R. F., Acc. Chem. Res., 2, 10 (1969).
231. Heck, R. F., J. Am. Chem. Soc., 85, 1220 (1963).
232. Heck, R. F., J. Am. Chem. Soc., 86, 5138 (1964).
233. Heck, R. F., J. Am. Chem. Soc., 87, 2572 (1965).
234. Heimbach, P., Angew, Chem., Int. Ed. Engl., 3, 648 (1964).
235. Heitsch, C. W., and J. G. Verkade, Inorg. Chem., 1, 392 (1962).
236. Heitsch, C. W., and J. G. Verkade, Inorg. Chem., 1, 863 (1962).
237. Hendricker, D. G., R. E. McCarley, R. W. King, and J. G. Verkade, Inorg. Chem., 5, 639 (1966).
238. Herber, R. H., R. King, and G. K. Wertheim, Inorg. Chem., 3, 101 (1964).

239. Hertz, C. H., and R. O. E. Davis, J. Am. Chem. Soc., 30, 1084 (1908).
240. Hieber, W., Brit. Pat. 873,153; C.A., 56, 8750c.
241. Hieber, W., and I. Bauer, Z. Naturforsch., 16b, 556 (1961).
242. Hieber, W., and I. Bauer, Z. Anorg. Allgem. Chem., 321, 107 (1963).
243. Hieber, W., I. Bauer, and G. Neumair, Z. Anorg. Allgem. Chem., 335, 250 (1965).
244. Hieber, W., and R. Breu, Chem. Ber., 90, 1259 (1957).
245. Hieber, W., and H. Duchatsch, Chem. Ber., 98, 1744 (1965).
246. Hieber, W., and H. Duchatsch, Chem. Ber., 98, 2530 (1965).
247. Hieber, W., and H. Duchatsch, Chem. Ber., 98, 2933 (1965).
248. Hieber, W., and J. Ellermann, Chem. Ber., 96, 1643 (1963).
249. Hieber, W., and J. Ellermann, Chem. Ber., 96, 1650 (1963).
250. Hieber, W., G. Faulhaber, and F. Theubert, Z. Anorg. Allgem. Chem., 314, 125 (1962).
251. Hieber, W., and V. Frey, Z. Naturforsch., 21B, 704 (1966).
252. Hieber, W., and V. Frey, Chem. Ber., 99, 2607 (1966).
253. Hieber, W., V. Frey, and P. John, Chem. Ber., 100, 1961 (1967).
254. Hieber, W., and W. Freyer, Chem. Ber., 92, 1765 (1959).
255. Hieber, W., and K. Heinicke, Z. Naturforsch., 16b, 553 (1961).
256. Hieber, W., and K. Heinicke, Z. Anorg. Allgem. Chem., 316, 305 (1962).
257. Hieber, W., M. Hoffler, and J. Muschi, Chem. Ber., 98, 311 (1965).
258. Hieber, W., and W. Klingshirn, Z. Anorg. Allgem. Chem., 323, 292 (1963).
259. Hieber, W., and R. Kramolowsky, Z. Naturforsch., 16b, 555 (1961).
260. Hieber, W., and E. Lindner, Z. Naturforsch., 16b, 137 (1961).
261. Hieber, W., and E. Lindner, Chem. Ber., 94, 1917 (1961).
262. Hieber, W., F. Lux, and C. Herget, Z. Naturforsch., 20b, 1159 (1965).
263. Hieber, W., J. Muschi, and H. Duchatsch, Chem. Ber., 98, 3924 (1965).
264. Hieber, W., and H. Tengler, Z. Naturforsch., 16b, 68 (1961).
265. Hieber, W., and H. Tengler, Z. Anorg. Allgem. Chem., 318, 136 (1962).

266. Hieber, W., and A. Zeidler, Z. Anorg. Allgem. Chem.,
     329, 92 (1964).
267. Hobday, M. D., and T. D. Smith, J. Chem. Soc. (A),
     1970, 1085.
268. Hoff, M. C., E. K. Fields, and R. W. Watson, U.S.
     Pat. 2,866,732; C.A., 53, P7989d.
269a. Hollander, J. M., and W. L. Jolly, Acc. Chem. Res.,
     3, 193 (1970).
269b. Holmes, R. R., and R. P. Wagner, J. Am. Chem. Soc.,
     84, 357 (1962).
270. Holmes, R. R., and R. P. Carter, Inorg. Chem., 2,
     1146 (1963).
271. Holmes, R. R., and R. P. Wagner, Inorg. Chem., 2,
     384 (1963).
272. Horrocks, W. D., and R. C. Taylor, Inorg. Chem., 2,
     723 (1963).
273. Houk, L. W., and G. R. Dobson, J. Chem. Soc. (A),
     1966, 317.
274a. Houk. L. W. and G. R. Dobson, Inorg. Chem., 5, 2119
     (1966).
274b. Hulley, G., B. F. G. Johnson, and J. Lewis, J. Chem.
     Soc. (A), 1970, 1732.
275. Huttemann, T. J., B. M. Foxman, C. R. Sperati, and
     J. G. Verkade, Inorg. Chem., 4, 950 (1965).
276. Ingemanson, C. M., and R. J. Angelici, Inorg. Chem.,
     7, 2646 (1968).
277. Jackson, H. G., R. F. Clark, and C. D. Storrs, Belg.
     Pat. 633,646; C.A., 61, 593c.
278. James, T. A., and J. C. McCleverty, J. Chem. Soc.
     (A), 1970, 850.
279. Jenkins, J. M., T. J. Huttemann, and J. G. Verkade,
     Adv. Chem., 62, 604 (1967).
280a. Jenkins, J. M., and B. L. Shaw, Proc. Chem. Soc.,
     1963, 279.
280b. Jenkins, J. M., J. R. Moss, and B. L. Shaw, J. Chem.
     Soc. (A), 1969, 2796.
281. Jenkins, J. M., and J. G. Verkade, Inorg. Chem., 6,
     2250 (1967).
282. Jenkins, J. M., and J. G. Verkade, Inorganic Syn-
     thesis, 11, 108 (1969).
283. Jennings, M. A., and A. Wojcicki, Inorg. Chem., 6,
     1854 (1967).
284. Jennings, M. A., and A. Wojcicki, Inorg. Chim. Acta,
     3, 335 (1969).
285. Jensen, K. A., B. Nygaard, G. Elisson, and P. H.
     Nielsen, Acta Chem. Scand., 19, 768 (1965).
286. Johnson, T. R., and J. F. Nixon, J. Chem. Soc. (A),
     1969, 2518.
287. Kabachnik, M. I., and Yu. M. Polikarpov, Dokl. Akad.
     SSSR, 115, 753 (1957) (Engl. transl.).
288. Kabachnik, M. I., and E. N. Tsvetkov, Dokl. Akad.

Nauk SSSR, 117, 1063 (1957) (Engl. Transl.).
289.   Kabachnik, M. I., and E. N. Tsvetkov, Zh. Obshch.
       Khim., 30, 3227 (1960); C.A., 55, 21067g.
290.   Kahn, O., and M. Bigorgne, C. R. Acad. Sci., Paris,
       Ser. C, 262, 906 (1966).
291.   Kalek, P., and R. Poilblanc, J. Organometal. Chem.,
       19, 115 (1969).
292a. Kasenally, A. S., J. Lewis, A. R. Manning, J. R.
       Miller, R. S. Nyholm, and M. H. B. Stiddard, J. Chem.
       Soc., 1965, 3407.
292b. Kardeshevski, V. V., Yu. M. Zino'ev, A. N. Medvedev,
       and V. A. Ginsberg, J. Gen. Chem., USSR, 39, 436
       (1969).
293.   Keiter, R. L., and J. G. Verkade, Inorg. Chem., 8,
       2115 (1969).
294.   Keiter, R. L., and J. G. Verkade, Inorg. Chem., 9,
       404 (1970).
295.   Kemmitt, R. D. W., D. I. Nichols, and R. D. Peacock,
       J. Chem. Soc. (A), 1968, 1898.
296.   Kemmitt, R. D. W., D. I. Nichols, and R. D. Peacock,
       J. Chem. Soc. (D), 1967, 599.
297.   King, R. B., Inorg. Chem., 2, 936 (1963).
298.   King, R. B., Inorg. Chem., 2, 1275 (1963).
299.   King, R. B., Inorg. Chem., 6, 30 (1967).
300.   King, R. B., J. Am. Chem. Soc., 90, 1412 (1968).
301.   King, R. B., and C. A. Eggers, Inorg. Chem., 7, 1214
       (1968).
302.   King, R. B., and T. F. Korenowski, J. Organometal.
       Chem., 17, 95 (1969).
303.   King, R. B., and K. H. Pannell, Inorg. Chem., 7,
       1510 (1968).
304.   King, R. B., and K. H. Pannell, Inorg. Chem., 7,
       2356 (1968).
305.   King, R. W., T. J. Huttemann, and J. G. Verkade, J.
       Chem. Soc. (D), 1965, 561.
306.   Knirik, K. S., A. D. Troitskaya, and R. R. Shagidullin,
       Tr. Kazan. Khim.-Tekhnol. Inst., 36, 116 (1967); C.A.,
       70, 25315q.
307a. Knoth, W. H., and R. A. Schuun, J. Am. Chem. Soc.,
       91, 2400 (1969).
307b. Knowles, W. S., and Q. E. Johnson, Chem. Ind., 1959,
       121.
308.   Kodama, G., and R. W. Parry, J. Inorg. Nucl. Chem.,
       17, 125 (1961).
309.   Kodama, G., and R. W. Parry, Inorg. Chem., 4, 410
       (1965).
310.   Kowala, C., and J. M. Swan, Austr. J. Chem., 19, 547
       (1966).
311a. Kraihanzel, C. S., and P. K. Maples, Inorg. Chem.,
       7, 1806 (1968).
311b. Kruck, T., Angew. Chem., Int. Ed. Engl., 6, 53 (1967).

312. Kruck, T., and K. Baur, Chem. Ber., $\underline{98}$, 3070 (1965).
313. Kruck, T., and M. Höfler, Angew. Chem., Int. Ed. Engl., $\underline{6}$, 563 (1967).
314. Kruck, T., M. Höfler, K. Baur, P. Junkes, and K. Glinka, Chem. Ber., $\underline{101}$, 3827 (1968).
315. Kruse, W., and R. H. Atalla, J. Chem. Soc. (D), $\underline{1968}$, 921.
316. Lambert, R. L., and T. A. Manuel, Inorg. Chem., $\underline{5}$, 1287 (1966).
317. Lane, A. P., and D. S. Payne, J. Chem. Soc., $\underline{1963}$, 4004.
318. La Prade, M. D., and C. E. Nordman, Inorg. Chem., $\underline{8}$, 1669 (1969).
319. Laube, B. L., R. D. Bertrand, G. A. Casedy, R. D. Compton, and J. G. Verkade, Inorg. Chem., $\underline{6}$, 173 (1967).
320. Laurent, J. P., and G. Jugie, C. R. Acad. Sci., Paris, Ser. C, $\underline{264}$, 20 (1967).
321. Laurent, J. P., G. Jugie, R. Wolf, and G. Commenges, J. Chim. Phys. Physiochim. Biol., $\underline{66}$, 409 (1969).
322. Lenzi, M., and R. Poilblanc, C. R. Acad. Sci., Paris, Ser. C, $\underline{263}$, 674 (1966).
323. Levine, R., Belg. Pat. 612,602; C.A., $\underline{57}$, 16436i.
324. Levine, R., Fr. Pat. 1,321,454; C.A., $\underline{59}$, 11293h.
325. Levison, J. J., and S. D. Robinson, J. Chem. Soc., (D), $\underline{1968}$, 1405.
326. Levison, J. J., and S. D. Robinson, J. Chem. Soc. (A), $\underline{1970}$, 96.
327. Levison, J. J., and S. D. Robinson, J. Chem. Soc. (A), $\underline{1970}$, 639.
328. Levshina, G. A., and A. D. Troitskaya, Tr. Kazan. Khim.-Tekhnol. Inst., $\underline{33}$, 21 (1964); C.A., $\underline{64}$, 15355f.
329. Levshina, G. A., and A. D. Troitskaya, Tr. Kazan. Khim.-Tekhnol. Inst., $\underline{34}$, 29 (1965); C.A., $\underline{67}$, 60491d.
330. Levshina, G. A., A. D. Troitskaya, and R. A. Shagidullin, Russ. J. Inorg. Chem., $\underline{11}$, 985 (1966).
331. Lewis, J., R. J. Irving, and G. Wilkinson, J. Inorg. Nucl. Chem., $\underline{7}$, 32 (1958).
332. Lewis, J., A. R. Manning, and J. R. Miller, J. Chem. Soc. (A), $\underline{1966}$, 845.
333. Lewis, J., A. R. Manning, J. R. Miller, and J. M. Wilson, J. Chem. Soc. (A), $\underline{1966}$, 1663.
334. Lewis, J., R. S. Nyholm, A. G. Osborne, S. S. Sandhu, and M. H. B. Stiddard, Chem. Ind. (London), $\underline{1963}$, 1398.
335. Lewis, J., and G. Wilkinson, J. Chem. Soc., $\underline{1961}$, 2259.
336. Lindet, M. L., Ann. Chim. (Rome), $\underline{11}$, 177, 190 (1887).
337. Lipscomb, W. N., Boron Hydrides, W. A. Benjamin, New

York, 1963.

338.   Luttinger, L. B., U.S. Pat. 3,131,155; C.A., <u>61</u>, P2060h.

339a.   Lynden-Bell, R. M., J. F. Nixon, and R. Schmutzler, J. Chem. Soc. (A), <u>1970</u>, 565.

339b.   Mann, B. E., C. Masters, and B. L. Shaw, J. Chem. Soc. (A), <u>1970</u>, 846.

339c.   Mathieu, R., M. Lenzi, and R. Poilblanc, Inorg. Chem., <u>9</u>, 2030 (1970).

340.   McAvoy, J., K. C. Moss, and D. W. A. Sharp, J. Chem. Soc., <u>1965</u>, 1376.

341.   McCleverty, J. A., N. M. Atherton, N. G. Connelly, and C. J. Winscom, J. Chem. Soc. (A), <u>1969</u>, 2242.

342.   McFarlane, W., J. Chem. Soc. (D), <u>1968</u>, 393.

343.   Magee, T. A., C. N. Matthews, T. S. Wang, and J. H. Wotiz, J. Am. Chem. Soc., <u>83</u>, 3200 (1961).

344.   Magnelli, D. D., G. Tesi, J. N. Lowe, and W. E. McQuiston, Inorg. Chem., <u>5</u>, 457 (1966).

345.   Maier, L., Angew. Chem., <u>71</u>, 574 (1959).

346.   Malatesta, L., J. Chem. Soc., <u>1955</u>, 3924.

347.   Malatesta, L., and M. Angoletta, Atti Accad. Nazl. Lincei, Rend., Classe Sci. Fis., Mat. and Nat., <u>19</u> (1955); C.A., <u>51</u>, 929c.

348.   Malatesta, L., and M. Angoletta, J. Chem. Soc., <u>1957</u>, 1186.

349.   Malatesta, L., and A. Araneo, J. Chem. Soc., <u>1957</u>, 3803.

350.   Malatesta, L., and C. Cariello, J. Inorg. Nucl. Chem., <u>8</u>, 561 (1958).

351.   Malatesta, L., and C. Cariello, J. Chem. Soc., <u>1958</u>, 2323.

352.   Malatesta, L., M. Freni, and V. Valenti, U.S. Dept. Commerce Office Tech. Serv., PB Rept., <u>155</u>, 141 (1960); C.A., <u>58</u>, 4140d.

353.   Malatesta, L., and A. Sacco, Ann. Chim. (Rome), <u>44</u>, 134 (1954); C.A., <u>48</u>, 13516c.

354.   Mannan, Kh. A. I. F. M., Acta Crystallogr., <u>23</u>, 649 (1967).

355.   Manning, A. R., J. Chem. Soc. (A), <u>1967</u>, 1984.

356.   Manning, A. R., J. Chem. Soc. (A), <u>1968</u>, 651.

357.   Manning, A. R., J. Chem. Soc. (A), <u>1968</u>, 1135.

358.   Manuel, T. A., Adv. Organometal. Chem., <u>3</u>, 181 (1965).

359.   Maples, P. K., and C. S. Kraihanzel, J. Am. Chem. Soc., <u>90</u>, 6645 (1968).

360.   Mark, V., C. H. Dungan, M. M. Crutchfield, and J. R. Wazer, Top. Phosphorus Chem., <u>5</u>, 227 (1967).

361.   Mather, G. G., and A. Pidcock, J. Chem. Soc. (A), <u>1970</u>, 1226.

362.   Mathieu, R., M. Lenzi, and R. Poilblanc, C. R. Acad. Sci., Paris, Ser. C, <u>266</u>, 806 (1968).

363.   Mathieu, R., and R. Poilblanc, C. R. Acad. Sci.,

Paris, Ser. C, 264, 1053 (1967).
364. Mathieu, R., and R. Poilblanc, C. R. Acad. Sci.,
     Paris, Ser. C, 265, 388 (1967).
365. Matthews, C. N., U.S. Pat. 3,177,983; C.A., 60,
     6870f.
366. Matthews, C. N., Brit. Pat. 946,674; C.A., 61,
     3148h.
367. Matthews, C. N., T. A. Magee, and J. H. Wotiz, J.
     Am. Chem. Soc., 81, 2273 (1959).
368. Mawby, R. J., F. Basolo, and R. G. Pearson, J. Am.
     Chem. Soc., 86, 3994 (1964).
369. Mawby, R. J., and C. White, J. Chem. Soc. (D), 1968,
     312.
370. Mays, M. J., and S. M. Pearson, J. Chem. Soc. (A),
     1968, 2291.
371. Mays, M. J., and S. M. Pearson, J. Chem. Soc. (A),
     1969, 136.
372. Meier, M., F. Basolo, and R. G. Pearson, Inorg.
     Chem., 8, 795 (1969).
373. Meriwether, L. S., and M. L. Fiene, J. Am. Chem.
     Soc., 81, 4200 (1959).
374. Meriwether, L. S., and J. R. Leto, J. Am. Chem.
     Soc., 83, 3192 (1961).
375. Miseno, A., Y. Uchida, M. Hidai, and I. Inomata, J.
     Chem. Soc. (D), 1968, 704.
376. Montelatici, S., A. Van der Ent, J. A. Osborn, and
     G. Wilkinson, J. Chem. Soc. (A), 1968, 1054.
377. Mooney, E. F., and B. S. Thornhill, J. Inorg. Nucl.
     Chem., 28, 2225 (1966).
378. Mueller, H., D. Wittenberg, and E. Scharf, Ger. Pat.
     1,204,669; C.A., 64, 3380c.
379. Muller, J., and K. Fenderd, J. Organometal. Chem.,
     19, 123 (1969).
380. Mullineaux, R. D., U.S. Pat. 3,110,747; C.A., 60,
     7504a.
381. Mullineaux, R. D., Ger. Pat. 1,220,843; C.A., 65,
     19749a.
382a. Mullineaux, R. D., U.S. Pat. 3,290,348; C.A., 66,
     65080h.
382b. Murray, R. W., and M. L. Kaplan, J. Am. Chem. Soc.,
     90, 537. 4161 (1968); 91, 5358 (1969).
383. Myers, V. G., F. Basolo, and K. Nakamoto, Inorg.
     Chem., 8, 1204 (1969).
384. Naumova, G. P., and A. D. Troitskaya, Tr. Kazan.
     Khim.-Tekhnol. Inst., 34, 33 (1965); C.A., 67,
     68161s.
385. Naumova, G. P., and A. D. Troitskaya, Tr. Kazan.
     Khim.-Tekhnol. Inst., 36, 101 (1967); C.A., 70,
     8528u.
386. Nesmeyanov, A. N., K. N. Anisimov, and N. E. Kolobova,
     Izvest. Akad. Nauk SSSR, Otdel. Khom, Nauk, 1962,

669 (Engl. transl.).

387a. Nesmeyanov, A. N., Yu. A. Chapovskii, B. V. Lokshin, I. V. Polovyonyak, and L. G. Makarova, Dokl. Akad. Nauk Nauk SSSR, 166, 213 (1966)(Engl. transl.).

387b. Nesmeyanov, A. N., and Yu. A. Chapovskii, Izvest. Akad. Nauk, SSSR, Ser. Khim., 1967, 223; C.A., 66, 95,162d.

388. Nesmeyanov, A. N., Yu. A. Chapovskii, and Yu. A. Ustynyak, Izvest. Akad. Nauk SSSR, Ser. Khim., 1966, 1814 (Engl. transl.).

389. Nesmeyanov, A. N., Yu. A. Chapovskii, and Yu. A. Ustynyak, Izvest. Akad. Nauk SSSR, Ser. Khim., 1966, 1815 (Engl. transl.).

390a. Nesmeyanov, A. N., Y. A. Chapovsky, and Y. A. Ustynyuk, J. Organometal. Chem., 9, 345 (1967).

390b. Nesmeyanov, A. N., L. G. Makarova, and I. V. Polovyanyuk, J. Organometal. Chem., 22, 707 (1970).

391. Nichols, D. I., J. Chem. Soc. (A), 1970, 1216.

392. Nishizawa, Y., Bull. Chem. Soc. Japan, 34, 1170 (1961).

393. Nishizawa, Y., Bull. Chem. Soc. Japan, 34, 1205 (1961).

394. Nishizawa, Y., Bull. Chem. Soc. Japan, 34, 1721 (1962).

395. Nishizawa, Y., and S. Nakagawa, Japan Pat. 4880; C.A., 59, 579h.

396. Nishizawa, Y., M. Nakagawa, and Y. Suzuki, Japan Pat. 21,321; C.A., 58, 4425e.

397a. Nixon, J. F., J. Chem. Soc. (A), 1967, 1136.

397b. Nixon, J. F., M. Murray, and R. Schmutzler, Z. Naturforsch., 25B, 110 (1970).

397c. Nixon, J. F., and A. Pidcock, in Annual Review of NMR Spectroscopy, Vol. 2, E. F. Mooney, Ed., Academic, New York, 1969, p. 345.

398. Nixon, J. F., and M. D. Sexton, Inorg. Nucl. Chem. Lett., 4, 275 (1968).

399. Nixon, J. F., and M. D. Sexton, J. Chem. Soc. (A), 1969, 827.

400. Nixon, J. F., and M. D. Sexton, J. Chem. Soc. (A), 1970, 321.

401. Nixon, J. F., and J. R. Swain, J. Organometal. Chem., 21, P13 (1970).

402a. Nolte, M. J., G. Gafner, and L. M. Haines, J. Chem. Soc. (D), 1969, 1406.

402b. Nordman, C. E., Acta Cryst., 13, 535 (1960).

403. Nöth, H., and H. Vetter, Chem. Ber., 96, 1298 (1963).

404. Nöth, H., and H. Vetter, Chem. Ber., 96, 1479 (1963).

405. Ogilvie, F. B., J. M. Jenkins, and J. G. Verkade, J. Am. Chem. Soc., 92, 1916 (1970).

406. Ogilvie, F. B., R. L. Keiter, G. Wulfsberg, and J. G. Verkade, Inorg. Chem., 8, 2346 (1969).

407. Olechowski, J. R., C. G. McAlister, and R. F.
     Clark, Inorg. Chem., 4, 246 (1965).
408a. Olechowski, J. R., C. G. McAlister, and R. F. Clark,
     Inorg. Synthesis, 9, 181 (1967).
408b. Oliver, A. J., and W. A. G. Graham, Inorg. Chem.,
     9, 2578 (1970).
409a. Orio, A. A., B. B. Chastain, and H. B. Gray, Inorg.
     Chim. Acta, 3, 8 (1969).
409b. Orio, A. A., U. Mazzi, and H. B. Gray, Proc. of the
     XIIIth I.C.C.C., Poland, Vol. 1, 342 (1970); results
     not recorded in the abstract were disclosed at the
     meeting.
410. Osborne, A. G., and M. H. B. Stiddard, J. Chem. Soc.,
     1964, 634.
411. Osborne, A. G., and M. H. B. Stiddard, J. Organometal.
     Chem., 3, 340 (1965).
412. Parker, D. J., J. Chem. Soc. (A), 1969, 246.
413. Parker, D. J., and M. H. B. Stiddard, J. Chem. Soc.
     (A), 1966, 695.
414. Parshall, G. W., in The Chemistry of Boron and its
     Compounds, E. L. Muetterties, Ed., Wiley, New York,
     1967, p. 617.
415a. Parshall, G. W., W. H. Knoth, and R. A. Schunn, J.
     Am. Chem. Soc., 91, 4990 (1969).
415b. Parshall, G. W., Acc. Chem. Res., 3, 139 (1970).
416. Payne, D. S., J. A. A. Mokuolu, and J. C. Speakman,
     J. Chem. Soc. (D), 1965, 599.
417. Payne, D. S., and A. P. Walker, J. Chem. Soc. (C),
     1966, 498.
418. Piacenti, F., M. Bianchi, E. Benedetti, and G.
     Sbrana, J. Inorg. Nucl. Chem., 29, 1389 (1967).
419. Pidcock, A., J. Chem. Soc. (D), 1968, 92.
420. Pidcock, A., R. E. Richards, and L. M. Venanzi, Proc.
     Chem. Soc., 1962, 184.
421. Pidcock, A., R. E. Richards, and L. M. Venanzi, J.
     Chem. Soc., (A), 1966, 1707.
422. Pidcock, A., J. D. Smith, and B. W. Taylor, J. Chem.
     Soc. (A), 1969, 1604.
423. Pidcock, A., J. D. Smith, and B. W. Taylor, Inorg.
     Nucl. Chem. Lett., 4, 467 (1968).
424. Pidcock, A., J. D. Smith, and B. W. Taylor, Inorg.
     Chem., 9, 638 (1970).
425. Pidcock, A., and B. W. Taylor, J. Chem. Soc. (A),
     1967, 877.
426a. Pidcock, A., and C. A. Waterhouse, Inorg. Nucl.
     Chem. Lett., 3, 487 (1967).
426b. Pidcock, A., and C. R. Waterhouse, J. Chem. Soc.
     (A), 1970, 2080.
427. Plastas, H. J., J. M. Stewart, and S. O. Grim, J.
     Am. Chem. Soc., 91, 4326 (1969).
428. Poilblanc, R., and M. Bigorgne, C. R. Acad. Sci.,

Paris, Ser. C, 250, 1064 (1960).
429. Poilblanc, R., and M. Bigorgne, C. R. Acad. Sci.,
      Paris, Ser. C, 252, 3054 (1961).
430. Poilblanc, R., and M. Bigorgne, Bull. Soc. Chim.
      France, 1962, 1301.
431. Polak, R. J., and T. L. Heying, J. Org. Chem., 27,
      1483 (1962).
432. Pollick, P. J., and A. Wojcicki, J. Organometal.
      Chem., 14, 469 (1968).
433. Pomey, E., C. R. Acad. Sci., Paris, Ser. C, 92, 794
      (1881).
434a. Pomey, E., C. R. Acad. Sci., Paris, Ser. C, 104,
      364 (1887).
434b. Powers, D. R., G. Faber, and G. R. Dobson, J. Inorg.
      Nucl. Chem., 31, 2970 (1969).
435. Quin, L. D., J. Am. Chem. Soc., 79, 3681 (1957).
436. Reckziegel, A., and M. Bigorgne, C. R. Acad. Sci.,
      Paris, Ser. C, 258, 4065 (1964).
437. Reckziegel, A., and M. Bigorgne, J. Organometal.
      Chem., 3, 341 (1965).
438. Reddy, G. S., and R. Schmutzler, Inorg. Chem., 6,
      823 (1967).
439. Reed, H. B., U.S. Pat. 2,686,209; C.A., 49, 15957b.
440. Reed, W. B., J. Chem. Soc., 1954, 1931.
441. Reetz, T., J. Am. Chem. Soc., 82, 5039 (1960).
442. Reetz, T., U.S. Pat. 3,104,253; C.A., 60, P557gh.
443. Reetz, T., U.S. Pat. 3,104,254; C.A., 60, P2768gh.
444. Reetz, T., U.S. Pat. 3,100,744; C.A., 60, 2756e.
445. Reetz, T., and W. D. Dixon, U.S. Pat. 3,119,853;
      C.A., 60, 11898b.
446. Reetz, T., and B. Katlafsky, J. Am. Chem. Soc., 82,
      5036 (1960).
447. Rick, E. A., and R. L. Pruett, J. Chem. Soc. (D),
      1966, 697.
448. Riedel, E. R., and R. A. Jacobson, U.S. A.E.C. report,
      IS-T-153.
449. Riess, J., and J. R. Van Wazer, Bull. Soc. Chim,
      France, 1966, 1846.
450. Riess, J. G., and J. R. Van Wazer, J. Organometal.
      Chem., 8, 347 (1967).
451. Riess, J. G., and J. R. Van Wazer, Bull. Soc. Chim.
      France, 1968, 3087.
452a. Robinson, S. D., J. Chem. Soc. (D), 1968, 521.
452b. Rodgers, J., D. W. White, and J. G. Verkade, J.
      Chem. Soc. (A), in press.
453. Rosenheim, A., and W. Levy, Z. Anorg. Allgem. Chem.,
      37, 394 (1903).
454. Rosenheim, A., and W. Levy, Z. Anorg. Allgem. Chem.,
      43, 34 (1905).
455. Ross, E. P., and G. R. Dobson, J. Chem. Soc. (D),
      1969, 1229.

456.   Rossi, M., and A. Sacco, J. Chem. Soc. (D), 1969,
       471.
457.   Rothenburg, R. A., and H. A. Willis, Spectrochim.
       Acta, Part B, 20, 759 (1964).
458.   Rudolph, R. W., R. C. Taylor, and R. W. Parry, J.
       Am. Chem. Soc., 88, 3729 (1966).
459.   Sacco, A., Ann Chim. (Rome), 43, 495 (1953); C.A.,
       48, 5012d.
460.   Sander, M., Angew. Chem., 73, 67 (1961).
461a. Sarkisyan, S. N., A. D. Troitskaya, and R. R. Shagi-
       dullin, Tr. Kazan. Khim.-Tekhnol. Inst., 34, 42
       (1965); C.A., 67, 87339x.
461b. Sartorelli, U., F. Canziani, S. Martinengo, and P.
       Chini, XIIIth I.C.C.C., Poland, Vol. 1, p. 144 (1970);
       results not recorded in the abstract were disclosed
       at the meeting.
462.   Schmutzler, R., Chem. Ber., 9, 2435 (1963).
463.   Schmutzler, R., Inorg. Chem., 3, 415 (1964).
464.   Schmutzler. R., J. Chem. Soc., 1965, 5630.
465.   Schmutzler, R., U.S. Pat. 3,242,171; C.A., 64,
       19683b.
466.   Schrauzer, G. N., V. P. Magweg, H. W. Finck, and W.
       Heinrich, J. Am. Chem. Soc., 88, 4604 (1966).
467.   Schrauzer, G. N., and R. J. Windgassen, Chem. Ber.,
       99, 602 (1966).
468a. Schrauzer, G. N., and R. J. Windgassen, J. Am. Chem.
       Soc., 88, 3738 (1966).
468b. Schroeder, H., J. R. Reiner, and T. Heying, Inorg.
       Chem., 1, 618 (1962).
469.   Schroll, G. E., U.S. Pat. 3,130,215; C.A., 61, P4396h.
470a. Schunn, R. A., Inorg. Chem., 9, 394 (1970).
470b. Schunn, R. A., Inorg. Chem., 9, 2567 (1970).
471.   Schuster-Woldan, H. G., and F. Basolo, J. Am. Chem.
       Soc., 88, 1657 (1966).
472.   Schutzenberger, P., Bull. Soc. Chim. France, 14, 177
       (1870).
473.   Schutzenberger, P., C. R. Acad. Sci., Paris, Ser. C,
       70, 1414 (1870).
474.   Schutzenberger, P., and M. Fontaine, Bull. Soc. Chim.
       France, 17, 386 (1872).
475.   Schutzenberger, P., and M. Fontaine, Bull. Soc. Chim.
       France, 17, 482 (1872).
476.   Schutzenberger, P., and M. Fontaine, Bull. Soc. Chim.
       France, 18, 101 (1872).
477.   Schutzenberger, P., and M. Fontaine, Bull. Soc. Chim.
       France, 18, 148 (1872).
478.   Schweckendiek, W. J., Ger. Pat. 841,589; C.A., 52,
       9192i.
479.   Sekul, A. A., and H. G. Sellers, Jr., U.S. Pat.
       2,964,575; C.A., 55, 14333i.
480a. Shechter, H., U.S. Pat. 3,187,062; C.A., 63, P6886c.

480b.   Seidel, W. C., and C. A. Tolman, Inorg. Chem., 9, 2354 (1970).

481.   Shell International Research Maatschappij, N. V., Neth. Appl. 6,506,276; C.A., 64, 11104e.

482.   Shrader, D., E. P. Ross, R. T. Jernigan, and G. R. Dobson, Inorg. Chem., 9, 1286 (1970).

483.   Shupack, S. I., and B. Wagner, J. Chem. Soc. (D), 1966, 547.

484.   Slaugh, L. H., and R. S. Mullineaux, U.S. Pat. 3,448,158; C.A., 71, 112406j.

485.   Smith, T. D., J. Inorg. Nucl. Chem., 15, 95 (1960).

486.   Spangenburg, S. F., and H. H. Sisler, Inorg. Chem., 8, 1004 (1969).

487.   Stalick, J. K., and J. A. Ibers, Inorg. Chem., 8, 1084 (1969).

488.   Stanclift, W. E., and D. G. Hendricker, Inorg. Chem., 7, 1242 (1968).

489.   Stanko, V. I., A. I. Klimova, and L. I. Zakharkin, Izvest. Akad. Nauk SSSR, Otdel. Khim. Nauk, 1962, 856 (Engl. transl.).

490.   Stelling, O., Z. Phys. Chem., 117, 161, 194 (1925).

491a.   Stephenson, T. A., J. Chem. Soc. (A), 1970, 889.

491b.   Stewart, R. P., and P. M. Treichel, J. Am. Chem. Soc., 92, 2710 (1970).

492.   Streuli, C. A., "Titrimetric Methods," Proc. Symp., Cornwall, Ont., Canada, 1961, 97; C.A., 57, 6866b.

493a.   Strohmeier, W. S., and R. Fleischmann, Z. Naturforsch., 24B, 1217 (1969).

493b.   Strohmeier, W., and F. Müller, Chem. Ber., 100, 2812 (1967).

494a.   Strohmeier, W., and F. Müller, Z. Naturforsch., 22B, 451 (1967).

494b.   Strohmeier, W., and F. J. Müller, Z. Naturforsch., 24B, 931 (1969).

495a.   Strohmeier, W., and T. Onoda, Z. Naturforsch., 23B, 1377 (1968).

495b.   Strohmeier, W., and T. Onoda, Z. Naturforsch., 24B, 1185 (1969).

496.   Szabo, P. L., G. Bor, L. Fekete, Z. Nagy-Magos, and L. Marko, J. Organometal. Chem., 12, 245 (1968).

497a.   Tate, D. P., J. M. Augl, and A. Buss, Inorg. Chem., 2, 427 (1963).

497b.   Tebbe, F. N., P. Meakin, J. P. Jesson, and E. L. Muetterties, J. Am. Chem. Soc., 92, 1068 (1970).

498a.   Ter Haar G., Sr., M. A. Fleming, and R. W. Parry, J. Am. Chem. Soc., 84, 1767 (1962).

498b.   Thompson, Q. E., J. Am. Chem. Soc., 83, 845 (1961).

499.   Thorsteinson, E. M., and F. Basolo, Inorg. Chem., 5, 1691 (1966).

500a.   Thorsteinson, E. M., and F. Basolo, J. Am. Chem.

Soc., 88, 3929 (1966).

500b.   Tolman, C. A., J. Am. Chem. Soc., 92, 2953 (1970).

500c.   Tolman, C. A., J. Am. Chem. Soc., 92, 2956 (1970).

500d.   Tolman, C. A., J. Am. Chem. Soc., 92, 4217 (1970).

501.   Treboganov, A. D., B. I. Mitsner, and E. P. Zinkevich, J. Org. Chem. USSR, 1, 1604 (1965).

502.   Treichel, P. M. and J. J. Benedict, J. Organometal. Chem., 17, P37 (1969).

503.   Troitskaya, A. D., Tr. Kazan. Khim.-Tekhnol. Inst., 23, 228 (1957); C.A., 52, 99516.

504.   Troitskaya, A. D., Tr. Kazan. Khim.-Tekhnol. Inst., 33, 16 (1964); C.A., 64, 13458g.

505.   Troitskaya, A. D., Russ. J. Inorg. Chem., 5, 585 (1961).

506.   Troitskaya, A. D., and T. B. Itskovich, Tr. Kazan. Khim.-Tekhnol. Inst., 18, 59 (1953).

507.   Troitskaya, A. D., and T. B. Itskovich, Tr. Kazan. Khim.-Tekhnol. Inst., 19-20, 79 (1954-55); C.A., 51, 11149a.

508.   Tsuka, S. O., and M. Rossi, J. Chem. Soc. (A), 1969, 497.

509.   Turbak, A. F., U.S. Pat. 3,072,703; C.A., 58, 10080h.

510.   Ugo, R., and F. Bonati, J. Organometal. Chem., 8, 189 (1967).

511.   Ugro, J. V., U.S. A.E.C., IS-T-168; Ugro, J. G., and Jacobson, R. A., submitted.

512.   Vallarino, L., Rend. Inst. Lombardo Sci., Pt. I Classe Sci., Mat. and Nat., 91, 391 (1957); C.A., 52, 18064h.

513.   Vallarino, L., J. Chem. Soc., 1957, 2287.

514.   Vallarino, L., J. Chem. Soc., 1957, 2473.

515.   Vallarino, L., J. Inorg. Nucl. Chem., 8, 288 (1958).

516.   Vandenbroucke, A. C., D. G. Hendricker, R. E. McCarley, and J. G. Verkade, Inorg. Chem. 7, 1825 (1968).

517.   Van Hecke, G. R., and W. de W. Horrocks Jr., Inorg. Chem., 5, 1960 (1966).

518.   Van Peski, A. J., and J. A. Van Melsen, U.S. Pat. 2,150,349; C.A., 33, 4779[7].

519.   Venezky, D. L., and R. B. Fox, J. Amer. Chem. Soc., 78, 1664 (1956).

520a.   Verkade, J. G., and C. W. Heitsch, Inorg. Chem., 2, 512 (1963).

520b.   Verkade, J. G., D. A. Allison, G. K. McEwen, T. Pecoraro, M. F. Traynor, and D. W. White, to be published.

521.   Verkade, J. G., R. W. King, and C. W. Heitsch, Inorg. Chem., 3, 884 (1964).

522.   Verkade, J. G., R. E. McCarley, D. G. Hendricker, and R. W. King, Inorg. Chem., 4, 228 (1965).

523.   Verkade, J. G., and T. S. Piper, Proc. Int. Conf.

Coord. Chem., 6th, Detroit, 1961, 634.

524. Verkade, J. G., and T. S. Piper, Inorg. Chem., 1, 453 (1962).

525. Verkade, J. G., and T. S. Piper, Inorg. Chem., 2, 944 (1963).

526. Vinal, R. S., and L. T. Reynolds, Inorg. Chem., 3, 1062 (1964).

527. Vol'pin, M. E., and I. S. Kolomnikov, Dokl. Akad. Nauk SSSR, 170, 997 (1966) (Engl. transl.).

528a. Vol'pin, M. E., and I. S. Kolomnikov, Izvest. Akad. Nauk SSSR, Ser. Khim., 1966, 1980 (Engl. transl.).

528b. Von Kutepow, N., H. Seibt, and F. Meier, Ger. Pat. 1,144,286.

528c. Vrieze, K., C. MacLean, P. Cossee, and C. W. Hilbers, Rec. Trav. Chim. Pays-Bas, 85, 1077 (1966).

529a. Walter, D., and G. Wilke, Angew. Chem., Int. Ed. Engl., 5, 897 (1966).

529b. Wasserman, E., R. W. Murray, M. L. Kaplan, and W. A. Yager, J. Am. Chem. Soc., 90, 4160 (1968).

530. Wawerski, H., and F. Basolo, J. Chem. Soc. (D), 1966, 366.

531a. Weiss, E., and W. Hubel, J. Inorg. Nucl. Chem., 11, 42 (1959).

531b. White, D. W., G. K. McEwen, and J. G. Verkade, Tetrahedron Lett., 1968, 5369.

532. Wilke, G., and G. Herrmann, Angew. Chem., Int. Ed. Engl., 5, 581 (1966).

533. Wilke, G., and P. Heimbach, Angew. Chem., Int. Ed. Engl., 6, 92 (1967).

534. Wilkinson, G., Z. Naturforsch., 9B, 446 (1954).

535. Yamashita, I., and M. Serizawa, Bull. Chem. Soc. Japan, 37, 1721 (1964).

536. Yamazaki, H., T. Nishido, Y. Matsumoto, S. Sumida, and N. Hagihara, J. Organometal. Chem., 6, 86 (1966).

537. Zgadzai, E. A., G. P. Naumova, and A. D. Troitskaya, Tr. Kazan. Khim.-Tekhnol. Inst., 36, 96 (1967); C.A., 69, 100661j.

538. Zingales, F., F. Canziani, and F. Basolo, J. Organometal. Chem., 7, 461 (1967).

539. Zingales, F., F. Canziani, and U. Sartorelli, Rend. Ist. Lombardo Sci. Lettere, A, 96, 771 (1967).

540. Zingales, F., A. Chiesa, and F. Basolo, J. Am. Chem. Soc., 88, 2707 (1966).

Chapter 4. Quaternary Phosphonium Compounds

PETER BECK

Org.-chem.Institut der Universität Mainz,
Germany

The substances discussed in this chapter are represented by the general formula R_4PX, where R may be a radical linked to phosphorus by carbon or another atom; X may be any compound, simple or complex, inorganic or organic, serving as an anion. Also included are the substances which bear the negative charge in one of the ligands. They are referred to as betaines.

All the substances which appear in the literature under the heading "phosphonium" have been taken into the list of compounds. No critical selection has been done, whether they belong to the tetra- or pentacovalent state of phosphorus. For the compounds with less than four carbon bonds to phosphorus it can often be assumed that there is an equilibrium between both states. Certain quaternary phosphonium salts are soluble in ethers. This may indicate that a pentacovalent state is valid even here.

A. SYNTHETIC ROUTES

I. PHOSPHINES AND ORGANIC HALIDES AND SIMILAR REACTANTS

The most widely used process leading to phosphonium compounds is the reaction of a tervalent phosphorus with a substance RX. R may be an H, an alkyl, or acyl radical. X can be a halogen or an acyl radical or an alkoxy substituent. In other words, esters, ethers, or anhydrides can react.

$$R_3P + RX \longrightarrow R_4\overset{+}{P} X^-$$

In the same manner halides of metaloids and metals are transposed. In this case the result is a quasi phosphonium compound (see also Chapter 2).

$$R_3P + NH_2X \longrightarrow R_3\overset{+}{P}NH_2 X^-$$

$$R_3P + R_2AsX \longrightarrow R_3\overset{+}{P}AsR_2 \; X^-$$

Quasi-phosphonium salts result, too, when the substituents at the phosphorus are alkoxy, aryloxy, or amine radicals.

$$(ArO)_3P + RX \longrightarrow (ArO)_3\overset{+}{P}R \; X^-$$

$$(R_2N)_3P + RX \longrightarrow (R_2N)_3\overset{+}{P}R \; X^-$$

Of course there can be different kinds of radicals attached to phosphorus.

If phosphine itself is used with organic halides phosphonium hydro halides form. These exist in an equilibrium as follows:

$$R\overset{+}{P}H_3 \; X^- \rightleftharpoons RPH_2 + HX$$

In the course of the reaction the whole spectrum of possible phosphonium salts arises. When the halide reactant is in excess the quaternary phosphonium salt is the final product. There is a marked difference in reactivity of the halides. Iodides are more reactive than bromides, and these in turn more than chlorides[440a] (see also Chapter 1, Section D). Fluorides are used for quaternization in special cases only. In general aryl halides do not react with phosphines. Special methods have to be used.

In simple cases the reaction is conducted by mixing of the reactants. The exclusion of oxygen is imperative, because of the autoxidability of the phosphines. With the lower alkyl halides the reaction might be quite violent. The biggest difference in reactivity is between methyl and ethyl halides. It is advisable to dilute with an inert solvent, such as ethers, petroleum ether or benzene. In this way the insoluble reaction product often is precipitated in an analytically pure state, while the unreacted substances remain in solution.

With the higher alkyl halides the reaction becomes sluggish and heating is necessary. Solvents of the polar type (formic acid, acetonitrile, nitromethane, dimethylformamide, dimethylsulfoxide), stimulate the reaction markedly. The quaternization may be conducted in a reflux apparatus or in a pressure bottle like the "German beer bottles." If there is more than one site of reaction in the molecule bis- to oligophosphonium-salts can be obtained.

Tertiary phosphines are more reactive than the analogous amines or arsines. In a concurrent reaction the phosphine is quaternized first. The electronic effects of a substituent in an aryl phosphine are lower than in a corresponding amine, because of the reduced mesomerism due to the larger bulk of the phosphorus atom. Nevertheless a p-ethoxy-substituted phenylphosphine reacts much faster with ethyl-iodide than a p-nitro-substituted one. Secondary halides in allyl position are subject to rearrangements during the quaternization reaction.[516]

$$\text{(cyclohexenyl)CH}_2\text{-Br} + Ph_3P \longrightarrow \text{(cyclohexenyl)CH-}\overset{+}{P}Ph_3 \qquad Br^-$$

In many cases alkyl halides with additional functional groups such as alcohols, aldehydes or ketones, acetals, esters, and acids react smoothly to the corresponding phosphonium salts. Nitro groups lead to side reactions because of their oxidizing properties. Phosphonium salts with phosphorus in the ring are accessible in two ways:

1. by alkylating cyclic phosphines
2. by a cyclizing alkylation reaction.[663]

$$\text{(aryl)}(PEt_2)(CH_2CH_2OMe) \xrightarrow{HBr} \text{(aryl)} \overset{Et_2}{P} (CH_2)(CH_2)$$

Tertiary phosphines and quaternary phosphonium salts can be synthesized starting with elementary phosphorus and alkyl halides, Grignard reagents, or alcohols. In other cases sulfur or halogen containing phosphorus derivatives are reduced with metals or Grignard reagents. But in every case it is assumed that the final step is an alkylating reaction of a tertiary phosphine generated in situ. Thus these reactions, as useful as they may be, are essentially the same method with respect to the arising phosphonium salt as already mentioned. Diarylthiophosphinites and alkyl halides under the conditions of the Arbusov reaction not only yield the tertiary phosphine sulfide but also a certain amount of diaryldialkylphosphonium salt. Tertiary phosphine sulfides and alkyl halides form quasi-phosphonium compounds with the alkyl group at the sulfur atom.

$$R_3P=S + R'X \longrightarrow R_3\overset{+}{P}-SR' \quad X^-$$

Heating halogenophosphoranes with alkyl halides, phosphonium salts, or quasi-phosphonium salts result, depending on the degree of substitution, with a perhalide anion. By reducing the perhalide anion with a suitable reagent the ordinary phosphonium salts result when the initial compound was a trisubstituted halogenophosphorane.

$$R_3PX_2 + MeJ \longrightarrow R_3\overset{+}{P}Me \quad JX_2^-$$

$$\overset{+}{R_3PMe}\ JX_2{}^- \xrightarrow{\text{reduction}} \overset{+}{R_3PMe}\ X^-$$

Similar compounds result from the Kinnear and Perren reaction, where phosphorus trichloride or primary halophosphines react with aluminum chloride and alkyl halides.

$$PX_3 + RX + AlX_3 \longrightarrow \overset{+}{RPX_3}\ AlX_4{}^-$$

The complex may be converted to a phosphonyl halide or a halophosphine by hydrolysis or reduction, respectively.

In di- or oligophosphines alkyl halides are capable of splitting the P-P bond. In the case of diphosphines the primary products are a tertiary phosphine and a halogenophosphine (see also Chapter 2).

$$R_2P-PR_2 + MeJ \longrightarrow R_2PMe + R_2PJ$$

In the second step the primary products are quaternized as follows:

$$R_2PMe + MeJ \longrightarrow \overset{+}{R_2PMe_2}\ J^-$$

$$R_2PJ + MeJ \longrightarrow R_2PMe\ J_2 \xrightarrow{MeJ} \overset{+}{R_2PMe_2}\ J_3{}^- \xrightarrow{red.} \overset{+}{R_2PMe_2}\ J^-$$

Starting with a symmetric diphosphine only one phosphonium salt results after reduction. With an unsymmetric diphosphine two different quaternary salts are obtained.

Very hygroscopic substances with all the properties of phosphonium salts are formed by reacting triethylphosphine with acyl halides. With 3,5-dinitrobenzoyl chloride the most stable adduct is obtained. The reaction of tertiary phosphines with carbonester chlorides is well known.

$$R_3P + ClCO_2R \longrightarrow \overset{+}{R_3PCO_2R}\ \ Cl^-$$

The same products are available from acylphosphides and alkyl halides.

$$R_2\overset{O}{\overset{\|}{P}}{-}CR + R'X \longrightarrow \overset{+}{R_2}\overset{O}{\overset{\|}{P}}{-}\underset{R'}{CR}\ \ X^-$$

These substances are of some interest because the former acyl radical is split off to the corresponding aldehyde by alkaline hydrolysis.

$$\overset{+}{R_3}\overset{O}{\overset{\|}{P}}CR'\ \ X^- \xrightarrow{NaOH} R_3PO + NaX + R'CHO$$

Pentaphenylphosphorane is transformed to the tetraphenylphosphonium cation by Lewis acids, hydrohalides, or by irradiation in a haloalkane.

$$Ph_5P + Ph_3B \longrightarrow Ph_4\overset{+}{P} \ Ph_4B^-$$

$$Ph_5P + HJ \longrightarrow Ph_4\overset{+}{P} \ J^- + PhH$$

$$Ph_5P \xrightarrow[RX]{h\nu} Ph_4\overset{+}{P} \ X^-$$

II. PHOPSHINES, ALCOHOL AND HYDROHALIDE

The same phosphonium salts as described in Section I are obtained with phosphines (primary, secondary, or tertiary), hydrohalides, and alcohols. It is also possible to conduct the reaction with phosphonium hydrohalides and an appropriate alcohol. In several cases the reaction may proceed more quickly and smoothly and the yields may be better than with the corresponding alkyl halides. The water arising during the reaction can be removed by azeotropic distillation. In special cases esters and ethers may be used instead of alcohols.

III. PHOSPHINES AND OXIRANES

Oxiranes react with phosphines to betaines or strong basic phosphonium hydroxides in the absence of acids in water-free mediums or in water. In the presence of acids or by later addition of them hydroxyalkyl phosphonium salts result.

$$R_3P + \ \overset{\triangle}{\underset{O}{\bigtriangledown}} \ \dashrightarrow R_3\overset{+}{P}\text{-}CH_2CH_2O^- \xrightarrow{HCl} R_3\overset{+}{P}\text{-}CH_2CH_2OH \ Cl^-$$

IV. COMPLEX-SALT METHOD

Tertiary phosphines, aryl halides, and Friedel-Crafts catalysts or salts or the transition metals form phosphoniur salts in good yields.

A mixture of triphenylphosphine, bromobenzene in moderate excess, and water-free aluminum chloride is refluxed above 200°C for several hours. Too high a temperature leads to difficulties in purifying the end product. Of course substances which suffer degradation by aluminum chloride are to be excluded. The reported replacement of the phenyl group by the anisyl group seems to be in error. Also the assertion of the autoxidability of the triphenylphosphine-aluminum chloride complex is erroneous.

$$Ph_3P + BrPh \xrightarrow{AlCl_3} Ph_4\overset{+}{P} \ Cl^-$$

Cu^{I}Cl, NiCl$_2$ or NiBr$_2$, CoCl$_2$ or CoBr$_2$ and ZnCl$_2$ are more suitable than AlCl$_3$. Tertiary phosphines with aliphatic and/or aromatic ligands can be quaternized with aryl halides. The reaction is best conducted by mixing the reactants, preferably with a solvent, e.g. acetonitrile, and by heating the mixture to at least 180°C for several hours in a sealed tube. The metal salt serving as catalyst has to be water-free. When the reaction is performed in a distilling apparatus with benzonitrile as a solvent; metal salts containing water of crystallization can be used. The water is distilled off azeotropically with an excess of benzonitrile at the beginning of the reaction. There are three known steps of the reaction.

$$2 \ Ph_3P\text{-}NiBr_2 \xrightarrow{\ 80° \ } (Ph_3P)_2NiBr_2$$

$$(Ph_3P)_2NiBr_2 + PhBr \xrightarrow{\ 115° \ } Ph_4P^+ \ Ph_3PNiBr_3^-$$

$$Ph_4P^+ \ Ph_3PNiBr_3^- + PhBr \xrightarrow{\ 180° \ } (Ph_4P)_2^{\ 2+} \ NiBr_4^{\ 2-}$$

In the end the phosphonium metal-salt complex is destroyed by water. The phosphonium halide is isolated in the usual way. If there are different halogen atoms in the metal salt and in the aryl compound used in the reaction, the phosphonium salt formed may have different anions causing difficulty in crystallization. In case of the anisyl halide, splitting of the ether linkage was observed when the reaction temperature was well above 180°C. The yields by this method are fair to excellent.

V. PHOSPHINES AND RADICALS OUT OF GRIGNARD REAGENTS

Ordinarily aryl halides, do not react with phosphines. There is one exception reported by Collie[183] who claims that triethylphenylphosphonium iodide can be prepared by a direct combination of triethylphosphine and iodobenzene.

Para- and orthohalogenophenols and arylamines do not form phosphonium salts with four P-C bonds but compounds of the quasi-phosphonium type. McDonald's[637] finding does not hold true.

$$p\text{-}XPhOH + Ph_3P \longrightarrow Ph_3PX_2 \xrightarrow{\ PhOH \ } Ph_3\overset{+}{P}\text{-}OPh \ X^-$$

In all cases, whenever an aromatic nucleus is to be linked to phosphorus, methods have to be used generating aryl ions, radicals, or substances which form these species during the reaction. With Grignard reagents and tertiary phosphines under normal conditions no reaction occurs. But when those mixtures are brought in contact with a stream of oxygen or air (less effective) the formation of the phosphonium ion takes place. The same reaction is

observed with lithium compounds. But the yields are not
as good. Phenyl sodium does not form phosphonium salts
under these conditions. It is also possible to produce
quaternary phosphonium salts directly, starting with phosphorus trichloride and a large excess of Grignard reagent
and oxygen without isolation of the tertiary phosphine
which is formed as an intermediate.

The mechanism of this reaction is not yet clear. It
is assumed that aryl radicals are formed, which in the
normal manner add to phosphorus. The formation of diphenyl
agrees with this assumption.

VI.   PHOSPHINES AND RADICALS PRODUCED BY IRRADIATING HALOGEN COMPOUNDS

Another means of activating aryl compounds to react
with tertiary phosphines is the irradiation of iodoaryl
compounds either with UV light[813] or γ-rays. Triphenylphosphine and iodobenzene for example, are irradiated in
bromobenzene as a solvent. The insoluble phosphonium salt
separates and is isolated by filtration. The photolysis
of tertiary phosphines alone also produces phosphonium
cations, when anions are in the reaction mixture, or anions
derived from alcohols.

$$Ph_3P \xrightarrow{h\nu} Ph_2P\cdot + Ph\cdot$$

$$Ph\cdot + Ph_3P \longrightarrow Ph_4P\cdot$$

$$Ph_4P\cdot + Ph_2P\cdot \longrightarrow Ph_4P^+ Ph_2P^-$$

$$Ph_4P^+ Ph_2P^- + MeOH \longrightarrow Ph_4P^+ OMe^- + Ph_2PH$$

VII.   COBALT-SALT METHOD

Another source of aryl radicals is the reduction of
Co(II) salts to Co(I) salts by Grignard reagents or lithium
alanate. The Co(I) species in turn attacks the aryl halide
producing a radical.

$$PhMgX + CoX_2 \longrightarrow PhCoX + MgX_2$$

$$2\ PhCoX \longrightarrow Ph_2 + 2\ CoX\cdot$$

$$CoX + PhX \longrightarrow CoX_2 + Ph\cdot$$

These formulas explain why the radical of the Grignard
reagents does not appear in the end product. For example,
triphenylphosphine, bromotoluene, and ethylmagnesium bromide form triphenyltoluylphosphonium salt. The yields are
fair.

VIII.   DIAZO METHOD

In contrast to the method mentioned above the generation of aryl radicals from neutral azocompounds leads to phosphonium salts in the presence of tertiary arylphosphines only.   An acetate-buffered solution of a diazonium salt in water is covered with an organic layer and stirred. The neutral azoacetate is extracted by the organic solvent and decomposes homolytically to radicals and nitrogen.   In the presence of a tertiary arylphosphine the phosphoranyl radical arises, which is oxidized by acyloxy radicals to the phosphonium cation.   Usually the phosphonium salts are isolated as halides or perchlorates.   Yields are fair to good.

$$Ph-N=NOAc \longrightarrow Ph\cdot + N_2 + \cdot OAc$$

$$Ph_3P + Ph\cdot \longrightarrow Ph_4P\cdot$$

$$Ph_4P\cdot + AcO\cdot \longrightarrow Ph_4P^+ \, AcO^-$$

$$Ph_4P^+ \, AcO^- + HClO_4 \longrightarrow Ph_4P^+ \, ClO_4^- + AcOH$$

IX.   TRIPHENYLPHOSPHINE AND DIPHENYLIODONIUM SALTS

When triphenylphosphine is refluxed with diphenyliodonium chloride or tetrafluoborate the result is a 40 to 50% yield of a tetraphenylphosphonium chloride or tetrafluoborate.   Usually the solvent is a higher alcohol (propanol). It is assumed that the phenyl group enters as a radical rather than as an ion.

X.   PHOSPHINES AND OXO COMPOUNDS

Carbonyl compounds such as aldehydes and ketones can react with primary, secondary, and tertiary phosphines in the presence of water or alcohol.   The resulting α-hydroxy alkyl phosphonium hydroxides transform to the salts if an acid is present during the reaction or by later addition of it.   The compounds decompose to the initial materials with alkali.

$$R_3P + OHC-R' + H_2O \longrightarrow \overset{+}{R_3P}-\underset{\underset{OH}{|}}{C}H-R' \, OH^-$$

XI.   ADDITION OF PHOSPHINES TO DOUBLE BONDS

A phosphine can be added to an olefinic double bond when it is strongly polarized.   Betaine like substances are the result. In the presence of mineral acids even the addition is possible on weaker polarized double bonds.[460] The same reaction is reported with acetylenes.

$$R_3P + \underset{\underset{R}{|}}{\overset{\overset{H}{|}}{C}}=CH\text{-}CO_2H + HBr \longrightarrow R_3\overset{+}{P}\text{-}\underset{\underset{R}{|}}{CH}\text{-}CH_2CO_2H \ Br^-$$

p-Quinone reacts similar with triphenylphosphine.

Phosphines can also add to systems with other double bonds, e.g., acid anhydrides (SO_3), carbon disulfide, and isothiocyanates. In all cases a betaine is the first product. Transformation to a phosphonium salt is maintained by common methods.

$$R_3P + SO_3 \longrightarrow R_3\overset{+}{P} - \underset{\underset{-O}{|}}{\overset{\overset{O}{\|}}{S}}=O$$

$$R_3P + CS_2 \longrightarrow R_3\overset{+}{P} - C\underset{S^-}{\overset{S}{\diagup}}$$

$$R_3P + S=C=NR \longrightarrow R_3\overset{+}{P} - \underset{\underset{-S}{|}}{C}=N\text{-}R$$

$$R_3P + \underset{O}{\overset{O}{\bigcirc}} \longrightarrow R_3\overset{+}{P}-\underset{OH}{\overset{^-O}{\bigcirc}}$$

Carbon dioxide does not react.

XII. PHOSPHINES AND CARBONIUM SALTS

Carbonium salts easily add to tertiary phosphines forming quaternary phosphonium salts.[495]

$$(Me_2NPh_2)\overset{\overset{R}{|}}{C}{}^{+} + X^- + Ph_3P \longrightarrow (Me_2NPh)_2\overset{\overset{R}{|}}{C}\text{---}\overset{+}{P}Ph_3 \ X^-$$

But triethoxycarbonium tetrafluoborate works as an alkylating reagent only.

$$[(EtO)_3\overset{+}{C}][BF_4{}^-] + R_3P \longrightarrow [R_3\overset{+}{P}Et][BF_4{}^-] + (EtO)_2CO$$

XIII. PHOSPHINES AND SULTONES

Sultones alkylate phosphines by opening their rings to form betaines. The reactants are heated together, with or without a solvent. The betaines may be transformed to phosphonium salts by addition of acids.

$$R_3P \; + \; \underset{\underset{O-S=O}{\overset{\displaystyle |}{|}}}{R-CH-CH_2} \longrightarrow \underset{\underset{H}{|}}{R_3\overset{+}{P}-\overset{\overset{R}{|}}{C}-CH_2-SO_3^-}$$

with O below $S=O$.

XIV. PHOSPHINE ALKYLIDENES AND POLAR REAGENTS

Phosphine alkylidenes, usually prepared from phosphonium salts by abstracting the α-hydrogen atom by bases, readily add polar reagents to yield phosphonium salts. With acids the corresponding phosphonium salts are obtained again.

Phosphine alkylidenes add alkyl halides to α-substituted phosphonium salts. If the halide is part of the same molecule in a favored position, ring closure occurs.

$$Ph_3P{=}CH \underset{CH_2-CH_2}{\overset{BrCH_2-CH_2}{\diagdown \diagup}} CH_2 \longrightarrow Ph_3\overset{+}{P} \text{---}\langle H \rangle \;\; Br^-$$

In the same way α-carbethoxyalkyl phosphonium ylids cyclize to ring ketones. Ylids with a carbonyl group in the β-position fix the alkyl group at the oxygen atom.

$$Ph_3P{=}CH{-}\underset{R}{\overset{|}{C}}{=}O \; + \; R'X \longrightarrow Ph_3\overset{+}{P}{-}CH{=}\underset{R}{\overset{|}{C}}{-}OR' \;\; X^-$$

Ylids add elementary halogens forming α-halogenated phosphonium salts.

Analogously trimethylbromosilane, bromostannane, or mercury dibromide add to phosphine alkylidenes producing phosphonium salts with the silyl, stannyl, or mercury group in the α-position. A phosphine and a phosphonium center in one molecule is obtained by addition of a halophosphine to an ylid.

$$R_3P{=}CH_2 \; + \; Ph_2PBr \longrightarrow R_3\overset{+}{P}{-}CH_2{-}PPh_2 \;\; Br^-$$

In solutions containing a phosphonium salt and an ylid, a transylidation takes place, if the ylid of the phosphonium salt is a weaker base than the ylid present.

$$Ar_3P{=}CR_2 \; + \; Ar_3\overset{+}{P}CHR'_2 \; X^- \longrightarrow Ar_3P{-}CHR_2 X^- + Ar_3P{=}CR_2'$$

In many cases phosphonium salts or betaines have the
characteristics of an intermediate which can rearrange,
e.g., the addition products of nitrenes, peroxy acids,
epoxides, and benzyne with phosphine alkylidenes. In case
of the cyclopentadienylide the addition of diazonium salts
is not in the α-position

$$Ph_3P=\text{⬠} \quad + \ ArN=\overset{+}{N} \ X^- \longrightarrow \ Ph_3\overset{+}{P}\text{⬠} \quad \quad X^-$$
$$\underset{N=N-Ar}{}$$

Addition of ylids to C=C double bonds has also been reported. Ylids and Lewis acids yield betaine like substances (see also Chapter 5A).

$$R_3P=CH_2 + BF_3 \longrightarrow R_3\overset{+}{P}-CH_2-\overset{-}{B}F_3$$

The above-mentioned reactions are useful in all cases
when the desired phosphonium salt is otherwise only accessible with difficulty. The reactions of phosphine imines
with the above-mentioned reactants proceed in essentially
the same way. The resulting products are quasi-phosphonium
salts, e.g.,

$$R_3P=NR' + R''X \longrightarrow R_3\overset{+}{P}-NR' \ X^-$$
$$\underset{R''}{|}$$

XV.   ALTERING OF THE ANION OF THE PHOSPHONIUM SALT

The methods described here do not serve to produce or
to alter the phosphonium ion. Only the anion is altered
by ion-exchange or complexing reactions with metal salts.
Other possible reactions are oxidation, saponification,
or the other reactions possible with an organic anion.

The anions may be exchanged for the hydroxide ion by
using bases. In the case of halide ions moist silver
oxide is applied frequently.

Depending on the nature of the cation the hydroxides
decompose in more or less time to phosphine oxides and
hydrocarbon, or to phosphines, water, and olefins.
Hydroxyphosphonium hydroxides form phosphine, carbonyl
compounds, and water. When the phosphonium hydroxides
are stable at least for a short time (the stability also
is dependent on the temperature), they can be converted
to salts with other anions by adding the appropriate acid.

$$R_4P^+ \ OH^- + HB \longrightarrow R_4P^+ \ B^- + H_2O$$

Tetraethylphosphonium iodide cannot be extracted from a
water solution by chloroform. But the tetraethylphosphonium
hydroxide readily is extracted and isolated as halide later.

A very common method is the reaction between two salts,
when one of the resulting salts is insoluble. This can
be the phosphonium compound or the inorganic salt.

$$Ph_4\overset{+}{P}\ X^- + KNO_3 \longrightarrow Ph_4P^+\ NO_3^-\downarrow + KX$$

$$Ph_4P^+SO_4^- + BaNO_2 \longrightarrow Ph_4P^+NO_2^- + BaSO_4\downarrow$$

The equations are examples for both cases. The iodides
usually may be converted to salts with other anions by the
appropriate silver salt. A very elegant method for exchanging anions is the passage of an aqueous solution of
the salt over an anion exchange resin charged with the
desired anion. The disadvantage is the time needed for
completion and the very diluted solutions resulting. They
have to be concentrated to isolate the product. The disadvantage mentioned first is easily overcome by automatic-
ally charging the column with the solution.

A less economical method uses repeated crystallizations from concentrated halide solutions. Thus the less
soluble tetraphenylphosphonium bromide can be transformed
to the more soluble chloride by applying sodium chloride.

Very few phosphonium alkoxides have been isolated in
the crystalline state. They are prepared by interaction
of the halides with alkaline alkoxides. Usually they are
quite unstable. When there are carbonyl groups in a
favorable position, betaines are formed instead of hydroxides.

Phosphonium salts ordinarily are stable to oxidation.
Therefore it is possible to oxidize the halide anion to
the perhalide by simply adding elemental halogen.

$$R_4\overset{+}{P}Br^- + J_2 \longrightarrow R_4P^+\ BrJ_2^-$$

Cyanides can be saponified to carboxylate anions.

A lot of complex anions are formed with salts of the
heavy metals, e.g.,

$$R_4\overset{+}{P}X^- + HgX_2 \longrightarrow R_4P^+\ HgX_3^-$$

Some of them have a low solubility in water and precipitate
when the reactants are mixed. Others form only under
strictly anhydrous conditions and decompose in water.

$$(R_4P)_2NiBr_4 \xrightarrow{\ H_2O\ } 2\ R_4PBr + NiBr_2$$

The formation of complex anions is analytically important
for the extraction of heavy-metal salts from aqueous solutions by organic solvents.

Some organic acids like picric acid and similar

compounds are used for the isolation of phosphonium salts with a low tendency of crystallization. Perchlorate, tetraphenylborate, dichromate, and nitrate are used for the same purpose. Phosphonium salts with these anions are of low solubility and are prepared by adding the sodium salt suitably dissolved in water. The possibility to exchange anions is of great importance for the preparation and resolution of chiral phosphonium salts.

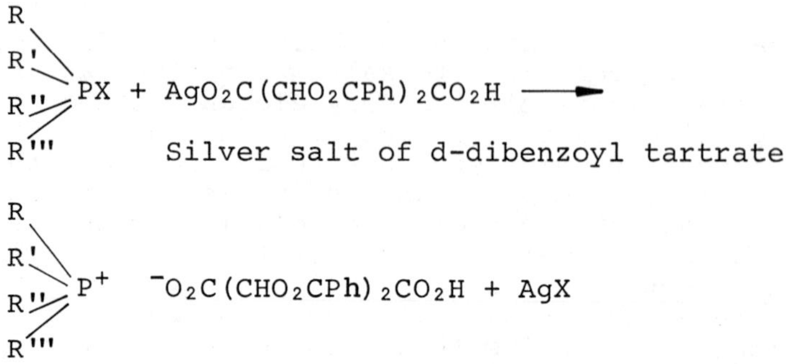

$$\overset{R}{\underset{R'''}{\overset{R'}{\underset{R''}{>}}}}PX + AgO_2C(CHO_2CPh)_2CO_2H \longrightarrow$$

Silver salt of d-dibenzoyl tartrate

$$\overset{R}{\underset{R'''}{\overset{R'}{\underset{R''}{>}}}}P^+ \; {}^-O_2C(CHO_2CPh)_2CO_2H + AgX$$

After repeated crystallizations to separate the diastereomers the halide is regained by the following reaction.

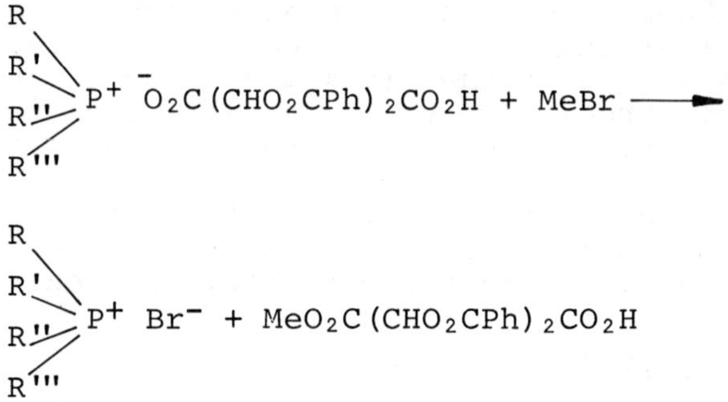

$$\overset{R}{\underset{R'''}{\overset{R'}{\underset{R''}{>}}}}P^+ \; {}^-O_2C(CHO_2CPh)_2CO_2H + MeBr \longrightarrow$$

$$\overset{R}{\underset{R'''}{\overset{R'}{\underset{R''}{>}}}}P^+ \; Br^- + MeO_2C(CHO_2CPh)_2CO_2H$$

XVI. ALTERING OF THE PHOSPHONIUM CATION

Altering of the phosphonium cations proceeds easily by the usual organic reactions, such as addition or substitution. For instance, the aromatic nucleus linked to phosphorus in a phosphonium compound can be considered as a substituted benzene (phosphoniono benzene). Electrophilic substitution reactions like nitration, sulfation, and halogenation introduce the substituents in the meta position. But in benzyl groups attached to phosphorus the position is substantially para.[514,726] Oxidation of a methyl group linked to a phenyl group at phosphorus leads to the expected carboxylic acid. Saponification of cyanide and esters proceeds normally; the same is true for

esterification reactions. In contrast to phosphine compounds, phosphonium salts do not poison metal catalysts,
at least not very seriously. Therefore nitro substituents
in quaternary phosphonium salts can be reduced to amino
groups by hydrogen in the presence of Raney nickel.
Ligands containing hydroxyl groups are readily transformed
to halides by thionylchloride or phosphorus pentachloride.
They can be acetylated by acid anhydrides.

$$(HOCH_2)_4PCl' \xrightarrow{\text{SOCl}_2} (ClCH_2)_4PCl; \quad (HOCH_2)_4PCl \xrightarrow{\text{Ac}_2O}$$

$$(ACOCH_2)_4PCl$$

Triphenylcarboxymethylphosphonium chloride decarboxylates,
when heated to triphenylmethylphosphonium chloride. In
view of practical application, besides the ylid formation,
the most important reactions at the organic radicals of
the phosphonium cation are condensation and polymerization
of hydroxy and vinyl groups respectively; e.g. tetrakis-
hydroxymethyl-phosphonium salts condense with phenols,
anilines, and ethylene-imines. The vinyl phosphonium compounds can be polymerized by common methods. Mixed polymers, e.g. with styrene, are also known.

XVII.   TRANSQUATERNIZATION

Tertiary phosphines are more readily quaternized than
comparable amines. Molecules containing one amine and
phosphine site together react with an alkylating reagent
like methyl iodide only at the phosphorus center. Another
example for this difference is the existence of tetraaryl-
phosphonium salts. Tetraphenylammonium salt is not known.
When a solution of an ammonium salt and a phosphine is
heated, the ammonium salt acts as an alkylating reagent on
the phosphine. The result is an amine and a phosphonium
salt (= transquaternization). It may be assumed that the
alkylating agent is the alkyl halide proper because of the
equilibrium between ammonium salt on one side and amine
and alkyl halide on the other.

$$R_4NX \rightleftharpoons R_3N + RX$$

$$RX + R_3P \longrightarrow R_4PX$$

There exists no such equilibrium with phosphonium compounds
under the reaction conditions. Therefore the reaction goes
to completeness even when the emerging amine is not volatile.

XVIII.   PHOSPHINES AND BENZYNE

In the presence of a hydrogen-donating substance benzyne and a tertiary phosphine form phosphonium salts.
These may be converted by usual metathesis.

$$Ph_3P + \text{[benzyne]} \xrightarrow{\text{fluorene}} Ph_4P^+ \quad ^- \text{[fluorenyl]}$$

XIX.   REDUCTIVE SYNTHESIS WITH LITHIUM ORGANO COMPOUNDS

Triarylphosphine oxides can react with lithium organo
compounds. The triaryl-lithiumoxy-phosphoranes form ordinary phosphonium salts with substances HX.

$$\text{[structure with P=O, Br]} + BuLi \longrightarrow \text{[structure with P, OLi]} \xrightarrow{HCl} \text{[structure with P}^+, Cl^-\text{]} + LiOH$$

XX.   QUASI-PHOSPHONIUM COMPOUNDS BY REACTING TERVALENT COMPOUNDS WITH HALOGEN

Phosphonium salts with one or more halogen-phosphorus
bonds in the cation, containing P-C-, P-O, or P-N-bonds,
may be prepared by addition of elemental halogen to the
tervalent compound, e.g.,

$$(R_2N)_2PX + X_2 \longrightarrow (R_2N)_2\overset{+}{P}X_2 \ X^-$$

If there is an alkoxy substituent in the molecule, the
products cleave at room temperature, forming alkyl halide
and the corresponding PO system. Aryloxy compounds are
also not very stable. But mild conditions during the reaction and careful evaporation allow the isolation of
products.
   The question still remains whether the substances thus
obtained are pentacovalent or ionic in structure. When
the anions are complex, for instance PCl_6^- or $SbCl_6^-$, one
can be quite sure that there exists a true onium compound.
Suitable solvents for the preparation of these quasi-
phosphonium salts are halogenoalkanes. The different
types, which can be prepared in this way are:

1. $R_n\overset{+}{P}X_{4-n} \quad Y^-$

2. $(ArO)_n P^+ X_4 - n$ Y^-

3. $(R_2N)_n P^+ X_4 - n$ Y^-

The first type is dealt with in Chapter 5B of this series. Of course there are mixed combinations of linkages, with P-C, P-O, and P-N, or P-S possible in one cation. Primary amines as ligands are attacked by halogen at the hydrogen atom.

XXI.   REACTIONS WITH PHOSPHORUS PENTACHLORIDE

Amines, phenols, phosphonyl and phosphinyl halides, phosphonic and phosphinic acids as well as their esters, phosphine oxides, and their alkyl-, aryl-, or amino-substituted derivatives form salts with phosphorus pentachloride. These are the same type as those referred to in Section XX.

If phenol itself is reacted with PCl_5, the ultimate product is $(PhO)_5P$, probably a phosphonium compound with a phenoxide anion.

$$(PhO)_4^+ \ OPh^-$$

When oxygen is to be replaced by halogen, somewhat vigorous conditions are necessary.

$$R_2PO_2H + PX_5 \longrightarrow R_2PX_2^+ \ X^-$$

Free primary or secondary amines or their hydrohalides may be used to produce the corresponding N-phosphonium salts.

$$R_2NH \xrightarrow{\ PCl_5\ } (R_2N)_n \ PX_4 - n^+ \ X^-$$

Tetraaminophosphonium salts behave like true phosphonium compounds. The anion is readily replaced by usual metathesis. They form hydroxides with alkali.

The products of lower substitution are isolated with the complex PCl_6^- anion.

$$R_2NPCl_3^+ \ PCl_6^-$$

$$Ph_2PCl_2^+ \ PCl_6^-$$

XXII.   QUASI-PHOSPHONIUM COMPOUNDS FROM PHOSPHINE OXIDES

The addition of phenols to triarylphosphine oxides at a temperature of about 100°C leads to the formation of aryloxyphosphonium hydroxides.

$$\overset{+}{Ar_3POPh}\ OH^-$$

Probably tertiary phosphine oxides and acids form phos-
phonium-like substances of the same structure as can be
expected in the hydrolysis of dihalogenophosphoranes in
strong acids.

$$R_3P=O\ +\ HX\ \longrightarrow\ \overset{+}{R_3POH}\ \ X^-$$

$$R_3PX_2\ +\ H_2O\ \longrightarrow\ \overset{+}{R_3POH}\ X^-\ +\ HX$$

With trialkyloxonium fluoborates the same compounds, with
alkyl instead of hydrogen, result.

$$R_3PO\ +\ R_3{}'O^+\ BF_4{}^-\ \longrightarrow\ \overset{+}{R_3POR}\ BF_4{}^-\ +\ R_2O$$

XXIII. TERTIARY PHOSPHINES AND QUINONE DIAZIDES

   When o- or p-quinonediazides react with tertiary phosphines betaine-like substances result with an azo link
between phosphorus and carbon atom. These substances are
colored from yellow to red, depending on the substituents.

B. ISOLATION OF PHOSPHONIUM COMPOUNDS

In most cases the reactants are brought together in a
suitable solvent. In the simplest case the resulting onium
salt is insoluble and precipitates during the reaction.
Sometimes the product is analytically pure. If not, the
salt is isolated by filtration, solved again, and the solution is mixed with another solvent in which the salt is
practically insoluble. This technique is used in most
cases. The solvents which have been applied are listed
in the table of compounds at the end of this chapter. In
other cases, the phosphonium salt is made less soluble by
metathesis of the anion. Most frequently used are iodide
ion, picric acid, and tetraphenylborate anion. This per-
formance is advantageous when the originally formed salt
is hygroscopic or difficult to crystallize. Often the
phosphonium salts come out of solution as an oil, especially
when they are slightly impure. This is often the case when
the radicals attached to phosphorus are different, and the
anion is organic in nature. A good method to overcome this
difficulty is treating the "oil" with absolute water-free

diethyl ether, acetone, or tetrahydrofurane repeatedly.
A certain amount of the solvent is put onto the "oil,"
stirred, and poured off. After repeating several times
there will be spontaneous crystallization in most cases.

C. ANALYZING PHOSPHONIUM COMPOUNDS

Mixtures of phosphonium compounds may be separated either
by thin layer chromatography or by electrophoresis. Quan-
titatively they may be determined by UV absorption of the
picrates as Schindlbauer proposed.[910] The tetramethyl-
phosphonium ion can be gravimetrically determined as the
chloro platinate.[21] Ross titrated phosphonium compounds
with perchloric acid[882] after complexing the anion with
mercury acetate.
When phosphonium salts are contaminated with phosphines,
these can be determined by Karl Fischer reagent.[432]
m-Dinitrobenzene was proposed by Hecker to detect the
phosphonium species by a color reaction.[434] Miscellaneous
methods are to be found in the following papers: Refs. 599,
614, 684, 423. The elements in a phosphonium compound are
determined by combusting it according to Liebig's method.
The tubes must have special fillings on account of the
phosphorus. The carbon evaluated by this method is usually
low. It is advisable to have a reference curve from analytically pure compounds.

D. TRANSFORMATION OF PHOSPHONIUM SALTS TO OTHER PHOSPHORUS COMPOUNDS

Other phosphorus compounds and the methods leading to them
are to be found in other chapters of this series. Therefore only a brief summary will be given here.

1. Strong heating of quaternary phosphonium compounds,
especially halides, to temperatures usually above 200°C,
results in a tertiary phosphine by losing one radical as
a halide. If the anion has oxidizing capabilities then a
tertiary phosphine oxide is obtained (see Chapters 1 and
6).

$$R_4PX \longrightarrow R_3P + RX$$

2. Phosphonium hydroxides generated in situ by moist
silver oxide, strong alkali, or alcoholates undergo thermal
decomposition to tertiary phosphine oxides, phosphines, or
betaines depending on the radicals linked to the phosphorus
atom (Chapter 1 and 6).
3. Reducing agents such as metals or metal hydrides
split the phosphonium salts to tertiary phosphines.

Phosphorus metal compounds react analogously (see Chapter 1).

$$R_4PX + 2 M \longrightarrow R_3P + RM + MX$$

$$Ph_4PCl + Ph_2PNa \longrightarrow 2 Ph_3P + NaCl$$

Electrolytic reduction of phosphonium cations is one of the most important reactions in generating optically active tertiary phosphines. The action of cyanide ion on phosphonium salts splits off one ligand as cyanide (see Chapter 1).

4. Aromatic metal organic compounds and aromatic phosphonium salts form pentacovalent phosphorus (see Chapter 5B).

$$Ph_4PX + PhLi \longrightarrow Ph_5P + LiX$$

The metal organic compound may be generated by splitting the phosphonium salt with sodium.

$$Ph_4PCl + 2 Na \longrightarrow Ph_3P + PhNa + NaCl$$

$$Ph_4PCl + PhNa \longrightarrow Ph_5P + NaCl$$

$$\overline{}$$

$$2 Ph_4PCl + 2 Na \longrightarrow Ph_3P + Ph_5P + 2 NaCl$$

5. Phosphonium salts with at least one hydrogen atom in the α position are transformed to phosphonium ylids by alkali metals, alkali amides, organoalkali compounds, and alcoholates (Chapter 5A).

$$\overset{+}{R_3}PCHR_2 + B^- \longrightarrow R_3P=CR_2 + HB$$

E. PHYSICAL, CHEMICAL AND MECHANISTIC ASPECTS OF PHOS-
    PHONIUM CHEMISTRY

Four-bonded phosphorus is the most familiar among phosphorus compounds. Formally, one can derive the fourth bond by reacting the nonbonding 3s electron pair with a Lewis acid. When the fourth substituent is neutral, phosphonium compounds result. If one of the ligands bears a negative charge, betaines or ylids form. Tetrahedral phosphorus can accept electrons from negatively charged particles into its 3d orbitals to form compounds with pentacovalent structures.

The base strength of PH_3 is lower than that of NH_3, because the lone pair of electrons in ammonia and phosphine have different hybridization states. Protonation to PH_4^+ involves hybridization from 3p to sp^3 atomic orbitals. Substitution of hydrogen in phosphine by organic radicals increases the basicity. The basicity of tertiary amines is similar to that of tertiary phosphines.

Phosphonium compounds with four different ligands can be resolved.[505,640,795,1004] The quaternization of tertiary phosphines follows the Taft equation.[136] (See also Refs. 779, 578, 480, 780 and Chapter 1). The steric requirements for the alkylating process are less than for amines. Phosphines are more powerful nucleophiles. This can be seen by the quaternization of Me_2P-NMe_2, which occurs at phosphorus. The strong nucleophilicity of Phosphorus is also responsible for the displacement of amine from an ammonium compound.

$$R_3P' + NR_4'^{+} X^{-} \longrightarrow R_3PR'^{+} X^{-} + NR_3'$$

Phosphites quaternize less readily and the aryl compounds more slowly than the alkyl compounds.[715] The trialkoxyphosphonium salts can be isolated under special conditions only.[250] Usually they stabilize to the well-known products of the Arbusov reaction by elimination of alkyl halides (see Chapter 18). The formation of the quasi-phosphonium intermediate is the rate determining step ordinarily. Tertiary phosphines add to unsaturated esters,[488] though they usually do not form stable products. But when mineral acids neutralize the negative charge, displaced by the attack of phosphorus,[462,460] a stable phosphonium salt forms. If the negative charge can be delocalized, betaines can be isolated.

In the presence of acids, carbonyl compounds and phosphines form true phosphonium salts with α-hydroxy groups.[710] Carbon dioxide cannot react with phosphines, whereas the adducts of carbon disulfide are well known. Their zwitterionic structure has been established by x-ray crystallographic determination.[762] The intermediate of the reaction of diaroyl peroxide with a tertiary phosphine is an aroyl phosphonium compound with an acyloxy anion.

$$\underset{\substack{\| \\ O}}{ArC}-O-O-\underset{\substack{\| \\ O}}{CAr} + R_3P \longrightarrow \underset{\substack{\| \\ O}}{ArC}-OPR_3^{+} \quad \underset{\substack{\| \\ O}}{O-CAr}^{-}$$

The reaction of hydroperoxides with phosphines results in the formation of phosphine oxides and alcohols. An ionic intermediate of phosphonium character was suggested.

$$t-BuOOH + R_3P \longrightarrow R_3\overset{+}{P}OH \quad \overset{-}{O}-t-Bu$$

The oxidation of phosphines to phosphine oxides by hypohalites going through a phosphonium intermediate has been confirmed by Denney, who isolated the phosphonium salt of a bridgehead hypohalite. In this case, final dealkylation is not possible.[233]

$$\text{(OCl-phenyl)} + (PhO)_3P \longrightarrow \text{(OP(OPh)}_3\text{-phenyl)} + Cl^-$$

The initial attack is taking place at the halogen site of the molecule. This is supported by the conversion of an optically active phosphine to the oxide of the opposite configuration.

$$Ph_3P: \ Cl\text{-}O\text{-}Bu\text{-}t \longrightarrow ClPPh_3 + t\text{-}BuO^- \longrightarrow$$

$$ClPPh_3 \longrightarrow t\text{-}BuOPPh_3 \ Cl^-$$
$$\underset{\displaystyle OBu\text{-}t}{|} \qquad \downarrow$$
$$t\text{-}BuCl + Ph_3PO$$

With p-quinones initial attack of the phosphine at the oxygen is probably the simplest explanation. But the mechanism is still in question because paramagnetic species can be detected in the reaction mixture.[828] The resulting compounds are quasi-phosphonium compounds. Only in one case has a Michael-type addition been confirmed.

$$Ph_3P + \text{(quinone)} \longrightarrow \text{(OH-phenyl-O}^-\text{)}\ -PPh_3$$

o-Quinones normally form pentacovalent structures. This was confirmed by ^{31}P NMR measurements.[89] Sulfur derivatives react like the corresponding oxygen compounds. Hilgetag[451] has been able to isolate phosphonium cations in the following way.

$$(MeO)_3P + MeSCl + SbCl_5 \longrightarrow MeSP(OMe)_3 \ SbCl_6^-$$

Trivalent phosphorus compounds and halogen together form products which seem to be in an equilibrium between the pentacovalent and ionic state.

$$R_3P + X_2 \longrightarrow R_3PX \ X^- \rightleftharpoons R_3PX_2$$

Optically active phosphines are racemized by this reaction. But in aqueous acetonitrile a phosphine oxide with inversed configuration is formed.[503]

Hexachlorocyclopentadiene is reduced to pentachlorocyclopentadiene by phosphines. The first step is the abstraction of Cl^+.

$$(RO)_3P' + \text{(C}_5Cl_6\text{)} \longrightarrow (RO)_3PCl \ \text{Cl}^-\text{(C}_5Cl_5\text{)}$$

Halocyclohexadienones react similarly, forming a phenolate anion. Ortho- and para-halophenols or haloanilines easily form quasi-phosphonium salts. Halocyclohexadienones or -imines might be the intermediates.[466]

There is some evidence that halogen acetylenes are attacked either at the halogen or at the carbon site. When the reaction proceeds in aqueous acetonitrile the products partly are dehalogenated acetylenes, phosphine oxides, and hydrogen halides. Acetylene phosphonium salts are stable under these conditions.[463]

$$Ph_3P + BrC\equiv CPh \longrightarrow Ph_3\overset{+}{P}Br \quad \overset{-}{C}\equiv CPh$$

Similar mechanisms are likely for α-bromosulfones and α-halonitriles. Initial attack of a tertiary phosphine at halogen is also assumed with tetrahalomethanes.

α-Halocarbonyl compounds have four potential sites of reaction with phosphines. They form either enol or true phosphonium salts. Evidence has been given that the carbonyl-C, the α-carbon atom, or the halogen is capable of the initial attack.

α-Ketophosphonium salts do not rearrange to enol phosphonium isomers. With phosphites instead of phosphines either the Perkov or the Arbuzov reaction occurs.

The arylation of phosphines by bubbling oxygen through a mixture of them with Grignard reagent, by homolytic cleavage of azo compounds, by attack of cobaltous chloride (CoCl) on arylhalides, or by UV irradiation of aryliodo compounds, have in common the generation of aryl radicals, which attack the phosphine, forming phosphoranyl radicals which in turn are oxidized to phosphonium cations.

$$Ar\cdot + R_3P \longrightarrow Ar\overset{\bullet}{P}R_3$$

$$Ar\overset{\bullet}{P}R_3 + X' \longrightarrow Ar\overset{+}{P}R_3 \ X^-$$

There is some evidence that the reaction of triphenylphosphine with bromoform to triphenyldibromomethylphosphonium bromide proceeds in a radical chain reaction.[138] Radicals may be produced directly by UV irradiation of phosphines. Phosphonium ylids readily react with electrophilic reagents to betaines. Frequently other reactions follow. This is especially the case with carbonyl compounds. The final products are olefins and phosphine oxides. But with alkyl halides, phosphonium salts with an

α-alkyl-substituted radical attached to phosphorus result. The reaction with hydrogen halides is reverse to the abstraction of an α-hydrogen atom by a base.

$$R_3\overset{+}{P}\text{-}CHR_2 + B^- \longrightarrow R_3P\text{=}CR_2 + HB$$

$$R_3P\text{=}CR_2 + H^+ \longrightarrow R_3\overset{+}{P}\text{-}CHR_2$$

Direct exchange reactions at the tetrahedral phosphorus are not satisfactorily established. An example might be the ester exchange reaction.

$$Y^- \quad \overset{\overset{X}{|}}{PR_3} \longrightarrow R_3\overset{+}{P}Y + X^-$$

$$(PhO)_3\overset{+}{PH} \xrightarrow{3R'OH} (R'O)_3\overset{+}{PH} + 3PhOH$$

The best known example for the addition-elimination mechanism is the hydroxide ion cleavage of phosphonium salts. The overall reaction is third order, first order with respect to phosphonium ion, and second order with respect to hydroxide ion.[459]

$$R_4\overset{+}{P} + OH^- \longrightarrow R_4P\text{-}OH$$

$$R_4POH\ OH^- \longrightarrow R_3PO + H_2O + R^-$$

The configuration at the phosphorus atom is inverted in this reaction as was shown with optically active phosphonium salts.[502] (See also Refs. 11, 16, 95, 127, 378, 782, 884, 998; see also Chapter 6).

With few exceptions the reactions giving rise to tetrahedral phosphorus are irreversible at room temperature. Normally under certain conditions and steric requirements only, the reverse reaction takes place.

Phosphonium salts will be dealkylated when they are heated to temperatures of about 200°C. Phosphines and either olefins or alkyl halides are the products. Both, nucleophilic displacement or β-elimination seem to occur.

It has been assumed that the reduction of phosphonium salts by LiAlH$_4$ proceeds by nucleophilic attack of the hydride ion.

Internal attack of carbanions results in ylids or cyclopropanes and phosphine (see also Chapter 1).

Heating alkyltriphenylphosphonium alkoxides leads to the formation of a phosphine and ether.[1036]

$$Ph_3PR^+ \ ^-OR \longrightarrow Ph_2PR + PhOR$$

Ylid addition to benzyne forms a betaine which breaks down to phosphine by nucleophilic displacement.

Physical properties of phosphonium compounds which have been investigated recently are summarized in the following references. For instance, the conductivity of phosphonium complexes, with tetracyanoquinodimethane was investigated.[981] The density of molten phosphonium halides is reported by Griffiths.[393] Deacon measured the solubility of $Me_3PCF_3^+ \ J^-$ in F_3CF.[229] The dissociation constants of α-mercurated phosphonium salts in solution are reported by Nesmeyanov.[750] Spectroscopic substituent constants in the visible and UV region are given in a paper by Schiemenz.[903] Laser-Raman spectra of PH_4^+ and their comparison with NMR and neutron-scattering results,[890] and the conclusion herefrom to hindered rotation of the cation,[890a] proton, and deuteron spin-lattice relaxations,[828] and constants of indirect nuclear spin coupling[268] have been reported by several authors.

F. APPLICATIONS

The applications of phosphonium compounds mentioned in this part are neither complete nor are the references cited in all cases the most important ones. It is a random selection to give the reader a survey impression of the purposes for which phosphonium salts may be used. In synthetic chemistry the phosphonium salts frequently are the precursors of the phosphonium ylids used for olefin synthesis. Many patents deal with tetrakis(hydroxymethyl)phosphonium salt and its derivatives alone or in condensation products.[14,565,857] Most patents claim that the substances are useful as flame retardants for textiles and paper.[41,58,230,339,475,721,856,1053] Other claims are the stabilization of polyacrylonitrile fiber to sunlight and heat,[30] or the condensation with organic dyes to wash-fast colors.[13] The uptake of colors by wool and hair is also improved by treating them with tetrakis(hydroxymethyl)phosphonium salt.[39] Shrink resistance of wool,[871] perspiration resistance of leather,[1066] and fat liquoring[563] are also

accomplished with treatment by this salt. The mono- and
bisphosphonium salts are proposed for various uses, such
as vulcanization accelerators,[808] treatment of felts,[123]
heat stabilizers of nylon,[445] flame retardants of plas-
tics,[149] antistatic textile finish,[179] emulsifiers,[851]
permanent-wave preparations,[362] to thicken pigment sus-
pension,[844] desinfectant detergents,[343] gelation inhibitors,[1051] water-repellent furnishes,[944] anti-icing additive for gasoline,[293] fungicidal soaps,[588] and gelling and
dispersing of organic liquids.[616]

Polymeric phosphonium salts, i.e. organic polymers
containing phosphonium units, are prepared in three different ways: (a) by condensing α-hydroxyphosphonium compounds with phenols or amines; (b) by polymerizing olefinic phosphonium salts; and (c) by introducing a phosphonium group into a polymeric substance.[329,363,509,957,
990,1043,1044,1046] Many phosphonium salts have been proposed and used as catalysts in polymerization reactions.[38,
185,245,357,749,824,1033] Other reactions leading to the
most diverse organic compounds are also catalyzed by them;
often the salts are complexed with heavy metal salts or
aluminum chloride.[50,119,291,338,577,603,862-864,942,1025,
1055]

Since phosphonium salts form a great number of compounds with heavy-metal salts which are extractable by organic solvents they are used for the determination of
metals such as: Ir,[745] Pt,[746,948] Mo,[747] Os,[744] Np,[773] and
Re.[1034] Further applications in the analytical field are
to be found in the following papers: Refs. 99, 945-947,
974, especially for bromide and iodide photometric determination,[57] determination of cobalt,[976] peroxidisulfate,[56]
and to evaluate the capacity of cation exchangers.[567]

In the photographic field, use is made of phosphonium
salts as mordants,[314] sensitizers,[261] developers,[244] and
against color contamination.[51] Anion exchange resins often
bear, as a working group, phosphonium units.[430,1016,1045]

Application of phosphonium salts in the bioligical
field has found much scientific and commercial interest.
Many of them have been used in plant growth regulation,[153,
706,817] and plant protection,[701,703,705,1065] outstanding
among them phosfon (2,2-dichlorobenzyltributylphosphonium
chloride).[31,154,594,1021,1065,1092] Much work has been
done in the developing of insecticidal,[163,354,587,690]
bactericidal,[735] nematocidal,[601,624] and fungicidal[93] compounds of phosphonium type and testing of their properties.
Their application in moth-proofing of textiles is well
known.[633,682] They have even been used against grass-
hoppers.[121] The qualities of phosphonium salts as gangli-
onic blockers,[365,366] their effects on the blood pressure,[920] as well as their properties as antineoplastic
agents,[1] their anti-influenza activity,[796] and their

astringent influence[924] have aroused medical interest.
The formation and decomposition of phosphonium salts
in vitamin A derivatives was thoroughly investigated.[335]
Phosphonium salts are excellent inhibitors of metal
corrosion in acid solutions in the absence of oxygen.[479,497,498,629] Phosphonium salts are cleaved electroyltically
to phosphines and hydrocarbones, via reactive intermediates,
which initiate other reactions. For the electrolytic reduction of nonconducting depolarizers they serve as sup-
porting electrolytes.[316,620,807,1052] Comparatively little
use has been made of phosphonium salts for organic synthesis directly out of the salt.[236,237,725,1002] In another reported synthesis, the intermediate formation of
ylids is assumed.[930,935,936]

G. LIST OF COMPOUNDS

The substances listed in Tables 1 to 20 are divided in four
main groups: (A) Quaternary phosphonium salts; (B) Diquaternary salts: (C) Betaines; and (D) Salts with less
than four carbon atoms attached to phosphorus (quasi-
phosphonium salts). Group A is subdivided into salts with
four equal ligands, with two equal ligands at a time, with
two equal and two different ligands, and with four different
ligands. The scheme for groups B, C, and D is analogous.
The compounds are ordered according to ascending numbers
of C and H, the other atoms are in the alphabetical order.
The same is true for the anions.
Aliphatic and aromatic groups form a series of their
own at a time. The aromatic ring linked directly to phosphorus is always after an aliphatic ligand. For instance,
a phenyl group comes after a hexadecyl group. The substances are listed under formulas not names. This is because names often are clumsy so that even a skilled chemist
has difficulty translating them into formulas, and there
are sometimes several quite different names used for the
same compound.
Some well-known ligands are given abbreviated: Me =
methyl, Et = ethyl, Pr = propyl, Bu = butyl, Ph = phenyl
and all the names which are derived from numbers, for
instance hexadec. = hexadecyl. The same abbreviations
have been used, when the ligands are between two phosphorus
atoms in the biquaternary salts: P—◯—P-, =P-Ph-P, or
P-Et-P = ethylene. The physical properties are to be
found under the following letters: m. = melting point,
b. = boiling point, IR = infrared spectrum, UV = ultraviolet spectrum, BMR, FMR, HMR, PMR = nuclear magnetic
resonance spectra, the first letter referring to the
nucleus. TLC = thin layer chromatography, Λ = conductivity,
$E_{1/2}$ = half-wave potential, χ = magnetic susceptibility,

n_D^{20} = refractive index, d_{20} = density, $[\alpha]$ = specific
rotation, ESR = electron spin resonance, σ = Hammet's
constant, k = rate constant, DK = dielectric constant,
pK_a value.

The Roman numbers give the method for preparation according to the number of the section on synthetic routes
(Section A). All substances have been registered as far
as they are registered in Chemical Abstracts as of July 1,
1970, when there was at least one data given. Of course
there not all available physical data is listed. For each
compound every constant was given once; the melting points
have been rounded off, and the highest one reported is
given, the same is true for the yields.

Table 1. R_4PX

Cation R_4P	Anion X	Method	m.	Solvent	Yield (%)	Other Data and Remarks
$(BrCH_2)_4P$	HSO_4	XV[24]	177			
	Br	XV[24], XV+ XVI[157]	236[157]		93[157]	IR, HMR[157]
$(ClCH_2)_4P$	HSO_4	XV[24]	189			
	CNS	[24]	102			
	$OPh(NO_2)-2,4,6$	XV[910]	113			
	Ph_2SnCl_2	XV[101]	124			
	Br	[24]	192		74	
	Cl	XV+ XVI[458]	192[458]		12[330]	IR[330]
	$UO_2(NO_3)_2Cl \cdot H_2O$	XV[23]	275	i-PrOH		Yellow
	$CdCl_3$	XV[23]	277	MeCN		
	$CuCl_3 \cdot H_2 \cdot H_2O$	XV[23]	212	i-PrOH		Yellow
	$HgCl_3 \cdot H_2O$	XV[23]	244	H_2O		
	$ZnCl_3$	XV[23]	285	H_2O		
$(JCH_2)_4P$	HSO_4	[24]	216			
	Cl	[24]	184			
Me_4P	$OPh(NO_3)-2,4,6$	XV[408]	328[408]	H_2O/EtOH[408]	100[408]	UV[910]
	Br	I[648a]	160–80[648a]	MeOH/Et_2O[648a]	90[648a]	PMR −25.1[673a]
	Cl	XV[408] I[730,648a]	406[49]	EtOH/Et_2O[648a]	89[730]	IR,[408] HMR,[878] PMR −24.4[648a] Hygroscopic[788]
	ClO_4	XV[408]	370	H_2O		
	$HgCl_3$	XV[408]	249 dec.	H_2O	100	

Table 1 (Continued)

Cation R_4P	Anion X	Method	m.	Solvent	Yield (%)	Other Data and Remarks
Me_4P	$PbCl_6$ Amalgam	XV[408]	250 dec.			ESR
	HgJ_3	Electrolytic[190] XV[227]	197[227]			IR,[224] UV,[227] lemon yellow[227]
	HgJ_4	XV[227]	300[227]			IR,[224] cream[227]
	Hg_2J_5	XV[227]	172[227]	EtOH[545]		IR,[224] UV,[227] deep yellow,[227] diamagnetic[229]
	J	II[352] I[256]	360[256]		33[256]	IR-Raman,[198] Λ[455]
$(HOCH_2)_4P$	OH	II[315] I[866]	Oil[866]			PMR,[1041]
	C_2O_4	XV	91[865]			
	$MeCO_2$					
	$OPh(NO_3)-2,4,6$ Cl	X[458]	150	AcOH	99[881]	PMR -25.2[673a]
	Br		151			PMR -25[673a]
	F	8[65]	98			
$(CH_2=CH)_4P$	Br	I[649]	105			
Et_4P	OH	1[44]	Solid			HMR,[679] hygroscopic
	CO_3	1[44]	240 dec.			Hygroscopic
	C_2O_4	6[23]	200 dec.			
	MeSP–S–CH CNHMe (O=…=O)	I[708]		PhH/Et_2O		IR, $n_D^{20} = 1.5458$; $d_{20} = 1.2507$
	$OPh(NO_3)-2,4,6$	XV[218]	212			Yellow

First cation (structure), ref [623]:

```
      O        Cl
      ‖        |
 MeSP–O–⟨ C6H3 ⟩
      |        |
      O        Cl
```
[623]

Cation	Anion	Prep. / Ref.	m.p. (°C)	Solvent	Yield (%)	Properties
(structure [623])	PhCO2	I[702]	71	CH2Cl2/Et2O		$n_D^{20}=1.5610$; $d_{20}=1.3204$
	BF4	I[495]	160			
			291			
	Br	I[352,478]	320, dec.[1018]			HMR[268]
		[623]	300 dec.			
	U^{VI}O2Cl4	XV[348]	>360			
	U^{IV}Cl6	I[348]	>300			
	HgJ3	XV[788]	117			
	HgJ4	XV[226]	275			
	J	I[6,478]	294[226]	EtOH/Et2O[531]	78[380]	IR[226]; IR, yellow[198]; IR–Raman,[198]; PMR –33[673a]
(MeCHOH)4P	J3	XIV+ XVI[1081]	66			
	MeCO2	II[352]	89			
		XV[226]	88			
	Br	I[654]				
		XV[867]				
		710				
	Cl	X[134]	115[134]	MeCN[135]	98[135]	
	J	367	64			
(H2C=CH–CHOH)4P	MeCO2	867	172			
	Cl	867	>260 dec.			
Pr4P	BF4	XV[886]	240	H2O		
	J	I[214]	200	EtOH[214]		UV[96]
			dec.[214]			
(EtCHOH)4P	Br	X[134]	105			
(MeCH=CH–CHOH)4P	MeCO2	X[867]	90			
	Cl	X[867]	88			

219

Table 1 (Continued)

Cation R_4P	Anion X	Method	m.	Solvent	Yield (%)	Other Data and Remarks
Bu_4P	Amalgam	Electrolytic[190]	dec.			ESR, PMR -33.9
	$MeSO_4$	I[589]	55		19	Very hygroscopic
	$Re(CO)_5$	XV[784]	Liquid			IR, red-orange
	$H_3B\ Re(CO)_5$	XV[784]		THF/Et_2O		IR, red-orange
	$OPh(NO_3)_2\text{-}2,4,6$	XV[592]	55	$H_2O/EtOH$		
	$(BuO)_2\overset{S}{\ddot{P}}\text{-}S$	XV[709]			86	$n_D^{20} = 1.15104$, $d_{20} = 1.033$
	BF_4	XV[576]	67			
	Br	I[380,849]	112[576]	Me_2CO/Et_2O[1014]	99[380]	PNM -34[673a]
	BrJ_2	XV[576]	107			
	JBr_2	XV[576]	131			
	Br_3	XV[576]	124			
	Cl	XV[576]	67			
	ClO_4	777	160	H_2O		
	J	I[592]	103[576]			PNM -32[673a]
		XV[576]	138			UV[96]
	J_3	XV[576]	86			
	S_2O_8	XV[155]	105			
$(PrCHOH)_4P$	Br	134	136			
$(Me_3SiCH_2)_4P$	J	I[952]				
$Pent_4P$	J	892				IR
$(Me_2CHCH_2CHOH)_4P$	J	X[367]	119			
Hex_4P	J	892				IR
$c\text{-}Hex_4P$	J	II[778]	310	EtOAc/EtOH		

Cation	Anion/Ligand	Ref	M.p.	Solvent		Methods
(PentCHOH)₄P	Cl	X[134]	123			
	J	X[134]	120			
(PhCH₂)₄P	OH	[622]	190 dec.			
	HSO₄	[622]	217		45[1030]	
	Br	IX[622]	216[622]			
		X[1030]				
	Cl	IX[622]	229[622]			
		I[478]				
	ClO₄	XV[777]	172	EtOH		
	J	IX[618]	191			
(HexCHOH)₄P	Cl	X[135]	123	PhH		
	J	X[134]	120			
Oct₄P	J	[892]				IR
(undec-CHOH)₄P	Cl	X[135]	109	PhH/EtOAc		ESR
Ph₄P	Amalgam	Electro-lytic[190]	dec.			IR
	HN₂O₆	XV[306]	142			
	BH₄	[433]	190			
	SCN	[255]	265			
	F⁻ MeSP=O S	XV[879]	208		82	IR
	F⁻ EtSP=O S	XV[879]	111		80	IR
	Mo^{III} Ni(SCN)₄·(PBu₃)₂	XV[303]	189		62	Blue
	(H₂C=C·CH₂)₂PdBr₂	XV[523]	125 dec.		82	Violet, red
	(H₂C···CH₂)₂PdCl₂	XV[372]	204			IR, HMR, Λ, yellow
		XV[372]	200			IR, HMR, Λ, yellow
	F₃C—S—Pt—S—CF₃ / F₃C—S—S—CF₃	XV[219]	174			
	(NO)₂Mo(NC—S / NC—S)₂	[627]				IR

221

Table 1 (Continued)

Cation R_4P	Anion X	Method	m.	Solvent	Yield (%)	Other Data and Remarks
Ph_4P	$Co^{III}(S_2C_2(CN)_2)_3$	XV[628]	224		-	IR, Λ
	$Mo^{IV}(S_2C_2(CN)_2)_3$	XV[628]	282	Me_2CO/H_2O	15	IR, HMR
	$W^{IV}(S_2C_2(CN)_2)_3$	XV[628]	269		35	IR, HMR
	$(F_5Ph)_2TlBr_2$	XV[223]	102		61	IR, x-ray
	$(F_5Ph)_2TlCl_2$	XV[223]	100		68	IR, x-ray
	$OPh(NO_2)_3\text{-}2,4,6$	XV[910]	197			
	$PhS-P(F)(=S)(O)$	XV[878]	180		83	
	$CdCl_3 \cdot PhNHCONH$	XV[793]	135			
	Ph_3SnCl_2	XV[101]	203			
	Ph_4B	XVIII[1069a] XV	309 dec.		78	IR, Λ
	TCNQ $[NC-C(CN)=C_6H_4=C(CN)-CN]^-$	XV[2]	228 dec.			

Ph₄P

	Ref	mp	Solvent		Other
BF₄	XV[439]	267 dec.	Me₂CO		IR,[1067] UV[912], E₁/₂[482], PMR -20.8[673a]
Br	VI[822]	350[751]	EtOH[650]	88[751]	
	IV[1091]	288[1078]	H₂O[698]	80[1079]	
	V[698]				
	VI[267]				
	XVIII+[1078]				
	XV[1078]				

$(Ph\text{-}D_3\text{-}2,4,6)_4P$
Ph₄P   PMR $-20.8, -23.2$[673a]

	Ref	mp	Solvent		Other
I	VIII[963,597]	298		45	
Br		286			
Br·2 H₂O					
BrJ₂	XV[576]	107			UV
Br₂Cl	XV[576]	117			UV
Br₂J	XV[576]	131			UV
Br₃	XV[576]	234			UV
HgBr₃	XV[226]	210	EtOH		IR
NiBr₄	[191]	273			
CeCl₄·8 H₂O	XV[896]	116	EtOH	82	
Cl	V[1062]	272[1071]	EtOH	79[1062]	E₁/₂[975]
	VI[1071]				
	XV[1071]				
CrO₃Cl	XV[898]		Petroleum ether/ CHCl₃		IR
CrO(O₂)₂Cl	XV[898]		Petroleum ether		IR
ClJ₂	XV[576]	102	EtOH		
HgCl₃	XV[1074]	216	EtOH		UV
LaCl₄	XV[896]	115	EtOH	82	
NdCl₄	XV[896]	118	EtOH	82	Purple

Table 1 (Continued)

Cation R_4P	Anion X	Method	m.	Solvent	Yield (%)	Other Data and Remarks
Ph$_4$P	$Co^{II}(N_3)_4$	XV[791]	164	Me$_2$CO/CCl$_4$		Blue
	$Hg^{II}J_3$	XV[226]	200	Me$_2$CO/EtOH		IR, yellow
	$Hg^{II}J_4$	XV[226]	225	Me$_2$CO		IR, yellow
	J	XVIII[1079]	337[493]	CHCl$_3$/PhH[813]	100[854]	IR,[375] ∧[455]
		XV[1078]				PMR −23.2, ppm[397]
		XX[854]				
		VI[813]				
	J_3	XV[576]	245			
	NO_3	XV[697]	284 dec.[255]	H$_2$O[692]	96[697]	IR,[306] UV ∧[306]
						E1/2[68]f
						IR, ∧
	OsO_3N	XV[733a]				
	$(NO_3)_2 \cdot Ce(NO_3)_2$	XV[695]	192	EtOH		
	$(NO_3)_2 \cdot La(NO_3)_2$	XV[695]	188	EtOH/CHCl$_3$		
	$(NO_3)_2 \cdot Nd(NO_3)_2$	XV[695]	198	EtOH/CHCl$_3$		Lilac
	$(NO_3)_2 \cdot Pr(NO_3)_2$	XV[695]	195	EtOH/CHCl$_3$		Greenish

Cation	Anion	Ref.	m.p. (°C)	Solvent	Yield (%)	Notes
(4-F₃C-Ph)₄P	Br	IV[481]	224		51	E$_{1/2}$[180]
(2-Me-Ph)₄P	Br	IV[481]			47[481]	
(3-Me-Ph)₄P	J	III[634]	175			
(4-Me-Ph)₄P	Br	IV[481]	203		95	
	J	III[634]	267[956]			
		IV[956]				
(4-Me₂N-Ph)₄P	J	XV[904]	328 dec.	H₂O/MeOH		Deep violet–brown
	J₃	VI[904]	239	EtOAc	83	
(1-Naphthyl)₄P	Br	IV[481]	251			
	J	II[692]	270			
(2-Naphthyl)₄P	Br	IV[481]	244		36	
(4-PhSO₂-Ph)₄P	Br	IV[481]	195		37	

Table 2. R₃R'PX

Cation R₃PR'		Anion X	Method	m.	Solvent	Yield (%)	Other Data and Remarks
R	R'						
ClCH₂	Et	BF₄	I[435]	176	EtOH/Et₂O	65	
	Pr	Cl	XVI[802]	197	EtOAc/MeOH		
	Ph	Cl	XVI[801]	236 dec.	MeOH/Et₂O/EtOAc		
Me	CH₂OH	Cl	I[400]	201			
		J	I[400]	250	EtOH		
	C₂F₄H	J	I[785] [442]	215 dec.			HMR
	CH=CH₂	J	XV + XVI[553]	150 dec.	CHCl₃/Pent.		IR
	CS₂Me	J	XV[313]				
	Et	OH	XV[313]	290			HMR,[679] hygroscopic, mass. 313
		OPh(NO₂)₃-2,4,6	XV[313],[397]				
		Br	[313]	300 dec.			PMR −28.0
		Cl	XVI[723]				IR
		PF₆	I[144]	315 dec.[530]	MeOH/Et₂O[108]	36	PMR −26[673a]
	CF₃COH·Me	J	I[401]	78			IR

Substituent	Anion	Ref.	mp (°C)	Solvent	Yield (%)	Notes
(CH₂)₃OH	Cl	I[798]	130			
CH₂CO₂Et	OPh(NO₂)₃-2,4,6	XV[313]	124			
	Br	I[1072,313]	160[1038]	Me₂CO[1038]	88	Hygroscopic[1038]
	Cl	I[313]	160 dec.			
(CH₂)₂O₂CMe	Cl	I[129]	126			
CH₂SiMe₃	J	I[917]	>300			
	PF₆	XV[722]	124	H₂O	70	IR, HMR
CHCO₂Et / Me	Br	I[88]				
(CH₂)₃O₂CMe	Cl	I[798]	94		72	
c-Hex	J	I[544]	263	EtOH/Et₂O	70	
(CH₂)₄O₂CMe	Br	I[798]	60		78	
	Cl	I[798]	122			
	ClO₄	XV[798]	93			
	PtCl₆	XV[798]	198 dec.			
CH₂Ph·NO₂-2	OPh(NO₂)₃-2,4,6	I[514], XV[331]	154[331]			
CH₂Ph-NO₂-3	OPh(NO₂)₃-2,4,6	I[514], XV[331]	176[331]			
CH₂Ph-NO₂-4	OPh(NO₂)₃-2,4,6	XVI[514], XV[331]	195[331]			
CH₂Ph	OPh(NO₂)₃-2,4,6	XV[876]	176			
	Br	I[313]	223[36]	MeCH(OH)iBu/petroleum ether[36]		
CH₂COPh	Br	I[12]	205[1038]	EtOH/Et₂O[1038]		IR,[12] pKa 8.60[556]
CD₂COPh	Br	XVI[12]		H₂O		IR
C(SiMe₃)₂	PF₆	XV[723]		H₂O		HMR
CH₂–⟨indol-3-yl⟩ (2-Me, N–H)	MeSO₄	XVII[436]	185	H₂O	70	Pale pink

Table 2 (Continued)

Cation R₃PR' R	R'	Anion X	Method	m.	Solvent	Yield (%)	Other Data and Remarks
Me	CH₂—[indolyl, N–H]	J	XV[436]	238	H₂O		IR
	(CH₂)₁₁Me	Br	RP(OR)₂ + RMgX[617]			6	IR
		Cl	I + XV[617]	87		100	IR, x-ray
		J	I[431]				IR, HMR, PMR −26 ppm
	[fluoren-9-yl]	Br	I[558]	221[558]	PhH/EtOH	57[558]	IR[1067]
		J	XIV[1076]	246 dec.	MeOH/Et₂O		
	CH₂CPh₂ (OH)	J	XIV[1081]	200 dec.	MeOH/Et₂O		
		J	I[419]	303	MeOH/H₂O		
	Ph–Cl-2 Ph	Amalgam	Electrolytic[190]				ESR
		OPh(NO₂)₃-2,4,6	XV[656]	133[656]	EtOH[656]		
		O₃SPh-Me-4	I[656]	161	Me₂CO		
		Br	I[10]				
			IV[481]	285[481]	EtOH/Et₂O[10]	60[481] $E_{1/2}$ 2.27[482]	UV[910]
		HgJ₄	I[151]	187	Me₂CO		

	J	I[531]	236[313]	EtOH/Et₂O[531]	85[531] UV,[912] PMR −21[673a]
Me⁺\P(OPh)₂J Ph⁄		XV[180]	−28		
PhP(SnEt₃)₂ + MeJ PhP thermal dec.					
Ph−NH₂−3	Cl	XVI[318]	253		
Ph−Me−4	Cl	I[202]	110		
	PtCl₆	XV[202]	230		
Ph(CO₂H)−2,4	J	I[202]	255		
Ph−Me₂−2,4	PtCl₆	XV[184]	258		
	Cl	I[184]	110		
	J	I[184]	265		
Ph−Me₂−2,5	HgJ₄	[552]	152		
	J	[552]	204		
Ph−Me₂−3,5	J	[184]	205		
Ph−Et−4	J	[552]	204		
Ph−NMe₂−4	HgJ₄	I[151]	171	Me₂CO	UV
Ph−Me₃−2,4,6	J	I[718]	264		
	J	[209]	232		
Ph−SiMe₃−3	J	I[117]	173	EtOAC	40
Ph−SiMe₃−4	J	I[117]	183	CHCl₃/petroleum ether	55
Ph−CH₂SiMe₃−3	J	[115]	160[115]	EtOAc[116]	80[115]
Ph−CH₂SiMe₃−4	J	[115]	186	EtOAc[116]	73[115]
Ph−OPh−4	J	I[115]	242		
Me	O₃SPh−Me−4	I[867]	83	i-PrOH	
	Ph−PO₃Me	I[590]	68		100
HOCH₂					

229

Table 2 (Continued)

Cation R_3PR'		Anion X	Method	m.	Solvent	Yield (%)	Other Data and Remarks
R	R'						
$HOCH_2$	Me	$H_2C=C(Ph)-PO_3Me$	I[590]			100	n_D^{20} = 1.5630
		$Me(CH_2)_{11}PO_3Me$	I[590]	34		100	PMR -27[673a]
		BPh_4	XV[378]	170	Me_2CO/PhH		
		J					
	Et	$EtPO_3Et$	I[590]			100	n_D^{20} = 1.4696
		$Me(CH_2)_{11}PO_3Et$	I[590]			100	n_D^{20} = 1.4801
	$CH_2NHCONH_2$	$OPh(NO_2)_3$-2,4,6-OH-3	XV[868]	192 dec.			
	$(CH_2)_2OH$	$OPh(NO_2)_3$-2,4,6-OH-3	III + XV[869]	121			
	$CHOHCH=CH_2$	O_2C-Me	XV[867]	172			
		Cl	XV[867]	>260 dec.			
	Pr	Br	I[803]			78	n_D^{20} = 1.5458
		Cl	X[799]				n_D^{20} = 1.5280, d^{20} = 1.2554
	$CHOHCH=CHMe$	O_2CMe	XV[867]	90			
		Cl	XV[867]	88			
	Bu	BPh_4	XV[378]	145	Me_2CO/PhH		n_D^{20} = 1.5460
		Br	I[771]				PMR -30[673a]
		Cl					PMR -28[673a]
		J					
	i-Bu	Cl	I[378]	52	Me_2CO		
	c-Hex	Cl	I[378]	88	Me_2CO/petroleum ether	68	PMR -28[673a]

R	Substituent	Ref	bp/mp	Solvent	Yield (%)	Properties
(CH₂)₂OP(OEt)₂ (O on P)	Br	I[867]	60			
CH₂Ph	Cl	I[803]			94	$n_D^{20} = 1.5805$; $n_D^{20} = 1.5265$
Hept	Br	I[805]				
Dodec	Br	I[867]				
Octadec	Br	I[254]	39			
CH₂O-octadec	Cl	[450]	61[450]	Et₂O[254]	25[254]	
			58			
CH₂CNH-octadec (O on C)	Cl	[450]	57			
	J	[450]	40			
Ph	OPh(NO₂)₃-2,4,6	XV[910]	78	MeOH		PMR, UV −17.7[673a] UV[912]
	Cl	I[378]	80[378]	MeOH/Me₂CO[378]		
		X[801]				
H₂C=CH	Me	I[649]	198	EtOH/Et₂O	100	HMR
	Et	I[649]	327			
	CH₂Cl	XV[798]	239			
	Me	[679]				
H₂C=CH	O=P(SMe)₂ (with –O)	I[708]	125	PhH/Et₂O		IR
Et	EtS–P(=O)(–O)–SMe	I[708]		PhH/Et₂O		IR, $n_D^{20} = 1.5352$, $d_{20} = 1.2241$
	MeNHCCH₂SP(=O)(–O)–SMe	I[708]	82	PhH/Et₂O		IR
	PrS–P(=O)(–O)–SMe	I[708]		PhH/Et₂O		IR, $n_D^{20} = 1.5355$, $d_{20} = 1.1510$

Table 2 (Continued)

Cation R_3PR'		Anion X	Method	m.	Solvent	Yield (%)	Other Data and Remarks
R	R'						
Et	Me	OPh(NO$_2$)$_3$-2,4,6	XV[313]	239			
		2,4,5-Cl$_3$PhO–P(=O)–SMe	XV[702]	88	Petroleum ether	71	
		4-NO$_2$-PhO–P(=O)–SMe	XV[702]			79	$n_D^{20} = 1.5702$, $d_{20} = 1.2756$, PMR -37[673a]
		J	I[6]	300			
	CH$_2$OH	OH·HgCl$_2$	XV[798]	159			
		J	X[435]	199			
	CH$_2$CN	Br	[534]	223	H$_2$O	60	pKa 7.5
	CH=CH$_2$	Br	XVI[826]	253	CH$_2$Cl$_2$/EtOAc	55	
	CH$_2$CO$_2$H	OPh(NO$_2$)$_3$-2,4,6	I[883]	122	Dioxane		
	CS$_2$Me	J	I[553]	96 dec.	CHCl$_3$/Pent		Ir
	CH$_2$CH$_2$Br	Br	I[454]	235 dec.	EtOH	42	
	CH$_2$CH$_2$OH	HgCl$_3$	XV[787]	164	EtOH		
		J	I[536] XVI[818]	237[818]	EtOH[818]	64[536]	
	CH$_2$CH=CHCl	PtCl$_6$	I + XV[883]	235 dec.	EtOH		Orange
	CH$_2$CH=CH$_2$	OPh(NO$_2$)$_3$-2,4,6	XV[883]	141	EtOH		Yellow
	CH$_2$C(=O)Me	Ph$_4$B	I[534]	196	DMF/EtOH		pKa 9.38

R	Anion	Ref	m.p.	Solvent	Yield	Notes
CHMe, CO₂H	Br	I[534]	70	EtOH/Et₂O		pKa 8.99, hygroscopic
CH₂O₂CMe	J	XV[798]	72			
CH₂CO₂Me	ClO₄, Ph₄B	XV[534]	171	DMF/EtOH		pKa 10.6
	Br	[534]	106	CH₂Cl₂/EtOAc		pKa 10.9
(CH₂)₂CO₂H	J	I[534]	97	Et₂O		Hygroscopic
CH₂CHCH₂Cl, OH	Monostyphnate	XV[883]	142	EtOH		Orange-yellow
	OPh(NO₂)₃-2,4,6	XV[883]	82	EtOH		
Pr, OH, CMe₂	J	I[883]	141			
	J	I[313]	178			
(CH₂)₃OH	J	X[473]	180			
CH₂CO₂Et	Cl	I[798]	142			
	Br	I[543]	83	Et₂O		
	J	I[543]	65	EtOH/Et₂O		Hygroscopic
		[861]	74			
(CH₂)₂O₂CMe, Bu	OPh(NO₂)₃-2,4,6, Br	XV[419]	82		100	
(CH₂)₄OH	Cl	I[798]	150			
CH₂AsOAsMe₂, Me	J	I[175]	66	Me₂CO	63	
		XV				
CHCO₂Et, Me	Br	I[88]	113			
(CH₂)₃O₂CMe	Cl	I[798]	86			
(CH₂)₂NMe₃Br⁻	Br	XVI[456]			47	Hygroscopic
EtO (cyclopentyl)	BF₄	XII[495]	60	Et₂O		

233

Table 2 (Continued)

Cation R_3PR' R	R'	Anion X	Method	m.	Solvent	Yield (%)	Other Data and Remarks
Et	CHEt CO2Et	J	5[43]	95	EtOH		
	(CH2)3CO2Et	J	5[43]	68	EtOH/Et2O	100	
	(CH2)4O2CMe	Br	I[798]	49		87	
	(CH2)2NEt2	J	I[539]	86 dec.		71	
	CH2Ph	Ph3SnBr	II[101]	129			
		Br	I[6]	179			
		Cl	[182]	178			
		PtCl6	XV[182]	78			
	CH2COPh	Br	5[34]	134	CH2Cl2/ EtOAc		pKa 8.17
	CH2Ph-Cl-2-Me-5	OPh(NO2)3-2,4,6	XV[218]	91	EtOH/H2O		Yellow
		J	I[218]	123	EtOH/Et2O		
	CH2Ph-Me-4	OPh(NO2)3-2,4,6	XV[218]	106	EtOH/Et2O		Yellow
		J	I[218]	133	EtOH/H2O		
	(CH2)2Ph	OPh(NO2)3-2,4,6	XV[313]	70			
	Oct	J	I[554]	94			
	C=NEt Ph	BF4	II[495]	50	EtOH/Et2O		
	EtO—(dihydroisobenzofuran structure)	BF4	II[495]	85	Et2O	90	
	Dec	O3SPh-Me-4	I[150]	61			

Structure	Anion/Group	Method [Ref]	m.p.	Solvent	Yield	Constants
naphthol-CH₂OH	Cl	R₃N⁺ R₃P⁺ HX[436]	154	EtOH/Et₂O	38	
OEt	BF₄	XII[495]	151	EtOH	60	
–CH₂–C=CMe / Me–N–C=O–Ph	J	XVII[436]	179	EtOH/Et₂O	88	
Dodec	O=P–S / MeS–P–OMe	XV[704]			91	$n_D^{20}=1.5142$, $d_{20}=1.0381$
	S=P–O / MeS–P–OMe	XV[587]				$n_D^{20}=1.5142$, $d_{20}=1.0381$
	S=P(OEt)₂–S	XV[704]			96	$n_D^{20}=1.5159$, $d_{20}=1.1111$
	Br	I[683]	121	EtOAc	100	IR, HMR, PMR –37, ppm
	Cl	I[431]	106		99	
	ClO₄	XV[150]	70			
CHPh₂	J	I[683]	111			
(CH₂)₂OCHPh₂	Br	I[491]	270	MeOH/Et₂O		
	J	I[818]	120	EtOH		
(CH₂)₂O₂CCH₂Ph-c-Hex-4	J	I[818]	123	Me₂CO/Et₂O		
Hexadec	J	I[683]	136	EtOAc		

Table 2 (Continued)

R	R'	Anion X	Method	m.	Solvent	Yield (%)	Other Data and Remarks
Et	Octadec	MeS–P(=S)–OMe, –O	XV^{187}	37		93	
		S=P(OEt)$_2$, –S	XV^{204}	34		65	
	Ph–Br–4	Br	5^{66}	138			
	Ph–Cl–2	J	I^{211}	182			
	Ph	J	I^{419}	178			
		Br	I^{211}	187^{211}	MeCOEt		
		Cl; U^{IV}Cl$_6$	I^{211}	<100	EtOH/ Et$_2$O^{546}		
		J	XV^{348}	151 dec.			
			I^{211,464}; IV^{494}	141^{494}	CHCl$_3$^{494}	95^{494}	
	Ph–Cl–2–Me–5	NO$_3$	6^{99}	50	EtOH		
		OPh(NO$_2$)$_3$–2,4,6	XV^{218}	121	EtOH		
		Br	2^{13}	181			
		J		188	MeCOEt		
	Ph–Me–2	J	I^{71}	162			
	Ph–Me–3	OPh(NO$_2$)$_3$–2,4,6	XV^{218}	88	EtOH/H$_2$O		
		J	I^{213}	151	EtOH		IR

Compound	Anion	Ref.	mp	Solvent	Yield (%)
Ph-Me-4	PtCl6	XV[203]	217		
Ph-OMe-4	J	I[711]	65		
Ph-Me2-2,4	PtCl6	XV[711]	148		
Ph-NMe2-4	J	I[711]	136		
PhCO2Et-4	J	I[718]	180		
(naphthyl)	OPh(NO2)3-2,4,6	XV[1019]	131		
	J	I[582]	209		
Ph-NMe2-4 Et	BF4	XV[495]	265	EtOH/H2O	70
Ph-Ph-4	OPh(NO2)3-2,4,6	XV[218]	89	Me2CO/H2O	
	J	I[218]	219	H2O	
Ph-OPh-4	J	I[218]	181		
Ph-PPh2-2	J	I[419]	214	EtOH/H2O	
Me	OPh(NO2)3-2,4,6	XV[656]	134	EtOH/H2O	
	O3SPh(NO2)2-2,4	I[656]	144	H2O	
		I[591]	229		
	Cl	I[656]	243[656]	AcOH[379]	42 ... 97[383]
	J	X[847]	149	AcOH	
CH2OH	Cl	I[358]	180	MeCN	
CH2CN	Cl	I[591]	242[591]	MeCN	90[492]
Et	Br	I[383]	269	BuOH	100
CH2CH=CH2	OPh(NO2)3-2,4,6	XV[847]	126	EtOH/Me2CO	
	Cl	I[847]	135	AcOH/Me2CO	
	J	I[383]	177	BuOH	99
Pr	J	I[611]	169	AcOH	
CH2CO2Et	Br	I[242]	169	MeCN	
Bu	J	I[611]	153	AcOH	62
c-Pent	Br	I[381]	83		
	J	I + XV[847]	176	H2O	
CH2Ph-Cl2-2,4	Br	I[380]	195	MeCN	
CH2Ph-NO2-4	Cl	I[242]	>300[380]	BuOH[380]	90
CH2Ph	Cl	I[242]	160	MeCN	97[591]
CH2COPh	Br	I[450]	54		
Octadec					

NC(CH2)2

Table 2 (Continued)

Cation R_3PR' R	R'	Anion X	Method	m.	Solvent	Yield (%)	Other Data and Remarks
MeCH=CH	Me	J	I[112]	305 dec.			
MeCOCH₂	Ph	BF_4	XVI[435]	96		90	
Pr	Me	MeS–P(=O)(–O)–OMe	XV[702]			44	$n_D^{20} = 1.4952$, $d_{20} = 1.1942$
		P–(SMe), O=P–O	I[708]	93	PhH/Et₂O		IR
		MeS–P(=O)(S)–OEt	XV[702]			36	$n_D^{20} = 1.4937$, $d_{20} = 1.1178$
		MeS–P(=O)(S)–SEt	I[708]		PhH/Et₂O		IR, $n_D^{20} = 1.5442$, $d_{20} = 1.2501$
		MeS–P(=O)(S)–OPr	XV[702]			35	$n_D^{20} = 1.4901$, $d_{20} = 1.0896$
		MeS–P(=O)(–O)–SCH₂CONHMe	I[708]		PhH/Et₂O		IR, $n_D^{20} = 1.5220$, $d_{20} = 1.2158$

R / Compound	Substituent	Method	m.p.	Solvent	Yield	Notes
	OPh(NO$_2$)$_3$-2,4,6	XV^{476}	81	EtOH/H$_2$O		
	MeS–P(S)(O)–OPh–Cl$_3$-2,4,5	XV^{702}	75		65	
	MeS–P(O)(S)–Ph–OPh–NO$_2$-4	XV^{702}		EtOH	80	$n_D^{20} = 1.5569$, $d_{20} = 1.2175$
CH$_2$CHBr$_2$	J	I^{214}	213			
	OPh(NO$_2$)$_3$-2,4,6	XVI^{883}	91	EtOH		
	Styphnate	XV^{883}	98			
CH$_2$CHCl$_2$	OPh(NO$_2$)$_3$-2,4,6	XVI^{883}	78	EtOH		
CH$_2$CO$_2$H	OPh(NO$_2$)$_3$-2,4,6	I + XV^{883} XVI + XV	104	EtOH		
CS$_2$Me	J	XV + XVI^{553}	71 dec.	CHCl$_3$/Pent		IR
Et	OPh(NO$_2$)$_3$-2,4,6	XV^{310}	64			
	MeS–P(S)(O)–OPh–Cl$_3$-2,4,5 J	XV^{702}			62	$n_D^{20} = 1.5495$, $d_{20} = 1.2989$
	MeS–P(S)(O)–OPh–NO-4	XV^{702}			44	$n_D^{20} = 1.5591$, $d_{20} = 1.1876$
	J	^{310}				
CH$_2$–CH=CHCl	OPh(NO$_2$)$_3$-2,4,6	XV^{883}	124	EtOH		
CH$_2$–CH=CH$_2$	OPh(NO$_2$)$_3$-2,4,5	XV^{883}	60	EtOH		
	Styphnate	XV^{883}	60	EtOH		Orange
	PtCl$_6$	XV^{883}	90	EtOH		
(CH$_2$)$_3$Br	OPh(NO$_2$)$_3$-2,4,6	XV^{883}	192	EtOH		
	Styphnate	XV^{883}	81	EtOH		
OH CH$_2$CHCH$_2$Cl	Styphnate	XV^{883}	81	EtOH		
	J	I^{883}	112	EtOH		
			145	EtOH/Et$_2$O	66	

Table 2 (Continued)

Cation R₃PR'						Yield	Other Data
R	R'	Anion X	Method	m.	Solvent	(%)	and Remarks
Pr	OH / CH₂CHMe	Styphnate	XV[883]	87	EtOH		
		PtCl₆	XV[883]	188	EtOH		
	CHCH₂=CHCl / Me						
	CH₂CH=CMe / Cl	OPh(NO₂)₃-2,4,6	XV[883]	75	EtOH		
	CH=CHCH₂Me	OPh(NO₂)₃-2,4,6	XV[476]	74	EtOH		HMR
	CH₂CH=CHMe	PtCl₆	XV[476]	192	EtOH		
	CH₂-CH-CH₂ / HN_N	OPh(NO₂)₃-2,4,6	XVI+ / XV[883]	79	EtOH		
	CH₂CO₂Et	Br	I[240]	96	EtOAC		
	Bu	OPh(NO₂)₃-2,4,6	XV[313]	76[313]	EtOH[476]		Pale yellow[476]
		PtCl₆	XV[476]	217			
		J	I[313]	239			
	CH₂CHMe / OMe	Styphnate	XV[883]	50			
	CHCO₂Et / Me	Br	I[88]	69			
	CH₂CHCH₂Cl / OEt	OPh(NO₂)₃-2,4,6	XVI+ / XV[883]	62	EtOH		

			M.p. (°C)	Solvent	Yield (%)	Notes
$CH{=}CHCH_2NMe_2$	$OPh(NO_2)_3\text{-}2,4,6$	$XV[L76]$	204	EtOH/ Me$_2$CO/ petroleum ether		HMR
$CH_2CH{=}CHNMe_2$	$OPh(NO_2)_3\text{-}2,4,6$	$XVI +$ / $XV[476]$	204 dec.	EtOH/ Me$_2$CO/ petroleum ether		HMR
$CH_2\underset{OEt}{CHMe}$	$OPh(NO_2)_3\text{-}2,4,6$	$XVI +$ / $XV[883]$	40	EtOH		
	Styphnate	$XVI +$ / $XV[883]$	67	EtOH		
	$PtCl_6$	$XVI +$ / $XV[883]$	117	EtOH		Orange
$(CH_2)_4OMe$	$PtCl_6$	$XV[476]$	194	EtOH		HMR, Deep orange, brown
$\underset{OH}{CH_2CHCH_2NMe_2}$ $+$	$OPh(NO_2)_3\text{-}2,4,6$	$XVI +$ / $XV[476]$	185	EtOH/ Me$_2$CO/ petroleum ether		HMR
$CH_2CH{=}CHNMe_3$ $+$	$OPh(NO_2)_3\text{-}2,4,6$	$XVI +$ / $XV[476]$	175	EtOH/ Me$_2$CO/ petroleum ether		HMR
$(CH_2)_3NMe_3$ $+$	$OPh(NO_2)_3\text{-}2,4,6$	$XVI +$ / $XV[476]$	186	EtOH/ Me$_2$CO/ petroleum ether	99	HMR, pale yellow
$\underset{OH}{CH_2CHCH_2NMe_3}$ $+$	$OPh(NO_2)_3\text{-}2,4,6$	$XVI +$ / $XV[476]$	194	EtOH/ Me$_2$CO/ petroleum ether		HMR, pale yellow

Table 2 (Continued)

Cation R_3PR'		Anion X	Method	m.	Solvent	Yield (%)	Other Data and Remarks
R	R'						
Pr	CH$_2$Ph	OPh(NO$_2$)$_3$-2,4,6	XV[765]	105	EtOH		UV
		Cl	I[765]	168	EtOH/Et$_2$O		
	Oct	AuCl$_3$	XV[313]	38			
	Dodec	MeS–P(=S)–OMe	XV[587]				$n_D^{20} = 1.5083$, $d_{20} = 1.1207$
		MeS–P(=O)–S	XV[704]			96	$n_D^{20} = 1.5088$, $d_{20} = 1.1207$
		(EtO)$_2$P(=S)–S	XV[587]				$n_D^{20} = 1.5098$, $d_{20} = 1.1102$
		Br	I[555]		Et$_2$O[555]	65[587]	Oil, $n_D^{20,555} = 1.4937$, n_D^{20}[587] $=$, d_{20}[587] $= 1.0377$[587]
	Octadec	MeS–P(=S)–OMe	XV[587]				$n_D^{20} = 1.5041$, $d_{20} = 1.0701$
		MeS–P(=O)–S	XV[704]			70	$n_D^{20} = 1.5041$, $d_{20} = 1.0701$
		(EtO)$_2$P(=S)–S	XV[704]			97	$n_D^{20} = 1.5100$, $d_{20} = 1.1024$

R	R'	Structure	Ref.		Solvent	Yield (%)	Spectra
i-Pr	Ph	Br	587	68		65	PMR −45.1, ppm
	Ph–Me–4	J	I[214]	132			
	Me	Br	I[214]	126			
		Br	397				
$H_2C=CCH_2$ $\vert$ Me	Me	J	I[208]	>360	EtOH/Et$_2$O		
		J	I[564]	151			
EtCH $\vert$ Me	Me	J	I[208]	149			
Bu	Me	MeS–P(=O)–OMe (O)	XV[702]			39	$n_D^{20} = 1.4931,$ $d_{20} = 1.0763$
		O=P(SMe)$_2$–O	I[702]				IR, $n_D^{20} = 1.5221,$ $d_{20} = 1.0713$
		MeS–P(=O)–OEt (O)	XV[702]			45	$n_D^{20} = 1.4935,$ $d_{20} = 1.0962$
		MeS–P(=O)–SEt (O)	I[702]				IR, $n_D^{20} = 1.5068,$ $d_{20} = 1.0663$
		MeS–P(=O)–OPr (O)	XV[702]			25	$n_D^{20} = 1.4872,$ $d_{20} = 1.0680$
		MeS–P(=O)–SPr (O)	I[702]				IR, $n_D^{20} = 1.5104,$ $d_{20} = 1.0597$
		MeS–P(=S)–OPhCl$_3$–2,4,5 (O)	XV[702]			85	$n_D^{20} = 1.5431,$ $d_{20} = 1.3005$

Table 2 (Continued)

Cation R_3PR'		Anion X	Method	m.	Solvent	Yield (%)	Other Data and Remarks
R	R'						
Bu	Me	$\begin{matrix}\ \ \ \ O\\ \ \ \ \ \|\|\\ MeS-P-OPhNO_2\text{-}4\\ \ \ \ \ \|\\ \ \ \ \ S\end{matrix}$	XV[702]	135[341]		75	$n_D^{20}=1.5462$, $d_{20}=1.2341$, HMR[389]
	CH_2CN	J	I[341]				HMR
	$CH=CH_2$	J	I[389]				
		Br	XVI[384]	152[384]	EtOAc/MeCN[384]	55[826]	
	CH_2CO_2H	Br	I[1067]			90	IR
	CH_2CH_2Br	Cl	I[677]				IR, HMR
		Br	I[826]	75	PhCl/Et$_2$O	72	
Et	Et	$\begin{matrix}\ \ \ \ O\\ \ \ \ \ \|\|\\ MeS-P-SCH_2CONHMe\\ \ \ \ \ \|\\ \ \ \ \ O\end{matrix}$	I[702]				$n_D^{20}=1.5258$, $d_{20}=1.0912$
		$\begin{matrix}\ \ \ \ O\\ \ \ \ \ \|\|\\ MeS-P-OPhCl_3\text{-}2,4,5\\ \ \ \ \ \|\\ \ \ \ \ O\end{matrix}$	XV[702]			58	$n_D^{20}=1.5402$, $d_{20}=1.1925$, PMR -35.5 ppm
		Br	[397]				PMR -36[673]
	$(CH_2)_2OH$	J	I[1210]	153			IR
		Ph_4B	XV[384]	124	EtOH		
		J	I[1067]				
	$CH=CHCO_2H$	Br	I[792]	185	CHCl$_3$/EtOAc	75	IR, HMR
	$CH_2CH=CH_2$	Br	I[584]	78	Et$_2$O	97	IR
		Cl	I[389]				HMR

R	X	Compound	m.p. (°C)	Solvent	Yield (%)	Notes
$CH=CHCO_2Me$	Br·PhH	I^{792}	78	$CHCl_3$/EtOAc/PhH	66	HMR
$CH_2CO_2CH=CH_2$	Cl	I^{426}	95		87	HMR
$CH_2C=CH_2$, Me	Cl	I^{389}				HMR
CH_2CO_2Et	Br	I^{1009}	99	CH_2Cl_2/Et_2O	53	HMR
$(CH_2)_2O_2CMe$	Cl	I^{677}	86	Et_2O^{384}		Very hygroscopic
	Ph_4B	XV^{384}	177^{384}	H_2O^{384}		
		XVI^{386}				
	Br	XVI^{385}	152^{385}	EtOAc/$MeCN^{385}$	87	
$(CH_2)_2OEt$	Ph_4B	XVI^{584}	174		85	
	Br	I^{171}, ^{389}	182			
$CH_2CH=CMe_2$	Br	XVI^{584}	165			
$(CH_2)_2-N-CH_2$, C=CH_2, Me-N (cyclic)	Br					
$(CH_2)_2NEt_2$	Ph_4B	$XVI+$, XV^{584}	155		63	
CH_2PhCl_2-2,4	$O=P(SBu)_2$, O	XV^{709}		PhH	86	$n_D^{27}=1.5104$, $d_4^{20}=1.033$
$O=P-OCH_2CH(CH_2)_3Me$, O-P-O, Et		XV^{709}		PhH	100	$n_D^{27}=1.4927$, $d_4^{20}=1.055$
CH_2PhCl_2-3,4	Cl	I^{287}	193	Dioxane	96	
CH_2Ph-NO_2-4	Br	I^{380}	168	PhH	91	
CH_2Ph	O_3SPh-Me-4	I^{589}	62	EtOH/Et_2O	83	
	Br	$XIVB^{836}$	153^{836}	CH_2Cl_2/Et_2O^{836}	72^{836}	HMR^{373}
	Cl	I^{237}	164^{287}	Et_2O^{287}		PMR -31.5^{5673a} IR^{1067} HMR^{389}

Table 2 (Continued)

Cation R₃PR'		Anion X	Method	m.	Solvent	Yield (%)	Other Data and Remarks
R	R'						
Bu	CH₂Ph	ClO₄ J	777 373	143	EtOH		HMR
		Ph₄B	XVI+[584] XV[556]	170 dec.	Chromatography	59	
	CH₂COPh	Br J	I[389] I[1067]				pK$_a$ 8.60 HMR IR
	CH₂COPh-NH₂-4	Cl	I[318]	130			
	CH₂Ph-Me-4	Cl	I[389]	155	Et₂O/ petroleum ether		HMR
	CH₂Ph-OMe-4	Ph₄B	I[557]	184			
	(CH₂)₂SPh	Ph₄B	XVI+[584] XV[584]	165	CHCl₃/PhH	74	
		Cl	I[19]	75	EtOAc	89	
	(CH₂)₂P(O)(CH₂CH₂CN)₂	Ph₄B	XVI+[584] XV[584]	176	MeCN	69	
	(CH₂)₂CHCMe(O)CO₂Et	Ph₄B	XVI+[584] XV[584]	163		90	

Compound	Reagent	m.p./b.p.	Method[ref]	Yield (%)	Solvent	Physical data
(CH₂)₂–N, N=CH (o-phenylene)	Ph₄B	155	XVI+ [584], XV[584]	80		
(CH₂)₂CPh–NHCOMe (O)	Cl	172	I[318]			
Me, Me / Ph (cyclobutene)	Br	225	XVI[584]	6		IR, UV, HMR
(CH₂)₂CMe₂, C=O, Ph	Ph₄B	152	XVI+ [584], XV[584]	95		
Dodec	MeS–P(=O)–S–OMe		XV[704]	97		$n_D^{20} = 1.5080$, $d_{20} = 1.1114$
	MeS–P(=S)–O–OMe		XV[587]			$n_D^{20} = 1.5060$, $d_{20} = 1.1114$
	S=P(OEt)₂–S		XV[704]	83		$n_D^{20} = 1.5875$, $d_{20} = 1.0901$
	Ph₃SnBrCl	118	II[101]	16		$n_D^{20} = 1.5115$[587],
	Br	33[727]	I[727]	65[587]		$d_{20} = 1.0501$[587]
(fluorene with H)	Br	194[558]	I[558]	47[558]	EtOH/Et₂O[558]	IR[1067]
CHCOPh, CH₂COPh	Cl·HCl	116	XIV[836]		CH₂Cl₂/Et₂O	HMR, PMR −35.7[673a]

Table 2 (Continued)

Cation R_3PR'		Anion X	Method	m.	Solvent	Yield (%)	Other Data and Remarks
R	R'						
Bu	Octadec	MeS–P(=O)(OMe)–S	XV[704]			89[703]	n_D^{20} = 1.5009,[704] d_{20} = 1.0822[704]
		MeS–P(=S)(OMe)–O	XV[587]				n_D^{20} = 1.5089, d_{20} = 1.0822
		S=P(OEt)$_2$–S	XV[704]			89	n_D^{20} = 1.5835, d_{20} = 1.1032
	CPh$_3$	Br	[587]	85		65	
		Cl	I[171]	125		42	
	(CH$_2$)$_2$CONHCO$_2$-octadec $\mid$ Ph	Br	I[324]	89			
	CH–CHCOCH$_2$Ph $\mid$ CH$_2$COPh	Br	XIV[836]				IR, HMR PMR −28.8[673a]
	Ph	Br	[1067]				IR
		J	IV[494] VI[813]	156[494]	CHCl$_3$[494]	75[494]	
i-Bu	Ph–Ph–4	J	IV[494]	128	CHCl$_3$	47	
	CH$_2$CO$_2$CH=CH$_2$	J	IV + XV[494]	114		70	
		Cl	I[426]	93		44	

R	Substituent	Anion	Type[ref]	mp (°C)	Solvent	Yield (%)	Other
t-Bu	Me	J	I[469]	>360	$EtOH/Et_2O$	86	
Me_3SiCH_2	Me	J	I[952]	126			PMR −40.0 ppm
c-Pent	Me	Br	I[597]				
c-Hex	Me	Br	I[77]	183	$CHCl_3/EtOAc$	94	
	Me	J	I[778]	182			
	CH_2CN	Br	5[34]	252	$EtOH/Et_2O$		pKa 9.98
	$CH=CH_2$	Br	XVI[846]	350	$MeCN/$dioxane		
	CH_2CO_2H	Br	I[94]	148[94]	CH_2Cl_2/Et_2O[94]	77[77]	
	CH_2COMe	Ph_4B	5[34]	199	$DMF/EtOH$		pKa 9.26
	CH_2COMe	Br	5[34]	196	$CH_2Cl_2/EtOAc$		pKa 8.73
	CH_2CO_2Me	Br	5[34]	160 dec.	$CH_2Cl_2/Et\,OAc$		
	CH_2CO_2Et	Ph_4B	XV[94]	193	$EtOH$		pKa 11
	$(CH_2)_2\text{-}N$〈triazole, N-Me〉	Br	XVI[20]	240	$MeCN$	85	IR
	CH_2Ph	Cl	I[77]	265	$CHCl_3/EtOAc$	90	
	CH_2COPh	Br	5[34]	229	$EtOH/Et_2O$	80	pKa 7.95
	$CH_2CH=CHPh$	Br	I[77]	242	$CHCl_3/EtOAc$		
Hex	$(CH_2)_2OH$	J	I[610]		$Et_2O/$petroleum ether		$n_D^{20} = 1.5225$, $d_{20} = 1.1650$
$PhCH_2$	CH_2CO_2Et	Ph_4B	XV[94]	124	CH_2Cl_2/Hex		
	CH_2CO_2Et	BH_4	XV[168]	132	$EtOH$		
	CH_2CO_2Et	$OPh(NO_2)_3\text{-}2,4,6$	XV[910]	157	$MeOH$		
	Me	Br	I[36]	229[36]	$EtOH$[36]	64[36]	UV
	Me	J	I[452]	170	$EtOH$		UV[912]

Table 2 (Continued)

Cation R_3PR'		Anion X	Method	m.	Solvent	Yield (%)	Other Data and Remarks
R	R'						
PhCH	Et	Cl	I[182]	110 dec.			
	$CH_2Ph-Br-3$	Br	I[638]	180	H_2O	88	
	$CH_3Ph-Cl-3$	Br	I[638]	178	H_2O	68	
	$CH_2Ph-Cl-4$	Cl	I[638]	229	H_2O	52	
	$CH_2Ph-F-3$	Cl	I[638]	224	H_2O	44	
	$CH_2Ph-F-4$	Cl	I[638]	228	H_2O	35	
	CH_2Ph-NO_2-3	Br	I[638]	191	H_2O	50	
	CH_2Ph-NO_2-4	Cl	I[638]	240	H_2O	13	
	$CH_2Ph-Me-3$	Br	I[638]	182	H_2O	85	
	$CH_2Ph-Me-4$	Cl	I[638]	277	H_2O	45	
	$CH_2Ph-OMe-4$	Cl	I[638]	184	H_2O	68	
	$CH_2Ph-t-Bu-4$	Br	I[638]	211	H_2O	76	
	Ph	Cl	I[478]	128[478]		85[464]	$E_{1/2}$[499]
		J	I[464]	199	$MeOH/H_2O$	100	
Hept	$(CH_2)_2OH$	J	I[610]		Et_2O/petroleum ether		$n_D^{20} = 1.5131$, $d_{20} = 1.1182$
$PhCO_2CH_2$	Ph	BF_4	XVI[435]	113			
Oct	$(CH_2)_2OH$	J	I[610]		Et_2O/petroleum ether		$n_D^{20} = 1.5096$, $d_{20} = 1.0905$
(fused bicyclic structure with Me and N–H)	CH_2CO_2Et	Ph_4B	XV[94]	79	CH_2Cl_2/hex		
	Me	J	I[719a]	171			

	R	Anion	Ref.	mp (°C) / solvent	Properties
Non	(CH$_2$)$_2$OH	J	I[610]		$n_D^{20} = 1.5031$, $d_{20} = 1.0573$
Ferrocenyl	Me	J	I[1005]	Infusible EtOH	4 Orange
Dec	CH$_2$CO$_2$Et	Ph$_4$B	XV[94]	72 CH$_2$Cl$_2$/hex	
Dodec	Me	Br	I[431]	99	IR, HMR, PMR, -32 ppm
4-Br-Ph	(fluorenyl-H structure) Pr	Br	I[559]	261 PhH/EtOH	
3-Cl-Ph		Br	I[412]	236 CHCl$_3$/EtOAc	
4-Cl-Ph	Me	J	I[905]	252 MeOH/H$_2$O	
	CH$_2$Ph–NO$_2$	Br	I[557]	260 PhH/CHCl$_3$	
	CH$_2$Ph–OMe-4	Br	I[557]	269 PhH/CHCl$_3$	
4-D-Ph	(structure-H)	Br	I[559]	273 PhH/EtOH	
Ph	Pr	Br	I[412]	240 CHCl$_3$/EtOAc	HMR Pale yellow
	CBr$_3$	Br	XVI[641], I[832]	246[832] MeNO$_2$[832]	
	CHBr$_2$	Br	XVI[641]	235[641] MeOH/ EtOAc[641], 100[832]	
	CHCl$_2$	OPh(NO$_2$)$_5$-2,4,6	I[138], VI[837]	159 EtOH/H$_2$O	29 IR, yellow
		Br	XV[874], I[1010]	MeOH/EtOAc	
		J	I[232]	129	
	CHJ$_2$	OPh(NO$_2$)$_3$-2,4,6	XV[961]	189 MeOH	
	CH$_2$Br	Ph$_4$B	XV	214	
		Br	I[1084]	242[961] H$_2$O[961], 75[1089]	UV[281], IR[281], PMR, -22.6 ppm[397]
			XVI[961]		

Table 2 (Continued)

Cation R_3PR'		Anion X	Method	m.	Solvent	Yield (%)	Other Data and Remarks
R	R'						
Ph	CD_2Br	Br	XVI[961]	237			IR, HMR
	CH_2Cl	$OPh(NO_2)_3\text{-}2,4,6$	XV[959]	192	MeOH/H_2O		
		Ph_3SnCl_2	II[101]	171		82	
		Ph_4B	XV[959]	222[1082]	Me_2CO[1082]		
		Br	XIV[960]	209	i-PrOH/EtOH		
		Cl	XVI[596]	267 dec.[596]	CH_2Cl_2/Et_2O[596]	77[435]	PMR -23.8 ppm[397]
	CD_2Cl	Cl	XVI[916]	261			IR, HMR
	CH_2J	Ph_4B	XV[961]	178	Me_2CO		
		J	I[961]	231[916]	PhH[961]		
	CD_2J	J	XVI[916]	231	MeOH/Et_2O		
	CH=NOH	Br	I[1042]	166	$MeNO_2$/EtOAC	65	
	CH_2NO_2	Br	I[1037]		$MeNO_2$		
	Me	B_9H_{12}	XIV[377]	166 dec.			BMR
		$B_{10}H_{13}$	XIV+ , XV[429]	134	$CHCl_3$/Et_2O	73	UV, light yellow
		$B_{10}H_{15}$	XV[285]	128 dec.		77	IR, BMR
		$B_{20}H_{18}$	[156]				UV, yellow
		$Ag(CN)_2$	XV[1008]	127	CCl_4/$CHCl_3$		
		$B_{10}Cl_{10}C_2Cl$	XV[927]	182			
		$(CO_2)_2$	XV[929]	130			
		$Co(CO)_4$	XV[929]	95			

Compound	Ref.	M.p./dec. (°C)	Solvent	Yield (%)	Color/notes
$Ni(CN)_4$	XV[929]	262	BuOH		Light yellow
$HFe(CO)_4$	XV[929]	90			Red
$Cr(NH_3)_2(SCN)_4$	XV[741]	177			Light yellow, green
$Fe(CN)_6$	XV	162			UV^{910}
$OPh(NO_2)_3$-2,4,6	XV[556]	134[656]	$EtOH^{656}$		
$O_3SPh(NO_2)_2$-2,4	I[656]	183	EtOH		
O_3SPh-Me-4	I[588a]	140		80	
$Co^{II}\left(\!-S\diagdown\diagup S-\!\right)_2$ (CN)$_2$	XV[1058]	215 dec.	MeCN/MeOH		Brown
$Co^{III}\left(\!-S\diagdown\diagup S-\!\right)_2$ (CN)$_2$	XV[1058]	>230	EtOH/MeCN		
$(CO)_4Mo\,I\ (-S\diagdown\diagup S-)$ (CN)$_2$	XV[186]	119[186]	Me_2CO/H_2O^{186}	43[186]	$IR,^{627}\ \diagdown^{627}$
$Ni^{II}\left(\!-S\diagdown\diagup S-\!\right)_2$ (CN)$_2$	XV[1058]	220 dec.[1058]	EtOH/MeCN[1058]		Orange-red[59]
$Ni^{III}\left(\!-S\diagdown\diagup S-\!\right)_2$ (CN)$_2$	XV[1058]	198 dec.[1058]	EtOH[1058]		Crystal structure,[340] black[1058]
$VO^{IV}\left(\!-S\diagdown\diagup S-\!\right)_2$ (CN)$_2$	XV[32]			40	IR, UV, olive green
$CuJ_2\,(F_3C\overset{O}{C}CH_2\overset{O}{C}Me)_2\,EtOH$	XV[361]				IR, dark green

Table 2 (Continued)

Cation R₃PR' R	R'	Anion X	Method	m.	Solvent	Yield (%)	Other Data and Remarks
Ph	Me	$\left[\text{C(CN)}_2\!=\!\bigcirc\!=\!\text{C(CN)}_2\right]_2$	XV[2]	231[2]		60[2]	PMR,[580] ESR,[561] ρ,[723] thermoelectricity,[1116] black[2]
		$\left[\text{C(CN)}_2\!=\!\bigcirc\!=\!\text{C(CN)}_2\right]_2$	XV[248]	180 dec.	MeCN	71	Blue-black
		$\left[(\text{CN})_2\text{C}\!=\!\bigcirc\!\bigcirc\!=\!\text{C(CN)}_2\right]_2$	XV[248]	446	MeCN[248]	76	UV,[248] black[248]
		Ph₃SnCl₂	II[101] [968]	146			
		Ph₃BCH₂CH=CH₂	XV[741]	168			
		Ph₄B	XV[964]	197	MeOH/Et₂O		
		Ph₃BSiPh₃	I[492]	>200 dec.	MeOH		
		Br		234[492]	MeOH/Et₂O[92]	98[676]	E½,[2][482] 1.862,[482] −22.767[3a] PMR, IR, HMR
	CD₃	Br	XVI[916]	230			
	¹⁴CH₃	Br	I[33]	229	PhH		
	Me	CdJ₄	XV[222]	174	Me₂CO	91	

	Reference	m.p.	Recryst. solvent	Yield	Methods/notes
Cl	XVI[165]	221[165]	EtOAc/petroleum ether[238]	92[238]	IR,[238] HMR[389]
	I[389] 777		EtOH		
ClO$_4$		163			
CuCl$_2$	XV[929]	143			
CuCl$_3$	XV[929]	120			
ZnCl$_3$	XV[1085]	194	2nHCl		
	XIV				
	I				
CuCl$_4$	XV[529]	180			Citron yellow
Fe^{III}Cl$_4$	II[741]	178			$\mu = 5.93$ Bohr magnetons, yellow UV
Mn^{II}Cl$_4$	XV[192]	216		62	
U^{VI}O$_2$Cl$_4$	XV[348]	316 dec.	MeOH		Orange-yellow UV
PtCl$_6$	XV[741] 347	242			
Th^{IV}Cl$_6$					
U^{IV}Cl$_6$	XV[348] 197	268 dec.			x-Ray diffraction pattern
Pt^{II}(SnCl$_3$)$_5$					
Hg^{II}J$_3$	XV[222]	146	Me$_2$CO/EtOH		
Hg^{II}J$_4$	XV[228]	172	Me$_2$CO/EtOH		
J	I[1070]	187[873]	H$_2$O[465]	91[465]	UV,[665] IR,[659] HMR[442]
	XVI[373]				PMR −21.0,[455] 673a
J	XV[465]	189[916]	Et$_2$O[33]	98[33]	pKa >>11[534]
	I[33]				

CD$_3$

255

Table 2 (Continued)

Cation R_3PR' R	R'	Anion X	Method	m.	Solvent	Yield (%)	Other Data and Remarks
Ph	$^{14}CH_3$	J	I[33]	187	Et₂O	100	
	CH₂OH	Ph₃SnCl₂	II[101]	138		49	
		BF₄	X[1082]	128	CH₂Cl₂/EtOAc	64	PMR -17.7 ppm,[397] UV[912]
		Br	X[961]	203	i-PrOH	69[435]	
		Cl	X[596]	192 dec.[596]	Et₂O[596]	90[435]	
		J	X[485]	170 dec.	EtOH/Et₂O	78	
	ÇHCHO HgCl	HgCl₃	XIV[754]	192	DMF		
	CH₂CN		[1105]	136			
		Br	XV[1052]	256[907]	PhH[907]		E1/2 1.301[482] 2.051[534]
	CH=CH₂	Cl	I[907] I[1052]	278[1037]	MeNO₂[1037]	83[1052]	pKa 7.44 UV[1037]
		Br	XVI[956]	189[956]	CHCl₃/EtOAc[956]	92[931]	
	CD=CH₂	Br	XVI[956]	186	CHCl₃/EtOAc		HMR, IR
	CH₂CHO	Br	XIV[895]	146	CHCl₃/EtOAc		
		Cl	I[1038]	212 dec.			

R	X	Type	m.p. (°C)	Solvent	Yield (%)	Spectra / Notes
CH_2CO_2H	Br	I[389]	157[916]			HMR[389], IR[1067], PMR −20[673a]
CD_2CO_2H	Cl	XVI[12]; I[238]	221[238]	EtOH/Et_2O[238]		IR; IR,[238] PMR −20 ppm[238]
	Br					
	Cl	XVI[677]			88[677]	ESR[632] after x-ray irradiation
$(CH_2)_2Br$	$Cr(NH_3)_2(CNS)_4$	[955]	136	Me_2CO/H_2O		
	Br	XVI[955]	298[337]	$CHCl_3$/EtOAc[955]	75[956]	HMR[955]
CD_2CH_2Br	Br	I[337]; XVI[956]	198	$CHCl_3$/EtOAc	80	HMR
$CHDCH_2Br$	Br	XVI[956]; [1105]	198		35	HMR
CH_2CONH_2	CNSe	I[171]	125			
	Cl	I[1037a]	206			
CHMe	Br		103 dec.	$MeNO_2$		
NO₂	$B_{10}H_{10}$	XV[449]	233 dec.	DMF/H_2O		
Et	$OPh(NO_2)_3$-2,4,6	XV[370]	135			
	O_3SPh	I[581]	125	EtOH/Et_2O	90	
	O_3SPh-Me-4	I[589]	93	EtOH/Et_2O	87	
	$[C(CN)_2\!-\!C_6H_4\!=\!C(CN)_2]_2$	XV[2]	223 dec.		41	Black
	$B_{12}H_{10}PPh_2$	XV[926]	242	Me_2CO	90	
	$Ph_3SnClBr$	II[101]	139		61	
	BF_4	I[495]	127	CH_2Cl_2/Et_2O	100	

Table 2 (Continued)

Cation R_3PR'		Anion X	Method	m.	Solvent	Yield (%)	Other Data and Remarks
R	R'						
Ph	Et	Br	I[1087] XV[228]	205[492]	EtOH/Et₂O[228]	96[1087]	UV,[65] HMR,[389] PMR -26.2 ppm[397] pKa>11,[534] E1/2 1.837, 1.970[482]
		CdBr₄ Re₄IIIBr₁₀	XV[222] XV[877]	109	EtOH		
		Cl	XIV[68] I + XV[169]	234[68]	CHCl₃/Et₂O[68]	89[82]	IR,[169] UV, Λ[169] HMR[169]
		ClO₄	XV[370]	164			
		HgIIJ₃	XV[228]	128	Me₂CO		
		HgJ₄	XV[228]	126			
		J	I[719]	167[493]	PhH[606]	80 100[606]	HMR[442] PMR -26[673a]
			XIV[493] XV[228] XIV[67] XV[370] 9[12]				
		NO₃		127			
	CHMe—OH	Cl					
	CH₂OMe	Ph₃SnCl₂	II[101]	136		55	UV

Compound	X	Ref.	mp (°C)	Solvent	Yield (%)	Method
CD_2OMe	Ph_4B	XV[1082]	198	Me_2CO/EtOH	86	HMR[442]
	Cl	I[1082]	201 dec.[1082]	$CHCl_3$/EtOAc[1082]		
$(CH_2)_2OH$	Cl	XVI[916]	194		88	IR, HMR
	$Ph_3Sn(CNS)_2$	II[101]	124			
	Br	I[956]	217[956]	PhH[956]	90[956]	HMR[956]
		XVI[928]				
CH_2SMe	Cl	I[9]	233	EtOH		
	ClO_4	XV[646]	192	MeOH	68	
	$PtCl_6$	XV[713]	222			
	$OPh(NO_2)_3$-2,4,6	XV[374]	102			
	Ph_4B	XV[1082]	211	Me_2CO/EtOH		
	BF_4	XV[374]	210			
	Cl	I[1082]	224[916]	$CHCl_3$/EtOAc[1082]		
CD_2SMe	Cl	XVI[916]	225	$CHCl_3$/EtOAc	85	IR, HMR
CH=CHCN-trans	Br	I[792]	224			IR, HMR
$CH_2C\equiv CH$	Br	I[26]	168[26]	EtOH[26] / EtOAc	58[291a]	UV[26]
CBr_2CO_2Me	Br	XIV[641]	172	MeOH/Et_2O	100	
$CH=CHCO_2H$-trans	Br	I[792]	174 dec.	$CHCl_3$/EtOAc	61	IR, HMR
CHCOMe	J	4[62]	271 dec.		99	
| HgCl	Cl	XIV[753]	208 dec.	$MeNO_2$	95	IR
$CHCO_2Me$ | HgCl	Cl	XIV[753]	190	$MeNO_2$/Et_2O	69	IR
CH_2COCH_2Cl	Cl	I[511]	210 dec.	MeOH/EtOAc	85	
$(CH_2)_2CN$	O_3SPh-Me-4	XVI[1052]	104	Me_2CO	91	

Table 2 (Continued)

Cation R₃PR'		Anion X	Method	m.	Solvent	Yield (%)	Other Data and Remarks
R	R'						
Ph	$(CH_2)_2CN$	Br	I[906]	218	$PhNO_2$	94	
		J	XVI[460]	187 dec.		85	
	C=CH₂ / Me	Br	XVI[955]	197		93	IR, HMR
	CHCOMe / SO_3H	Cl	9[55]	243			
		OH	XIV[798]	176			
	CH=CHMe	Br	XVI[584]	235[584]	PhH/hex[584]	82[955]	IR,[584] HMR[584]
	Ch=CHMe-trans	Cl	6[46]	234	Me_2CO		HMR
		ClO_4	XVI[646]	221	H_3PO_4	78	
	c-Pr	Br	I[939]	188		99	IR, HMR
	$CH_2CH=CH_2$	O_3SPh-Me-4	XVI I[589]	146	$EtOH/Et_2O$	98	
		$Ph_3SnClBr$	XV[101]	117			
		Br	I[584]	225[584]	PhH[380]	100[380]	IR,[584] PMR -21.4[397] $E_{1/2}$ 1.626, 1.775, 2.055, 2.468[482] UV[162]
		Cl	I[584]	234[584]	PhH[427]	64[427]	

Compound	Anion	Ref	m.p.	Solvent	Yield	Data
$CHMe$ $-CHO$ CH_2COMe	Cl	XIV[1038]	231	EtOH/Et$_2$O		IR,[12] E1/2[681]
	OPh(NO$_2$)$_3$-2,4,6	XV[716]	166			pKa 6.80[534]
	Br	I[12]	227[12]	CHCl$_3$/Me$_2$CO[12]		PMR -20.3[397]
CD_2COMe	Br	XVI[12]	237	D$_2$O		IR
CH_2COMe	Cl	I[492]	dec.[492]	EtOH/Et$_2$O[492]		UV,[830] IR,[830] HMR[442]
CD_2COMe	PtCl$_6$	XIV[1038] XV[716]	198			IR, HMR
CH_2COMe	Cl	XVI[916]	237			UV
$CH_2-CH-CH_2$ (O)	J	XVI[830]	207	H$_2$O		
	J	I[105]	260 dec.	Et$_2$O	30	IR
$CHMe$ $-CO_2H$	Cl	9[49]				IR
CH_2O_2CMe	BF$_4$	XVI[435]	174		96	
CH_2CO_2Me	BF$_4$	XIV[759]	136		25	
	Br	I[520]	164 dec.[534]	PhH[520]	97[81]	PMR -20.3,[397] pKa 8.81,[534] E1/2[681]
CD_2CO_2Me	Br	XIV[81]	156			IR, HMR
CH_2CO_2Me	Cl	XVI[916]	149			IR
$(CH_2)_2CO_2H$	Cl	9[50] I[510]	198[510]	MeOH/ EtOAc[510]	67[238]	IR,[238] PMR -24 ppm[238]
$(CH_2)_2CO_2H$	PtCl$_6$·H$_2$O J	XIV[510] XVI[460]	199 dec. 161 dec.	EtOH/H$_2$O	78	IR
$(CH_2)_3Br$ $CHCHCH_2OH$ $-I-$ $BrOH$	Br J	I[159] I[414]	229[159] 138	Xylene[159] EtOH	63[337] 60	PMR -24[673a] UV

261

Table 2 (Continued)

Cation R_3PR'		Anion X	Method	m.	Solvent	Yield (%)	Other Data and Remarks
R	R'						
Ph	$CH_2\overset{O}{\overset{\|}{P}}(CH_2Br)_2$	OPh(NO₂)₃-2,4,6	XV[204]	182			
		Br	I[204]	226	MeOH/PhH	87	IR, HMR
	$CH_2\overset{O}{\overset{\|}{P}}(CH_2Cl)_2$	OPh(NO₂)₃-2,4,6	XV[204]	165			IR
		Br	XIV[204]	225			
		Cl	I[204]	230	MeOH/PhH	83	IR, HMR
	$CH_2\overset{O}{\overset{\|}{P}}(CH_2J)_2$	J	I[204]	228	H₂O	85	IR, HMR
		B₁₀H₁₀	XV[449]	226 dec.	DMF/H₂O		
		O₃SMe	I[589]	263	EtOH/Et₂O	87	
		CoII(CNS)₄	XV[946]	130			
		PdII(CNS)₄	XV[946]	133			
		FeII(CNS)₆	XV[946]	185 dec.			
		O₃SPh-Me-4	I[589]	141	EtOH/Et₂O	80	
		Ph₃SnClBr	XV[101]	114		50	
	Pr	Br	[62]	229[62]			PMR −24.1 ppm[397]
i-Pr		PtIVBr₆	XV[946]	215 dec.			
		PdCl₆	XV[946]	225			
		J	I[719]	202			
		BiIIIJ₄	XV[946]	75			
		PtIV(N₃)₆	XV[946]	86			
		CoIII(CNS)₄	XV[946]	141			

	Deriv.	Struct.	M.p. (°C)	Solvent	Yield (%)	Notes
	$Pd^{II}(CNS)_4$	XV[946]	180			
	$Fe^{III}(CNS)_6$	XV[946]	155 dec.			
	Br	I[389]	239[916]			HMR,[389a]; $E_{1/2}$ 1.833, 2.092,[499] −30.9[397] PMR
$CDMe_2$, i-Pr	Br	XIV[916]	239	CH_2Cl_2/EtOAc	88	
	$Bi^{III}J_4$	XV[946]	89			
	$Pt^{IV}Br_6$	XV[946]	225 dec.			
	$Pd^{II}Cl_4$	XV[946]	273 dec.			
	J	I[1087]	195[1087]	EtOH/Et_2O[1087]	94[1087]	HMR[442]
$CH(CD_3)_2$	J	I[33]	197	MeCN	77	
CHMe, −OMe	$Pt^{IV}(N_3)_6$	XV[946]	126			
	$Cl·2\ H_2O$	I[194]	88	$CHCl_3$/PhH	88	
$CHCH_2OH$, −Me	Br	I[956]	156		25	HMR
CH_2OEt, CH_2CHMe, −OH	J	9[56]	172			
	Ph_3SnCl_2	II[101]	117			
	Br	XVI[916]	221	i-PrOH	54, 50	IR
$(CH_2)_2OMe$	ClO_4	XVI[646]	185			
	Br	III[47]	216	MeOH	71	
	J	III[122]	138			
CH_2CHCH_2OH, −OH	J	I[414]		EtOH	60	Pale yellow, UV
CH_2SEt	$OPh(NO_2)_3$-2,4,6	XV[374]	103			
	Ph_3SnCl_2	II[101]	160		83	

Table 2 (Continued)

Cation R_3PR'

R	R'	Anion X	Method	m.	Solvent	Yield (%)	Other Data and Remarks
Ph	CH_2SEt	BF_4	XV[374]	159	$CHCl_3$/EtOAc	55	
	$(CH_2)_2SMe$	Cl	I[102a]	150	EtOH		
	$(CH_2)_3NH_2^+$	Br·HBr	XVI[159]	260			
	CH_2SMe_2	$OPh(NO_2)_3$-2,4,6	XV[374]	156			
		BF_4	XV[374]	234			
	CH–CH₂ / O=C–O–CH₂	Br	I[319]	196	MeOH/EtOAc	38	
	$CH{=}C(CF_3)_2$	Cl	XIV[91]	153	PhH		HMR, FMR PMR −17.3 ppm
	$CH{=}C(CO_2H)_2$	BF_4	XV[462]	261 dec.	Et_2O[915]	55	
		Cl	I[915]	310 dec.[915]		99[915]	IR[580]
	$CH_2C{\equiv}CCH_2Br$	Br	I[915]	179	Et_2O	83	
	$CH{=}CHCH{=}CH_2$	ClO_4	XV[323]	135	H_2O	90	
	$CH{=}CHCO_2Me$	Br	I[792]	158	$CHCl_3$/EtOAc		IR, HMR
	–CH–CH₂ / O=C–O–CH₂	Cl	I[792]	151 dec.	$CHCl_3$/EtOAc	74	IR, HMR
		Br	I[319]	196	MeOH/EtOAc	38	

Substituent	Halide	Ref.	m.p. (°C)	Solvent	Yield (%)	Method
$CHCH_2CO_2H$ ⎪ CO_2H	Br	XVI[460]	140 dec.		88	
	Cl	XVI[510]	125 dec.	Me_2CO/H_2O/HCl	99	
$CH_2CH=CHCH_2Br$	J	I[460]	104 dec.		69	
$(CH_2)_3CN$	Br	I[132]	180	H_2O	84	
$(CH_2)_2CONHCO_2H$	Br	I[1052]	218	EtOH		
$C=CHMe$ ⎪ Me	Br	I[324]	99	EtOAc	60	
	Br	XVI[956]	204	$CHCl_3$/EtOAc	81	HMR, IR
$CH=CHEt$	Cl	[956]	165			
CH–CH_2 ⎪ CH_2–CH_2	ClO$_4$	XVI[646]	171	H_3PO_4	92	
	Br	XVI[940]	280	THF/DMF	67	HMR
$CH_2C=CH_2$ ⎪ Me	Cl	I[427]	218[812]	H_2O[812]		
$CH_2CH=CHMe$	Br	I[107]	245	EtOH	84	
$CH_2CH=CHMe$	Cl	I[586]	214[323]	EtOAc[323]	80[323]	HMR[586]
$CD_2CH=CHMe$	Cl	I[940]	232	THF	90	HMR
$CH_2CH=CHMe$	ClO$_4$	XV[323] / I	178	EtOH	62	
$CH_2CH{<}^{CH_2}_{CH_2}$	Br	I[940]	183	EtOAc	83	
$CD_2CH{<}^{CH_2}_{CH_2}$	Br	XVI[940]	180	$CHCl_3$/EtOAc	97	HMR, TLC
$(CH_2)_2CH=CH_2$	Br	XVI[940]	228	CH_2Cl_2/EtOAc	66	IR, HMR
$CHCOMe$ ⎪ Me	Br	I[956]	185	H_2O	22	IR, HMR

Table 2 (Continued)

Cation R₃PR'		Anion X	Method	m.	Solvent	Yield (%)	Other Data and Remarks
R	R'						
Ph	CH₂OCH₂CH=CH₂	Cl	I[171]	175			
	CH₂CH=CHCH₂OH	Br	II[47]	214			
	CH₂CH=CHCH₂OH-cis	Cl	I[125]	198	CH₂Cl₂/ EtOAc/ MeOH		IR
	CHMe—CO₂Me	Br	I[356]	178[950]			
	CHCH₂CO₂H—Me	J	XIV[80]	138	EtOAc	75	
		Br	XVI[460]	178 dec.		70	
	CH₂CO₂Et	CNSe	[1105]	160 dec.	CHCl₃/ EtOAc		
		O₃SPh-Me-4	XIV[797]	155 dec.			
		Ph₃SnCl₂	II[101]	177		77	
		Ph₄B	XV[1083]	173	PhH[356]	27	
		Br	I[356]	158[356]		92[894]	
		Cl	I[187]	87 dec.	CH₂Cl₂/ CCl₄[187]	89[187]	IR,[12] HMR,[384] pKa 8.81[534] IR,[187] E1/2[681]
				144 dec. after drying over P₂O₅			

R	Substituent	Type[ref]	m.p. (°C)	Solvent	Yield (%)	Properties
CH_2CHMe | CO_2H	J	XVI[713,460]	165, 178 dec.		81	
$(CH_2)_3CO_2H$	Br	I[238]	234	EtOH/Et₂O	42	IR, PMR, -23 ppm
$(CH_2)_4Br$	Cl	I[726]	207[728]	CHCl₃/Me₂CO[728]	99[728]	PMR -24.2 ppm[397]
	Br	XV[728]	89	MeOH		Yellow
	Br_3	I[337]	200	PhH	6	
	Cl	XV[728]	174	H₂O		
	J	XV[728]	198	CHCl₃/Me₂CO		
	NO_3	XV[449]				
Bu	$B_{10}H_{10}$	I[589]	167 dec.	DMF/H₂O	86	Λ, X, IR, UV, pale green
	O_3SMe	XV[328]	252	EtOH/Et₂O	47	green UV
	$Mn^{II}(CNS)_4$	XV[910]	119	Me₂CO		
	$OPh(NO_2)_3$-2,4,6	I[583]	98	MeOH	96	
	O_3SPh-Me-4	I[101]	141	EtOH/Et₂O	65	
	$Ph_3SnBrCl$	I[380]	127, 249[380]	BuOH[380]	91[380]	$E_{1/2}$ 1.160, 1.766, 2.151,[482] PMR -24[673a]
	Br	II[47], IV[1091]				
CD_2Pr	Br	XVI[916]	240		80	IR, HMR
	$Br_2U^{IV}Cl_4$	XV[220]	254			Λ, X, IR, UV, green
Bu	$Br_2U^{VI}O_2Cl_2$	XV[221]	206	MeCN	60	Λ, X, UV, IR, Yellow
	$Br_2U^{VI}O_2J_2$	XV[221]	210	MeCN	50	Λ, X, UV, IR, bright red

Table 2 (Continued)

Cation R₃PR′		Anion X	Method	m.	Solvent	Yield (%)	Other Data and Remarks
R	R′						
Ph	Bu	Br₂UVIO₂	XV[221]	238	MeCN	80	Λ, X, UV, IR
		Br₆UIV	XV[220]	256		80	Λ, X, UV, IR, green
		Hg₂Br₇	XV[228]	111	Me₂CO/EtOH		
		Cl	I + XVI[79][77][77]	220	CHCl₃	77	
		ClO₄IV		186	EtOH		
		Cl₆UIV	XV[220]	254		90	Λ, X, UV, IR
		J	[912]				UV
		J₄UVIO₂	XV[221]	241	MeCN	70	Λ, X, UV, IR, dark red
	Et —CH —Me	Br	I[956]	230	CHCl₃/EtOAc	79	PMR −30.2[397]
	i-Bu	J·H₂O OPh(NO₂)₃-2,4,6	XVI[301] XV[910]	208 142	H₂O MeOH	>90	UV
		Br	I[78]	235[78]	PhH[78]	95[78]	E1/2 1.720[482], 2.039[482], HMR[442]
		J	I[719]	176[719]			UV[912]

	Ph₃P·NiBr₃ Ph₄B	Ref.	m.p.	Solvent	Yield	Notes
t-Bu		XV[191] [954]	221 219	BuOH MeNO₂/ MeOH	11	HMR,[954] E1/2 1.657,[482] 2.041[397]
	Br	I[954]	241 dec.[954]	CHCl₃/ EtOAc[954]	70[954]	
$CHCHMe$ | $MeOH$	Cl	XIV[954] I[492]	231 201[301] dec.	HCO₂H CHCl₃/ EtOAc[301]		PMR −34.7,[499] E1/2
	J	XVI[301]				
	Br	XVI[956]	244	CHCl / EtOAc	54	HMR
CH_2CHEt | OH	J	[956]	234	MeOH	88	
	ClO₄	XVI[646]	173			
CH_2CHMe | OMe	Br	[956]	195	MeOH/EtOAc	78	HMR
$(CH_2)_2OEt$	Br	XVI[931] I[171]	182[171]	EtOH/ Et₂O[931]	72[931]	IR[931]
$(CH_2)_3NHMe$	Br·HBr	XVI[159]	279	EtOH		
$CH_2\overset{+}{S}Et$ | Me	BF₄	XV[374]	223			
$CH_2\overset{+}{Sb}Me_3$	Br	I[953]	93			
$CH_2\overset{+}{Sb}Me_3$	Cr(NH₃)₂(SCN)₄ J, Br	I[952] XIV[398]	107 142			
CH_2SiMe_3	Br HgBr₃	XIV[398] XIV[398]	175 134			

Table 2 (Continued)

Cation R_3PR'		Anion X	Method	m.	Solvent	Yield (%)	Other Data and Remarks
R	R'						
Ph	CH_2SiMe_3	J	I[965]		EtOH/ EtOAc	98	IR
	CH_2SnMe_3	Me_3SnBr_2	XV[957]	152 dec.[957]	Et_2O[957]		
		$Cr(NH_3)_2(CNS)_4$	XIV[398]	123[398]	H_2O[957]		
		$HgBr_3$	XIV[398]	165	H_2O		
			XIV[957]				
		J	VII[484]	253 dec.	H_2O	50	
	CH_2–(pyridine ring, N; CH₃)	J	I[474]	182	EtOH		Yellow-brown
	CH_2–(furan ring, O)–NO_2	Br	I[934]	269 dec.	Et_2O		IR, HMR
	CH_2–(furan ring, O)	Cl	I[1095]	277[560]	PhH[1095]	70[1095]	IR[560]
	CH_2–(thiophene ring, S)	Br	I[972]	262			
	CH_2–(selenophene ring, Se)	Cl	I[1095]	276 dec.		53	

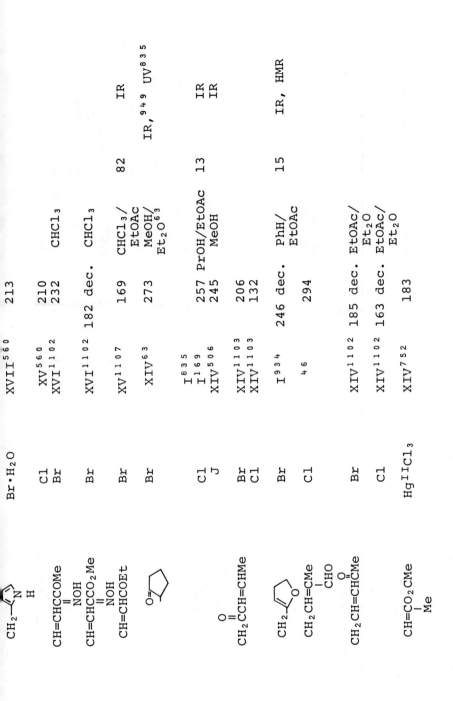

R		Ref.	mp (°C)	Solvent	Yield (%)	Spectra
CH₂=(2-pyrrolyl, N–H)	Br·H₂O[560]	XVII[560]	213			
CH=CHCCOMe ‖NOH	Cl	XV[560]	210			
	Br	XVI[1102]	232	CHCl₃		
CH=CHCCO₂Me ‖NOH	Br	XVI[1102]	182 dec.	CHCl₃		
CH=CHCOEt	Br	XV[1107]	169	CHCl₃/EtOAc	82	IR
(cyclopentanone, O=)	Br	XIV[63]	273	MeOH/Et₂O[63]		IR,[949] UV[835]
O=CCH=CHMe	Cl	I[835] I[169]	257	PrOH/EtOAc	13	IR
	J	XIV[506]	245	MeOH		IR
CH₂=(furanone)	Br	XIV[1103]	206			
	Cl	XIV[1103]	132			
CH₂CH=CMe CHO	Br	I[934]	246 dec.	PhH/EtOAc	15	IR, HMR
CH₂CH=CHCMe O=	Cl	[46]	294			
CH₂CH=CHCMe	Br	XIV[1102]	185 dec.	EtOAc/Et₂O		
	Cl	XIV[1102]	163 dec.	EtOAc/Et₂O		
CH=CO₂CMe Me	HgICl₃	XIV[752]	183			

271

Table 2 (Continued)

272

Cation R_3PR'		Anion X	Method	m.	Solvent	Yield (%)	Other Data and Remarks
R	R'						
Ph	$CH_2CH=CHCO_2Me$	Br	I[131]	166[131]	MeCN/ EtOAc[131]	72[131]	IR[949]
	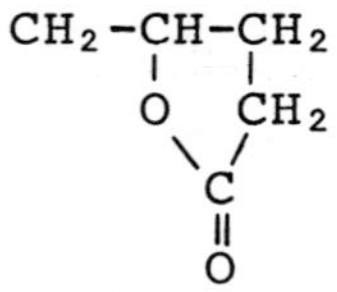 $CH_2-CH-CH_2$ $O-C$ CH_2 $O=C$ $C=O$	J	I[733]	218	Me₂CO/ petroleum ether		
	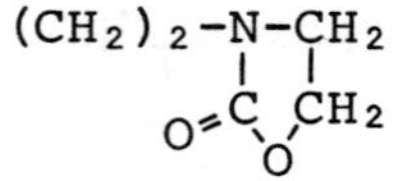 $(CH_2)_2-N-CH_2$ $O=C-O$ CH_2	Cl	[171]	74			
	$CH=CHCHMe_2$	CNS	XV[1107]	137			
		Br	XV[1107]	167			IR, HMR
		Cl	I[1107]	195	CHCl₃/ EtOAc	100	
		J	XV[1107]	175			
	 ⬠	Br	I[834]	261[834]	H₂O[834]		IR,[1067] UV,[834] PMR -30.7 ppm[397]
	$CH_2CH=CMe_2$	Br	XIV[834] I[42]	242[888]	H₂O[42]		
	$CH_2CH=CHCH_2Me$	Br	I[102]	202	Et₂O	90	HMR[389]

Structure	X	Code	No.	Solvent	Yield	Methods
CH₂CH=CHCH₂Me–trans	Br	I[61]	199	PhMe		IR
	Cl	I[61]	188	PhMe		
(CH₂)₃CH=CH₂	Br	I[427]	134			
CH=COEt / Me (O)	J	XIV[830]	163	MeOH/EtOAc	90	UV, IR
CH₂CCHMe₂	Cl	XVI[1106]	198	CH₂Cl₂/EtOAc	60	HMR
CH₂ (O)	Br	I[934]	232	EtOAc	75	IR, HMR, TLC
CHCO₂Et / Me	Br	I[894]	157	PhH		
CHCH₂Me / CO₂Me	J	XIV[81]	133	CHCl₃/EtOAc	46	
CHCO₂Et / Me	Br	I[356]	199			
CH₂CHO₂CMe / Me	BF₄	XIV[170]	150	PhH/Et₂O/EtOAc	100	
(CH₂)₂CO₂Et	Br	I[520]	199[520]	EtOH/H₂O[9]		
(CH₂)₃CO₂Me	Br	[1003]	177			
Me / CEt / Me	J	3[97]				PMR –36.7 ppm
CH(Et)₂	Br	3[97]				PMR –30.1 ppm
CHPr / Me	J	XIV[82]	172	H₂O	51	
CH₂CMe₃	BF₄	XIV[642]	168	CHCl₃/EtOAc	100	IR, HMR
	J	I[965]			87	
(CH₂)₂CHMe₂	Ph₃SnBrCl	II[101]	150		73	

Table 2 (Continued)

Cation R$_3$PR'

R	R'	Anion X	Method	m.	Solvent	Yield (%)	Other Data and Remarks
Ph	(CH$_2$)$_2$CHMe$_2$	Br	I[517]	152	MeOH/Et$_2$O	58	
		J	I[65]	174	EtOH/H$_2$O		UV
	CH$_2$OCMe$_3$	Cl	II[1068]	145 dec.	PhH	85	
	CH$_2$OBu	Ph$_4$B	XV[1069]	154	EtOH		
		Cl	I[1069]	185 dec.	CHCl$_3$/Et$_2$O	76	
	(CH$_2$)$_2$OCHMe$_2$	Br	II[931]	198 dec.	Me$_2$CO/Et$_2$O	50	IR
	CHEt–CO$_2$Et	Br	I[172]	158 dec.	EtOAc/ EtOH/Et$_2$O		
	CH$_2$OCH$_2$ (1,3-dioxane) 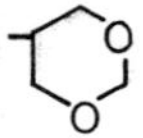	J	XV[196]	179	CHCl$_3$/ EtOAc		
	CH$_2$SCMe$_3$	Ph$_4$B	XV[374]	162	EtOH		
	(CH$_2$)$_3$NMe$_2$ +	Br·HBr	XVI[159]	280	EtOH		
	CH$_2$SEt$_2$	BF$_4$	XV[374]	194	CHCl$_3$/ EtOAc	88	HMR
	(CH$_2$)$_2$SiMe$_3$	J	XIV[965]				
	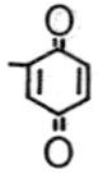 (cyclohexadienone)	Br	XVI[485]	305 dec.	EtOH/ Et$_2$O		
	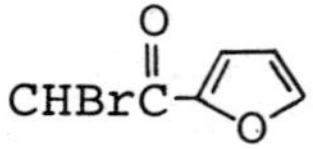 (furyl) CHBrC=	Br	XIV[394]	247	CCl$_4$		

Structure	X	No.	mp	Solvent	Yield	Method
(2-thienyl)C(=O)CHBr	Br	XIV[394]	246	CCl$_4$		
(2-thienyl)C(=O)CHJ	Br	XIX[396]	211	CHCl$_3$		
(2-furyl)C(=O)CH$_2$	Br	I[972]	275[972]	PhH[680]	99[972]	
CH$_2$CH=C–C≡CH (Me)	Br	I[518]	153	EtOAc/Et$_2$O		
(N-Me-pyrrolyl)CH$_2$–	Br	XVII[560]■	216			IR
(dimethyltriazolyl)CH$_2$–	Cl	XV[560]	210			IR
CH$_2$–	Cl	I[899]	195	PhH	90	
CH=CHCO$_2$Me / CO$_2$Me	Br	[462]	143 dec.		63	
CH$_2$C≡CPr	Br	I[140]	168	PhH/MeOH		
CH$_2$CH=CHCH=CHMe	Br	[107]	151		90	
(cyclohexenyl)	J	I[1040]	225	CHCl$_3$/EtOAc	90	HMR
CH=CHCPr(=O)	Cl	I[1107]	140	CHCl$_3$/EtOAc	100	IR, HMR
(cyclohexanonyl)	Cl	I[169]	238	EtOH/EtOAc		IR, HMR

Table 2 (Continued)

Cation R₃PR'		Anion X	Method	m.	Solvent	Yield (%)	Other Data and Remarks
R	R'						
Ph	(2-methylcyclohexanone)	J	XIV[506]	235		81	IR
	(cyclohexanone) + (2-methylcyclohexanone) — Mixture	Br	I[113]	105	MeCN		IR
	(2-methylcyclohexanone)	Cl	XI[169]	195	Bu₂O	90	HMR
	CH₂C=CHCO₂Me, Me	Br	I[215] II[47]	179[215]	i-PrOH[215]	81[215]	
	CH₂CH=CMe, CO₂Me	Br	I[307]	178	BuOH/EtOAc		
	CH₂C(O)CH₂CO₂Et	Br	I[951]	175 dec.	EtOH/Et₂O	99	UV
	CH=CHCCHMe₂, NOH	Cl Cl	I[738] XIV[1102]	140 210	PhH/Et₂O	70	

Substituent	X	Type[ref]	mp	Solvent	Yield	Spectral data
C=CHNMe$_2$ / CO$_2$Me	BF$_4$	XVI[642]	204			
(CH$_2$)$_3$–N (ring, O)	Cl	I[171]	70			
(benzene ring) H	Br	I[78]	255[78]	CHCl$_3$/EtOAc[78]	94[78]	PMR –26.6,[397] E1/2 1.859, 2.092[482]
CH$_2$C=CMe$_2$ / Me	Br	I[923]	192	Et$_2$O/PhH		
CH$_2$CH=CHPr	Br	I[141]	143[141] 178 trans[1047]	PhH[141]		
(CH$_2$)$_2$CH=CMe$_2$	J	I[22]	142	Et$_2$O	88	
(CH$_2$)$_3$C=CH$_2$ / Me	J	I[22]	161	Et$_2$O	89	
(CH$_2$)$_4$CH=CH$_2$	Br	I[427]	165		95	
CHCCHMe$_2$ / MeO	SCN	I[104]	140			
CH=C(OEt)$_2$	J	XIV[3]	162 dec.	EtOAc/ EtOH/Et$_2$O		
CHEt / CO$_2$Et	Br	I[172]	158 dec.			
CH$_2$CH=CCH<OMe / OH, Me	Cl	XVI[336]	113			
(CH$_2$)$_5$CO$_2$H	J	I[505]	Oil	PhH	100	
CH$_2$OCH$_2$ (dioxane ring)	J	XV[196]	179	CHCl$_3$/ EtOAc		IR, HMR, TLC

Table 2 (Continued)

Cation R_3PR'		Anion X	Method	m.	Solvent	Yield (%)	Other Data and Remarks
R	R'						
Ph	$CH{=}C(SEt)_2$	J	XIV[69]	163		95	
	$CH_2CONH(CH_2)_3OMe$	Cl	I[171]	235			
	$CHC(Me)_3$ | Me	J	I[965]		$CHCl_3$/EtOAc	87	IR, HMR
	$CHPr$ | Et	Br	[397]				PMR -30.3 ppm
	$(CH_2)_3CHMe_2$	Br	I[1094]	231			
	Hex	$Ph_3SnBrCl$	II[101]	93	$CHCl_3$/Et_2O	80	
	$CH_2CH(OEt)_2$	Br	I[301]	202[301]	$CHCl_3$/EtOAc[301]		PMR -24.4[397]
	$(CH_2)_2OBu$	Cl	I[171]	74			
	$(CH_2)_6OH$	Br	XVI[931]	170	BuOH/Et_2O	50	IR, HMR
	$C(SEt)_2$ | Me	J	I[414]	133	Me_2CO	50	UV
		J	XIV[70]	185			
	$CH_2OPh{-}Cl_3{-}2,4,5$	Cl	I[706]	197	PhMe	47	
	$CH_2Ph{-}Cl_2{-}2,4$	Cl	I[581]	246	PhH/EtOH	93	
		$Cl·Br_2$	I[171]	114			
	$CH_2Ph{-}Cl_2{-}3,4$	Cl	I[171]	306			
	$CH_2Ph{-}Br{-}3$	Cl	3[99]	291			
		Br	I[492]	278			
	$CH_2Ph{-}Br{-}4$	Cl	I[560]	278			IR

Cation substituent	Anion	Prep.	mp (°C)	Solvent	Yield (%)	Spectra
CH₂Ph-Cl-3	Cl	I[171]	240			
CH₂Ph-Cl-4	Ph₃SnCl₂	II[101]	152			
	Br	I[605]	273	EtOH	68	
	Cl	I[560]	289[560]			IR[1067]
	NO₃	I[605]	273	EtOH		
CH₂OPh-Cl-4	Cl	I[706]	198	EtOH	54	
CH₂SPh-Cl-4	BF₄	XV[374]	131	PhMe		
CH₂Ph-F-4	Cl	I[560]	310			
CH₂Ph-NO₂-2	Cl	I[605]	230 dec.	EtOH/Et₂O	68	
	NO₃	I[605]	269 dec.	EtOH/CHCl₃/Et₂O		
CH₂Ph-NO₂-3	Br·CHCl₃	I[605]	269 dec.[605]	EtOH/H₂O[605]	68	IR,[1067] PMR[397] −24.0 ppm[397]
	NO₃	I[605]	230 dec.	EtOH/Et₂O		
CH₂Ph-NO₂-4	Br	I[557]	269	PhH[557] CHCl₃[557]	97[885]	PMR −23.8 ppm[397] IR,[1067] HMR[389]
	Cl	I[585]	278[585]	CCl₄[585]	58[585]	
	NO₃	XV[605]	158 dec.			
(benzene ring)	ClO₄	I + XV[164]				HMR, UV
CH₂Ph	H(NO₃)₂	XV[306]	72			
	BH₄	XV[168]	295	CHCl₃	45	IR
	B₁₀H₁₀	XV[449]	235 dec.			
	Br	I[505] XIV[836] Ph₃P + R₄AsX[756]	295[836]	EtOH[605]	84[606]	IR[1067] HMR,[373] PMR[397] −23.5 ppm[397]

Table 2 (Continued)

Cation R₃PR'		Anion X	Method	m.	Solvent	Yield (%)	Other Data and Remarks
R	R'						
Ph	CH₂Ph	Br_3	XV[836]	144	CH_2Cl_2/Et_2O		HMR, PMR −21.7 ppm
		$U^{VI}O_2Br_4$	XV[221]	317	MeCN	80	UV, Λ, orange yellow
		$U^{IV}Br_6$	XV[221]	260	MeCN	70	Λ, green
		$Re^{III}_3Br_{10}$	XV[877]				UV, Λ
		$Ce(NO_3)_3 \cdot 3\ H_2O$	XV[693]	161	EtOH		
		Cl	I[399]	338[399]	CHCl₃/petroleum ether[399]	100[387]	IR[1067] UV[162] HMR[507]
			II[47]				pKa >>11,[534] σ[399]
			XIV[1083]				E1/2[399] 1.826,[399] K[780]
CD₂Ph	CH₂Ph	Cl	XVI[916][777]	333	EtOH		IR, HMR
CH₂Ph		ClO_4	XV[221]	237			UV, Λ, yellow
		$U^{VI}O_2Cl_4$	XV[221]	318	MeCN	70	
		$U^{IV}Cl_6$	XV[221]	259	MeCN	70	Λ, blue green
		$Re^{III}_3Cl_{11}$	XV[877][719]				UV, Λ Red
		Cr_2O_7	VII[493]	172 dec.			HMR[373] IR[1067]
		J	I[425]	256[493]			E1/2[499]

Compound	Ref.	m.p.	Solvent	Yield	Notes
La(NO₃)₃·6 H₂O	XV[693]	159[306]	EtOH	69	[306]
NO₃	XV[306]	212[306]	H₂O[696]	94[696]	
Nd(NO₃)₃·6 H₂O	XV[693]	165	EtOH		
Pr(NO₃)₃·6 H₂O	XV[693]	163	EtOH		
SCN	719	189			
Ni(CS₃)₂	XV[305]	186			
$Pt^{II}\left(\!\!\begin{array}{c} S \\ S \end{array}\!\!C{=}CHNO_2\right)_2$	XV[304]				IR, ν, x-ray
Ni(CN)₄	XV[929]	270			
$Re_3^{III}Br_{10}\cdot$ pyridine	XV[877]				UV
OPh(NO₂)₃-2,4,6 O₃SPh-Me-4	XV[719] I[589]	148 173	EtOH/Et₂O	92	
Ni(S...O...O...S)₂·(SnCl₄)₂	XV[193]	219			
Ni(S...O...O...S)₂·SnCl₄	XV[193]	163			
Pd(S...O...O...S)₂·(SnCl₄)₂	XV[193]	219			
Pd(S...O...O...S)₂·SnCl₄	XV[193]	168			
Ph₃SnBrCl	II[101]	195		78	
Ph₃SnCl₂	II[101]	205		67	
Ph₄B	XV[1083]	228			
Cl	977	76	Et₂O	75	
CH₂NHO₂SPh-Cl-4					

Table 2 (Continued)

| Cation R₃PR' | | Anion | | | | Yield | Other Data |
R	R'	X	Method	m.	Solvent	(%)	and Remarks
Ph	CH_2CONH–[pyridinium]	Cl	I[342]	218	$MeNO_2$	45	
	CH_2OPh	Cl	I[1069]	215	Me_2CO/Et_2O	80	
	$CH-C(=O)-Me$, [thiophene, S]	J	XIV[395]	217			
	$CH-C(=O)-Me$, [furan, O]	J	XIV[395]	220			
	CH_2SPh	Ph₄B	XV[374]	165			
		Cl	I[706]	210			
	CH_2NHSO_2Ph	Cl	I[977]	58		35	
	$CH_2C\!\equiv\!C\!-\!C\!=\!CHMe$ (H H Me)	Br	[48]	151	Et_2O	75	
	$CH_2CH=CCH=CHCHO$, Me	Br	I[941]	200 dec.	$CH_2Cl_2/EtOAc$		UV
	$CH=CH-CO_2CMe$, =CHMe	Br	XVI[1107]	189	$CH_2Cl_2/EtOAc$	39	IR
	$CH_2=$[dihydropyran]$-CH_2OH$	Cl	I + XVI[79]	254	$CHCl_3$	67	

	Anion	Prep.	m.p.	Solvent	Yield (%)	
CH=CH-C-CO$_2$Pr $\parallel$ NOH	Br	XIV[1102]	209 dec.			
CH=CH-C-CO$_2$-i-Pr $\parallel$ NOH	Br	XIV[1102]	210			
(cyclohexene-CH$_2$)	Br	I[515]	219	Me$_2$CO	98	
(cycloheptanone, Me)	J	XIV[506]	206	MeOH	58	
CH$_2$CH=CH-C̈Pr (O)	Br	XIV[1102]	99			
CH$_2$CH=CH-C̈CHMe$_2$ (O)	Br·CH$_2$Cl$_2$	XIV[1102] XIV[1102]	137 83	H$_2$O EtOAc/ Et$_2$O	59	
MeO $\mid$ -C-Me $\mid$ CO$_2$Et	J	XIV + XV[75]	142	MeCN/ Et$_2$O	86	UV
CH=CCH$_2$CO$_2$Et $\mid$ OMe	J	XIV[951]	111			
CHCH=CCO$_2$Me $\mid$ $\mid$ Me OMe	Br	XIV[1102]		EtOAc/ Et$_2$O		
CHCH$_2$CO$_2$Et $\mid$ O$_2$CMe	Cl	XVI[951]	dec.		96	UV
NHEt (triazine) NHEt NHEt	Cl	I[969]	210	EtOH		
NHEt C=CMe CO$_2$Me	BF$_4$	XVI[642]	103			IR

Table 2 (Continued)

Cation R₃PR' R	R'	Anion X	Method	m.	Solvent	Yield (%)	Other Data and Remarks
Ph	$-CH$ $\overset{NEt}{\underset{CO_2Me}{\|\|}}$CMe	BF₄	XVI[642]	218		58	IR
	(cyclohexyl, H, Me, CH₂)	J	[397]				PMR −35.1
	(cyclohexyl, H, CH₂)	Br	I[78]	225	Petroleum ether	98	
	$(CH_2)_5CH=CH_2$	Br	I[1003]	167[106]	PhH/Et₂O[1003]	75[1003]	
	(dioxane ring, O=CHMe₂, Me, Me)	BF₄	XII[495]	160	EtOH	80	
	(CH₂, H, Me)	J	I[105]	195	Et₂O	37	
	$(CH_2)_3CCH_2CO_2H$, Me	Br	I[635]	158	PhMe/Et₂O	78	IR

Structure	Halide	Reference	m.p. / Oil	Solvent	Yield (%)	Notes
$(CH_2)_4CO_2Et$	J	I[506]	Oil	PhH	97	IR, HMR, TLC, pale yellow
$CH_2OCH_2CH{-}CH_2$ (—O—C(Me)(Me)—O—)	Cl	I[549]	180		47	
$(CH_2)_3-$ [Me-dioxane]	Br	I[635]	215	PhMe/PhH	85	
$C{=}C(SEt)_2$ / Me	J	[69]	186		80	
$(CH_2)_2-$ [piperidine, H N-H]	Br	$Ph_3P(CH_2)_2PPh_3$ + LiN[ring] I[1026][1070]	178	Me_2CO	27	
CH—Pent / Me	Br	I[1026]	170	THF	60	
Hept	Br	I[427]	171	Xylene	74	
$(CH_2)_3-N{\:}NH$	Br·HBr	XVI[152]	278			
$CBr_2CPh{-}Br{-}4$ (C=O), $C{\equiv}CPh$	Br / J	XIV[394] / XV+[63]	216 / 181	$PhNO_2$/PhH / H_2O	98	
$CHBrCPh{-}Cl{-}4$ (C=O)	Br	XIV[394]	233			
$CHJCPh{-}Br{-}4$ (C=O)	Br	XIV[395]	210	$CHCl_3$		
$CHBrCPh{-}NO_2{-}4$ (C=O)	Br	XIV[394]	182			

Table 2 (Continued)

Cation R_3PR'		Anion X	Method	m.	Solvent	Yield (%)	Other Data and Remarks
R	R'						
Ph	$CBr_2\overset{O}{C}Ph$	Br	XIV[394]	207[1013]	CHCl₃/PhH[1013]		IR[1013]
	$CHBr\overset{O}{C}Ph-Br$	Br	XIV[394]	240			
	$CH_2\overset{O}{C}ONHPh-Br_3-2,4,6$	Cl	I[342]	95	MeNO₂	60	
	$CHJ\overset{O}{C}Ph-Cl-4$	Br	XIV[396]	208	CHCl₃		
	$CH_2\overset{O}{C}ONHPh-Cl_3-2,4,6$	Cl	I[342]	107	MeNO₂	79	
	$C(HgJ)_2\overset{O}{C}Ph$	Cl	XIV[752]	205			
	$CHJ\overset{O}{C}Ph-NO_2-4$	Br	XIV[396]	180	CHCl₃		
	$CHBr\overset{O}{C}Ph$	Br	XIV[394]	255			
	$CH_2\overset{O}{C}Ph-Br-4$	Br	I[12]	239[950]	EtOH/Et₂O[12]		IR[950]
	$CD_2\overset{O}{C}Ph-Br-4$	Br	XVI[12]	D₂O			
	$CH_2\overset{O}{C}ONHPh-Br_2-2,4$	Cl	I[342]	191		88	IR
	$CH\overset{O}{C}Ph$ HgCl	Cl	XIV[753]	208	MeNO₂	100	IR

Structure	X	Method[ref]	mp	Solvent	Yield (%)	Notes
CH₂C(=O)Ph-Cl-4	Br, J	I[256], XIV[355]	264[258], 249	PhH[258]	97[258]	IR[950]
CH₂C(=O)NHPh-Cl₂-2,4	Cl	I[342]	176	MeNO₂	84	
CH₂C(=O)NHPh-Cl₂-2,5	Cl	I[342]	280 dec.		70	
CH₂C(=O)NHPh·Cl₂-3,4	Cl	I[396]	221	MeNO₂	89	
CHJ C(=O)Ph	Br	XIV[396]	216	CHCl₃		
(benzoxazinone structure)	Cl	I[247]	186 dec.			
CH₂-N (cyclohexane fused)	Cl	I[175]	148			
CH₂C(=O)Ph-NO₂-4	Br	I[519], Ph₃P +	126	THF	40	
CBr₂CH₂Ph	Br	CBr₄ + PhCH₂Br[832]	172	CH₂Cl₂/Et₂O		
CH₂CONHPh-Cl-2	Cl	I[342]	217		68	
CH₂CONHPh-Cl-4	Cl	I[342]	220		70	
CH₂CONHPh-NO₂-3	Cl	I[342]	221		90	
CHPh	Cl	XVI[6+2]	223	H₂O	72	
CHO						
CH₂COPh	SCN	XV[1100]	193	CH₂Cl₂/EtOAc		
	SeCN	[1105]	160 dec.			
	BF₄	XIV[759]	249		31[980]	IR,[830] pKa 6.03[534]
	Br	XIV[930]	284 dec.[492]	EtOH/Et₂O[492]	95[980]	UV[830]
		I[492]				

287

Table 2 (Continued)

Cation R_3PR'		Anion X	Method	m.	Solvent	Yield (%)	Other Data and Remarks
R	R'						
Ph	$CD_2\overset{O}{C}Ph$	Br	XVI[12]				IR[12]
		Cl	XIV[1038] I[716]	270[916]	D_2O[1]	27[1038]	IR[1011]
	CH_2COPh	Cl	XVI[916]				IR, HMR
		J	XIV[830]	259	MeOH/EtOAc		UV, IR
	$CH_2Ph\text{-}CHO\text{-}2$	NO_2	[716]	183			
		Cl	I[128]	233	MeOH/Et_2O	98	IR, HMR
	$CH_2Ph\text{-}CO_2H\text{-}4$	Br	[373]				HMR
		Cl	[1067]				IR,[1067]
	$CH_2Ph\text{-}CH_2Br\text{-}2$	J	I[373]				HMR[373]
		Br	I[388]	253	EtOH	92	
	$CH_2Ph\text{-}CH_2Br\text{-}3$	Br	I[337]	239	Xylene	38	
	$CH_2Ph\text{-}CH_2Br\text{-}4$	Br	I[337]	240	Xylene	94	
	$CHCONH_2$ (—Ph)	Br	XVI[1037] XI[460]	276[1037]	HBr dil./EtOH[1037]	81[460]	
	$CH_2CONHPh$	Cl	I[342]	240		94	
	$CHPh$ (—CH_2NO_2)	Br	XI[460]	191 dec.		77	

R	X	Prep. (ref.)	M.p. (°C)	Solvent	Yield (%)	Notes
$CH_2CONHPh$-CH-3 (=O)	Cl	I[342]	260		87	
$CH_2SO_2NHC(=O)Ph$	ClO_4	XIV[246]	228 dec.	Explos.	97	
	Cl	I[342]	340 dec.	DMF	71	
$CH_2CONHPh$-SO_3H-4	Cl	I + XVI[79]	241	$CHCl_3$	59	
CHPh(Me)	Br	I[389]	253	Petroleum ether/Et_2O		HMR
CH_2Ph-Me-2	Br	I[389]	271[389]	Petroleum ether/Et_2O[389]	100[337]	HMR[389]
CH_2Ph-Me-3	Br	I[636]	276[636]	DMF[636]	76[636]	IR[1067], HMR[389]
CH_2Ph-Me-4	$B_{10}H_{10}$	XV[449]	210 dec.	DMF/Et_2O		
CH_2Ph-Me						Very hygroscopic
$CH_2CONHPh$-SO_2NH_2-4	Cl	I[342]	275 dec.	$CHCl_3$/Et_2O	68	
	Br	II[931]	132		95	
CHOPh(Me)	Cl	I[1069]	192	EtOH/Et_2O	96	
$CH_2CHOHPh$	Br	XIV[1084]	190	MeOH/EtOAc	21	
$(CH_2)_2OPh$	Br	I[931]	138	PhOH/Et_2O	100	IR, HMR
CH_2Ph-OMe-3	ClO_4	XVI[646], [399]	186	MeOH	43	
	Cl		262			
CH_2Ph-OMe-4	Br	I[557]	236[557]	PhH/$CHCl_3$[557]		HMR[389], IR[1067]
	Cl	I[585]	258[79]	$CHCl_3$[79]	99[79]	IR[560], E½ 1.845, σ[399]

I + XV[79]

Table 2 (Continued)

Cation R₃PR'		Anion X	Method	m.	Solvent	Yield (%)	Other Data and Remarks
R	R'						
Ph	CH=C(-OEt)(2-thienyl)	J	XIV[395]	171			
	CH₂SO₂Ph-Me-4	Br	I[1012]	267	CH₂Cl₂/PhH	62	
	CH₂SCH₂Ph	Ph₄B	XV[374]	187			
		BF₄	XV[374]	133			
	(CH₂)₂SPh	Br	XVI[931]	148	CHCl₃/THF	100	
	CH₂NHSO₂Ph-Me-4	Cl	I[977]	67	Et₂O	75	
	CH=CHC=CMe₂—O₂CMe	Br	XVI[1107]	219	CH₂Cl₂/EtOAc	73	IR, HMR
	O₂CMe	Cl	XVI[1107]	166			
	CH=CHCCO₂Bu =NOH	Cl	XIV[1102]	208 dec.			
	CH₂CCH₂CH₂C-N(morpholine)	O₂CMe	XIV[165]	178		75	
	CH₂CH=(cyclohexyl)H	Br	I[418]	164			
	CH₂C=CC=CHCH₂OMe H H Me	Br	I[1080]	155 dec.			IR

Structure		Code	mp	Solvent	Yield	
$\overset{O}{CClCPr}$ $-Pr$	Cl	XIV[116]	137	CH$_2$Cl$_2$/ EtOAc	84	
CH$_2$CH=CHCBu cis (=O), HO (CH$_2$)$_2$	Cl	XIV[112]	163		50	
	Br·H$_2$O	I[417]	110 dec.			
CH$_2$OCH$_2$CHCH$_2$O$_2$CMe $-O_2CMe$ $-OH$	Cl	I[549]	157			
CH$_2$CH=CCOCHMe$_2$ $-Me$	Cl	XVI[338]	87	THF		
(CH$_2$)$_3$O	Br	[103]	238 dec.	PhH		
(CH$_2$)$_2$CHC MeMe	Br	I[595]	218	PhH		
	J	XV[595]	158			
(CH$_2$)$_3$ Me	Br	I[775]	219[775]	PhMe/ PhH[635]		
	J	I[776]	210	PhH		
CH$_2$OCH$_2$ OMe Me	Cl	I[549]	180		47	UV

Table 2 (Continued)

Cation R_3PR'		Anion X	Method	m.	Solvent	Yield (%)	Other Data and Remarks
R	R'						
Ph	CH_2CONPr_2	Cl	I[171]	165			
	NEt₂ (S–S–OMe)	Cl	I[969]	148	EtOH	89	
	$CH_2\overset{O}{\overset{\|}{C}}NHPh-CO_2H-3$	Cl	I[342]	172		68	
	$CH_2\overset{O}{\overset{\|}{C}}NHPh-CO_2H-4$	Cl	I[342]	142		75	
	$CH_2Ph-CH=CH_2-4$	Br	I[276]	246	EtOH/Et₂O	10	
		Cl	I[350]	226	MeOH/Et₂O	60	
	$CH_2CH=CHPh$	Br	I[337]	240		100	
		Cl	I[636]	224[636]	EtOH/Et₂O[636]	91[636]	HMR,[389] UV[162]
	$CH_2O(CH_2)_2OPh-Cl_2-2,4$	Cl	I[355]	196			
	$CH_2O(CH_2)_2OPh-Cl_2-3,4$	Cl	I[355]	157			
	$\underset{\|}{\overset{O}{\overset{\|}{C}}HCMe}$ Ph	Cl	XIV[1038]	172	EtOH/H₂O	47	
	$CH=CMe$ OPh	BF₄	XIV[759]	191	MeNO₂	15	
	$\overset{O}{\overset{\|}{C}}HCPh$ Me	Br	I[113]	245	EtOH/H₂O	24	IR, HMR

R	Anion	Ref.	m.p. (°C)	Solvent	Yield (%)	Method
$CH_2C(=O)Ph$-Me-4	J	XIV[7,395]	240		83[258]	IR[950]
$CH_2C(=O)CH_2Ph$	Br	I[716]	261[716]			
	Cl	I[716]	231			
	$PtCl_6$	XV[716]	211 dec.			
	J	XIV[395]	255			
	NO_3	[716]	183			
$CH_2C(=O)CH_2Ph$	Cl	I[507]	204 dec.	CH_2Cl_2/Et_2O	85	IR, HMR
$(CH_2)_2C(=O)Ph$	Br	I[391]	180[391]	$CHCl_3$[391]	69[391]	IR[1067]
	J	XVII[336]	160	H_2O	77	
$CH(Ph)CO_2Me$	BF_4	XIV[759]	210		45	
$CH(Ph)CH_2CO_2H$	Br	I[757]	204[757]	$CHCl_3$[759]	61[757]	
	Br	XV[460]	245 dec.		71	
$CH_2CO_2CH_2Ph$ (O=)	Cl	I[389]				HMR
$CH_2C(=O)Ph$-OMe-4	Br	I[258]	222	PhH	91	
	Cl	I[319]	165	EtOAC	78	
	J	XIV[395]	246			
CH_2Ph-OMe-2-CHO-5	Cl	I[537]	208[637]	DMF[637]	97[637]	IR[1076]
CH_2PH-CO_2Me-3	Br	I[188]	225	EtOH	92	
CH_2Ph-CO_2Me-4	Cl	I[148]	240[148]	EtOH/Et_2O[148]	92[148]	HMR[389]
CH_2Ph-OH-2-CH_2Br-3-Me-5	Br	I[337]	243	PhH	97	
CH_2Ph-Me-2-OH-3-Cl-4-Me-6	Cl	I[1057]	256	MeOH/EtOAc	90	
	ClO_4	XV[1057]	224	H_2O		
$(CH_2)_3OPh$-Cl-4	Br	I[706]	171	PhMe	89	

Table 2 (Continued)

Cation R₃PR'		Anion X	Method	m.	Solvent	Yield (%)	Other Data and Remarks
R	R'						
Ph	$(CH_2)_3OPh$-Cl-4	Cl	I[706]	129	PhMe	59	
	$CH_2O(CH_2)_2SPh$-Cl-4	Cl	I[355]	179			
	C—NEt / Ph	BF₄	XII[495]	205	CH_2Cl_2/Et₂O		
	$CH_2CONHPh$-Me-2	Cl	I[342]	170		49	
	$CH_2CONHPh$-Me-3	Cl	I[342]	147		55	
	$CH_2CONHPh$-Me-4	Cl	I[342]	214		94	
	$CH_2CONHPh$-OMe-4	Cl	I[342]	230		91	
	$CH_2CONHPh$-Me-OH	Cl	I[342]	285 dec.		68	
	CHEt / Ph	BF₄	XIV[642]	263	MeNO₂	98	
	$CHCH_2Ph$ / Me	Ph₄B	XV[956]	179	Pyridine		
		Br	XIV[956]	162	CHCl₃/EtOAc	80	HMR
	$(CH_2)_3Ph$	Ph₃SnClBr	II[101]	125	CHCl₃/PhH/petroleum ether	57	UV, IR
	$CH=CPh$ / OEt	J	XIV[830]	178			
	CHCHCMe (O=, ring)	Br	XVI[1107]	236	CH_2Cl_2/EtOAc	84	IR

Compound	Halogen	Ref	mp	Solvent	Yield	Notes
(CH₂)₃OPh	Br	I[706]	177	PhMe	85	HMR
	Cl	I[706]	172	PhMe	62	
	Br	I[932]	207	PhH	80	
CH₂CHCH₂OPh (OH)	Br	I[106]	173	Et₂O	60	
CH₂CH=CH(CH₂)₄CH=CH₂						
[cyclobutane] C=C–, PrOH	Br	XIV[1100]	172	CHCl₃/EtOAc		
[cyclohexane] CH₂OH; CH₂CH=	Br	I[417]	218	CHCl₃/THF		
CH₂C(=O)CC(Me)₂CCHMe₂	Cl	XIV[612]	184 dec.	PhH	97	UV, HMR
Cl–CCC(Me)₃	Cl	XIV[1106]	165	CH₂Cl₂/EtOAc	91	
C(=O)– CH₂CH₂Me	Cl	I[969]	209	EtOH	83	
[triazole] –N–N–NEt₂	Cl	I[969]	208	EtOH	98	
[ring] –N=N– NH-i-Pr						
[ring] –N=N– NH-i-Pr						
CH₂–CH=CH–Hex	Br	I[110]	170			
CH₂C≡CCH(OEt)₂	Br	I[652]	204			
CH₂CH=CCH(OEt)₂ (Me)	Br	XI[651]	244	EtOH	70	
[tetrahydropyran] (CH₂)₄O	J	I[104]	181	Me₂CO/Et₂O		

Table 2 (Continued)

Cation R₃PR'		Anion X	Method	m.	Solvent	Yield (%)	Other Data and Remarks
R	R'						
Ph	(CH₂)₆CO₂Et	J	I[506]	Oil		90	IR, HMR, TLC
	CBrJC(=O)-benzofuran	Br	XIV[971]	181			UV
	CBr₂C(=O)-benzofuran	Br	XIV[971]	167			UV
	naphthoquinone-CH₃	J	XVI[485]	171	MeOH		
	CHBrC(=O)-benzofuran	Br	XIV[971]	234			UV
	CH₂C(=O)-benzofuran	Br	I[971]	289			
	CH=CPh-Br-4 / O₂CMe	Cl	XIV[257]	>150 dec.			
	CH-CPh-Cl-4 / O₂CMe	Cl	XIV[257]	>150 dec.			
	CH=CHCH=CHPh-NO₂-4	Br	XVI[755]	166	DMF	45	

Group	Anion	Ref.	mp	Solvent	Yield	Notes
$CH=CPh-NO_2-4$, O_2CMe	BF_4	XIV[170]	180	PhH/Et$_2$O/EtOAc	100	
$C=CHPh-OH-2$, $CH=CH_2$	Br	XVI[938]	265	MeOH/CHCl$_3$	75	IR, UV, HMR, yellow
$CH_2\overset{O}{C}CH=CHPh$	Br / Cl	XIV[1103] / I[511]	240 / 200	EtOH/EtOAc	32	
$\underset{O=CMe}{CHCPh}$	Cl	XIV[753]	260	THF	52	
$CH=CPh$, O_2CMe	BF_4	XIV[170]	163	PhH/Et$_2$O/EtOAc	100	
$\overset{O}{=}$	Cl	XIV[257]	>150 dec.		87	
$CHBrCPh-Me_2-2,4$, $CH=CPh-Br-4$, OEt	Br / J	XIV[394] / XIV[395]	186 / 173	CCl$_4$		
$CH_2CH=CPh-Cl-4$, Me	Cl	I[349]	217	PhMe		
$CH_2CH=CHCH_2Ph-Cl-4$, $CH=CPh-Cl-4$, OEt	Cl / J	I[349] / XIV[395]	180 / 169	PhMe		
$(CH_2)_4OPh-Cl_3-2,4,5$	Br	I[706]	150	PhMe	87	
CH_2 (indole, N-Me)	J	XVII[436]	218	EtOH	70	
$CH_2CONHPh-CMe-4$	Cl	I[342]	218		94	
$CH_2CONHPh-CH_2CO_2H-4$	Cl	I[342]	160		85	

297

298

Table 2 (Continued)

Cation R_3PR' R	R'	Anion X	Method	m.	Solvent	Yield (%)	Other Data and Remarks
Ph	$CH_2CH=CHPh-Me-4$	Cl	I[586]	145	PhMe		HMR
	$CH_2CH=CHCH_2Ph$	Cl	I[349]	240	MeNO_2	60	
	$CH_2CONHPhSO_2NHCOMe$	Cl	I[342]	177			
	CH=CPh / OEt	J	XIV[395]				
	CHCPh-Me-4 / C=O / Me	Br	I[973]	225	PhMe		
		J	XIV[395]	249			
	$CH_2CPh-Me_2-2,4$ / C=O	Br	9[50]				IR
	$(CH_2)_3CPh$ / C=O	Cl	I[392]	185[392]	Me_2CO/petroleum ether[392]	33[392]	IR[1067]
	CH–C–CO_2Me / H / Ph / C=O	Br	II[73]		CHCl_3/EtOAc		
	CHCPh–OMe-4 / Me	Br	I[973]	229	PhMe		

Structure	Anion	Type	m.p.	Solvent	Yield (%)	Notes
$CHCH_2OPh$–CO_2Me	J	XIV[392]	232	EtOAc	84[933]	HMR,[933] IR[937]
	Br	XIV[67]	185			
$(CH_2)_3OPh$–CHO-2	Br	I[933]	162[933]	$PhCl$/Et_2O[933]		
$(CH_2)_3OPh$–CHO-4	Br	I[933]	140	Me_2CO	58	IR, HMR
CH_2CHCH_2OPh–CHO-2, OH	Br	III[932]	214	MeCN	80	IR, HMR
$C{=}C(SMe)_2$, Ph	J	XIV[69]	255		96	
$(CH_2)_4OPh$–Cl-4	Br	I[706]	171	PhMe	89	
$C{=}CHNMe_2$, Ph, Me	BF_4	XVI[642]	182			
CCH_2CONH_2, Ph	J	XVI[1037]	276 dec.	EtOH		
$CH_2CONHPh$–OEt-2	Cl	I[342]	209		75	
$CH_2CONHPh$–OEt-4	Cl	I[342]	239		87	
$CH_2CONHPh$–Me-OMe	Cl	I[342]	218	$MeNO_2$	76	
CH_2Ph–Me_3-2,3,4	Br	I[189]	222	i-PrOH/petroleum ether	99	
CH_2Ph–Me_3-2,3,6	Br	I[189]	243	i-PrOH/petroleum ether	97	
CH_2Ph–Me_3-2,4,6	Br	I[142]	233	EtOH	90	
$CH_2CONHPh$–NMe_2-4	Cl	I[342]	229	$MeNO_2$	50	
CH_2–	Br	I[593]	262	EtOH/Et_2O	84	

Table 2 (Continued)

Cation R_3PR'		Anion X	Method	m.	Solvent	Yield (%)	Other Data and Remarks
R	R'						
Ph	CHBrĊPh-Me-4 (Ċ=O)	Br	XIV[394]	237			
	CHĊPh-Br-4 / Me (Ċ=O)	Br	I[973]	252	PhMe		
	(Ċ=O)	J	XIV[395]	223			
	CHBrĊPh-OMe-4 / HgCl (Ċ=O)	Br	XIV[394]	209			
	CCO₂Me / Ph	Cl	XIV[753]	190[752]		64[753]	IR[752]
	CHĊPh-Cl-4 / Me (Ċ=O)	Br	I[973]	256	PhMe		
	(CH₂)₃OPh-Cl₃-2,4,5	J	XIV[395]	212			
	CH₂O(CH₂)₂OPh-Cl₃-2,4,5	Cl	I[706]	236	PhMe	13	
		Cl	I[355]	170			
	CHJĊPh-Me-4 (Ċ=O)	Br	XIV[396]	217	CHCl₃		
	CH₂– (indole, N–H)	O₃SOMe	XVII[437]	185		70	

Structure	X	Ref / Method	mp	Solvent	Yield (%)	IR
$C{=}CHCONH_2$ | Ph	J	[462]	276 dec.		75	
Oct	$O_3SPh{-}Me{-}4$	I[569]	70	$EtOH/Et_2O$	59	
CH_2N (phthalimide)	Cl	I[171]	250			
	J	XVII[436]	264 dec.	H_2O	90	
$O{=}CBr_2CPh{-}Me{-}4$	Br	XIV[394]	193	$PhNO_2/PhH$		
$O{=}CBr_2CPh{-}OMe{-}4$	Br	XIV[394]	201	$PhNO_2/PhH$		
triazine NH_2 $\cdots$ $N{-}NHPh{-}Cl{-}4$	Cl	I[969]	282	EtOH	71	
$O{=}CCH{=}CHPh$	Cl	[1067]				
$C{=}CHCHO$ | Ph	J	[462]	85 dec.		87	IR
$O{=}CH{=}CHCPh$	Br	XV[1107]	180	$CH_2Cl_2/EtOAc$	79	IR
$C{=}CHCO_2H$ | Ph	J	[462]	202 dec.		90	
$O{=}CHCHCEt$ (cyclopentadienyl)	Br	XVI[1107]	199 / 247 dec.	$CH_2Cl_2/EtOAc$	62	IR

Table 2 (Continued)

Cation R₃PR'		Anion X	Method	m.	Solvent	Yield (%)	Other Data and Remarks
R	R'						
Ph	CHCHCEt (‖O), cyclopentadiene ring	J	XVI[1107]	224	CH₂Cl₂/EtOAc		UV
	CH₂CH=CCH=CHCH=CMe, —Me, CHO	Br	I[941]	238 dec. 203 dec.	CH₂Cl₂/EtOAc		
	(CH₂)₃Ph-OMe-3	Br	I[742]	135	PhCl/Et₂O		
	(CH₂)₄OPh	Br	I[706]	178	PhMe	79	
	CHCH(SMe)₂, -Ph	J	XIV[70]	224			
	(CH₂)₂Ph-NHEt	Br	I[318]	205			
	CH₂, Me–C–Me bridged ring	Br	I[206]	239	PhH	53	
	CH₂CH=CCH=CHCH=CMe₂, —Me	Br	II[1023]	184	Me₂CO	45	
		Cl	I + XVI[79]	144	CHCl₃	80	
	O=, CH₂, Me–C–Me ring	Br	I[206]	304	PhH	10	Addition of Ph₃PR'Br → 76% yield

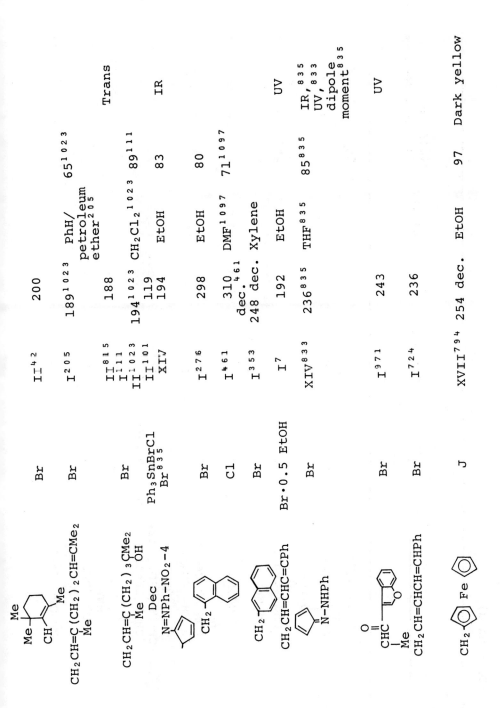

	Anion	Ref.	m.p.	Solvent	Yield %	Notes
[cyclohexene, Me, Me, CH, Me]	Br	II[42]	200			Trans
$CH_2CH=C(CH_2)_2CH=CMe_2$ / Me	Br	I[205]	189[1023]	PhH/petroleum ether[205]	65[1023]	
$CH_2CH=C(CH_2)_3CMe_2$ / Me OH Dec	Br	II[815]	188			IR
	$Ph_3SnBrCl$ Br[835]	I[111]	194[1023]	CH_2Cl_2[1023]	89[111]	
		II[1023]	119	EtOH	83	
N=NPh-NO2-4		II[101] XIV	194	EtOH		
[naphthalene-CH_2]	Br	I[276]	298	EtOH	80	
[naphthalene-CH_2, N-]	Cl	I[461]	310 dec.[461]	DMF[1097]	71[1097]	
$CH_2CH=CHC=CPh$	Br	I[353]	248 dec.	Xylene		
[cyclopentadiene] $CH_2CH=CHC=CPh$	Br·0.5 EtOH	I[7]	192	EtOH		UV
N-NHPh	Br	XIV[833]	236[835]	THF[835]	85[835]	IR,[835] UV,[833] dipole moment[835]
[benzofuran] $CHC=O$ / Me	Br	I[971]	243			UV
$CH_2CH=CHCH=CHPh$	Br	I[724]	236			
[ferrocene] CH_2–Fe	J	XVII[794]	254 dec.	EtOH	97	Dark yellow

303

Table 2 (Continued)

Cation R_3PR'		Anion X	Method	m.	Solvent	Yield (%)	Other Data and Remarks
R	R'						
Ph	$C(Ph)=CHCO_2Et$	J	[462]	174 dec.		80	
	$CH=CPh\text{-}Me\text{-}4$, O_2CMe	Cl	XIV[257]	150 dec.			
	$CH(CO_2Me)CH_2C(=O)Ph$	Br	I[83]	95 dec.[83]	$CHCl_3/i\text{-}Pr_2O$[83]	79[83]	
	$C(CO_2Me)_2$, Ph	J	XIV[73] I[748]	183	EtOH	85	UV
	$CH_2CONHPh\text{-}CO_2Et\text{-}4$	Cl	I[342]	60			
	$CH_2CH=CCH_2Ph$, Me	Cl	I[349]	199	PhMe	91	
	$CH_2CH=CH(C\equiv C)_2CMe_3$	Br	I[108]	194 dec.	Et_2O	70	
	$(CH_2)_3CH=CHPh$	Br·1.5 H_2O	XVI[728]	208			
	(cyclopentadienylidene), NNHPh	Br	XVI[833]	204			UV, IR
	$CH=COPr$, Ph	J	XIV[395]	158			
	$CH=CPh\text{-}Me\text{-}4$, OEt	J	XIV[395]	158			

Structure	X	Ref	m.p.	Solvent	Yield (%)	Method/Notes
(CH$_2$)$_4$CPh, O= CH$_2$Ph-t-Bu, O=	Br Cl	I[86] I[399]	167 234	Et$_2$O	71	E$_{1/2}$ 1.842, σ
CHCHCPr (O=)	Cl	XVI[1·07]	210	CH$_2$Cl$_2$/EtOAc	32	IR
CHCHCHMe$_2$ (O=)	Br	XVI[1107]	236 dec.	CH$_2$Cl$_2$/EtOAc	44	
CH$_2$CH=CCH=CHCH$_2$COMe, Me, Me	Br	II[1023]	168	CH$_2$Cl$_2$/EtOAc	40	
(NEt$_2$, N-ring, NEt$_2$)	Cl	I[965]	127	EtOH		
CH$_2$CH=C(CH$_2$)$_3$CMe$_2$, Me, OMe, OMe	Br	II[1023]	168	EtOAc	80	
(CH$_2$)$_9$CO$_2$Me	J	I[49a]	108	PhMe/EtOAc	80	
(CH$_2$)$_{10}$CO$_2$H	J	I[64]	114			
(CH$_2$)$_{11}$OH	Br	I[414]	90	Me$_2$CO	77	UV
CBr$_2$C (O=) naphthalene	Br	XIV[394]	209	PhNO$_2$/PhH		
CHBrC (O=) naphthalene	Br	XIV[394]	256			
CH$_2$C (O=) naphthalene	Br	XIV[396]	218	CHCl$_3$		
CBrC (O=) -Et, benzofuran	J	XIV[971]	169			UV

Table 2 (Continued)

Cation R₃PR'		Anion X	Method	m.	Solvent	Yield (%)	Other Data and Remarks
R	R'						
Ph	CCl₂C(=O)–Et, benzofuran-O	J	XIV[971]	215			UV
	CJ₂C(=O)–Et, benzofuran-O	J	XIV[971]	205			UV
	CH₂C(=O)–ferrocenyl	J	XIV[794]	200 dec.	Et₂O		Brick red
	CH₂– naphthalene-OMe	J	XVII[436]	226	MeOH/Et₂O	80	
	CH₂C(=O)– naphthalene	Br J	I[258] XIV[395]	251[258] 238	PhH[258]	60[258]	IR[950]
	CH=C–OEt, benzofuran-O	J	XIV[971]	253		92	UV

Substituent / Structure	Anion	Ref	m.p.	Solvent	Yield (%)	Method
(ring) CH_2, C=O, N-Ph, N-Me, Me	J	XVII[436]	193	H_2O	67	
$CHOCH_2CO_2Et$, O=CPh	NO_3	XV[435]	173 dec.		90	UV
	Cl	XVI[951]	135 dec.			UV
$C=CHPh-NMe_2-4$, O=CMe	Cl	XVI[603]			25	UV
(dimethoxy-tetrahydronaphthalene) CH_2, OMe, OMe	Br·2 HBr	I[1067]	218 dec.			
	Cl·HCl·2 H_2O	I[1060]	197 dec.			
$CH_2C≡CCH_2Ph$, MeMe	Cl	I[349]	200	PhMe		
$CH_2CH=CCH_2Ph-Me-4$, Me	Cl	I[349]	202	PhMe		
$CH_2CH=CCH=CHCH=CHCHO$, Me Me	Br	I[941]	202 dec.			UV
$CH=CPh-Me_2-2,4$, OEt	J	XIV[395]	174			
$CH_2CH=CCH_2Ph-OMe-4$, Me	Cl	I[349]	198	PhMe		
$(CH_2)_5CPh$, O=	J	XV[392]	158	EtOH/Et_2O	60	

Table 2 (Continued)

Cation R₃PR'		Anion X	Method	m.	Solvent	Yield (%)	Other Data and Remarks
R	R'						
Ph	(CH₂)₂– [dioxolane-Ph]	J	I[109]	207	PhH/Et₂O	88	
	C=CSCH₂Ph / MeSEt	J	XIV[69]	179		85	
	Me OMe [tetrahydroisoquinoline, CH₂, N–H]	Br·H₂O	I[1060]	218			
	(CH₂)₆OPh	Br	I[414]	149	Me₂CO	89	UV
	CH₂Ph-Me₃-3,4,6-OMe-2,5	Cl	I[999]	233	EtOH/Et₂O	95	
	CH₂CH=CCH=CHCH=CMe	Br	I[941]	197	CH₂Cl₂/EtOAc	93	UV
	Me CO₂Et [tetrahydropyran]	Br	I[516]	250	EtOAc/MeOH		
	C=C(SEt)₂	J	XIV[70]	219			
	(CH₂)₁₀O₂CMe	Br	I[49a]	97			
	(CH₂)₁₀CO₂Me	J	I[49a]	132[49a]	PhH[409]		
	C=C(SEt)₂ / Hex	J	XIV[69]	220		78	

Compound	Group	Type [ref]	mp (°C)	Solvent	Yield (%)	Properties
Dodec	MeSPOMe (S=, O)	XV[704]		Et₂O	94	$n_D^{20} = 1.5868$, $d_{20} = 1.0601$
	$\overset{S}{\underset{S}{\overset{\|}{P}}}$(OEt)₂	XV[704]		MeNO₂		$n_D^{20} = 1.5886$, $d_{20} = 1.1001$
[3,6-dibromofluorene]	Br	I[548]	93		84	
[2,7-dibromofluorene]	Br	I[925]	288 dec.	PhH/EtOH	91	
[dibromobenzofluorene]	Br	I[559]	264	PhH/EtOH		
	Br	I[559]	261			
[acridine]	Br	I[1050]	82 dec.			
[3-nitrofluorene]	Br	I[559]	281 dec.	PhH/CHCl₃		
[fluorene]	Br	I[811] XIV[1083]	310 dec.[491]	EtOH/Et₂O[491]	50[491]	IR[1067]
CH₂Ph-(SO₂Ph-Br-p)-4	Br	I[973]	308	EtOH		
CH₂Ph-(SO₂Ph-Cl-p)-4	Cl	I[171]	220			
CHPh-NO₂-4, Ph	Br	I[275]	245 dec.	EtOH/Et₂O		
CHPh₂	Ph₃SnCl₂	II[101]	205		54	

Table 2 (Continued)

Cation R_3PR'		Anion X	Method	m.	Solvent	Yield (%)	Other Data and Remarks
R	R'						
Ph	CHPh$_2$	Br	I[491]	230 dec.[491]	MeOH/H$_2$O[491]	50[491]	IR[1067] UV,[912] PMR −21.4 ppm[397] $E_{1/2}$ 1.271, 2.067[482]
		Cl	I[384] XIV[1083] XIV[1017] XVI[643]	280[1083]		34[1083]	HMR,[389] IR[1067]
	C=NNHPh−NO$_2$−4 NHNHPh−NO$_2$−4	BF$_4$		218		71	UV, dark red
	CHPh −OPh O= CHCPh	Ph$_4$B	XVI + XV[1068]	167	CH$_2$Cl$_2$/ Et$_2$O	5	
		J	XIV[794]	176	Me$_2$CO/ H$_2$O		IR, deep red
	CHC−Me O= (naphthalenyl)	Br	I[973]	237	PhMe		
		J	XIV[395]	216			

Structure	Anion	Ref	mp (°C)	Solvent	Yield (%)	Notes
CH(OPh)$_2$	Cl	I[1068]	130 dec.	CH$_2$Cl$_2$/Et$_2$O EtOH	82	pKa 10.72[534]
CH$_2$Ph-SO$_2$Ph-4	Br	I[983]	255			
CH$_2$PPh$_2$ (O=)	Ph$_4$B	XIV[533]	214	DMF/MeOH		
CH$_2$PPh$_2$	Cl	XIV[533]	255[533]	DMF/MeOH		
	Ph$_4$B	XV[953]	186	DMF/MeOH		
	(HgBr$_3$)$_2$	XV[953]	201	DMF/MeOH		
	HgBr$_3$·HgBr$_2$	XV[953]	195	DMF/MeOH		
	Cl	[534]				pKa >>11
Me–C(=O)–C–Et (benzofuran)	J	XIV[971]	148			UV
O=CPh CHCH$_2$CO$_2$Et	Cl	XIV[951]	135 dec.		96	UV
C=CHPh (Me)	Br	XVI[463]	205		95	UV
CCH$_2$Ph PEt$_2$	Cl	XIV[532]	249			
CH(CH$_2$)$_2$Ph PEt$_2$ PEt$_2$	BF$_4$	XIV[532]	169			
Tridec	Br	I[360]	84	PhH/Et$_2$O		
(fluorene-CN)	Br	I[559]	279	PhH/CHCl$_3$		
(anthrone)	Br	I[1109]	257	MeNO$_2$	78	

Table 2 (Continued)

Cation R₃PR'		Anion X	Method	m.	Solvent	Yield (%)	Other Data and Remarks
R	R'						
Ph	CHCPh-Ph-4 (=O) / HgCl	Cl	[752]	150			
	CClCPh (O=) / Ph	Cl	XIV[790]	Oil	CHCl₃		IR
	CCl₂CONPh₂	ClO₄ / Cl	XV[790] / XIV[1013]	230	MeOH/H₂O / CHCl₃		IR / IR
	(anthracenyl, Me)	Br / Cl	XIV[72] / XIV[72]	223 / 215	PhH	85	
	(fluorenyl)	Br	I[559]	254	PhH/EtOH		
	Ph-CH₂Br-2 / CH₂ NO₂	Br	I[411]	157	Me₂CO	19	IR
	Ph-CH₂Br-2-NO₂-4 / CH₂ NO₂	Br	I[72]	257 dec.	MeOH/Et₂O	79	

Structure	Anion		mp (°C)	Solvent	Yield (%)	Notes
$\overset{O}{\overset{\|}{C}}$HCNPh$_2$ $\|$ Cl	Cl	XIV[1013]	185 dec.	CHCl$_3$/PhH	93	HMR
$\overset{O}{\overset{\|}{C}}$HCPh $\|$ Ph	Br	I[389]				
Br (fluorene, OMe)	Br	I[559]	265	PhH/CHCl$_3$		
CH=CPh SPh	Br	XVI[463]	196			
CH$_2$CPh=NPh	Br	XVI[463]	222		80	
$\overset{O}{\overset{\|}{C}}H_2$CNPh$_2$	Cl	I[1013]	200 dec.	PhH	83	
CH$_2$CONHPh–N=NPh–4	Cl	I[342]	244	MeNO$_2$	38	
CH$_2$Ph–OCH$_2$Ph–4	Cl	I[44]	242	PhH	100	
CO$_2$Me $\|$ C=CCH$_2$CO$_2$Me $\|$ O=CPh	BF$_4$	XIV[441]	175	Et$_2$O/CH$_2$Cl$_2$		IR, UV
$\overset{O}{\overset{\|}{C}}$HPPh$_2$ $\|$ Me	J	XIV[533]	159[533]			pKa >11[534]
CH$_2$C= (cyclohexenone, H) $\|$ Ph	J	XVI[463]	225			IR
C=C(SEt)$_2$ $\|$ CH$_2$CH$_2$Ph	J	XIV[69]	217		85	

Table 2 (Continued)

Cation R_3PR'		Anion X	Method	m.	Solvent	Yield (%)	Other Data and Remarks
R	R'						
Ph	CH=CPh-Br-4 O_2CPh	J	XIV[257]	150		87	
	CH=CPh-Cl-4 O_2CPh	Cl	XIV[257]	150			
	(anthracenyl)CH_2	Cl	I[276]	282	EtOH/Et$_2$O	75	Light yellow
	CH$_2$Ph-C=CPh-4	Br	I[279]	288	EtOH/Et$_2$O	36	
	(fluorenyl)CH$_2$CH=	ClO$_4$	XV[317]	231	Me$_2$CO/ Et$_2$O	54	UV
	CH=CPh O_2CPh	Cl	XIV[257]	>150			
	CH(CPh)$_2$ (O)	Cl·HCl	XIV[836]	74 dec.	CH$_2$Cl$_2$/ Et$_2$O		IR, PMR -25.7, -21.2 ppm

Group	Anion	Ref	m.p.	Solvent	Yield	
(fluorenyl-type, O=/NHCMe structure)	Br	I⁵⁵⁹	264	PhH/EtOH		UV
$CH_2Ph-CH=CHPh-3$	Br	I³³⁷	260	PhH	46	
$CH_2Ph-CH=CHPh-4$	Br	I²⁷³	279	EtOH/Et₂O	87	
	Cl	I²⁷¹	287	PhH/Et₂O	80	
$CH_2CH=CPh_2$	Br	I²⁷⁴	238	Xylene/Et₂O	93	
	Cl	I⁷⁹	241	CHCl₃	86	
CH_2–C₆H₄–$CH=CHPh-NO_2-4$	Br	I⁹⁷⁸	240		69	
$CH(CONHPh)_2$	Cl	XVI¹⁰³⁷	207	EtOH/HCl		
$CHCH_2\overset{O}{C}Ph$ | Ph	Br	I³⁹²	150	Me₂CO	76	
$CH_2\overset{O}{C}CHPh_2$	Br	XIV¹⁰⁷³	191 dec.	EtOAc/CH₂Cl₂	85	
$CHCH=CH_2$ | PPh_2	Ph₄B	XV⁹⁵³	132	DMF/MeOH		
$CH_2O(CH_2)_2OPh-Ph-4$	HgBr₄	XV⁹⁵³	134	DMF/MeOH		
	HgBr₅	XV⁹⁵³	137	DMF/MeOH		
	Cl	I³⁵⁵	149			
C₆H₄=$CHPh-NMe_2-4$	ClO₄	XIV²⁶⁰	250	EtOH		
CMe_2 | PPh_2	Ph₄B	XIV⁻⁵³²	183			

Table 2 (Continued)

Cation R₃PR'		Anion X	Method	m.	Solvent	Yield (%)	Other Data and Remarks
R	R'						
Ph	CH=CHCH= [benzothiophene, Et] O=CMe	ClO₄	XVI[608]			30	UV
	CH₂CH=CCH=CH–CH=CCH=CHCH=CMe Me — Me — CHO	Br	I[941]	162 dec.			UV
	CN=NPh–OMe-2 NNHPh–OMe-2	BF₄	XVI[643]	212	MeOH/Et₂O	86	UV, brown-red
	Et C=CNHC=NEt PhMe Me	BF₄	XVI[642]	154			IR
	CH₂CH=CCH=CH [cyclohexene, Me] Me Me Me	HSO₄	II[45]	175			
		Br Cl	II[897] II[887]	135 142	THF/Et₂O		
	Me CH₂CH=CCH=CH [cyclohexane, H, Me, Me] Me Me	HSO₄	II[774]	175			Yellow

R group	X	Ref.	mp (°C)	Solvent	Yield (%)	Notes
$\overset{Me}{\underset{PPr_2}{C}}$—$CH_2Ph$	Cl	II[774]	113			
	Cl	XIV[532]	225			
$CH_2CH{=}C(CH_2)_2CH{=}C(CH_2)_2CH{=}CMe_2$ (Me, Me, CMe_2; $P(c\text{-}Hex)_2$)	Br	I[215]	128	EtOAc	53	
	Ph_4B	XIV[532]	211			
$(CH_2)_2CH(CH_2)_3CH(CH_2)_3CHMe_2$ (Me, Me)	Br	I[691]	92 (3R, 7R)	Me_2CO/Et_2O	72	HMR, hygroscopic
			81 Racemic	Me_2CO/Et_2O	69	IR, HMR
Pentadec	Br	I[201]	92	PhH/Et_2O		
$CH_2Ph{-}$ [2-Ph-oxazole, O, N, Ph-4]	Br	I[280]	284	$EtOH/Et_2O$	67	Light yellow
$CH{=}CPh{-}Me{-}4$, O_2CPh	Cl	XIV[257]	>150 dec.			
$CH_2{-}$[C₆H₄]${-}CH{=}CHPh{-}Me{-}4$	Br	I[273]	286	$EtOH/Et_2O$	83	Greenish yellow
$\overset{O}{\overset{\|}{C}}$ $CH{\ddot C}CHPh_2$, Me	Br	XIV[1c73]	233 dec.	$CHCl_3/EtOAc$	77	
$CH_2{-}$[C₆H₄]${-}CH{=}CHPh{-}OMe{-}4$		I[978]	390 dec.	DMF		
$C{=}CSCH_2Ph$, PhSMe	J	XIV[69]	191		95	

318

Table 2 (Continued)

Cation R_3PR'		Anion X	Method	m.	Solvent	Yield (%)	Other Data and Remarks
R	R'						
Ph	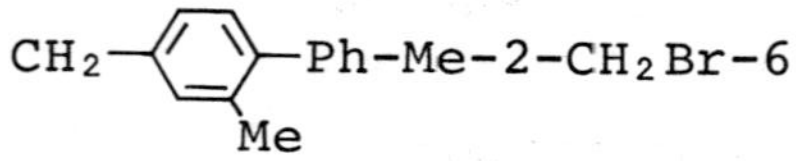 CH$_2$—(2-Me-6-CH$_2$Br-phenyl) Ph-Me-2-CH$_2$Br-6, Me	Br	I[72]	213	PhH	93	
	CHCH$_2$PPh$_2$ / —CO$_2$Me	J	XIV[532]	212			
	(CH$_2$)$_4$PPh$_2$	Br	XVI[728]	172	CHCl$_3$		
	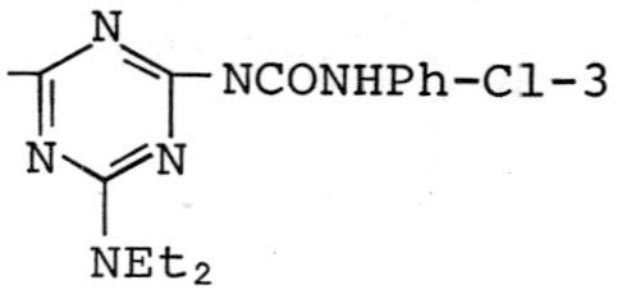 —NCONHPh-Cl-3	Cl	I[969]	169	EtOH		
	triazine ring with NEt$_2$	Br	I[683]	202	EtOAc		Hygroscopic
	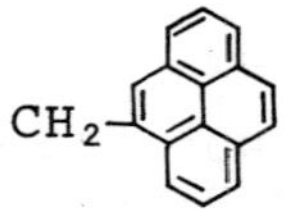 Hexadec	Br	I[353]	265 dec.			
	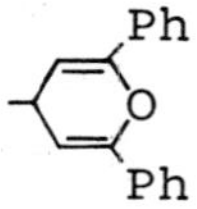 pyrene—CH$_2$; 2,6-diphenylpyran (Ph, Ph)	ClO$_4$	I[604]	164	MeNO$_2$/ EtOAc	99	

Structure	Anion	Ref	mp	Solvent	Yield	Method
CH$_2$CH=CHCH= (fluorenyl)	ClO$_4$	XV[≤17]	207	MeOH/Me$_2$CO	25	UV, light yellow
CH$_2$Ph–CH=CHCH=CHPh-3	Br / J	I[337] / I[733]	269 / 236	PhH/Et$_2$O	10 / 17	
CH$_2$ / Ph Ph / O=, lactone	Cl	XVI[608]			44	UV
C=CHPh–NMe$_2$-4 / O=CPh / MeO / CCCH$_2$Ph$_2$ / Me	Br	XIV[1073]	259	MeOH/Et$_2$O	91	
CH$_2$CHPPh$_2$ / CO$_2$Et	J	XIV[1073]	219	EtOAc/CH$_2$Cl$_2$	74	
	J	XIV[532]	219			
CN=NPh–NMe$_2$-4	BF$_4$	XVI[643]	180		84	UV, black
NNHPh–NMe$_2$-4 / (CH$_2$)$_{11}$OPh	J	I[414]			23	UV
O‖ CH$_2$CCH$_2$CH$_2$	OH	X–V[87]	Resinous			IR
Me Me / H H / Me, O (decalone) / Me O‖ CCCPh$_2$ / MeMe	J	XIV[1073]	273 dec.	CHCl$_3$/MeOH/EtOAc	64	

319

Table 2 (Continued)

Cation R₃PR'		Anion X	Method	m.	Solvent	Yield (%)	Other Data and Remarks
R	R'						
Ph	(2-morpholino-cyclohex-1-enyl) CH=CPh	Br	XVI[463]	278 dec.		81	
	CH₂CO₂-hexadec	Br	I[350]	223		65	
	Octadec	Br	[587]	91			
	(pyrenyl)	ClO₄	I[859]	185 dec.	MeCN/Et₂O	94	
	CPh₃	Br	XV[1071]	290	H₂O		
		Cl	I[171]	245	MeCN		
	CH₂–(C₆H₄)–Ph-Ph-4 (o)	Br	I[279]	122	EtOH/Et₂O	35	
	CH₂–(C₆H₄)–Ph-Ph-4 (m)	Br	I[279]	315	EtOH/Et₂O	61	
	CH₂–(C₆H₄)–Ph-Ph-4 (p)	Br	I[278]	291	EtOH	66	
	CBr=PPh₃	Br	XIV[831]	278	CH₂Cl₂	70	IR

Ligand	Anion	Structure	m.p. (°C)	Solvent	Yield (%)	Characterization
CJ=PPh₃	J	XIV[30]	300 dec.		94	IR, yellow
C̈HPPh₂ / PhO	Cl	XIV[533]	256[533]		pK$_a$ 9.16[534]	
CH=PPh₃	BF₄	XV[282]	211 dec.	EtOH/Et₂O	96	UV, thermo-chromic
	Ph₃BBu	XVI[686]	170			
	Ph₄B	XV[689]	205[689]	Me₂CO/Et₂O[281]	78[281]	IR, UV, HMR, BMR, PMR[689]
	BF₄	XIV[533] XVI[686] XV[689]	260[689]	EtOH[281]	83[281]	IR, UV, HMR, PMR,[689]
	Br	XV[831]	274[689]	CH₂Cl₂/EtOAc[686]	80[832]	IR, UV, HMR, PMR,[689]
	CoBrCl₂	XVI[689,283]	228	CH₂Cl₂		
	FeBrCl₂	XVI[283]	224			
	PdBrCl₂	[283]	253			
	ZnBrCl₂	[283]	235			
	FeBrCl₃	[283]	230			
	CuBr₂	XIV[283]	195			
	FeBr₂Cl₃	[283]	205			
	CuBr₃	[283]	257			
	CuBr₃	[283]	175			
	ZnBr₃		258			
	Cl	XIV[281]	274	CH₂Cl₂/Et₂O	14	IR
	FeJCl₃	[283]	165			
	FeCl₄	I + XVI[283]	230	CH₂Cl₂		
	J	XV[689]	253[689]	MeOH/Hex[281]	96[281]	UV, IR,[689] HMR, PMR[689]

Table 2 (Continued)

Cation R_3PR'		Anion X	Method	m.	Solvent	Yield (%)	Other Data and Remarks
R	R'						
Ph	$CH=PPh_3$	J_3	XV[689]	199	MeOH	77	
	$CHPPH_2$ —Ph	Cl	XIV[534]	130 dec.	CH_2Cl_2/Et_2O		pKa >>11
	$S{=}CHPPh_2$ —Ph	Ph_4B	XIV[533]	201			
	CH_2GePh_3	Cl	XIV[533]	268[533]			pKa 9.01[534]
		Ph_4B	XV[957]	84			Light sensitive
	$O{=}\;CH_2CHCH_2CPh\text{-}Me_3\text{-}2,4,6$ —Ph	Br	XIV[957]	121	Et_2O		
		Br	XIV[333]	218 dec.[333]	EtOAc[333]		IR,[333] HMR[334]
	$(CH_2)_2C{=}CHCH{=}CHCH{=}$ (2,4,4,6-Me₄-cyclohexyl) —Me	Br	I[1020]	88	PhH/petroleum ether		UV
	$CHCO_2$-Hexadec —Me	Br	I[356]	203			
	CH_2O-Octadec	Cl	I[171]	77			

CH=CPPh$_2$ | Ph	Br	XVI[463]	195		72	
CH$_2$Ph–CH$_2$PPh$_2$–4 CHCH$_2$Ph | PPh$_2$	Br Cl	I[489] XIV[532]	277 249	H$_2$O	61	
MeO | =O C–CCHPh–Me$_3$–2,4,6 | MePh	Br	XIV[1073]	207 dec.	EtOH/ Et$_2$O	62	
(cyclohexenyl, Me, Me Me Me)	HSO$_4$	II[896a]	194	MeCN		UV
CH$_2$Ph=CCH=CHCH=CCH=CH |Me |Me	Br Cl	II[+70] II[–9+1] II[660]	180 194[941]	Me$_2$CO[660] EtOAc[660]	UV 18[660]	UV,[660] yellow
CH$_2$CH=C(CH$_2$)$_2$CH= | Me	Br	I[215]	127	EtOAc		
C(CH$_2$)$_2$CH=C(CH$_2$)$_3$CMe$_2$ |Me |Me	Br·EtOH	I[277]	155	EtOH/Et$_2$O	30	
(imidazole, Ph, Ph, Ph, N–N) CH$_2$–	Br	I[269]	264	EtOH	35	UV
CH$_2$Ph–CH=CPh$_2$–4	Br	XVI[489]	261	EtOH/H$_2$O	53	

Table 2 (Continued)

Cation R_3PR'		Anion X	Method	m.	Solvent	Yield (%)	Other Data and Remarks
R	R'						
Ph	$CH_2\overset{O}{C}CH=PPh_3$	Cl	XVI[350]	220	CH_2Cl_2/monoglyme Et_2O	80	
	$\underset{CO_2Me}{C}=PPh_3$	Cl	XIV[92]	195			
		J	XV[92]	219 dec.[687a]	Et_2O[687a]		IR, HMR, PMR −21.4 ppm[687a]
	$CH=CHCH=PPh_3$	Ph_4B	XIV[687a]	204			UV
		Br	XV[755]	217			UV
		J	XVI[755]	232			UV
	$\underset{CS_2Me}{C}=PPh_3$	J	XV[755]	152			
	[benzoselenole ring, Et, Se]		XV[92]		NH_4OH	34	
	$\underset{O=CCH_2Ph}{C}=CHCH=$ OEt	ClO_4	XVI[608]			35	UV
	$C(CH=CHPh\text{-}OMe\text{-}4)_2$	BF_4	XII[495]	185	Me_2CO	90	
	$\underset{PPh_2}{CCH_2Ph}$	Cl	XIV[532]	229			

Structure	X	No.	mp (°C)	Solvent	Yield (%)	Spectra
decalin derivative: ring-junction Me, CH₂C(Me)–Me, OCH₂Ph, 2 H $\text{CH}_2\text{C}(\text{Me})\text{Me}$	Br	I^{720}	260 dec.	PhMe		IR
$(\text{CH}_2)_2\text{C=CHCH=CHCH=CHCH=CH}$– (2,6,6-trimethylcyclohexenyl), Me	HOBF₃	809			90	UV
$\overset{\mid}{\text{C}}\text{=PPh}_3$, $\text{HOC(CF}_3)_2$	Br	II^{810}	213 dec.	Diglyme	83	HMR, FMR, PMR −21.6 +144.8 ppm
oxazole ring (2-(Ph-Ph-4), 5-(4-CH₂–phenyl))	PF₆	91	261	EtOH/Et₂O	64	
$\text{CH}_2\overset{\text{O}}{\overset{\|}{\text{C}}}\overset{\mid}{\text{C}}\text{=PPh}_3$, C=O, Cl	Br	I^{280}	195 dec.			
$\text{CH}_2\overset{\text{O}}{\overset{\|}{\text{C}}}\overset{\mid}{\text{C}}\text{=PPh}_3$, CN	Cl	I^{165}				
$\text{C=C(SCH}_2\text{Ph})_2$, $\overset{\mid}{\text{Ph}}$	J	XIV^{626}	220			
$\text{C=C(SCH}_2\text{Ph})_2$, $\overset{\mid}{\text{Ph}}$	Cl	XIV^{70}	186			
$\text{C=C(SCH}_2\text{Ph})_2$, Ph	Cl	XIV^{69}	187		90	
anthracene (9-substituted), $\text{CH}_2\text{Ph–CH=CH}$	Br	I^{276}	285	EtOH/Et₂O	55	

Table 2 (Continued)

Cation R_3PR'		Anion X	Method	m.	Solvent	Yield (%)	Other Data and Remarks
R	R'						
Ph	CH₂-[phenyl]CH=CHPh-CH=CHPh-4	Br	I[273]	264	EtOH/Et₂O[269]	75[273]	UV,[269] yellow greenish[273]
	C=COCH₂Ph (Ph; CH₂CPh=O)	Br	XIV[836]	230	CH₂Cl₂/Et₂O	65	IR, HMR
	CH₂CC=PPh₃ / CO₂Me	Br	XIV[626]	190			
		Cl	XIV[626]	220			
	[benzothiophene structure: Et, OMe, SMe, S, CH₂CC=CHCH, Ph, O]	Cl	XVI[608]			33	UV
	CH₂-[biphenyl] Ph-Ph-4	Br	I[279]	217	EtOH/Et₂O	25	

Structure	Anion	Compound	m.p.	Solvent	Yield	Notes
CH$_2$—⟨Ph-Ph-4⟩	Br	I[279]	298	EtOH/Et$_2$O	61	
CH$_2$—⟨Ph-Ph-4⟩ (biphenyl)	Br	I[273]	295	DMF	76	
Ph—CHCH$_2$Ph—PPH$_2$	Ph$_4$B	XIV[532]	216			
O=CH$_2$CCO$_2$—t-Bu, PPh$_3$	Cl	XVI[165]	218		90	
CH$_2$—CH=CH—⟨Ph-Ph-4⟩	Br	I[276]	274	EtOH/Et$_2$O	69	Light yellow
O=CH$_2$CCO$_2$Ph, PPh$_3$	Cl	XVI[165]	185	CHCl$_3$/EtOAc	38	
CH(CH=)$_2$ fluorenylidene, PPh$_3$	ClO$_4$	XV[517]	232	Me$_2$CO/Et$_2$O	38	UV
CH—cyclopentadienyl, PPh$_3$	ClO$_4$	[241]				UV
PPh$_3$	J	XIV[260]	249	EtOH		UV

Table 2 (Continued)

328

Cation R_3PR'		Anion X	Method	m.	Solvent	Yield (%)	Other Data and Remarks
R	R'						
Ph	C=CHCH=CHC=PPh₃ with —CO₂Et, —CO₂Et groups	ClO₄	XVI + XV[608]	142	EtOH	37	UV
	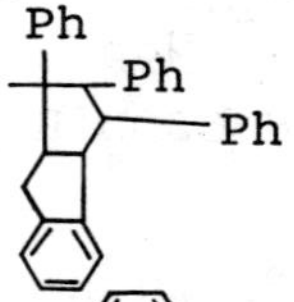 (Ph, Ph indenyl/cyclopentane fused bicyclic structure)	Br	II[619]	189	MeCN	39	UV, HMR
	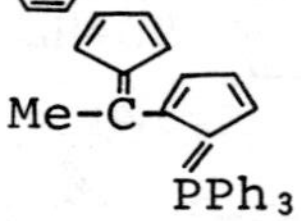 Me—C (indenyl) PPh₃	ClO₄	XIV[260]	160	EtOH		UV
	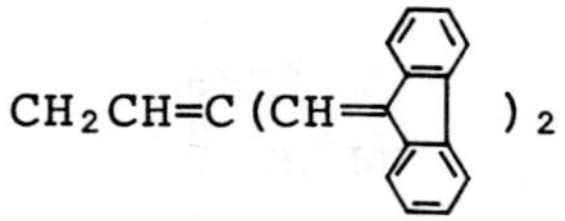 (fluorenylidene)₂ structure)₂	ClO₄	XV[317]	253	CH₂Cl₂/Et₂O	18	UV, light yellow
	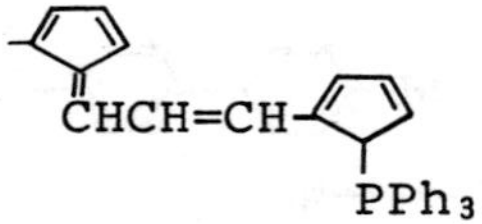 CH₂CH=C(CH ...)	ClO₄	XIV[260]	244	EtOH		UV
	CHCH=CH PPh₃ ; C=PPh₃ / O=PPh₂	Cl	XVI[90]	296	EtOH/diglyme	96	IR, PMR −23.2, −26.9 ppm

Compound	Anion	Type	mp	Recryst. solvent	Yield (%)	Notes
C=PPh₃ | PPh₂	Ph₄B	XV[50]	252			
	BF₄	XV[50]	261	MeCN/diglyme	96	IR, HMR, PMR −26.0; +1.5 ppm
	Cl	XIV[90]	253 dec.			IR
	HgCl₃	XV[50]	230	DMF	92	HMR, PMR −26.5, −1.4, +114.2 ppm
	PF₆	XV[90]	264	MeCN/EtOH	38	
C=PPh₃ | S=PPh₂	J	XV[90]	264	EtOH		
	Cl	XV⁻[90]	309			IR, HMR, PMR −24.7, −44.0 ppm
CH₂Ph—O—CHC·C=PPh₃ | CO₂Et	Br	XIV⁻[65]	160	CHCl₃/monoglyme	48	
[cyclopentadienylidene]CHCH=CHCH=CH[—cyclopentadienyl]—PPh₃	ClO₄	XIV[260]	>260[240]	EtOH[260]		UV[240]
Ph–Cl₂-3,4	Cl	[1050]	296			
Ph–Br-4	Br	VI[267]	134			
	J	VIII[483]	210	H₂O	25	
Ph–Cl-2	Cl	[1050]	243			
	J	VIII[483]	242	H₂O	40	Pale yellow
Ph–Cl-4	II[101]		161		68	
	Ph₃SnBrCl					
	BF₄	VI[322]	198	DMF	61	
	Cl	VI[267]	279[1050]		90[267]	
	J	VIII[483]	219[267]	H₂O[483]	55[267]	

Table 2 (Continued)

Cation
R_3PR'

R	R'	Anion X	Method	m.	Solvent	Yield (%)	Other Data and Remarks
Ph	Ph-(OH)$_2$-2,5-Cl-4	OPh(NO$_2$)$_3$-2,4,6	XV[828]	225	MeOH/H$_2$O		UV
	Ph-NO$_2$-2	Cl	XVI[828]	297	MeOH/EtOAc		IR, UV
		BF$_4$	VI[819]	209	Me$_2$CO		
		J	VIII[483]	230 dec.	H$_2$O	40	Yellow
	Ph-NO$_2$-3	BF$_4$	VI[822]	194	Me$_2$CO	51	
	Ph-NO$_2$-4	J	VIII[483]	215 dec.	H$_2$O	40	Red
		Ph$_3$SnBrCl	II[101]	201		81	
		Br	I[636]	275	Xylene	95	
		J	VIII[483]	228 dec.[483]	H$_2$O[483]	40[483]	Orange[906]
	Ph-OH-2	J	[76]	284		35	IR
	Ph-OH-3	J	XVI[487][482]	270	MeOH/H$_2$O		
	Ph-OH-4	Ph$_3$SnCl$_2$	II[101]	201		54	IR
		Br	XV[813]	279[813]			
			XIV[467] I[637]				
		J	VI[813]	270[467]	EtOH/ H$_2$O[467]	92[467]	IR[1067]
	Ph(OH)$_2$-2,5	OPh(NO$_2$)$_3$-2,4,6	XVI[467]				
		Cl	XIV[828]	188[828]	MeOH/H$_2$O		UV
			XIV[828]	297[828]	MeOH/ Et$_2$O[828]		IR, UV[828]

Compound	Anion	Structure	mp (°C)	Solvent	Yield (%)	Data
Ph-NH$_2$-3	J	I[8] XVI[485]	270 dec.	EtOH/H$_2$O	93	E1/2 1.801, 1.916, 2.144, 2.437[482]
Ph-NH$_2$-4	J, Br	XVI[487], XVI[487]	315 dec., 316[487]			
Ph-CF$_3$-4	J, Br	XVI[487] [482]	321 dec.		70	E1/2 1.410, 2.077, 2.323, 2.482
Ph-CO$_2$H-4	J, Br	IV+[94] [432]	218	CHCl$_3$	65	E1/2 1.630, 2.346
Ph-Me-2	J, Br	IV+[94], IV+[94]	298[494], 285[494]	CHCl$_3$[494], CHCl$_3$[494]	52, 91[494]	E1/2 1.678[482], 2.061[482], IR,[1078] E1/2[499]
Ph-Me-3	J	XV+[494]	294[1078]			
	Ph$_4$B	XVIII[1078] XV[1079]	254	Me$_2$CO/H$_2$O		
	Br	IV[494]	196[494]	CHCl$_3$[494]	95[494]	E1/2 1.684[482], 2.066[482]
Ph-Me-3 and -4 1:1	J	VIII[483] V[1079]	202[1079]	H$_2$O[483]	40[483]	
	Ph$_4$B	XVIII+ XV[1079]	246	Me$_2$CO		
Ph-Me-4	Ph$_4$B	XV[1079]	246			IR

331

Table 2 (Continued)

Cation R_3PR'		Anion X	Method	m.	Solvent	Yield (%)	Other Data and Remarks
R	R'						
Ph	Br	Br	IV[494]	228[267]	$CHCl_3$[494]	85[494]	E1/2 1.602 2.073[482]
		Cl	IV[494] VI[267]	194	$CHCl_3$	68	
		J	VI[813] V[1079] I[1071]	212[1071]	H_2O[1071]	78[1071]	IR[1079]
	Ph-OMe-2	J	XIV[76]	266	MeOH/EtOAc	76	HMR
	Ph-OMe-3	J	VII[484]	204	H_2O	50	
	Ph-OMe-4	$Ph_3SnBrCl$	II[101]	134	76		
		BF_4	VI[822]	196	Me_2CO	49	
		Br	IV[494]	218[494]	$CHCl_3$[494]	49[494]	
		J	VII[484] IV[813] XV[494] VIII[467]	217[813]		80[467]	E1/2 1.835, 1.988, 2.128[482] IR[1067]
	Ph-CH=CH$_2$-2	OPh(NO$_2$)$_3$-2,4,6	XV[125]	155			IR
	Ph-OEt-2-Cl-4-OH-5	J	XIV[828]	217	MeOH/EtOAc		IR, UV
	Ph-Me$_2$-3,4	J	XV[1079]	218	MeOH/Et$_2$O	74	
	Ph-(OMe)$_2$-2,5	J	VII[484]	202	H_2O	50	Yellowish

Substituent	Anion	Method	mp (°C)	Solvent	Yield (%)	Physical data
Ph-OEt-2-OH-5, Ph-NMe₂-4	J Br	XIV[828][482]	237	MeOH/EtOAc		IR, UV $E_{1/2}$ 1.777, 2.091
Ph-CO₂Et-3 Ph-CO₂Et-4 Ph-i-Pr-4	Cl BF₄ J Br	IV[494] VI[822] VIII[483] IV[494]	184 191 204 dec. 226[494]	CHCl₃ Me₂CO H₂O CHCl₃[494]	78 47 40 94[494]	$E_{1/2}$ 1.654[482] 2.056[482]
Ph-Me₃-2,4,6	Br	IV[494]	273	CHCl₃	28	
![naphthyl]	Ph₄B	V + XV[1079]	268	MeCN		IR
	Br	IV[494]	280[494]	CHCl₃[494]	48[494]	$E_{1/2}$ 1.439, 1.781, 2.054, 2.326, 2.488[482]
![naphthyl]	J	VIII[483] XV[494]	278[483]	H₂O[483]	40	
	Ph₄B	XVIII + XV[1079] V[482]	215	Me₂CO	68	IR
	Br					$E_{1/2}$ 1.554, 1.653, 2.049, 2.207, 2.325, 2.494
Ph(OEt)₂-2,5	Cl J	IV[494] XVI[828]	298[1050] CHCl₃[494] 212	MeOH/EtOAc	74[494]	IR, UV
![Ph-NO₂-4] Ph-NO₂-4	Br	[482]				$E_{1/2}$ 0.454, 1.502, 1.831, 2.118

Table 2 (Continued)

Cation R_3PR'		Anion X	Method	m.	Solvent	Yield (%)	Other Data and Remarks
R	R'						
Ph	Ph-Ph-4	Br	IV[694]	255[494]	H_2O[694]	89[494]	$E_{1/2}$ 1.527, 1.655, 2.067, 2.279, 2.490[482]
		J	XV[813] VII[484] VI[813] [482]	241[484]	H_2O[484]	50[484]	IR[1067]
	Ph-SO_2Ph-4	Br					$E_{1/2}$ 1.227, 1.677, 1.954, 2.083, 2.350, 2.490
	Ph-$CHPh_2$-4	Br Cl	[482] I[468] II[468] IV[494] I[764]	242[494]	$CHCl_3$[494]	92[468]	$E_{1/2}$ 1.668 PMR[468]
Ph-OH-2	Me	J		298 dec.	HJ/H_3PO_2/H_2O		
Ph-OH-3	Me	SO_4	XVI XIV[613]	150	NaOH/H_2SO_4 dil.		
Ph-OH-4	Me	Cl	XV[764]	87[658]	HCl dil.		
Ph-Me-2	Me	OPh(NO_2)$_3$-2,4,6	XV[656]	170[658]	EtOH[656]	70	
		HgJ_3	XV[659]	214			TLC[914]

		X	Prep	mp (°C)	Solvent	Yield (%)	Notes
Ph—Me—3	Me	J	I[225],[91+]	234[225]	Et_2O[225]		IR[655] TLC
		HgJ_3	XV[225]	95[225]			
		J	I[225]	168[225]	Et_2O[225]		UV,[911] { IR,[659] DK[911]
Ph—Me—4	Me	Br	I[559]	255	PhH/EtOH		
		$OPh(NO_2)_3$-2,4,6	XV[656]	134[656]	EtOH[656]		TLC[914]
		Br	I[1086]	221	$BuOH$/Et_2O	97	
		$Cl \cdot 2\ H_2O$	I[712]	80			
		$PtCl_6$	XV[712]	245			Red
		J	I[732]	221		94[732]	UV,[911] IR,[959] DK[911]
	Et	$J \cdot H_2O$	I[732]	100			
		O_3SPh	I[589]	145	EtOH/Et_2O	73	
		Br	I[1099]	226	$CHCl_3$/EtOAc	80	
		Br	I[732]	187		73	
		$J \cdot H_2O$	I[732]	>100			
	$CH_2C(=O)Me$	Br	I[712]	210 dec.			
		Cl	I[712]	245 dec.			
		$PtCl_6$	I[712]	220			
		J	I[712]	189			
	Pr	J	I[712]	182			
	i—Pr	J	I[712]	184			
	i—Bu	J	I[712]	104			
	$CH_2C(=O)Ph$	Br	I[712]	248			
		Cl	I[712]	226			
		$PtCl_6$	I[712]	240			
		J	I[712]	236			
	CH_2CHPh—OH	Ph_4B	XIV + XV[1086]	170 dec.	Me_2CO/Et_2O	92	

Table 2 (Continued)

Cation R₃PR'

R	R'	Anion X	Method	m.	Solvent	Yield (%)	Other Data and Remarks
Ph–Me-4	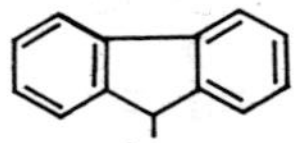 [fluorenyl structure]	Br	I[559]	268	PhH/EtOH		
	Ph(NO₂)₂-2,4	OPh(NO₂)₃-2,4,6	XV[656]	191	EtOH/Me₂CO/petroleum ether		
	Ph	Br	[481]	284[481]		59[481]	$E_{1/2}$ 1.727, 2.071[482]
	Ph-i-Pr-4	J	XI[483]	196	H₂O	40	
		Br	[481]	241[481]		47[481]	$E_{1/2}$ 1.713, 2.004, 2.455[482]
Ph–OMe-2	Me	Br	I[1086]	234	H₂O	95	
		J	I[583]	212		100	
	Et	J	XIV[1086]	225 dec.	EtOH/Et₂O	70	
Ph–OMe-3	[fluorenyl structure]	J	I[613]	160	H₂O		
	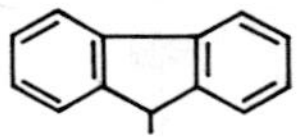	Br	I[559]	116	PhH/EtOH		
Ph–OMe-4	Me	Br	I[1086]	214	EtOH/Et₂O	91	
		J	I[389]	219[764]	H₂O[764]	95[764]	HMR,[389] UV[902]

R	X⁻	Ref.	mp (°C)	Solvent	Yield (%)	Other data
CH₂CN	Cl	XV[1086], I[731][432]	204	MeOH/Et₂O	78	$E_{1/2}$ 1.934, 2.109, 2.364
Pr	Br					
CH₂CO₂Et	Ph₄B	XV[34]	148	CH₂Cl₂/Hex		
CH₂Ph-NO₂-4	Br	I[557]	227[1032]	PhH[1032] 97[1032]		
CH₂Ph-OMe-4	Br	I[557]	254	PhH/CHCl₃		
CH₂CHPh—OH	Ph₄B	XIV[1086] + XV[1086]	211 dec.	BuOH	68	
[9H-fluoren-9-yl structure]	Br	I[559]	232	PhH/EtOH		
Ph-(NO₂)₂-2,4	OPh(NO₂)₃-2,4,6	XV[656]	179	EtOH/petroleum ether		$E_{1/2}$ 1.848, 2.006, 2.128
Ph	Br	[482]				
Ph-Me₂-2,4 / Me	J	I[712]	231			
Et	PtCl₆	XV[712]	252			
Ph-Me₂-2,5 / Me	J	I[712]	225			
Et	J	I[712]	169			
PhCH₂OMe / Me	J	I[712]	220			
Ph-(OMe)₂-2,4 / Me	J	I[621]	203	EtOH	99	
Dodec	Br	I[821]	192	EtOH		
Ph-NMe₂-4 / Me	Br	I[821]	92	MeNO₂		
Et	J	[482]	302	DMF/EtOAc		
	Br	I[904]	296 dec.[904]	H₂O[904]	81[904]	$E_{1/2}$ 1.697, UV[901]
CH₂Ph-NO₂-4	Br	I[904]	295 dec.	EtOH	66	Yellow, $E_{1/2}$ 1.279, 1.846
CH₂Ph	Br	I[904][482]				

338

Table 2 (Continued)

Cation R₃PR'		Anion				Yield	Other Data
R	R'	X	Method	m.	Solvent	(%)	and Remarks
Ph-Me₃-2,4,5	Me	J	I[712]	291			
Ph-Me₃-2,4,6	Me	J	I[712]	269			
Ph(OMe₃)-2,4,6	Me	J	I[820]	175			
	i-Pent	Br	I[820]	197			
	Non	Br	I[820]	101			
Ph-(OMe)-2,4,6	Dodec	Br	I[820]	106			
	Hexadec	Br	I[820]	69			
	Octadec	Br	I[820]	98			
Ph-Ph-2	Me	J	I[49]	250 dec.			
Ph-Ph-4	Me	J	I[1090]	135 dec.			
	CH₂CH=CH₂	Br	I[1090]	195			
	CH₂CO₂Et	Cl	I[1090]	164			
	CH₂Ph	Br	I[1090]	277			
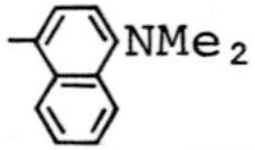 [1-methyl-naphthalenyl-NMe₂]	Me	J	I[1031]	194 dec.			
Me₂N-[Ph-4]	Me	J	I[1031]	348			

Table 3. $R_2R'_2PX$

| Cation $R_2PR'_2$ | | Anion X | Method | m. | Solvent | Yield (%) | Other Data and Remarks |
R	R'						
ClCH₂	Et	BF₄	XVI[435]	170		90	
	c-Hex	Cl	XVI[435]	249		89	
	Ph	Cl	XVI[435]	221[435]	MeOH/Et₂O[800]	85[435]	
Me	CH₂OH	J	I[400]	170			HMR, PMR −32[673a]
	Et	J	I[679]				
	(CH₂)₂CN	J	I[383]	194	MeOH	66	
	Pr	J	I[531]	204	EtOH/Et₂O	76	
	Bu	J	I[531]	168	EtOH/Et₂O	88	
	t-Bu	J	I[468]	>365	Et₂O		
	c-Hex	J	I[544]				
			XVI + I[542]	218[542]	EtOH/Et₂O[542]	65[542]	
	CH₂Ph	Br	I[36]	141	i-Bu-CHOH-Me/ petroleum ether	85	
	Dodec	J	I[431]	49		61	IR, HMR, PMR −29.5 ppm
	CH₂-CPh₂ OH	J	XIV[1081]	200 dec.	MeOH/Et₂O		
	Ph	OPh(NO₃)₃-2,4,6	XV[447][656]	119[447]	EtOH[656]		UV[910]
		O₃S-Ph-Me-4		166	EtOH/Me₂CO		

Table 3 (Continued)

R	R'	Anion X	Method	m.	Solvent	Yield (%)	Other Data and Remarks
Me	Ph	Br	I[10]	215[481]	$EtOH/Et_2O$[10]	71[481]	PMR -22.1,[397] TLC,[914] $E_{1/2}$ -2.087[482]
		$U^{VI}O_2Cl_4$	XV[348]	255			
		$PtCl_6$	7[17]	218			
		$U^{IV}Cl_6$	I[348]	177			
		J	I[207], XVI + I[967], XVI[648]	256[207]	$EtOH$ Et_2O[648]	82[544]	Λ,[455] PMR -22[397]
	Ph-Me-2	$O_3S\text{-}Ph(NO_2)_2\text{-}2,4$	I[656]	172	EtOH		
		J	[656]	227	EtOH		
	Ph-OMe-2	$O_3S\text{-}Ph(NO_2)_2\text{-}2,4$	[656]	162	H_2O		
		$OPh(NO_2)_3\text{-}2,4,6$[656]		164	$EtOH/H_2O$		
		J	I[656]	218[656]	H_2O[656]	100[583]	
HOCH₂	Bu	Br	I[805]	134			
	i-Bu	Cl	X[378]	169	Me_2CO	90	
	c-Hex	Cl	XVI·I[435]				$n_D^{20} = 1.5278$
	Hept	Br	I[805]				$n_D^{20} = 1.5082$
	Ph	$OPh(NO_2)_3\text{-}2,4,6$	9[10]	144	$AcOH/Et_2O$[800]		
		Cl	X[800]	165[800]			
Et	Bu	Br	XVI + I[29]	146	i-PrOH/EtOAc	88[378]	UV[912]

R	X	Method	m.p.	Solvent	Yield	Notes
c-Hex	J	I[542]	134			
CH_2Ph	Br	XVI + I[29]	234			
Ph	Br	XVI + I[124]	195[124]	EtOAC	21[124]	E1/2 2.031[82]; Red
Ph	$PtCl_6$	717	218			
Ph	J	I[772]; XVI + I[465]; VII[487]; IV[494]	209[465]	$CHCl_3$[494]	98[494]	
Ph-Me-3-Cl-6	$OPh(NO_2)_3$-2,4,6	XV[218]; I[218]	175; 218	EtOH		Yellow
MeCHOH	J	X[800]	139	MeOH/Et_2O		
$MeCO_2CH_2$	Cl	XVI +[435]; XV[435]	123	EtOH	70	
Ph	ClO_4					
Ph	BF_4	XVI +[435]; XV[435]	103			
Pr	J	IV[494]	164	H_2O	65	
$Me-CH-CH_2$ / $-OH$	Br	XVI +[848]; III[848]	197	i-PrOH	19	
$HO(CH_2)_3$	Cl	X[1035]	140 dec.	EtOAC		
Ph	Br	I[419]	154[419]	EtCOMe/ EtOAC[419]		IR[1067]
i-Bu	J	I[27]	183			
t-Bu	Cl	I[468]	264	MeOH/Et_2O	67	
$Me(CH_2)_2CHOH$	Cl	X[1035]	127 dec.	EtOAC		
$PhCH_2$	Br	I[5]	252	HBr dil.	55	
Ph	Cl	I[478]	262	H_2O	62	
Ph	Cl	X[1035]	165[1035]	EtOAC/ Et_2O[1035]	98[800]	
Ph-Me-4	Br	[481]	200[481]		61[481]	E1/2 1.605[82]; 2.092[82]

341

Table 3 (Continued)

Cation $R_2PR'_2$		Anion X	Method	m.	Solvent	Yield (%)	Other Data and Remarks
R	R'						
Ph	Ph-Me-4	J J·H_2O	VIII[483] [483]	201 110	H_2O H_2O		
	Ph-OMe-4	Br	[481]			67[481]	E 1.790 $1.954_{\ell 2}$[482]

Table 4. $R_2R'R''PX$

Cation $R_2PR'R''$			Anion X	Method	m.	Solvent	Yield (%)	Other Data and Remarks
R	R'	R''						
ClCH₂	Et	Ph	BF₄	XV[435]	105		85	
			Cl	XVI[435]	163		66	
Me	CH₂J	Ph-Me-4	J	I[203]	158			
	CH=CH₂	Ph	Br	XVI[826]	167	MeCN	87	
	CH₂CO₂H	Ph-Me-4	Cl	I[712]	172 dec.			
			PtCl₆	XV[712]	220			
	(CH₂)₂Br	Ph	Br	I[369]	173			
		Ph-Me-4	Br	I[203]	194	i-BuCHMe/		
	Et	Pent	J	I[37]	108	EtOAc	81	
		Ph	J	I[37]	149	OH i-PrOH		
		Ph-NMe₂-4	J	I[718]	199			
		Ph-Me₃-2,4,6	J	I[209]	168		94	
	(CH₂)₂CN	Ph	OPh(NO₂)₃-2,4,6	XV[345]	123	H₂O		Yellow
			J	I[346]	171	EtOH		
	CH₂-CH=CH₂	Ph	Br	I[600]	113			
Me	CH₂CMe(O)	Ph	Cl	I[1038]	107	Me₂CO		
	(CH₂)₂CO₂H	Ph	J	I[345]	167	EtOH		
	Pr	Ph	Br	I[505]	118			
	Me₂COH	Ph	J	X[464]	117	Et₂O		
	CH₂CO₂Et	Ph	Br	I[1038]	124	Me₂CO/PhH	71	

Table 4 (Continued)

Cation $R_2PR'R''$			Anion X	Method	m.	Solvent	Yield (%)	Other Data and Remarks
R	R'	R''						
Me	CH_2CO_2Et	Ph–Me–4	Cl	I[712]	153			
			$PtCl_6$	XV[712]	200			
	Bu	CH_2Ph	Cl	I[571]	147			
	t–Bu	Ph	J	I[468]	160	$MeOH/Et_2O$		
	$CH_2As(Me)OAsMe_2$	Ph	Cl	I[175]	116	Me_2CO		
			J	XV[175]	147	Me_2CO		
	$(CH_2)_2$–N⟨ ⟩(Me)	Ph	Br	XVI[584]	183		99	IR, tan
	CH_2Ph–NO_2–4	Ph	Cl	I[318]	219			
	CH_2Ph	Ph	J	I[554]	72			
		Oct	Cl	I[554]	176			
		Dodec	J	I[554]	49			
		Hexadec	Cl	I[554]	189			
			J	I[554]	66			
		Ph	Br	I[1038]	158			
			Cl	I[699]	101			
			J	I[1035]	243			
		Ph–NMe_2–4	Cl	901		EtOH	83	UV
	$CH_2C(=O)Ph$	Ph		556				pK_a 7.75

R	Ar	Anion	Structure	M.p.	Solvent	Other data
			I[12]	177[1038]	$CHCl_3$/PhH[1038]	IR,[12] pK values[12]
$CD_2C(=O)Ph$	Ph	Br	XVI[12]			IR,[12] pK values
$CH_2C(=O)Ph$	Ph-NMe_2-4	Br	I[1039]	249 dec.		
$CH_2C(=O)Ph$-NH_2-4	Ph	Cl	I[318]	242		HMR
MeCHPh	Ph	Br[309]	I[699]	172		
$(CH_2)_2Ph$	Ph-Me-4	$PtCl_6$	XV[202]	226		
$(CH_2)_2$N(Et)Ph	Ph	Br	I[318]	113	MeOH	
Dodec	Ph	Br	I[555]	138		
		Cl	I[555]	156		85
$Ph(NO_2)_2$-2,4	Ph	$OPh(NO_2)_3$-2,4,6	XV[656]	235	EtOH/H_2O	Hygroscopic
Ph-Br-2	Ph	J	I[545]	228	EtOH	
Ph-Cl-2	Ph	J	I[545]	205	EtOH	
Ph-Cl-4	Ph	$OPh(NO_2)_3$-2,4,6	XV[655]	134	EtOH	
Ph	Ph-OH-3	$OPh(NO_2)_3$-2,4,6	XV[657]	180	EtOH	
		J	I[557]	174	EtOH/Et_2O	
Ph-OMe-2	Ph	$O_3SPh(NO_2)_2$-2,4	I[556]	98	H_2O	
		$OPh(NO_2)_3$-2,4,6	XV[656]			
Ph-OMe-3	Ph	J	I[656]	198	EtOH	
		J	I[657]	129		
Ph-OMe-4	Ph	O_3SPh-Me-4	I[453]	132	Me_2CO	
		J	I[453]	135	Me_2CO/MeJ	IR

346

Table 4 (Continued)

Cation R₂PR'R"			Anion X	Method	m.	Solvent	Yield (%)	Other Data and Remarks
R	R'	R"						
Me	Ph	Ph–NMe₂-4	Br	I[1039]	252	CHCl₃/PhH		
			J	I[1039]	234	CHCl₃/PhH		
HOCH₂	Br	Ph	OPh(NO₂)₃-2,4,6 Cl	XV[910]	105	MeOH		UV
				[912]				
Et	Me	$\overset{O}{\overset{\|\|}{C}}$Me	J	I[538]	122	EtOH/Et₂O		Very hygroscopic
		(CH₂)₂OH	J	I[536]	253	EtOH		
		CH₂CHMe–OH	J	I[541]	173	EtOH/Et₂O	72	
		(CH₂)₃OH	J	I[541]	173	EtOH/Et₂O		
		(methylthiophene)	J	I[893]	122	EtOH/Et₂O		
		CH₂CO₂Et	J	I[543]	41	EtOH/Et₂O		Very hygroscopic
		–C(=O)–(thiophene)	J	I[535]	73	Et₂O	58	Yellow
		(CH₂)₂CO₂Et	J	I[543]	99	EtOH/Et₂O		

Group		Ref.	mp	Solvent	Yield	Notes
HO— (ring, H, Me) CHEt	J	I[541]	110	EtOH/Et₂O		
CO₂Et	J	I[543]	82	EtOH		
(CH₂)₃CO₂Et	J	—[543]	102	EtOH/Et₂O		
O=C—CPh	J	I[538]	68	Me₂CO/Et₂O CH₂Cl₂/EtOAc		Hygroscopic
C=CHPh	J	XIV[532]	181			IR
Me (CH₂)₂Ph— CH₂OMe-2	OPh(NO₂)₃-2,4,6 J	XV[54]	67	EtOH		Yellow
		I[54]	107			
O=C (naphthalene)	J	I[535]	118 dec.	Et₂O	78	
C=CHPh Ph	J	I[528]	174	EtOH	85	
C=C—CO₂H PhPh	J	I[528]	230 dec.	EtOH	32	
C=N—C=NMe Ph Ph	J	I[524]	178	EtOH	71	
CPh₃ Ph-Br-2	OPh(NO₂)₃-2,4,6 J	I[547] XV[419]	205 dec. 116	EtOH MeOH/H₂O		
Ph-Br-4	J	I[419]	123	EtOAc/Et₂O		
	J	I[711]	97			

Table 4 (Continued)

Cation $R_2PR'R''$ R	R'	R''	Anion X	Method	m.	Solvent	Yield (%)	Other Data and Remarks
Et	Me	Ph-Cl-4	J	I[711]	135			
		Ph	OPh(NO₂)₃-2,4,6	XV[656]	74	EtOH		
		PhOH-4	J	I[699]	110			
		Ph-CF₃-4	J	XVI[212]	168	EtOH		
		Ph-Me-3-J-6	J	I[151]	108	Me₂CO		
		Ph-Me-2	J	I[421]	190			
		Ph-Me-4	J	I[213]	163			
			J	I[202]	137			
		Ph-OMe-4	PtCl₆	XV[711]	142			
			J	I[711]	91			
		Ph-Me₂-2,4	PtCl₆	XV[711]	202			
			J	I[711]	90			
		Ph-Me₂-2,5	PtCl₆	XV[552]	137			
			HgJ₃	XV[552]	105			
			J	I[552]	137			
			J₃	XV[552]	85			
		Ph-Et-4	PtCl₆	XV[711]	195			
			J	I[711]	135			
		Ph-OEt-4	PtCl₆	XV[711]	208			
			J	I[711]	60			
		Ph-NMe₂-2	OPh(NO₂)₃-2,4,6	II[671]	111	H₂O		
			J	I[671]	128	MeOH/Et₂O		Yellow

Substituents	X	Ref.	M.P.	Solvent	Color / Notes
Ph-NMe-4	J	I[711]	186		
Ph-Me3-2,4,5	J	I[711]	160		
Ph-Me-2,4,6	J	I[711]	125		
Ph-(CH2)2OMe-2	PdBr3	XV[663]	178	EtOH	Orange
		I[665]	96		
(1-methylnaphthyl)	J	I[711]	125		
Ph-(CH2)3OMe-2	OPh(NO2)3-2,4,6 / J	XV[54]	99	EtOH	Yellow
Ph-AsEt2-2	J	I[295]	148	H2O	
Ph-CH2AsOAsMe2, Me	J	I[295]	162	EtOH	
Ph-Cl-2-Ph-4	O3SPh-Me-4	I[213]	151	Me2CO/Et2O	
Ph-OPh-4	J	I[213]	163		
Ph-Me-4	Cl	XV[712]	96		
	PtCl6	XV[712]	157		
	Br	I[699]	152		
CH2CO2H, Ph		XV[712]	127		
CH2-CH=CH2	OPh(NO2)3-2,4,6 / PtCl6	XV[712]	178	EtOH/Me2CO	
CH2CMe(=O), Ph-Me-4	Cl	I[571]	148		
Bu, CH2Ph	Br	I[1076]	201	MeOH/Et2O	Hygroscopic
(fluorenyl) CH2Ph, Ph	Ph3SnBrCl	XV[101]	114		93[836]
Ph	Br	XIV[836]	170[568]	CH2Cl2/Et2O[836]	71[836] HMR,[836] PMR −32.7[673a]
Cl		[568]	194		

Table 4 (Continued)

Cation $R_2PR'R''$			Anion X	Method	m.	Solvent	Yield (%)	Other Data and Remarks
R	R'	R''						
Et	CH_2Ph	Ph-Me-4	Cl	I[572]	104	EtOAc	70	
	$CH_2C(=O)Ph$	Ph-Me-4	$PtCl_6$	[712]	173			
	CHPh-CH_2CO_2H	Ph	$J \cdot H_2O$	XI[460]	100 dec.	H_2O	61	
	CH_2Ph-CH_2OMe-2	Ph	J	I[666]	113	$EtOH/Et_2O$		
	Dodec	Ph	$MeS\text{-}P(=O)(OMe)\text{-}S$	XV[704]			81	$n_D^{20} = 1.5372$, $d_{20} = 1.0520$
			$MeS\text{-}P(=S)(OMe)\text{-}O$	XV[587]				$n_D^{20} = 1.5372$, $d_{20} = 1.0520$
			$(OEt)_2P(=S)\text{-}S$	XV[587]			92	$n_D^{20} = 1.5423$, $d_{20} = 1.1054$
			Br	[587]	109	$EtOAc/Et_2O$	65	
			$Br \cdot EtOAc$	I[555]	62	$MeOH/Et_2O$	75	
	CHPh-NHPh	Ph	Br	XI[461]	182 dec.	Et_2O	69	

R	R′	X	Method [ref]	mp (°C)	Solvent	Yield (%)	Properties
$\overset{O}{\parallel}$CHCPh / CH_2CPh=O	Ph	Cl·H_2O	XIV[836]	214			IR, PMR −34.1[673a]
Octadec	Ph	MeS–$\overset{O}{\overset{\parallel}{P}}$–OMe	XV[704]			95	n_D^{20} = 1.5296, d_{20} = 1.0860
		MeS–$\overset{S}{\overset{\parallel}{P}}$–OMe	XV[587]				n_D^{20} = 1.5296, d_{20} = 1.0860
		Br	[587]	82		98	
$\overset{Ph}{\mid}$C=C–OCH_2Ph / CH_2CPh=O	Ph	Br	XIV[836]	96	CH_2Cl_2/Et_2O/PhH	70	IR, HMR, PMR −28[673a]
Ph-Cl_3-2,4,5-$(OH)_2$-3,6	Ph	Cl	I[842]	187	MeOH/EtOAc		IR, UV
Ph-$(NO_2)_2$-2,4	Ph-$(NO_2)_2$-2,4	OPh$(NO_2)_3$-2,4,6	XIV[656]	207			
2-naphthyl		J	IV + XV[494]	127		80	
Ph-Ph-4		J	IV + XV[494]	145			
Et		J	I[383]	210	EtOH	73	
$(CH_2)_3$CN		Br	I[383]	116	BuOH	78	
Ph	NC$(CH_2)_2$ / Me	J	I[664]	115	Me_2CO	96	
Et	Pr / Me	J	I[214]	137[214]	EtOH/Et_2O[531]	85[531]	

Table 4 (Continued)

Cation R₂PR'R"			Anion	Method	m.	Solvent	Yield (%)	Other Data and Remarks
R	R'	R"	X					
Pr	Me	Ph-Me-4	J	I[14]	82			
		Ph-OMe-4	J	I[50]	60			
		Ph-Me₂-2,5	PtCl₆	X[552]	141			
			HgJ₃	X[552]	90			
			J	I[52]	105			
		Ph-Et-4	PtCl₆	X[550]	195			
		Ph-OPh-4	J	I[13]	126			
	Bu	CH₂Ph	Cl	I[71]	155[570]			
	CH₂Ph	Ph	Cl	I[70]	134			
i-Pr	Me	Ph-OPh-4	J	I[08]	203			
	Et	Ph	Br	I[97]				PMR −42.2 ppm
	Pr	Ph	Br	I[97]				PMR −39.9 ppm

CH =C−CH₂
   |
   Me
	Me	Ph-Br-4	CdJ₃	X[564]	178			

			HgJ₃	X[564]	67			
			J	I[64]	174			
		Ph	CdJ₃	I[64]	114			
			HgJ₃	I[64]	133			
			J	I[64]	188			
		Ph-Me	HgJ₃	X[564]	79			
			J	I[64]	94			
		Ph-OMe-4	CdJ₃	X[564]	132			
			HgJ₃	X[564]	71			

R	R′	X	Ref.	m.p. (°C)	Solvent	Yield (%)	Other
Bu	Ph-Me$_2$-2,5	J	I[564]	135			
		CdJ$_3$	XV[564]	159			
		HgJ$_3$	XV[564]	71			
	Ph-Et-4	J	I[564]	161			
		HgJ$_3$	XV[564]	83			
		J	I[564]	153			
Me	$-$C(=O)$-$(1-naphthyl)	J	I[535]	141 dec.	CHCl$_3$/Et$_2$O	60	
	Ph	J	I[210]	168[210]	EtOH Et$_2$O[531]	77[531]	IR,[1067] HMR[389]
	Ph-Me-4	J	I[210]	131			
	Ph-OMe-4	PtCl$_6$	XV[550]	196			
		I	I[550]	86			
	Ph-Me$_2$-2,5	PtCl$_6$	XV[552]	215			
		J	I[552]	93			
		J$_3$	XV[552]	70			
	Ph-OPh-4	J$_3$	I[213]	227			
	CH$_2$CO$_2$Et	Br	I[866]				n_D^{20} = 1.5010
		Cl	I[866]				n_D^{20} = 1.5020
	CH$_2$P(=O)(OBu)$_2$	Cl	I[866]				n_D^{20} = 1.4880
	CH$_2$P(=O)(OHex)$_2$	Cl	I[210]				n_D^{20} = 1.4696
Et	Ph	J	5[70]	147			
CH$_2$Ph	Ph	Cl	I[210]	156			
Me	CH$_2$OH	Ph$_4$B	XV[578]	176	Me$_2$CO/PhH		
	Ph	J	I[214]	167			
	Ph-Me$_2$-2,5	J	I[552]	120			
i-Bu	i-Pr	J	I[469]	360	MeOH/Et$_2$O EtOH		
t-Bu	Ph	J	I[670]	134	Me$_2$CO/Et$_2$O	22	

Table 4 (Continued)

Cation R₂PR'R"								
R	**R'**	**R"**	**Anion X**	**Method**	**m.**	**Solvent**	**Yield (%)**	**Other Data and Remarks**
(cyclopentadienyl ring)	Me	Ph–Cl–4	J	I[1007]	173	PhH		
	Me	Ph–OMe–4	J	I[1007] [397]	87	i-PrOH		
c-Pent	Me	Ph	Br					PMR −36.4
	Et	Ph	Br					PMR −37.3 [397]
	Pr	Ph	Br					PMR −35.6 [397]
	Me	Ph	J	I[214]	91			
Pent	Me	Ph–OMe–4	PtCl₆	XV[550]	153			
	Me	Ph–Me₂–2,5	PtCl₆	XV[552]	151			
i-Pent	Me	Ph	J	I[214]	182			
MeCHCH₂ — Et	Me	Ph	J	I[214]	150			
		Ph–Me–4	J	I[214]	131			
		Ph–Me–4	J	I[214]	150			
c-Hex	Me	$\underset{\text{C-Me}}{\overset{\text{O}}{\|}}$	J	I[537]	127			Decomposition by H₂O
		Et	J	I[542]	127			
		$\underset{\text{COEt}}{\overset{\text{O}}{\|}}$	J	I[522]	92	EtOH/Et₂O	99	
		CH₂CO₂Et	J	I[543]	113	EtOH/Et₂O		
		(CH₂)₂CO₂Et	J	I[543]	106	EtOH/Et₂O		

Column header (far right): $E_{1/2}$

R / structure	R'	X	Ref.	m.p.	Solvent	%
(cyclohexyl, H_3C/H)$(CH_2)_4SbEt_2$		J	I[541]	173	EtOH/Et$_2$O	
C=CHPh, Ph		J	I[526]	71		91
		J·MeJ	I[526]	135	MeOH/Et$_2$O	84
		J	I[528]	194	EtOH	66
C=C–CO$_2$H		J	I[526]	220 dec.		96
C=N–C=N–Me, PhPh		J	I[524]	189	EtOH	51
Ph—Ph, CPh$_3$	Ph	J	I[547]	203	EtOH/Et$_2$O	99
	Ph	J	I[551]	67		
Hex	Me, Ph	Br	I[640]	134	Me$_2$CO	
	CH$_2$Ph, Ph	J	I[214]	146		
	Me, Et	J	I[214]	116		
	Et, Ph	Br	I[143]	134		
Me$_2$CH(CH$_2$)$_3$	Me, Et	J	I[37]	123	i-PrOH/EtOAc	84
PhCH$_2$	Ph	Ph$_3$SnCl$_2$	XV[101]	121		
		Br	I[492]	121		22
		J	I[37]	207	EtOH/EtOAc	82
		NO$_3$	XV[699]	162		
CH=CH$_2$	Ph	Br	XVI[928]	189	CHCl$_3$/EtOAc	84
Et	Pr	Br	I[143]	99		
	Ph	Br	I[459]			
(CH$_2$)$_2$OH	Ph	J	I[659]	150	EtOH/Et$_2$O	
CH$_2$CO$_2$Et	Ph	Cl	I[492]	153	MeOH/Et$_2$O	
Bu	Ph	Br	I[143]	147		
t-Bu	Ph	Br	I[492]	159		
	Ph	Br	I[143]	207		

355

Table 4 (Continued)

Cation $R_2PR'R''$			Anion X	Method	m.	Solvent	Yield (%)	Other Data and Remarks
R	R'	R"						
PhCH₂	t-Bu	Ph	J	I[468]	214			
	(CH₂)₂OPh	Ph	Br	I[928]	93	CHCl₃/Et₂O	60	
	CHPh₂	Ph	Br	I[143]	248	EtOH		
	Ph-Cl-2	Ph	OPh(NO₂)₃-2,4,6	XV[217]	136			
	Ph	Ph-Me-4	Br	IV[494]	219	CHCl₃	94	
		PhOMe-4	Br	IV[500]	220		50	
	Me	Ph	J	I[551]	87			
Hept	CH₂OH	CH₂CO₂Et	Br	[806]				$n_D^{20} = 1.4964$
		CH₂P(=O)(O-Oct)₂	Cl	[806]				$n_D^{20} = 1.4771$
	(CH₂)₂P(=O)(O-Oct)₂		Cl					$n_D^{20} = 1.4855$
Oct	Me	Ph	PtCl₆	XV[551]	102			
			J	[551]	8			
Ph-Cl-2	Me	Et	O₃SPh(NO₂)₂-2,4	I[656][55]	165	EtOH		Yellow
Ph	CF₃	Me	J	I[55]	123			
	CH₂Cl	Et	BF₄	XVI[435]	129		85	Waxy
	Me	CH₂OH	J	I[376]	223 dec.	MeOH/PhH	63	
		CH=CH₂	J	I[928]	119[928]	CHCl₃/EtOAc[928]	80[827]	HMR[928]

Substituent	Anion	Ref.	m.p.	Solvent	Notes
Et	OPh(NO₂)₃-2,4,6	XV[656]	117	EtOH	
	Ph₄B	XV[962]	145	MeOH	
	PtCl₆	XV[717]	220		
	Br	I[962]	170[962]	CHCl₃/EtOAc[962]	99[492] HMR[962]
	J	XV[962]	183[962]	CHCl₃/petroleum ether[376]	
CH₂OMe	J	I[376] I[1035]	144	CHCl₃/EtOAc	
(CH₂)₂OH	J	I[536]	122	EtOH	
C≡CMe	J	I[158]	139[158]	EtOH	84 47[158] HMR[984]
(CH₂)₂CN	J	I[664]	163	Et₂O	
CH₂CH=CH₂	Br	I[699]	161		
	J	540	145 dec.	EtOH/Et₂O	62
CH₂$\overset{\text{O}}{\overset{\|}{C}}$Me	Cl	I[1038]	170	CHCl₃/EtOAc	
CO₂Et	J	I[544]	114	EtOH/Et₂O	
Pr	J	I[699]	153		
i-Pr	Br	397			
	J	376	225	CHCl₃/petroleum ether	PMR −30.9 ppm
CH₂CHMe—OH	J	I[541] III[850] 1038	163[850]	i-PrOH[850]	
CH₂CO₂Et	Br	I[542]	119	Me₂CO	
	J		114	EtOH/Et₂O	

Table 4 (Continued)

Cation $R_2PR'R''$			Anion X	Method	m.	Solvent	Yield (%)	Other Data and Remarks
R	R'	R''						
Ph	Me	Bu	J	I[376]	172[376]	CHCl₃/ petroleum ether[376]		IR,[1067] HMR[389]
		t-Bu	Ph₄B	XV[954]	169	MeOH		HMR
			Br	I[954]	207	CHCl₃/ EtOAc		
		[pyridyl ring, with N]	J	I[655]	165	EtOH		IR
		[methylcyclopentadienyl ring]	J	I[670]	141	EtOH/Et₂O		
		C=C–CMe₂–OH	J	I[1006]	178			Dark orange
		c-Pent	J	I[158]	151		55	
		[methylcyclohexenyl ring]	Br	3[97]				PMR –29.8 ppm
		[cyclohexane ring, HO– and H and methyl]	J	XIV[1040]	228	CHCl₃/ EtOAc		HMR
			J	I[541]	218	EtOH/Et₂O		
		CH₂Ph	OPh(NO₂)₃– 2,4,6	XV[669]	118	EtOH		

Substituent	Type	Compound	m.p. (°C)	Solvent	Yield (%)	Notes
$CHPh$–OH	Br	I^{669}	248^{669}	$CHCl_3$/PhH^{1038}	61^{97}	
		I^{1035}	243^{1035}	$EtOH^{1035}$		
		I^{298}	153	$MeOH$/Et_2O		
$C{=}CPh$	J	I^{158}				UV
$CH{=}CHPh$	J	I^{462}	164		49	
$CH_2\overset{O}{C}Ph$	Br	I^{565}	197	$CHCl_3$/$EtOAc$		pK$_a$ 6.51
		I^{12}				
$CD_2\overset{O}{C}Ph$	Br	XVI^{12}				IR, pK$_a$ value
$C{\equiv}C$–(cyclohexenyl)	J	I^{158}	124	$EtOH$/Et_2O	33	UV
$(CH_2)_2Ph$	J	I^{528}	120	$EtOH$/H_2O	47	
CH_2CHPh–OH	J	I^{562}	188			
$(CH_2)_2OPh$	J	I^{928}	128	$CHCl_3$/$EtOAc$		HMR
$(CH_2)_2NPh$–Me		I^{126}				
CH_2Ph–CH_2OMe-2 $CHPh$	J	I^{566}	130	$MeOH$/Et_2O		
$\overset{O}{C}NMe_2$	J	I^{97}	237	$EtOH$	54	
$C{=}CHPh$–CO_2Et	J	I^{462}	105 dec.	$BuOH$/Et_2O	50	
CH_2–ferrocenyl	J	I^{794}	201	Me_2CO		Golden

359

Table 4 (Continued)

Cation $R_2PR'R''$			Anion X	Method	m.	Solvent	Yield (%)	Other Data and Remarks
R	R'	R''						
Ph	Me	$C{\equiv}CPPh_2$	J	I[424]	156	Me_2CO/Et_2O	90	
		$C{=}CHPh$ $-Ph$	J	I[528]	170	EtOH	55	
		$CH-CH_2Ph$ $-Ph$	J	I[528]	234	EtOH	75	
		$CH-CHPh$ $-Ph$ $-OH$	J	I[562]	222	$CHCl_3$		
		$(CH_2)_2PPh_2$ $=O$		126				HMR
		Me $CH{=}PPh_2$		126				
		$C{\equiv}C-CPh_2$ $-OH$	J	I[158]	165		52	HMR
		CPh_3	J	I[547]	242 dec.	$EtOH/Et_2O$		
		$Ph(NO_2)_2-2,4$	$OPh(NO_2)_3-$ 2,4,6	XV[656]	182	$Me_2CO/EtOH$		Brown, hygroscopic
			$O_3SPh-Me-4$	I[656]	166			
		$Ph-Cl-3$	J	I[1040]	170	$CHCl_3/EtOAc$		

R	X	Method [ref]	mp (°C)	Solvent	Yield (%)	Notes
Ph-CH$_2$Br-2	J	XVI[645]	192	Me$_2$CO	98	IR
Ph-OMe-3	Ph$_4$B	XV[613,902]	146			UV[901]
Ph-OMe-4	Br					
Ph-CH$_2$OMe-2	Br	I[645]	158		100	
Ph-NMe$_2$-4	J	I[904]	173[904]	H$_2$O[904]	100[904]	
Ph-OPh-2	J	I[76]	198	MeOH/EtOAc	72	
Ph-CH$_2$Ph-2	J	I[1059]	297			
Ph-CHPh-2, Me	J	I[1059]	312			
Ph-CHPh-2, Pr	J	I[1059]	258			
Ph-OMe-2-CHPh-6, Pr	J	I[1099]	197			
⟨benzene ring⟩ Ph-Ph-2	J	I[1077]	313	MeOH	88	
CH$_2$OH / CH$_2$Ph	Br	I[1035]	220	CHCl$_3$/EtOAc		
CH$_2$PPh$_2$	Cl	X[403]	154 dec.	PhH	4	
Ph-t-Bu$_2$-3,5-OH-4	Cl	X[678]	215 dec.	EtOH/Et$_2$O		
CH$_2$CN / Ph-NMe$_2$-4	Cl	I[781]	279	CHCl$_3$/PhH	80	
CH=CH$_2$ / CH$_2$Ph	Br	XVI[928]	211	CHCl$_3$/Et$_2$O	62	HMR
CH$_2$CHO / Ph-NMe$_2$-4	Cl	I[1038]	172			
CH$_2$CO$_2$H / Ph-CH$_2$Br-2	Br	XVI[645,397]	152	CHCl$_3$/EtOAc	42	
Et / i-Pr	Br					PMR -36.2 ppm
Bu	J	I[607,397]	153			
c-Pent	Br					PMR -34.8 ppm

Table 4 (Continued)

Cation $R_2PR'R''$			Anion X	Method	m.	Solvent	Yield (%)	Other Data and Remarks
R	R'	R''						
Ph	Et	$(CH_2)_2NEt_2$	J	I[539]	139 dec.[539]	EtOH/Et$_2$O[539]	81[539]	PMR -30.2 ppm[321]
		CH_2Ph	Ph$_3$SnClBr Cl	XV[101] I[772]	178 243		100	
		$C\equiv CPPh_2$	J	I[424]	126	Me$_2$CO/Et$_2$O	84	
		Octadec	O=P(OEt)$_2$ S	XV[704]			87	$n_D^{20} = 1.5501$, $d_{20} = 1.1094$
		Ph(NO$_2$)$_2$-2,4	OPh(NO$_2$)$_3$-2,4,6	XV[656]	136	EtOH/ Me$_2$CO/ petroleum ether		
		Ph-NMe$_2$-4	Br	I[1039]	193	Me$_2$CO/ petroleum ether		
		(naphthalene)	J	IV[494]	183	H$_2$O	95	
	CH$_2$OMe	Ph-t-Bu$_2$-3,5-OH-4 CH$_2$Ph	BF$_4$ Br	I[734] I[1035]	202 173	EtOH/Et$_2$O CHCl$_3$/ EtOAc	1	
	(CH$_2$)$_2$CN CH$_2$CH=CH$_2$	Ph-NO$_2$-4 CH$_2$Ph	J J	XV[906] I[478]	89 208	H$_2$O		

Compound	X	Ref.	mp/temp	Solvent	Yield (%)	Notes
(CH2)2OPh	Br	I[928]	141	CHCl3/Et2O	87	
Ph-NMe2-4	Br	I[1039]	232	H2O	81	
O‖CH2CMe Ph-NMe2-4	Cl	I[1039]	204	CHCl3/EtOAc		PMR −32.5 ppm
CH2CO2Me (CH2)2OPh	Br	I[928]	107	MeOH/Et2O		
Ph-CO2H-4	Br	[909]	160 dec.			
Ph-CO2Me	Br	[909]	146	H2O		
Pr	Br	[397]				
CH2CO2Et Ph-NMe2-4	Br	I[1039]	84	CHCl3/Et2O	80	HMR
(CH2)2CO2Me CH2Ph	Br	XVI[928]	157			
CH2Ph-NO2-4 Ph-CO2H-4	Br	I[909]	290 dec.	MeOH	79	Light yellow
Ph-OMe-4	Br	I[1032]	240		72	Light yellow
Ph-CO2Me-4	Br	[909]	239 dec.	EtOH	85	Light yellow
Ph-NMe2-4	Br	I[904]	230	PhH		
CH2Ph [cyclopentadienyl Mn(CO)3] CH2CHPh–OH	Br	I[860]	195 dec.			
	Br	I[562]	195	CHCl3/EtOAc	28	
(CH2)2OPh	Br	I[928]	95	CHCl3/EtOAc[1039]		HMR
Ph-NMe2-4	Br	I[1039]	227[1039]			
CH=CHPPh2	Cl	XV[904]				
	Cl	I[904]	248	H2O	90	HMR, PMR
	Br	I[3]	218	MeOH/Me2CO		
Ph-CH2Br-2	Br	XVI[645]	247		100	
Ph(CH2)2Br-2	Br	[644]	225			
Ph-CH2OMe-2	Cl	I[645]	254		96	
O‖CH2CPh (CH2)2OPh	Br	I[928]	223			

Table 4 (Continued)

Cation $R_2PR'R''$			Anion X	Method	m.	Solvent	Yield (%)	Other Data and Remarks
R	R'	R''						
Ph-Cl-2-Me-5	Me	Et	$OPh(NO_2)_3$-2,4,6	XV[656]	117	EtOH		
			$O_3SPh(NO_2)_2$-2,4	XV[656]	218	EtOH		
			O_3SPh-Me-4	I[656]	234	Me_2CO		
			J	I[656]	203			
Ph-Me-2	Ph	Ph-Ph-4	J	XVIII+ XV[1078]	271			IR
Ph-Me-3	$(CH_2)_3$Br	Ph	Br	I[420]	191	EtOH/EtOAc		
Ph-Me-4	Me	$Ph(NO_2)_2$-2,4	$OPh(NO_2)_3$-2,4,6	XV[656]	78 dec.	EtOH/H_2O		IR
	Et	Ph-Cl-4	J	I[712]	177			
	CH_2Ph	Ph-Cl-4	J·2 H_2O	I[712]	257			
	Ph-NO_2-4	Ph	J	VIII[483]	143 dec.	H_2O		Light yellow
Ph-OMe-2	Me	Et	$OPh(NO_2)_3$-2,4,6	XV[615]	136 dec.			
			J	I[583]	157	EtOH/Me_2CO		
			NO_3	I[615]	154			
Ph-OMe-4	CH_2Ph-NO_2-4	Ph	Br	I[1032]	62	EtOH	53	
		Ph-NMe_2-4	Br	I[1031]	238	EtOH		
Ph-Me_2-3,5	$(CH_2)_3$Br	Ph	Br	I[420]	203	EtOH/EtOAc		

			Ref.	mp	Solvent	Yield	Color
Ph-NMe$_2$-4	Me	Cl·H$_2$O	[7,1,2]	72		100	UV
Ph-Cl-4		PtCl$_6$	XV[7,1,2]	235 dec.			Yellow
	Ph	J	I[7,1,2]	135			
CH$_2$Ph-NO$_2$-4	Ph	Br	I[9,4]	166	H$_2$O		
	Ph	Br	I[9,4]	253	H$_2$O		
	Ph-OMe-4		I[1,3,1]	254	EtOH		

Table 5. RR'R''R'''PX.

Cation RR'PR''R'''				Anion X	Method	m.	Solvent	Yield (%)	Other Data and Remarks
R	R'	R''	R'''						
Me	Et	CH₂-CH=CH₂	Pr	$O_2C(HCO_2-CPh)_2CO_2H$	XV[574]	124			$[\alpha]_D^{20} = -66.40$ MeOH
			Ph	$O_2C(HCO_2-CPh)_2CO_2H$	XV[574]	130			$[\alpha]_D^{20} = -78.82$ MeOH
		Pr	Ph-Me-4	Br	I[573]	97		61	
			Ph	Br	I[573]	75			
			Ph	J	I[505]	62			$[\alpha]_D = 4.6$ c. 1.732 MeOH
		CH₂CO₂Et	Ph-Me-4	$O_2C(HCO_2-CPh)_2CO_2H$	XV[574]	149			$[\alpha]_D^{20} = -57.9$ MeOH
			Ph	$O_2C(HCO_2-CPh)_2CO_2H$	XV[574]	122			$[\alpha]_D^{20} = -88.84$ MeOH
		Bu	PhMe	Br	I[573]	71		71	
			CH₂Ph	Br	I[573]	8		75	
		t-Bu	Ph	Br	I[492]	134			
		Pent	CH₂Ph	J	I[468]	206	MeOH/Et_2O	90	
				Br	I[37]	129	PrOH/petroleum ether	61	
				J	I[37]	91	i-PrOH/petroleum ether	84	

CH_2Ph	Ph	$MeCO_2 \cdot H_2O$	XV^{773}	104	EtOAc	Extremely hygroscopic $[\alpha]_D^{25} = -0.7$ c. 1.940 MeOH, +0.9 c. 5.65 MeOH
		OBu	XV^{783}			
		$O_2C(HCO_2-CPh)_2CO_2H$	XV^{609}	142^{609}	$PrOH^{640}$	$[\alpha]_D^{25} = -54.0$ c. 1.281 $MeOH^{609}$ $[\alpha]_D^{25} = +54.0$ c. 1.100 $MeOH^{609}$
		Br	I^{40}	141^{40}	$MeOH/Et_2O^{40}$ 83^{574}	$R(-)$ $[\alpha]_D^{20} = 23.07$ MeOH
		Cl	XV^{574} I^{573}	168^{573}	$Xylene^{640}$ 77^{573}	$E1/2$ -2.118^{499}
		J	I^{699}	167^{699}	Et_2O^{640} 90^{640}	
			XV^{609}			$[\alpha]_D^{25} = +24.0$ c. 0.824 $MeOH^{609}$ $[\alpha]_D^{25} = -23.8$ c. 0.927 $MeOH^{609}$
Ph-Me-4		Br	I^{574}	141	42	$[\alpha]_D^{20} = +24.42$ MeOH
		Cl	I^{573}	57	78	$[\alpha]_D^{20} = +20.87$ MeOH
	Ph	J	XV^{574} XV^{574}	112		$[\alpha]_D^{20} = 22.20$ MeOH
$(CH_2)_2Ph$	Ph	$O_2C(HCO_2-CPh)_2CO_2H$	XV^{574}	130		$[\alpha]_D^{20} = -63.29$ MeOH

Table 5 (Continued)

Cation RR'PR''R'''				Anion X	Method	m.	Solvent	Yield (%)	Other Data and Remarks
R	R'	R''	R'''						
Me	Et	Dodec	Ph	Br	I[575]	76			
		Octadec	Ph	Br	I[575]	105		60	
		Ph	Ph-Me-4	O$_2$C(HCO$_2$-CPh)$_2$CO$_2$H	XV[640]	143		53	$[\alpha]_D^{25}$ = -70.0 c. 1.120 MeOH
				Camphorsulfonate	[1056]	128			
				J	XV[640]	187			$[\alpha]_D^{25}$ = -0.00 MeOH
						185			$[\alpha]_D^{25}$ = 0.00 MeOH
			Ph-OMe-4	J	I[212]	114			
			Ph-Me$_3$-2,4,5	PtCl$_6$	[712]	186			
	(CH$_2$)$_2$CN	CH$_2$Ph	Ph	L(+)O$_2$C(HCO$_2$-CPh)$_2$CO$_2$H	XV[1093]	129 dec.	PrOH	67	c. 2.002 MeOH
				L(-)	XV[1093]	127 dec.	PrOH		$[\alpha]_D^{27}$ = -69 c. 2.07 MeOH
		Ph-Br-2	Ph	OPh(NO$_2$)$_3$-2,4,6	XV[345]	88	EtOH		
		Ph-Cl-2	Ph	OPh(NO$_2$)$_3$-2,4,6	XV[345]	93	EtOH		

R	R′	R″	X⁻ / group	Ref.	No.	Solvent	$[\alpha]$ / notes
Ph	Ph-CN-4		$OPh(NO_2)_3$-2,4,6	XV[345]	158	EtOH	$[\alpha]_D^{20} = -9.2$; c. 1.19 MeOH; R(−),
$(CH_2)_2CO_2H$	Ph-Me₂-3,5	Ph	J	I[345]	104	EtOH/c-Hex	$[\alpha]_{578}^{20} = -89.5$; c. 0.62 MeOE; $[\alpha]_{546}^{20} = -104$; R(−),
CH_2-CH=CH₂	Pr	Ph	Ph_4B	4[80]	160	EtOH	5
t-Bu	Ph		$O_2C(HCO_2\text{-}CPh)_2CO_2H$	XV[40]	128	EtOH	$[\alpha]_{578}^{20} = -19.9$; c. 0.627 MeOH; $[\alpha]_{546}^{20} = -23.8$; $[\alpha]_D^{20} = 15.1[480]$
			Br	XV[40]	232	MeOH/Et_2O	
				I	142[505]		
CH₂Ph	Ph		Br	I[505]	175	MeOH/Et_2O	$[\alpha]_{578}^{20} = 2.5$; c. 1398 MeOH; $[\alpha]_{546}^{20} = 3.2$; 32
Pr	Bu		J / Br	I[8,16][40]	129	MeOH/Et_2O	
	Ph-Me-4	Ph	J	I[435]	151	MeOH/EtOAC	HMR·R(−); $[\alpha]_{578}^{20} = -3.8$; c. 1.05 MeCH; $[\alpha]_{546}^{20} = -4.3$; R(+), 34
t-Bu	Ph		Br	XV′[40]	228	MeOH/Et_2O	$[\alpha]_{578}^{20} = -28.5$; c. 1.354 MeOH; $[\alpha]_{546}^{20} = -32.5$; R(+)−
(cyclohexen-3-yl)	Ph		Br	I[40]	170	MeOH/Et_2O	
c-Hex	Ph		Br	XVI[40]	172	MeOH/Et_2O	$[\alpha]_{578}^{20} = -3.7$; c. 0.82 MeOH; $[\alpha]_{546}^{20} = -3.7$

Table 5 (Continued)

Cation RR'PR"R'''				Anion X	Method	m.	Solvent	Yield (%)	Other Data and Remarks
R	R'	R"	R'''						
Me	Pr	CH$_2$Ph-Cl-2	Ph	Br	I[40]	156	MeOH/Et$_2$O	69	$[\alpha]^{20}_{578}$ = 28.1 c. 0.604 MeOH, $[\alpha]^{20}_{546}$ = 33.2 HMR,
		CH$_2$Ph-Cl-4	Ph	Br	I[40]	Oil	MeCN/Et$_2$O		$[\alpha]^{20}_{578}$ = 36.6 c. 1.14 MeOH, $[\alpha]^{20}_{546}$ = 42.5 S(+)
		CH$_2$Ph	Ph	Br	I[40]	178	MeOH/Et$_2$O	79	$[\alpha]^{20}_{578}$ = 7.05 c. 0.852 MeOH, $[\alpha]^{20}_{548}$ = 8.22 HMR,
		(CH$_2$)$_2$Ph	Ph	Br	I[40]		MeCN/Et$_2$O		$[\alpha]^{20}_{578}$ = -5.7 c. 1.41 MeOH, $[\alpha]^{20}_{546}$ = -6.7
		MeO$_2$C-CHPh	Ph	Cl	I[504]	66 dec.	MeCN/Et$_2$O	94	Very hygroscopic, $[\alpha]_{436}$ = -2.2 c. 1.873 MeOH, $[\alpha]_D$ = 4.1 c. 3.146 MeOH
Me—C-Ph—CO$_2$Me			Ph	J	XIV[504] XV I	65 dec.	MeOH/Et$_2$O	42	

R	R'	R''	Anion	Ref.	No.	Solvent		Rotation
(9-methylfluorene structure)	Ph		Br	I[40]	196	MeOH/Et₂O	30	$[\alpha]^{20}_{578}$ = 26.6, c. 1.599 MeOH, $[\alpha]^{20}_{546}$ = 30.3
CHPh₂	Ph		ClO₄	I + XV[40]	128	MeOH/Et₂O	51	$[\alpha]^{20}_{578}$ = -4.6 S(-), c .0862 MeOH, $[\alpha]^{20}_{546}$ = -5.8
i-Pr	PhOMe-4		J	I[212]	114	MeOH/Et₂O[40]		R(+) $[\alpha]^{20}_{578}$ = 87.6 c. 0.879 MeOH,[40] $[\alpha]^{20}_{546}$ = 101.6; HMR[143]
	Ph		Br	I[40]	160[40]			
CH₂CO₂Et	CH₂Ph	Ph	Br	I[505]	163	MeOH/Et₂O[40]		R(-) $[\alpha]^{20}_{578}$ = -27.7 c. 2.022 MeOH,[40] $[\alpha]^{20}_{546}$ = -31.9, HMR[143]
Bu	CH₂Ph	Ph	Br	I[40]	207[40]			
t-Bu	CH₂Ph	Ph	J	I[492]	167	EtOH	14[40]	HMR,[143] $[\alpha]^{20}_{578}$ = -21.9 c. 0.824 MeOH, $[\alpha]^{20}_{546}$ = -24.3 $[\alpha]^{20}_{589}$ = -84.5 $[\alpha]^{20}_{546}$ = -118
		Ph	Br	I[505]	245[143]			
			J	231				

Table 5 (Continued)

Cation RR'PR"R'''				Anion X	Method	m.	Solvent	Yield (%)	Other Data and Remarks
R	R'	R"	R'''						
Me	t-Bu	$PhCH-C(=O)Ph$	Ph	J	I[468]	175	H_2O	72	
		$PhCHCH_2C(=O)Ph$	Ph	J	I[468]	180	MeOH/Et$_2$O		
		Ph	Ph–CHPh$_2$-4	J	XV[468]	122	H_2O	64	
	(CH$_2$)$_4$OH	Ph	Ph–Me$_2$-2,5	J	I[345]	146	EtOH/Et$_2$O		
			Ph–Me$_2$-3,5	J	I[658]	146			IR
	(cyclohexenyl)	CH$_2$Ph	Ph	Br	I[143]	206[143]	MeOH/Et$_2$O[40]	35	HMR,[143] R(+) $[\alpha]^{20}_{578} = 42.4$ c. 0.766 MeOH[40] $[\alpha]^{20}_{546} = 48.9$
	c-Hex	CH$_2$Ph	Ph	Br	XVI[40]	284	MeOH/Et$_2$O		HMR, R(+), $[\alpha]^{20}_{578} = 79.25$ c. 0.978 MeOH, $[\alpha]^{20}_{546} = 99.5$
	CH$_2$Ph	CH$_2$Ph–NO$_2$-4	Ph	Br	I[505]	174			
		$CH_2C(=O)Ph$	Ph	Br	I[505]	171			

R1	R2	X	Method	mp	Solvent	Notes
Ph-CH-Me	Ph	Br	I[143]	197	EtOH/Et2O	
CH2Ph-2	Ph	J	I[667]	193		
CH2OMe-2	Ph	Br	I[505]	112		HMR
(CH2)4OPh	Ph	Br	I[143]	237		
[fluorene structure]	Ph	Br	I[492]	125		HMR
CHPh2	Ph	Br	I + XV[143]	257		HMR
CPh3	Ph	Br	908			
Ph-Cl-4	Ph	Cl	908			
		J3	908			
Ph	Ph-Me-4	d-Br-camphor sulfonate	816	129		
		Camphor-sulfonate	816	134		
		Br	816	211		
		J	816	215		
Ph-OMe-4	Ph	OPh(NO2)3-2,4,6	XV[669]	94	EtOH	Yellow
		Cl				
Ph-Me3-2,4,6	Ph	Br	IV[143]	162	MeOH/Et2O	HMR
[naphthalene structure]	Ph	Br	908			HMR
		Cl	908			
		J3	908			
		NO3	908			
CH2Ph- (CH2)2Br-2 (CH2)2Ph- CH2OMe-2						
Ph	Ph-Me-4	Br	I[471]	185	EtOH/Et2O	HMR
Ph-Br-4	Ph	J	I[471]	167	EtOH	

373

Table 5 (Continued)

Cation RR'PR"R'''				Anion X	Method	m.	Solvent	Yield (%)	Other Data and Remarks
R	R'	R"	R'''						
Me	$(CH_2)_2Ph-CH_2OMe-2$	Ph	Ph-OMe-4	J	I[471]	121	EtOH		
Et	$CH_2-CH=CH_2$	Pr	Ph	$O_2C(HCO_2-CPh)_2CO_2H$	XV[574]	125			$[\alpha]_D^{20} =$ -75.14 MeOH
		Bu	Ph-Me-4	Br	I[573]	131		69	
			CH_2Ph	Br	I[573]	94		65	
			Ph	Br	I[570]	87		69	
			Ph-Me-4	Br	I[573]	73		67	
		Pent	Ph	Br	I[573]	63		67	
		Hex	Ph	Br	I[573]	58		70	
		CH_2Ph	Ph	Br	I[573]	51			
				Br-camphorsulfonate	XV[570]	148	EtOAc		
		Hept	Ph	Br	I[505]	154		69	
		$CH_2C(=O)Ph$	Ph	Br	I[573]	41			
		Oct	Ph	J	I[566]	166		70	
		CH_2CO_2Et	Ph	Br	I[573]	47		66	
			Ph	Br	I[573]	72			
	Pr	Bu	Ph-Me-4	$O_2C(HCO_2-CPh)_2CO_2H$	XV[574]	139			$[\alpha]_D^{20} =$ -65.65 MeOH
		CH_2Ph	Ph	Cl	I[573]	163		76	
			Ph-Me-4	Cl	I[573]	146		81	

374

R¹	R²	R³	X⁻	Ref	mp	Solvent	Yield	$[\alpha]_D^{20}$
Octadec	Ph	Ph-Me-4	Br	I[574]	52			
Octadec	Ph	Ph-Me-4	$O_2C(HCO_2\text{-}CPh)_2CO_2H$	XV[574]	122			$[\alpha]_D^{20} = -60.61$
CH_2CO_2Et	Bu	Ph	Br	I[573]	83		69	
CH_2CO_2Et	Pent	Ph	Br	I[573]	93		69	
CH_2CO_2Et	Hex	Ph	Br	I[573]	69		71	
CH_2CO_2Et	CH_2Ph	Ph	Br	I[143]	142			
CH_2CO_2Et	Ph	Ph-Me-4	Cl	I[573]	125		76	
CH_2CO_2Et	Ph	Ph-Me-4	Cl	I[573]	139		75	
CH_2CO_2Et	Ph	Ph-Me-4	$O_2C(HCO_2\text{-}CPh)_2CO_2H$	XV[574]	134	MeOH/Et₂O	77	$[\alpha]_D^{20} = -65.52$ MeOH
Bu	CH_2Ph	Ph	Cl	I[573]	133			
Bu	Dodec	Ph	Br	I[575]	56			
Bu	Octadec	Ph	Br	I[575]	74			
Bu	Ph	Ph-Me-4	$O_2C(HCO_2\text{-}CPh)_2CO_2H$	XV[574]	131			$[\alpha]_D^{20} = -68.10$ MeOH
Pent	CH_2Ph	Ph	Br	I[640]	121	Me₂CO	89	
Pent	Dodec	Ph	Cl	I[573]	139		75	
Pent	Octadec	Ph	Br	I[575]	70			
Pent	Hept	Ph-Me-4	Br	I[575]	77		78	
Hex	CH_2Ph	Ph	Cl	I[573]	99			
CH_2Ph	$CH_2C(O)Ph$	Ph	Cl	I[568]	166			
CH_2Ph	Oct	Ph	Cl	I[573]	81		65	
CH_2Ph	Ph	Ph-Me-4	Br	I[1056]	216			
CH_2Ph	Ph	Ph-Me-4	J	I[1056]	192			
Hept	Dodec	Ph	Br	I[575]	90			
Hept	Ph	Ph-Me-4	$O_2C(HCO_2\text{-}CPh)_2CO_2H$	XV[574]	127			$[\alpha]_D^{20} = -56.08$ MeOH
$CH_2Ph\text{-}(CH_2)_2Br\text{-}2$ / Pr	Ph	Ph-Me-4	Br	I[471]	146	Et₂O		
$CH_2\text{-}CH{=}CH_2$	CH_2Ph	Ph	Br	I[570]	153			

Table 5 (Continued)

Cation $RR'PR''R'''$				Anion X	Method	m.	Solvent	Yield (%)	Other Data and Remarks
R	R'	R''	R'''						
$CH_2-CH=CH_2$	Bu	CH_2Ph	Ph	Br	I[570]	102			
Pr	Bu	CH_2Ph	Ph	Cl[908] (J)	I[435]	138		34	HMR
CH_2Ph	Ph-Br	Ph	Ph-OMe-4	$O_2C(HCO_2-CPh)_2CO_2H$	XV[500]	102		90	
Ph	Ph	Ph-OMe-4	(naphthalene ring structure)	Br	[500]	259 dec.			$[\alpha]_{578} = 11.0$, c. 0.137, $PhNO_2$:DMF = 1:2 HMR,[908]
				Cl	XV[500]	253 dec.[560]		82[500]	$[\alpha]_D = 15.7$, c. 1.4492, MeOH HMR
(naphthalene ring structure)			Ph-Ph-4	Cl	[908]				
$CH_2C(=O)Ph$-Cl-4	Ph	Ph-Me-4	Ph-OMe-4	Br	I[212]	199			
Ph	Ph-Me-4	Ph-Me-4	Ph-Ph-4	Ph_4B	XV[1079]	207	Me_2CO/ H_2O		
Ph-Me-2	Ph-Me-4	Ph-Me-4	Ph-Ph-4	J	XVIII + XV[1079]	271	H_2O	55	IR

Table 6. Salts with Phosphorus in the Ring

Cation RR'PR"			Anion X	Method	m.	Solvent	Yield (%)	Other Data and Remarks
R	R'	R"P						
ClCH$_2$	Me	[dibenzophosphole ring, P]	Cl	XVI[17]				HMR
JCH$_2$	Me	[dibenzophosphole ring, P]	OPh(NO$_2$)$_3$-2,4,6	XV[17]	214	EtOH		
		[phenoxaphosphine ring, O···P]	J	I[17]	235	EtOH	83	HMR
	Ph	[dibenzophosphole ring, P]	J	I[17]	241	EtOH		HMR
		[phenoxaphosphine ring, O···P]	J	I[17]	219[17]	EtOH[17]	75[17]	HMR[18]
Me	Me	[dibenzophosphole ring, P]	J	I[17]	298	EtOH		HMR
		[bicyclic cage: Me–C–O, CH$_2$, O, Me, P]	J	I[2,9,7]	243	i-PrOH		
CH$_2$OH		[dibenzophosphole ring, P]	Br	[18]				HMR

Table 6 (Continued)

Cation RR'PR''			Anion X	Method	m.	Solvent	Yield (%)	Other Data and Remarks
R	R'	R''P						
Me	CH₂OH	(dibenzophosphole)	Cl	X[17]	151 dec.	i-PrOH		HMR
	Et	(dihydrophosphindole)	OPh(NO₂)₃⁻ 2,4,6	XV[54]	121	EtOH		
		(dibenzophosphole)	J	XV[54] [18]	184	EtOH		HMR
			J					
	CH₂OMe	(dibenzophosphole)	Br	XVI + XV[17]	170	CHCl₃/ EtOAc		
			Cl	I[17]	167	Et₂O		HMR
	i-Bu	(cage phosphorus structure)	J	I[297]	202	Me₂CO		
	CH₂Ph	(dibenzophosphole)	Br	I[17]	246			

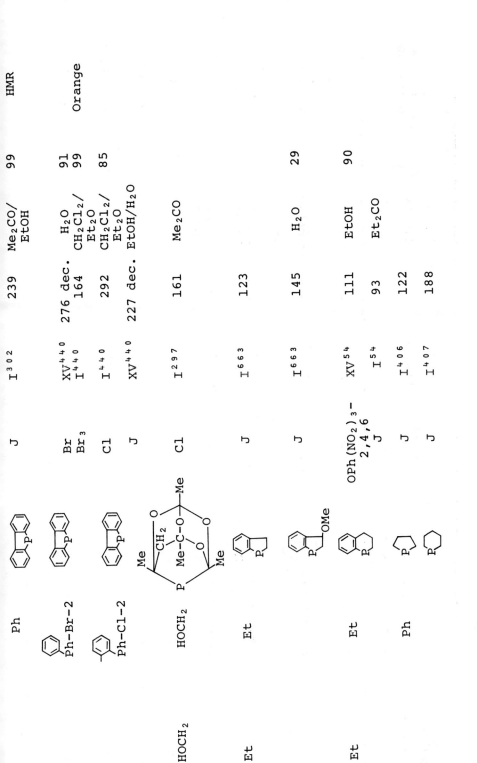

R	X	Ref.	m.p.	Solvent	Yield	HMR
Ph	J	I^{302}	239	Me$_2$CO/EtOH	99	
Ph–Br-2	Br, Br$_3$	XV^{440}, I^{440}	276 dec., 164	H$_2$O, CH$_2$Cl$_2$/Et$_2$O	91, 99	Orange
Ph–Cl-2	Cl	I^{440}	292	CH$_2$Cl$_2$/Et$_2$O	85	
	J	XV^{440}	227 dec.	EtOH/H$_2$O		
HOCH$_2$	Cl	I^{297}	161	Me$_2$CO		
HOCH$_2$	J	I^{663}	123			
Et	J	I^{663}	145	H$_2$O	29	
Et	OPh(NO$_2$)$_3$-2,4,6, J	XV^{54}, I^{54}	111, 93	EtOH, Et$_2$CO	90	
Ph	J	I^{406}	122			
	J	I^{407}	188			

379

Table 6 (Continued)

R	R'	R"P	Anion X	Method	m.	Solvent	Yield (%)	Other Data and Remarks
Et	Ph-Me-4	(P in 6-membered ring)	J	I[407]	163			
	Ph-CH$_2$CH=CH$_2$-2	(P in benzo-fused ring)	OPh(NO$_2$)$_3$-2,4,6	XV[661]	130	MeOH/EtOH		Yellow
		(P in benzo-fused ring)	J	XV[661]	168	EtOH		
	Ph-(CH$_2$)$_2$Br-2	(P in benzo-fused ring)	Br·H$_2$O	[661]	110	EtOH / EtOAc		
	Ph-(CH$_2$)$_3$OMe-2	(P in benzo-fused ring)	Br·H$_2$O	XVI[661]	144	EtOH/ EtOAc		
Pr	Ph	(P in 5-membered ring)	J	I[406]	153			
Bu	Bu	(P in 5-membered ring)	Cl	[134a]				PMR -46
Bu	Bu	(P in 5-membered ring with two OH)	Cl	[673a]				PMR -46
Bu	Bu	(P in 5-membered ring with two OH)	Cl	[673a]				PMR -20

Table (rotated, landscape orientation):

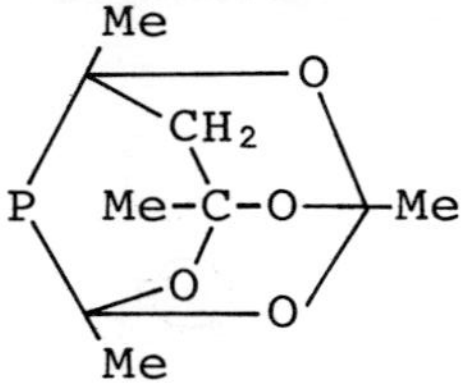

R	R'	Structure	J	Ref.	m.p.	Solvent	Yield	PMR
Bu	Bu	Me cage (Me-C-O, CH₂, P, O)	J	I²⁹⁷	209	THF		
Bu	Bu	Ph-Br-2 (dibenzophosphole)	Br₃	I⁴⁴⁰	132	CH₂Cl₂/Et₂O	83	
	Ph-Br-2		J Cl	XV⁴⁴⁰ 134a	186	EtOH/Et₂O		
i-Bu	i-Bu	(phospholane)	Cl	673a				PMR −39
i-Bu	i-Bu	(phospholane with OH, OH)	Cl	134a				PMR −23
c-C₆H₁₁	C-C₆H₁₁	(phosphinane)	Cl	134a				PMR −20
PhCH₂	Ph-Br-2	(dibenzophosphole)	Br	I⁴⁴⁰	283 dec.	EtOH/Et₂O	87	
	Ph	Me Me / P / Me Me (phosphetane)	Br	I⁴²⁸	220			
		CH₂Ph / Et—As / P (dibenzo)	Cl	I³⁰²	303	Et₂O	94	
			OPh(NO₂)₃-2,4,6	216	180			
			OPh(NO₂)₃-2,4,6	216	206			
			Br	216	296			

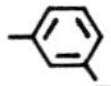

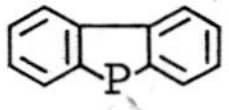

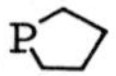

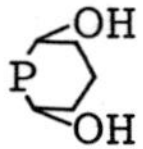

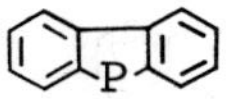

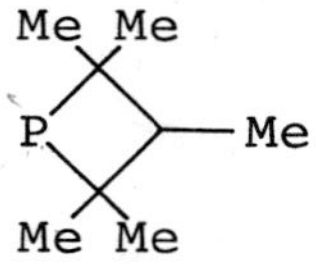

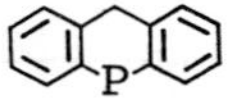

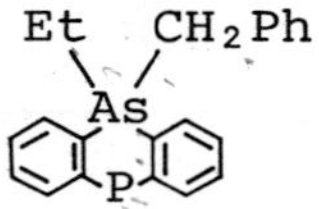

Table 6 (Continued)

Cation RR'PR"			Anion X	Method	m.	Solvent	Yield (%)	Other Data and Remarks
R	R'	R"P						
Oct	Oct	(structure: ring with OH, OH)	Cl	673a				PMR −44
Oct	Oct	(structure: ring with OH, OH)	Cl	673a				PMR −25
Ph-Br-4	Ph	(structure)	d-Camphor sulfonate	471	206			
			$OPh(NO_2)_3$-2,4,6	471	137			
Ph-Cl-4	Ph-Me-4	(structure)	Br	471	227			
	Ph	(structure: Me, O, Me)	Br	I + XVI[982]	243	PhH/$(Me_2N)_3PO$	72	
		(structure: Ph, O)	Br	I + XVI[302]	245	PhH/$(Me_2N)_3PO$	25	

Structure				Method		Solvent	Yield	Notes
(Ph, Me)	Ph	Ph	Br	I + XVI[982]	245	PhH/(Me$_2$N)$_3$PO	86	
Ph-Br-4	Ph	Ph	Br	I + XVI[982]	306	PhH/(Me$_2$N)$_3$PO	78	
Ph-NO$_2$-4	Ph	Ph	Br	I + XVI[982]	305	PhH/(Me$_2$N)$_3$PO	85	PMR 16.5
	Ph	Ph	Cl	673a				
	Ph	Ph	Br	I + XVI[982]	316	PhH/(Me$_2$N)$_3$PO	91	
Ph-OH-2			d-Br-Camphor sulfonate	471	137			Resolved
			Camphor sulfonate	471	174			
Ph-OH-4			Br	471	287			
			Br	472				$[M]_D = +32.9$
Ph-C≡CH	Ph	Ph	Br	I + XVI[982]	241	PhH/(Me$_2$N)$_3$PO	47	

Table 6 (Continued)

Cation	Anion X	Method	m.	Solvent	Yield (%)	Other Data and Remarks
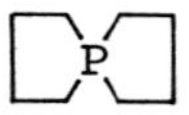	Cl	II[133]	94	Pent OH/HCl		
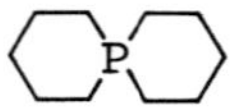	Cl	II[133]	167			
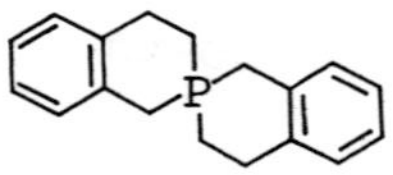	OPh(NO$_2$)$_3$-2,4,6	XV[661]	39	H$_2$O		Yellow
	J	XV[661]	294 / 246			d [M]$_D$ = 66 CHCl$_3$ c. 0.520 / l [M]$_D$ = -65 CHCl$_3$ c. 0.668 (dl)
			246			
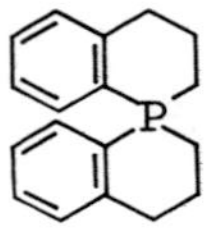 (-)	(-)Menthoxy-acetate	[661]	78	EtOAc		[M]$_D$ = -140 EtOH c. 0.812

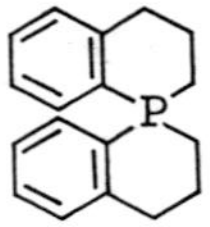

384

			mp	Solvent		Rotation / spectra
(±)						
(−)						
(+)Camphor-sulfonate	661	171	EtCMe (C=O)			
(+)Br-Camphor-sulfonate	661	185	EtCMe (C=O)			
(+)Camphor-sulfonate	661	150				
	438	247 dec.				$[\alpha]^{24.5}_{578} = 1265 \pm 15$ c. 0.7 DMF, -1265 ± 15 $[M]^{24.5}_{578} = -10,410 \pm 120$ $-10,410 \pm 120$
Br	440	398	EtOH/Et$_2$O	79		IR
Cl	440	>400	EtOH/EtOAc	19		IR
J	439					PMR -24

Table 6 (Continued)

Cation	Anion X	Method	m.	Solvent	Yield (%)	Other Data and Remarks
bis-cyclo						
R						
NC(CH₂)₂	O₂CMe	I[20]	149	i-PrOH/Me₂CO		
	Br	XV[20]	303			
	J	XV[20]	320			
	Cl	XVI[20]	280	HCl/H₂O		
(CH₂)₂CO₂H Bu	O₂CMe	I[20]	195			
	Br	XV[20]	316			
c-Hex	J	XV[20]	340 dec.	Me₂CO		
	O₂CMe	[20]	95			
	J	[20]	317			
(CH₂)₂OBu	J	XV[20]	271			
Dec	O₂CMe	[20]	144	EtOAc/i-PrOH		
bis-cyclo						
Ph		6 7 3 a				PMR −14.4

R	R'	X	Compound[ref]	m.p. (°C)	Recrystn. solvent	Yield (%)	Remarks
Me	CH₂Ph	OPh(NO₂)₃-2,4,6-	[216]	281			
	Ph	Br	[216]	387			
		OPh(NO₂)₃-2,4,6-	[216]	269			
Et	CH₂Ph	OPh(NO₂)₃-2,4,6-	XV[216]	330			Two isomeric forms
		J		299			Two isomeric forms
		Br	I[216]	241			
				345			
		J	I[216]	318			Two isomeric forms
				326			
CH₂Ph	Ph	OPh(NO₂)₃-2,4,6-	XV[662]	320			
		Br		<100 dec.			
			I[662]	375[662]	EtOH/Et₂O[662]		HMR,[284] IR,[284] TLC[284]
bis-bicyclo	Me–P⟨⟩P–Me (Et)	OPh(NO₂)₃-2,4,6-	XV[662]	270 dec.	H₂O		Yellow
		J	I[662]	375 dec.			
CH₂Ph	Et–P⟨⟩P–OH	Br	I[825]	220 dec.	MeOH/EtOAc	46	UV
		ClO₄	XV[825]	>200 dec.			IR, HMR

387

Table 6 (Continued)

Cation	Anion X	Method	m.	Solvent	Yield (%)	Other Data and Remarks
bis-bicyclo						
(Et—P—Et structure)	OPh(NO$_2$)$_3$-2,4,6	XV[216]	272			
	Br	I[216]	325			
(P—P—P structure)	OPh(NO$_2$)$_3$-2,4,6	XV[216]	278			
	J	I[216]	305			

Table 7. Diquaternary Salts

| Cation R₃PR'PR₃'' | | | Anion | Method | m. | Solvent | Yield | Other Data |
R	R'	R''	X				(%)	and Remarks
Me₃P	$(CH_2)_2$	PMe₃	$B_{12}H_{12}$	XV[736]	600			
Me₂EtP	(1,3-disubstituted benzene ring)	PMe₂Et	$OPh(NO_2)_3^-$ 2,4,6 J	XV[668]	151	H_2O		
Me₂P(CH₂)₂CN	$(CH_2)_2$	PMe₂–(CH₂)₂CN	Br	I[668] I[383]	308 249	H_2O BuOH	 75	
Me₂PhP	$(CH_2)_3$ $(CH_2)_2$ $(CH_2)_4$	PPhMe₂	Br J J	I[383] I[492] I[492]	175 298 217	 MeOH	98 83 	
	(1,3-bis-CH₂ benzene ring)		$OPh(NO_2)_3^-$ 2,4,6 J	XV[666]	181	EtOH		Yellow
MeEt₂P	(1,3-disubstituted benzene ring)	PEt₂Me	$OPh(NO_2)_3^-$ 2,4,6 J	XV[668] I[668]	253 311	DMF/H_2O 		
	(2,2'-biphenyl structure)		$OPh(NO_2)_3^-$ 2,4,6 J J	XV[448] I[448] I[15]	239 280 255	Me₂CO EtOH 		

Table 7 (Continued)

Cation $R_3PR'PR_3''$ R	R'	R''	Anion X	Method	m.	Solvent	Yield (%)	Other Data and Remarks
$\begin{array}{c}CH_2Ph\\ \mid\\ Me-P\\ \mid\\ Ph\end{array}$	$(CH_2)_2$	$\begin{array}{c}Me\\ \mid\\ P-CH_2Ph\\ \mid\\ Ph\end{array}$		[126]				HMR
$MePh_2P$	CH_2	PPh_2Me	Br	I[492]	272	MeOH	93	
			J	I[452] [126]	300	MeOH		
	$-CH=CH-$		J	I[448] [126]	322	MeOH		HMR
	$(CH_2)_2$		$OPh(NO_2)_3$-2,4,6	XV[913]	205			HMR
	$(CH_2)_3$		J	I[160] [126]	306	MeOH		HMR
	$\begin{array}{c}CH_2-CH\\ \mid\\ Me\end{array}$		J	I[448]	287	EtOH		
	$\begin{array}{c}CH_2CHCH_2\\ \mid\\ OH\end{array}$		J	540	135 dec.	EtOH/Et$_2$O	78	
	Fe (ferrocenediyl)		J	I[1007]	223 dec.			
	$\begin{array}{c}(CH_2)_2P(CH_2)_2\\ \mid\\ Ph\end{array}$	PPh_2Me	$OPh(NO_2)_3$-2,4,6	XV[448]	193	Me$_2$CO/H$_2$O		
			J·H$_2$O	I[448]	162	MeOH/Et$_2$O		

Structure	Phosphine	Anion	Method	mp	Solvent	Yield	Notes
Me CH_2CCH_2 CH_2PPh_2 CH_2 $(CH_2)_2$ CH $CH=CH_2$	PPh_3	J	I[448]	310	$MeNO_2$		HMR
		Ph_4B	$XIV + XV$[126]	194	DMF/MeOH		
		Ph_4B	$XIV + XV$[953]	137	DMF/MeOH		HMR
		Hg^IBr_4	$XIV + XV$[953]	134	DMF/MeOH		
(m-tolyl)	PPh_2Me	J	I[1115] [126]	199	EtOH		
(p-tolyl)		J	I[443]	334 dec.	EtOH		
$(HOCH_2)_3P$	$P(CH_2OH)_3$	Br	I[867]	107	MeOH		
$HOCH_2PBu_2$ $CH_2NHCNHCH_2$ CH_2	Bu_2PCH_2OH	Picrolonate J	XV[868] I[804]	158 dec.	PhMe	76	$n_D^{20} = 1.5487$
$HOCH_2PHept_2$ $(CH_2)_2$ CH_2	$Hept_2PCH_2OH$	Br J	I[804] I[804]		PhMe PhH	88 100	$n_D^{20} = 1.5440$ $n_D^{20} = 1.5146$
$HOCH_2PPh_2$ CH_2	Ph_2PCH_2OH	Br Cl	I[804] X[403]	166	c-Hex	96	$n_D^{20} = 1.5180$
$(CH_2)_2$	$PCH=CH_2$ Ph_2	J Br Br	I[804] I[804] I[928]	58 62 325	EtCHO	100 100	

Table 7 (Continued)

392

R	R'	R''	Anion X	Method	m.	Solvent	Yield (%)	Other Data and Remarks
Et_3P	$(CH_2)_2$	PEt_3	$OPh(NO_2)_3$-2,4,6	XV[454]	208	MeOH		Yellow
	$(CH_2)_3$		Br	I[454]	260 dec.	EtOH	92	
	$(CH_2)_4$		Br	I[454]	260 dec.	EtOH	88	
			Br	I[150]	158			
	$CH_2CH=CH$	PPr_3	$OPh(NO_2)_3$-2,4,6	XV + XVI[883]	151	EtOAc		
	$(CH_3)_3$		$OPh(NO_2)_3$-2,4,6	XV[883]	138	EtOH		
	CH_2CHCH_2, OH	PPr_3	$OPh(NO_2)_3$-2,4,6	XV[883]	151	EtOH		Yellow
			$OPh(NO_2)_3$-2,4,6	XV[883]	213 dec.	EtOH/H₂O		
			$OPh(NO_2)_3$-2,4,6-OH-3	XV[883]	143	EtOH/H₂O		
	$CH_2\overset{Me}{C}CH_2$, PEt_2	PEt_3	$OPh(NO_2)_3$-2,4,6	XV[448]	213	H₂O		
	$(CH_2)_{10}$	PEt_3	J	I[448]	357 dec.	H₂O		
			Br	I[262]	226	CHCl₃/Et₂O		
Et_2PhP	$(CH_2)_2$	$PPhEt_2$	$OPh(NO_2)_3$-2,4,6	I[454]	188 dec.			

Phosphine / group	Anion	Ref.	m.p. (°C)	Solvent	Yield (%)	Colour
$(CH_2)_4$	Br	I^{454}	257 dec.	EtOH	78	Yellow
$C_6H_4(CH_2\cdots CH_2)$	J	I^{529}	210	EtOH	74	
$Me_2NPhPEt_2$	$OPh(NO_2)_3\text{-}2,4,6$	XV^{566}	159	Me_2CO/EtOH		Yellow
$Et_2PPhNMe_2$; $(CH_2)_2$	$OPh(NO_2)_3\text{-}2,4,6$	XV^{671}	183	MeOH		Yellow
	Br	I^{671}	221	Me_2CO/MeOH		
$(CH_2)_3$	$OPh(NO_2)_3\text{-}2,4,6$	XV^{671}	158	MeOH		Yellow
	$J\cdot H_2O$	XV^{671}	186	H_2O		
$C_6H_4CH_2$	$OPh(NO_2)_3\text{-}2,4,6$	XV^{671}	141	MeOH		
Pr_3P	Br	I^{671}	223	Me_2CO		Orange–yellow
PPr_3	$OPh(NO_2)_3\text{-}2,4,6$	$XV +$ / XVI^{883}	156	EtOH/EtOAc		
$(CH_2)_3$	$OPh(NO_2)_3\text{-}2,4,6$; $PtCl_6$	$XV +$ / XVI^{883}; XV^{883}	128			
$C_6H_4CH{=}CHCH_2$	$OPh(NO_2)_3\text{-}2,4,6$	XV^{883}	226	EtOH		Orange
CH_2CHCH_2–OH	$OPh(NO_2)_3\text{-}2,4,6\text{-}O\text{-}3$	XV^{883}	214 dec.	EtOH/H_2O		
	$OPh(NO_2)_3\text{-}2,4,6$	XV^{883}	135	EtOH/H_2O		
$CH{=}CHCH_2$–Me	J	XV^{883}	277	EtOH		
$(CH_2)_2$; $(CH_3)_3$	$OPh(NO_2)_3\text{-}2,4,6$	XV^{883}	156	EtOH		
$[NC(CH_2)_2]_3P$; $P[(CH_2)_2CN]_3$	Cl	XV^{476}	104			
	Br	I^{683}	>300	MeCN	93	
	Br	I^{382}	215^{382}		87^{683}	

Table 7 (Continued)

Cation $R_3PR'PR_3''$			Anion X	Method	m.	Solvent	Yield (%)	Other Data and Remarks
R	R'	R''						
Bu₃P	(CH₂)₂	PBu₃	J	XV[845] + XVI[845]	194	EtOH/MeCN	92	
	(CH₂)₂O(CH₂)₂– O(CH₂)₂		Ph₄B	XV[584] + XVI[584]	174		87	
			Br	XVI[584]	225	MeCN/Et₂O	54	IR, yellow
			Cl	I + XVI[1014]	234		54	
	C≡CH / Me₂NPh / CO₂Et		Cl	I + XVI[1014]	234	CHCl₃/ EtOAc	54	IR, yellow
	(CH₂)₂C(CH₂)₄ CO₂Et / CO₂Et		Ph₄B	XV[584] + XVI[584]	137		77	
	(CH₂)₂ [fluorenyl] (CH₂)₂		Br	XVI[584]	225	MeCN/Et₂O	92	
	 [p-tolyl]		Br	IV[494]	288	H₂O	85	
	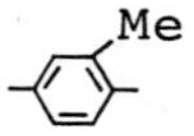 [methyl-substituted phenyl, Me]		Br	IV[494]	213	H₂O	39	
	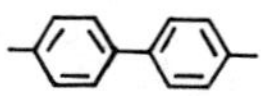 [4,4'-dimethylbiphenyl]		Br	IV[494]	295	H₂O	90	

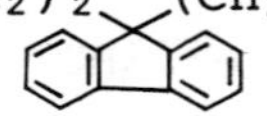

394

R₃P	Bridge	PR₃′	Anion	Method[ref]	mp	Solvent	Yield	Spectra
c-Hex₃P	(ring: Cl₂C–S–CCl₂, S, Cl)	P-c-Hex₃	Cl	I[292]	141	EtOH/H₂O		
(PhCH₂)₃P	CH₂–C₆H₄–CH₂ (para)	P(CH₂Ph)₃	Br	I[264]	300		97	UV
	(CH₂)₂		OPh(NO₂)₃-2,4,6 Br	XV[452]	220 dec.	EtC(=O)Me		UV, HMR
	(CH₂)₄			I[452]	275	MeOH		
			OPh(NO₂)₃-2,4,6 Cl	XV[910]	161			
(PhCH₂)₂PPh	(CH₂)₂	PhP(CH₂Ph)₂		[912]				
				[226]				
			OPh(NO₂)₂-2,4,6 Br	XV[452]	290 dec.	Me₂CO		Yellow
				I[452]	300 dec.[452]	MeOH/Et₂O[452]	94[492]	HMR[928]
			OPh(NO₂)₃-2,4,6·EtOH Br	XV[667]	122	Me₂CO/EtOH		
PhCH₂PPh₂	(o-C₆H₄)(CH₂)(CH₂)	Ph₂PCH₂Ph	Br	I[4]				IR, HMR
	CH=CH		Br	I[3]	333	MeOH/MeCN		HMR, PMR −3.8 ppm
	(CH₂)₂	PhPCH₂PPh₂	Br	[126] / I[405]	159	CHCl₃/Et₂O		HMR / IR
Ph₃P	CH₂	PPh₃	Br	I[415]	310[283]	MeOH/EtOAc[415]	80[831]	IR,[831] pKa 6.87[687]; 2.78,[534] UV[687]; PMR −18.4[6 7/3a]

Table 7 (Continued)

Cation R₃PR'PR''₃								
R	R'	R''	Anion X	Method	m.	Solvent	Yield (%)	Other Data and Remarks
Ph₃P	CH₂	PPh₃	Br	XIV[31]	312	EtOAc/MeOH	43	IR
			Br·0.5 H₂O	I[21]				
			Cl	I[44]	256	PhCN		
			Cl·H₂O	XV[81]	240	MeOH/EtOAc	68	
	CH=CH		J	XV[81]	306 dec.	MeOH	58	
			Br	I + XV I[792]	305	CHCl₃/EtOAc	62	IR, HMR
	(CH₂)₂			III				PMR −24.7 ppm
			Br	I[1070]	308 dec.[1070]	CHCl₃/c-Hex[1070]	100[931]	IR,[1067] HMR[386]
				II[31]		i-PrOH		
			Cl	I[845]	287			
	CH₂HgCH₂	PPh₃	Ph₄B	II[398]	101			
			HgBr₃	XIV[398]	123[398]			
	CH₂OCH₂		Ph₄B	XV[252]	220	MeOH[957]		
			Br	I[252]	292[251]	DMF/H₂O	30	
	CH₂SCH₂		Br	XV[249]	302	CHCl₃[252]	90[252]	
	CH=CHCH—HgBr		HgBr₃	XIV[755]	198	H₂O	40	UV

Structure	Anion	Method[ref]	M.p.	Solvent	Yield (%)	Spectral / other data
$CH_2C(=O)CH_2$	Cl	I[239]	260	$CHCl_3$/Et_2O	86	
$(CH_2)_3$	Cl·H_2O, ClO_4	XIV[312], XV[966]	274, 289[486]	MeOH	88	
	Br	I[486]	335[486]	DMF/petroleum ether[486]	94[486]	PMR −23.2 ppm,[397] $E_{1/2}$ 1.307, 1.712, 2.054[482] UV
$CH_2CH(OH)CH_2$	Br	XVI[755]	292			
$CH_2C{=}CCH_2$	Br	I[915]	220 dec.	$MeNO_2$/Et_2O	95	
$CH_2CH{=}CHCH_2$	Br	I[653]	260[653]	EtOH/EtOAc[653]	41[132]	
$(CH_2)_4$	Cl	II[47], I[636]		i-PrOH Et_2O[636]	65[636]	IR[1067]
	Br	I[486]	296[486]	MeOH/Et_2O[486]	100[486]	$E_{1/2}$ 1.798, 2.058[482]
$^{14}CH_2(CH_2)_2{}^{14}CH_2$ / $(CH_2)_2O(CH_2)_2$	Br·H_2O	I[728]	105 dec.	H_2O		
	Br_3	XV[728]	213	H_2O/Br_2	91	
	Br	I[1088]	296	$CHCl_3$	75	
	Br	I[252]	281			
$CH_2Sn(Me)_2CH_2$	Me_2SnBr_4	XIV[398]	135			
$Cr(NH_3)_2(SCN)_4$	Ph_4B	XV[398]	115			
		XV[398]	78			
	$HgBr_2Cl$	XV[398]	139			

Table 7 (Continued)

Cation R₃PR'PR₃''			Anion X	Method	m.	Solvent	Yield (%)	Other Data and Remarks
R	R'	R''						
Ph₃P	CH₂$\overset{O}{\overset{\|}{C}}$CH₂ / CO₂Me	PPh₃	Cl	XVI[165]	167	c-Hex/monoglyme	94	
	(CH₂)₅		Br	I[427]	262[602]	EtOH/i-Pr₂O[427]	66[427]	E1/2[482] 1.807[482]
			Br·H₂O	[486]	120	MeOH/Et₂O DMF		
	(CH₂)₂OCH₂O(CH₂)₂		Br	I[263]	235	EtOH/i-PrOH		
	CH=CHCH₂OCH₂OCH₂CH=CH		J	I[263]	242			
	$\overset{O}{\overset{\|}{C}}$HCCH₂ / CO₂Et		Cl	XVI[165]	162	CH₂Cl₂/EtOAc		
	(CH₂)₂O(CH₂)₂O(CH₂)		Br	I[263]	195	EtOAc		
	(CH₂)₃O(CH₂)₃		J	I[263]	242	EtOH/i-PrOH		
	CH(CH₂)₄ / Me		Br	I[602]	70	BuOH/Et₂O		
	(CH₂)₆		Br	I[427]	335[427]	CHCl₃/Me₂CO[427]	82[427]	E1/2 1.822[482]
	(p-tolyl)	PPh₃	J / Br	I[414] IV[494] I[1029]	292 >300[1029]	EtOH Me₂CO/Et₂O[1029]	18	2.000[482]

	Anion	Prep.	mp	Solvent	Yield (%)	Notes
OH Me (cyclohexanol)	Br	I[1028]	184	Me$_2$CO/Et$_2$O		
Me HO (cyclohexanediol)	Br	I[337]	298	EtOH/Et$_2$O	95	
CHCH$_2$—CO$_2$-i-Pr (C=O)	Cl	XVI[165]	178	c-Hex/monoglyme		
C≡CH Ph	J	XI + XV[462]	205	PrNO$_2$/EtOAc	59	
CH$_2$— CH$_2$ (benzocyclobutene)	Br	I[98]	231	CH$_2$Cl$_2$/Et$_2$O	99	UV
	Br	I[390]	340[387]	CHCl$_3$[387]	89[387]	HMR,[389] IR[390]
	AuCl$_3$	786	163			
	PtCl$_6$	786	235			
	J$_2$HJ	786	247			
CH$_2$ CH$_2$ (m-xylylene)	Br	I[739]	275		83	
	Co(SCN)$_4$	XV[326]				IR[325]
	Cu(SCN)$_4$	XV[326]				IR[325]
	Ni(SCN)$_4$	XV[326]				IR[325]
	Co(SeCN)$_4$	XV[327]				IR, blue
CH$_2$ CH$_2$ (p-xylylene)	Mn(SeCN)$_4$	XV[327]				IR, pale green
	Fe(SeCN)$_4$	XV[327]			85	IR, pale brown

400

Table 7 (Continued)

Cation $R_3PR'PR_3'$			Anion X	Method	m.	Solvent	Yield (%)	Other Data and Remarks
R	R'	R"						
Ph_3P	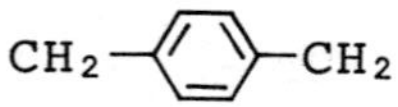 CH_2–C₆H₄–CH_2	PPh_3	$Ni(SeCN)_4$	XV^{327}	Nonmelting	EtOH[337]	80	IR, yellow
			Br	I^{337}	360[337]		89[337]	PMR -22.8 ppm[397]
	$(CH_2)_2$–C₆H₄(–CH_3)	PPh_3	Cl	I^{147}	400[147]	DMF[147]	95[147]	HMR[389] IR[1667]
	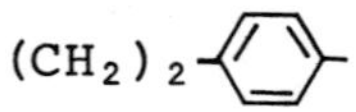 CH_2PCH_2 / Ph		Br	I^{1029}	207	Me_2CO/Et_2O		
			Ph_4B	$XIV + XV^{953}$	128	DMF/MeOH		
	CHTeCH– / –CO_2 –CO_2Et / –Et		Br	XIV^{797}	212 dec.	$PhNO_2$/EtOAc		Orange-yellow
	$(CH_2)_8$		Br	[482]				
	$(CH_2)_4O(CH_2)_4$		J	I^{602}	211	BuOH/EtOAc		$E_{1/2}$ 1.732, 1.971
	O= / CHCCH₂ / CO_2Ph		Cl	XVI^{165}	125	CH_2Cl_2/EtOAc	91	
	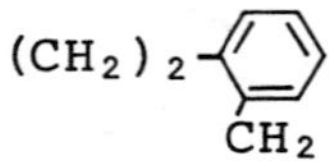 $(CH_2)_2$–C₆H₄–CH_2		Br	I^{74}	262	PhH	99	

Structure	X	Prep.	m.p.	Solvent	Yield (%)	HMR
(2,4,5-trimethylbenzyl), CH$_2$	Cl	[389]				
(3,4-dimethoxybenzyl), CH$_2$, OMe OMe	Cl	I[737]	271			
PPh$_3$; dimethylnaphthyl–HO–naphthyl dimethyl	Br	I[1028]	200			
	Br	I[1029]	300			
(CH$_2$)$_3$, CH$_2$ (ortho phenylene)	Br	I[74]	286 dec.		95	
CH$_2$C=CHC≡CCH=CCH$_2$, Me, Me	Br	I[1022]	256	MeOH/EtOAc	90	
CH$_2$C=CHCH=CHC=CCH$_2$, Me, Me	Br	I[1022]	280	MeOH/EtOAc	90	
(CH$_2$)$_2$C(=O)(CH$_2$)$_4$C(=O)(CH$_2$)$_2$, (CH$_2$)$_{10}$	ClO$_4$, Br	I + XV[1063], I[60]	135, 232[60]	BuOH/Et$_2$O[60]		E1/2 1.795, 1.916, 2.073[482]
(CH$_2$)$_2$[O(CH$_2$)$_2$]$_4$	Br, Br	I[263], IV	177, 360[494]	H$_2$O[494]	78[494]	E1/2 1.155, 1.269, 1.667, 1.477, 2.069, 2.435[482]

Table 7 (Continued)

| Cation R₃PR'PR₃' | | | Anion | | | | Yield | Other Data |
R	R'	R''	X	Method	m.	Solvent	(%)	and Remarks
Ph₃P	CH₂—naphthalene—CH₂	PPh₃	Br	I[74]	311 dec.	PhH	98	
	CH₂—naphthalene—CH₂ (CH₂)₁₂		Br	I[1029]	300			
			Br	I[602]	228[602]	BuOH/ EtOAc[602]		$E_{1/2}$ 1.664 1.944[482]
	—CH— O=PPh₂ —CH— PPh₂		Cl	XVI[687]	269			
			Ph₄B	XV[687]	252	EtOH/MeCN	83	
			BF₃Cl	XV[687]	172	EtOH/MeCN	89	
			BF₄	XV[687]	261	EtOH/MeCN	48	
			Cl	XIV[687]	254 dec.	MeCN/ diglyme	96	
			PF₆	XV[687]	264	EtOH/MeCN	38	
			J	XV[687]	264	EtOH/MeCN	69	
	CH₂—biphenyl(NO₂)—CH₂		Ph-2-Oxo-10-bornane sulfonate	XV[411]	170			IR, $[\alpha]_{546}^{22}$ = 210 c. 1.00 EtOH

	Anion	Method	mp (°C)	Solvent	Yield (%)	Notes
	Ph-3-Br-2-Oxo-8-bornane sulfonate	XV[411]	168 dec.	EtOAc		$[\alpha]_{546}^{186} = 48.0$ c. 1.00 EtOH
	OPh(NO₂)₃- 2,4,6	XV[411]	135 dec.	EtOH		Yellow
	Br·0.5 H₂O	I[411]	265	Me₂CO	68	
	J	XV[411]	280	EtOH		IR
	OPh(NO₂)₃- 2,4,6	XV[411]	209	EtOH/H₂O		
	Br	I[74]	305	PhH	98	
	Br·H₂O	I[411]	323	EtOH/ EtOAc	78	
	Br	I[74]	284	DMF/PhH	95	
	Ph₄B	XV[90]	252			
	BF₄	XVI[90]	257 dec.	MeOH		IR, PMR HMR, PMR -25.8, 22.7, 143.9 ppm
	PF₆	XV[90]	310	Et₂O		HMR, PMR -25.6, -22.6
	J	XIV[90]	293 dec.	EtOH	90	
	J₃	XV[90]	127			
	Br	I[74]	288 dec.	PhH	97	
	Cl	I[740]	298		85	
	Br	I[279]	360	DMF	23	

Table 7 (Continued)

Cation R₃PR'PR₃'

R	R'	R''	Anion X	Method	m.	Solvent	Yield (%)	Other Data and Remarks
Ph₃P CH₂–⟨benzene⟩–C≡C–⟨benzene⟩–CH₂ PPh₃	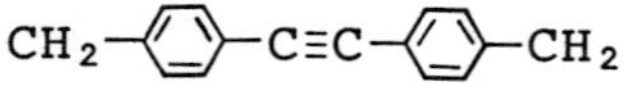		Br	I[279]	360	EtOH/Et₂O	23	
C=CHCH=C (–Ph, Ph, CH₂, CH₂)			J	I[462]	399 dec.		23	
–CH=CH–⟨benzene, CH₂⟩	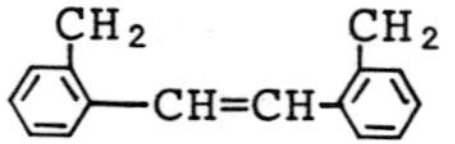		Br	I[1015]	327 dec.[1015]	DMF/PhH/Et₂O[272]	85[1015]	
CH₂CH₂ CH₂CH₂	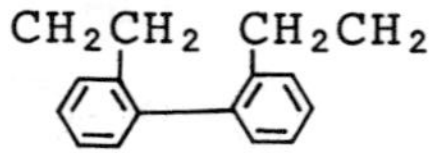		Br	I[74]	278	DMF/Et₂O	76	
⟨biphenyl–biphenyl⟩			J	XV[74]	276 dec.	EtOAc/CH₂Cl₂		Hygroscopic
–C=CHCCH=C– (=O, Ph, Ph, –C–)			Br	II[462]	95 dec.		66	
Ph₂PN=N–Ph			BF₄	XIV[90]	255	EtOH/EtOAc	86	
CH₂–⟨biphenyl⟩–CH₂			Br	I[279]	>360	DMF	30	PMR –38.9, –24.1 ppm

-C=
PhCH₂PPh₂ →

Structure	Ph₄B / J	XV[90] / I[...]	m.p.	Solvent	Yield	HMR, PMR / notes
	Ph₄B	XV[90]	206 dec.	DMF/MeOH	88	HMR, PMR −27.0, −28.5 ppm
	J	XIV + XV[90]	230	MeOH		UV
CH₂(⬡-C≡C-)₂⬡-CH₂	Br	I[269]	349 dec.	EtOH/Et₂O	38	UV
CH₂(⬡-CH=CH)₂⬡-CH₂	Br	I[269]	363 dec.	EtOH/Et₂O	40	UV, yellow
CH₂⬡CH=CH⬡CH=CH⬡CH₂-	Br	I[508]	360[508]	PhH/PhMe[508]	56[387]	
(3-MePh)₂P–Ph (CH₂)₃ P(PhMe-3)₂–Ph	Br	I[420]	314	H₂O	30	
(4-MePh)₃P (CH₂)₅ P(PhMe-4)₃	Br	I[602]	80			
(MeO-)₃P (CH₂)₂ P(-OMe)₃	Br	I[820]	201			
	J	I[820]	214			
(CH₂)₆	Br	I[820]	118			
(CH₂)₁₀	J	I[820]	178			
	Br	I[820]	158		44	
(CH₂)₂₀	J	XV[820]	105			
	Br	I[820]	104			

Table 7 (Continued)

Cation Tris	Anion X	Method	m.	Solvent	Yield (%)	Other Data and Remarks
(HOH$_2$C)$_3$PH$_2$CHN—[triazine]—NHCH$_2$P(CH$_2$OH)$_3$, NHCH$_2$P(CH$_2$OH)$_3$	O$_2$N—[pyrazolone with O$^{(-)}$, N–Ph–NO$_2$-4, Me]	XV[868]	164			Yellow
MePPh$_2$, Ph$_2$P=N–N [triazine, Me, Me]	J	I[447]	246	MeOH/Et$_2$O		
PhCC[(CH$_2$)$_2$PBu$_3$]$_3$ (O=)	Br	XVI[584]	203	PhCMe/Et$_2$O (O=)	80	
Ph$_3$PCH$_2$PCH$_2$PPh$_3$ (Ph, Me, O=)	Ph$_4$B	XIV + XV[953]	177	DMF/MeOH		
Cl, PPh$_3$ / Ph$_3$P, PPh$_3$ [dithiolane-S–S with Cl, Cl]	Cl	I[292]	148 dec.	MeOH/H$_2$O	83	

Structure	Anion	Ref.	M.p. (°C)	Solvent	Yield (%)
(benzene ring) CH₂PPh₃ / CH₂PPh₃ / CH₂PPh₃	Br	I[265]	280 dec.		
Ph₃PH₂C—(benzene)—CH₂PPh₃ , CH₂PPh₃	Br	I[265]	228 dec.		
tetra (benzene ring) PPh₃ / Ph₃P , CH₂PPh₃ / CH₂PPh₃	Br	IV[494]	295	CHCl₃	90
Ph₃PCH₂—(benzene)—CH₂PPh₃ , CH₂PPh₃	Br	I[265] [270]	>300		
Si(Ph–CH₂PPh₃–4)₄	Br	[270]	255	EtOH/Et₂O	62

Table 8. Betaines

	Method	m.	Solvent	Yield (%)	Other Data and Remarks
$(+)$ H Me$_3$P–CPhNO$_2$-4 | C$(-)$ | (CN)$_2$	XV[490]	161			
H | –C–Ph | C$(-)$ | (CN)$_2$	XV[490]	166			
Ph(CO$_2$H)$_2$-2,4 | CO$_2$ $(-)$	[184]	160			
Ph(CO$_2$H)$_2$-3,5 | CO$_2$ $(-)$	[184]	115			
CH$_2$CH$_2$CO$_2$ $(-)$ Et$_3\overset{+}{P}$CS$_2$ $(-)$	XI[870, 672]	147	EtOH		
–C=NPh | S $(-)$	XI[490]	63	THF/petroleum ether	40	X-ray diffraction
–CHPh | NC–C–NO$_2$ $(-)$	XI[490]	151 dec.			

	Ref.	mp (°C)	Solvent	Color
-CHPhCl$_2$-2,4 | (-)C(CN)$_2$	XI^{490}	155		
-CHPhCl-2 | C(CN)$_2$ (-)	XI^{477}	145	PhH	
-CHPhCl-3 | C(CN)$_2$ (-)	XI^{477}	143	PhH	
-CHPh-Cl$_4$ | C(CN)$_2$ (-)	XI^{477}	143	PhH	
-CHPhNO$_2$-2 | C(CN)$_2$ (-)	XI^{477}	92	PhH	
-CHPhNO$_2$-3 | C(CN)$_2$ (-)	XI^{477}	141	PhH	
-CHPhNO$_2$-4 | C(CN)$_2$ (-)	XI^{477}	145	PhH	Orange-red
-CHPh | C(CN)$_2$ (-)	XI^{477}	147 dec.		
-CHPh | NC-C-CO$_2$H (-)	XI^{477}	121		
-CHPhCN-4 | C(CN)$_2$ (-)	XI^{477}	120	PhH	

Table 8 (Continued)

	Method	m.	Solvent	Yield (%)	Other Data and Remarks
Et₃P⁺–CHPhMe-4					
$\overset{\mid}{\underset{(-)}{C}}(CN)_2$	XI[477]	149	PhH		
	XI[490]	296			
–CHPhOCMe-4 (=O)					
$\overset{\mid}{\underset{(-)}{C}}(CN)_2$	XI[477]	133	PhH		
–CHPhNHCMe-4 (=O)					
$\overset{\mid}{\underset{(-)}{C}}(CN)_2$	XI[490]	218			
–CHPh					
NC–C–CO₂Et	XI[490]	122			
(−)					
PhNO₂					
–CHC–Ph	XI[490]	131 dec.			
(−)					
–CHPh					
$\overset{\mid}{\underset{(-)}{C}}(CO_2Et)_2$	XI[490]	76			

Structure		Temp.		Yield	Notes
$-CHPh$ (indanone, $-$)	XI^{490}	217 dec.	60		Yellow to orange
$-CHPh$, $-C=C-O$, $HN-O-Ph$ ($-$)	XI^{490}	145		100	Pale green
(+) $Bu_3P-CH_2Ph-SO_3-4$; ($-$) $-CHPh-NO_2-3$, $-C(CN)_2$	I^{858}, XI^{490}	202, 145	H_2O	65	
($-$) $-CHPh$, $-C(CN)_2$ ($-$)	XI^{490}	124			
(dimethylphenyl dioxy, $-$)	$6\,3\,0$				ESR
(methylnaphthalene dioxy, $-$)	$6\,3\,0$				ESR

Table 8 (Continued)

	Method	m.	Solvent	Yield (%)	Other Data and Remarks
[naphthoquinone structure: Bu₃P(+), Me, O(−)]	630				ESR
[benzene structure: Bu, Bu, Me, O(−) O(−)]	630				ESR
[benzene structure: Ph, Ph, O(−) O(−)]	630				ESR
c-Hex$_3$P-C=NPh (+)S(−)	R$_3$P + RNCS[527]	127	Petroleum ether	86	Yellow
Ph$_3$P-CH$_2$BF$_3^-$ (+)	XIV[398]	218[958]	Me$_2$CO/H$_2$O[398]		
-CH$_2$CO$_2$ (−)	713	124			
-CH$_2$CO$_2$ (−)	XV + XVI[9]	186[9]	EtOH/Et$_2$O[238]	87[238]	IR,[238] PMR -23 ppm[238]

Structure	Ref.	mp	Solvent	Yield	Methods
$-CHCO_2Me$, $SO_3^{(-)}$	XIV[758]	199 dec.	THF/MeNO$_2$	67	IR,[295] HMR,[295] FMR[295]
$-(CH_2)_3SO_3^{(-)}$	XIII[410]	320	MeOH/PhMe		
$-C-C=O$, $^{(-)}O-C-CF_2$	[295]	235[294]	Me$_2$CO/H$_2$O[295]	29[295]	IR, HMR
$O^{(-)}$, $-CH=C-CHF_2$	XIV[295]	210 dec.	EtOH/Me$_2$CO/H$_2$O		IR, HMR
lactone ring $^{(-)}$	I[167]				IR
$-(CH_2)_3CO_2^{(-)}$	XV[238]	227	EtOH/Et$_2$O	38	HMR, FMR, IR
fluorinated cyclopentanedione ring, $S^{(-)}$	[295]	181	EtOH/H$_2$O	35	IR
$CH_2S-C=C(CN)_2$; S^{253}—$C(CN)_2$—S / CH_2 + Ph$_3$P		245	MeCN	89	IR, HMR, UV
cyclopentadienyl (methyl) $^{(-)}$	[834]	229	PhMe	60	IR, UV, yellow
$-C-CH_2-CO_2^{(-)}$, CO_2Et	XVI[510]	142	EtOAc		IR

Table 8 (Continued)

Structure	Method	m.	Solvent	Yield (%)	Other Data and Remarks
$Ph_3P-CH_2PhSO_3$ (+)(−)	I[858]	>350	AcOH/H$_2$O	51	
−CH=CPh, O$^{(−)}$	716	183			
−CH−CPh, SO$_3$ (−), O$^{(−)}$	XIV[798]	227	MeNO$_2$	91	
Ph$_3$P + [structure: N[246]=CPh, SO$_2$, H$_2$C−O]	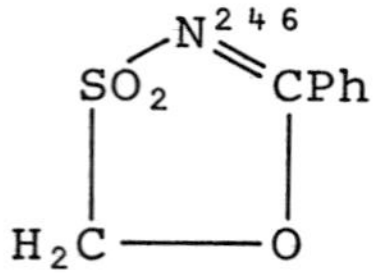	242 dec.	DMF	90	
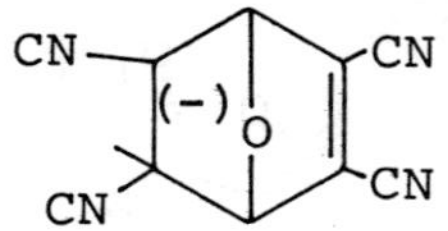 −CH$_2$SO$_2$N=C−Ph	XI[1059]	193	MeCN	100	IR, dark blue, deep red in MeCN solution
[bicyclic structure with CN groups, O, (−)]	XV[755]	145 dec.		65	
CH=CH−CH=CH−Ph−O−4 (−), [ring structure] N=N−Ph	905	238	PhH/MeOH		UV

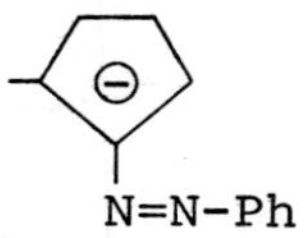

414

Substituent	Method	mp (°C)	Solvent	Yield (%)
$-CH_2-CH_2-$[1-naphthyl], $O^{(-)}$	XIV[1097]	221 subl.	MeOH/THF	40
$-C-Ph_2$, $SO_3^{(-)}$ ·MeOH, $O^{(-)}$	XIV	195		
$-CH_2-C=CPh_2$, $O^{(-)}$	XIV[1073]	176	PhH/c-Hex	12
$-CH-CHPh$, Me N–Ph $(-)$	XIV[82]	107 dec.	Petroleum ether	87
[ring–O] N=N–PhNEt$_2$–4·2 HBr $(-)$	833a	181	MeOH/EtOAc	
	XIV[1073]	199	PhH/MeOH	
		182	EtOAc/ petroleum ether	45
$-CH-C=CPh_2$, Me O $(-)$	82	116 dec.	Petroleum ether	89
$-CH-CH-Ph$, $(-)$N Me; Ph–Me–4	XIV[1073]	139	CH$_2$Cl$_2$/ EtOH	83
$-C(Me)_2C=CPh_2$, O $(-)$	XIV[82]	104 dec.	Petroleum ether	66
$-CH-CPh$, Pr NPh $(-)$ / CH–CH–Ph, Pr NPh–Me–4 $(-)$	XIV[82]	116	Petroleum ether	65

Table 8 (Continued)

	Method	m.	Solvent	Yield (%)	Other Data and Remarks
$(+)\text{MeO}^{(-)}$ $\text{Ph}_3\text{P-C-C=C=C-Ph-Me}_3\text{-2,4,6}$ $\quad\quad\mid$ $\quad\quad\text{MePh}$ -C=PPh_3	XIV[1063]	145	$\text{CHCl}_2/$ MeOH	57	
C=O $\text{S}(-)$ -C=PPh_3	XIV[687a]	152			
C=O $\text{O}(-)$ -C=PPh_3	XIV[687a]	140 dec.	Et_2O	96	IR
C=S $\text{S}(-)$ $(-)$ BPh_3	XIV[687a]	132			Light yellow
-C=PPh_3	XIV[689]				IR, HMR, BMR, PMR, UV
-PhO^-	XV[1067]	282	$\text{MeOH}/$ H_2O^{76}		IR[1067]
$\text{-PhOH-5-O}^{(-)}\text{-2}$	[828]	262^{828}	PhH^{828}		IR,[829] UV,[828] ESR[631]

	XI[485]	dec.[485]	MeOH/Et₂O[485]	88[485]	ESR,[630] yellow

Naphthalene structure with $O^{(-)}$, methyl, OH, $O^{(-)}$ substituents — 161 dec.; MeOH/Et₂O[485]; 88[485]; ESR,[630] yellow

Substituent	XI[485]	dec.[485]
$(Ph-Me-4)_3P$—$CH=C$—Me, $O^{(-)}$ $(-)$	7^{12}	107
—$CH=CPh$, $O^{(-)}$ $(+)$	7^{12}	177
$(Ph-Ph-2)_3P$—CH_2CO $(-)$ $(+)$	I^{1090}	109
$(ClCH_2)PCH_2CO_2$ $(-)$, Ph—Me—4	7^{12}	206
$Et_2P-CH=CMe$ $(+)$, $O^{(-)}$, Ph—Me—4	7^{12}	75

417

Table 9. Salts with P–H Bonds

Cation Quasi PH	Anion X	Method	m.	Solvent	Yield (%)	Other Data and Remarks
PH_4	Br	7[31]				Molecular orbital wave function for XH_n molecules χ[199] HMR,[308] ESR after x-ray irradiation[674] PMR +2.8[673a]
	J	XV[1079] [199]	207[1079]	Me_2CO/H_2O[1079]	79[1079]	
Me_3PH	Cl					
Et_3PH	$UO_2^{VI}Br_4$	XV[221]	230	EtOH	40	IR, UV, Λ, Orange-yellow
	$UO_2^{VI}Cl_4$	XV[221]	255	EtOH	50	IR, UV, Λ, yellow
	$U^{IV}Cl_6$	XV[221]	150	MeCN	50	Blue-green
Pr_3PH	$U^{VI}O_2Br_4$	XV[221]	181	EtOH	40	IR, UV, Λ, Orange-yellow
	$U^{VI}O_2Cl_4$	XV[221]	210	EtOH	50	IR, UV, Λ, yellow
	J	I[778]	205	EtOH/EtOAc		
	J	I[778]	180			
$c-Hex_3PH$	Cl					PMR ±0[673a]
Ph_3PH	$U^{IV}Br_6$	XV[221]	252	MeCN	80	Λ, green
	$U^{VI}O_2Cl_4$	XV[221]	280	MeCN	90	IR, UV, Λ
	$U^{IV}Cl_6$	XV[221]	249	MeCN		Λ, blue-green
$(4-HO-Ph)_3PH$	Br	I[764]	220 dec.			
	J	I[764]	200 dec.			

Table 10. Salts with P-As Bonds (see also Chapter 2)

Cation R₂AsP	R'₃	Anion X	Method	m.	Solvent	Yield (%)	Other Data and Remarks
Me₂AsP	Me₃	J	I[118]	270	EtOH/H₂O		
	Me₂Ph	Br	I[118]	178	EtOH		
		Cl	I[118]	115	Me₂CO		
		J	I[118]	147	EtOH	62	
	MePh₂	J	I[118]	115	EtOH		
	Et₃	Br	I[118]	142	EtOH		
		Cl	I[118]	73	EtOH/Et₂O		
		J	I[118]	132	EtOH		
	Et₂Ph	Br	I[118]	142	PrOH		
	Pr₃	J	I[118]	100	PrOH		
Et₂AsP	Me₂Ph	J	I[118]	112	EtOH		
Ph₂AsP	Et₃	J	I[118]	85	EtOH		
Et—Bu>AsP	(OEt)₃	J	I[569]	182	EtOH/Et₂O	100	

Table 11. Salts with P-Br Bonds (see also Chapter 5B)

Cation	Anion X	Method	m.	Solvent	Yield (%)	Other Data and Remarks
Br₂PEt₂	Br	[49]				IR
BrPPh₃	BrJ₂	I[84]	79	MeCN/Et₂O		Red
	Br₂J	I[84]	101	MeCN/Et₂O		Orange
	Br₃	I[84]	117	MeCN/Et₂O		Orange
	J₃	XV[84]	127	MeCN/Et₂O		Red
BrPPh₂ | HC=PPh₃ | CH=CH₂	HgBr₄	XIV + XV[953]	137	DMF/MeOH		
Br₂P(NMe₂)₂	Br	I[769]	207 dec.	MeCN/Et₂O	76	
	Ph₄B	XV[770]	243	MeCN		
BrP(NMe₂)₃	Br	I[770]	244 dec.		95	
BrP(OPh)₃	Br₃	XV4[16]	105	PhNO₂/Et₂O	60	

Table 12. Salts with P-Cl Bonds (See also Chapter 5B)

Cation	Anion X	Method	m.	Solvent	Yield (%)	Other Data and Remarks
PCl_4	ClO_4	[52]				PMR −87.1 ppm
	JCl_2	XXI[52]	180 dec.	$MeNO_2$		PMR −88.6 ppm
	$AlCl_4$	[52]				PMR −88.5 ppm
	BCl_4	XXI[52]		$MeNO_2$		PMR −88.1 ppm
	JCl_4	XXI[52]	221 dec.	$MeNO_2$		PMR −89.3 ppm, orange
	$N(SO_2Cl_2)_2$	[52]		$POCl_3$		PMR −87.0 ppm
	$SbCl_6$	I[889]	>300[889]	CH_2Cl_2[889]	84[889]	HMR,[889] ∧[889] IR,[889] ∧[889] PMR −87.9[52]
Cl_3PF	$SbCl_6$	I[889]	>300	CH_2Cl_2	74	IR, HMR, ∧
$Cl_3\underline{P}$						
Cl_3P R̲ Me	Cl	XXI[49]	132 dec.		94	
Cl_3P Et	Cl	II[600]	230 dec.		99	
	PCl_6	XXII[1111]	163 dec.		65	
$Ph-NO_2-3$	PCl_6	XXII[1111]	155 dec.		73	Very hygroscopic, yellowish
$Ph-NO_2-4$	PCl_6	XXII[311]	200 dec.			
Ph	$SbCl_6$	I[889]	290	CH_2Cl_2	90	IR, HMR, ∧

Table 12 (Continued)

Cation		Anion X	Method	m.	Solvent	Yield (%)	Other Data and Remarks
Cl_3P	R						
Cl_3P	Ph-Me-4	PCl_6	XXII[311]	158			
	NMe_2	PCl_6	XXII[184]	242 dec.			
	$N{=}C(OMe)_2$	Cl	I[243]	119	PhH		
	NEt_2	PCl_6	XXII[184]	232 dec.			
	$N{=}C(OEt)_2$	$SbCl_6$	I[889]	247	CH_2Cl_2	67	HMR, IR, ∧
		Cl	I[243]	46	PhH		
	NPr_2	PCl_6	XXII[184]	220 dec.			
	$-N{-}i{-}Bu_2$	PCl_6	XXII[184]	168 dec.			
	OEt $-N{=}CPh{-}Br{-}4$	Cl	I[243]	131	PhH		
	OEt $-N{=}CPh_2$	Cl	I[243]	90	PhH		
	$-OPh$	$SbCl_6$	I[889]	265	CH_2Cl_2	81	IR, HMR, ∧

Cl_2PRR'

Cation	R	R'	Anion X	Method	m.	Solvent	Yield (%)	Other Data and Remarks
Cl_2P	Me	Me	Cl	[49]				
	Ph-Br-4	Ph	PCl_6	XXII[970]	156			IR

422

ClPRR'R''

R	R'	R''		ref.	mp/bp	solvent	%	data
	Ph	Ph-Cl-4	PCl_6	XXII[970]	146			
	Ph	Ph-NO$_2$-4	PCl_6	XXII[970]	168			
	Ph	Ph	PCl_6	XXII[970]	178			IR, HMR, ∧
		Ph-Me-4	$SbCl_6$	I[889]	168	CH_2Cl_2	67	
		Ph-OMe-4	PCl_6	XXII[970]	124			
			PCl_6	XXII[970]	102			
			Cl	I[34]	61			
			Cl		b_{11}[158]			
ClP	Et	Et	(CH$_2$)$_2$OBu	Cl	II[1049]	b_{11}[44]		49	$n_D^{20} = 1.5018$, d_{20} 1.1018
	Et		CH$_2$Ph	Cl	I[1048]	92			
				Cl	I[1048]	114			
	Ph	Ph	Ph	$SbCl_6$	I[889]	172	CH_2Cl_2	59	IR, HMR, ∧
	NMe$_2$	NMe$_2$	NMe$_2$	$AlCl_4$	XV[770]	260	Et_2O	90	
				$FeCl_4$	XV[770]	220	Et_2O	90	
	NEt$_2$	NEt$_2$	NEt$_2$	$SbCl_6$	I[889]	230 dec.	CH_2Cl_2	91	IR, HMR, ∧
	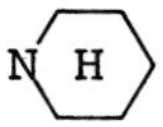 (piperidino) N–H	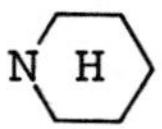 (piperidino) N–H	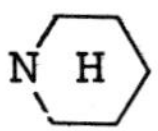 (piperidino) N–H	$SbCl_6$	[114]	183			
	OPh-Me$_2$-2,6	OPh-Me$_2$-2,6	OPh-Me$_2$-2,6	Cl$_3$P(OPh-Me$_2$-2,6)$_3$	I[891]	124	PhCl		

423

Table 13. Salts with P-F Bonds

Cation	Anion X	Method	m.	Solvent	Yield (%)	Other Data and Remarks
$FPMe_3$	Cl	$XV^{49,922}$		Et_2O	50	IR, Raman FMR,[922] HMR,[855]
$FP(NMe_2)_2$ Me	$MePF_5$	[921]				PMR -71.4, 126.4 ppm[922]
$FP(NMe_2)_2$ Ph	$PhPF_5$					$\wedge$, HMR, FMR, PMR -56.0, 136.0 ppm

Table 14. Salts with P–Hg Bonds

Cation	Anion X	Method	m.	Solvent	Yield (%)	Other Data and Remarks
MeHgPEt$_3$	Br	I^{200}	90	Et$_2$O		IR
EtHgPMe$_2$ — Ph	Br	I^{200}				$\wedge$

Table 15. Salts with P-J Bonds

Cation	Anion X	Method	m.	Solvent	Yield (%)	Other Data and Remarks
JPPh$_3$	J$_3$	XXI[84]	123	MeCN		Purple
JP(NHMe)$_3$	NO$_3$	XV[770]	130 dec.	MeCN		
JP(NMe$_2$)$_3$	J	XXI[770]	228 dec.	MeCN/Et$_2$O	96	
JP(N-i-Bu$_2$)$_2$ Me	J	I[184]	132			Light yellow
JP(OPh)$_3$	J	XX[841]	68	PhCl		Brown
	J$_3$	[891]	76	PhCl		Red
JP(OPh-Me$_2$-2,6)$_3$	J	I[891]	90	PhCl		Black

Table 16. Salts with P-N Bonds

Cation	Anion X	Method	m.	Solvent	Yield (%)	Other Data and Remarks
$(NH_2)_3PPh$	Cl	I[332]	164	Me_2CO/EtOH/Et_2O	19	
$(NH_2)_3PSMe$ Me	J	I[918]	161 dec.			
$(NH_2)_2P$ Me, Me	Cl	I[989]	192	EtOH/Me_2CO	60	
Et, Et	Cl	I[989]	106	EtOH	19	
Bu, Bu	Cl	I[989]	114	$CHCl_3$	25	
Ph, Ph	Cl	I[332]	207	EtOH/Me_2CO		
H_2NP NEt_2, NEt_2	Cl	I[422]	148	PhH	43	
Me, Dodec	Cl	I[852]	75	CH_2Cl_2/PhH	100	IR
Et, $(CH_2)_2CN$	Cl	I[43]	77			
Et, $(CH_2)_2CN$	$OPh(NO_2)_3$-2,4,6	XV[987]	154			
Et, $(CH_2)_2CN$	$PtCl_6$	XV[987]	210 dec.			
$CH_2CH{=}CH_2$, $CH_2CH{=}CH_2$	O_3S–[anthraquinone]–SO_3	XV[987]	181			
Bu, Bu	$PtCl_6$	XV[987]	192			
Bu, Bu	O_3S	XV[992]	89			
Bu, Bu	$OPh(NO_2)_3$-2,4,6	XV[43]	71			
Bu, Bu	Cl	I[992]	62			
Bu, Bu	$PtCl_6$	XV[43]	140			

427

Table 16 (Continued)

Cation				Anion X	Method	m.	Solvent	Yield (%)	Other Data and Remarks
H₂NP	Bu	Bu	Bu	PF_6	XV[992]	72			
H₂NP	CH₂Ph	CH₂Ph	CH₂Ph	$OPh(NO_2)_3$-2,4,6	XV[987]	180			
				Cl	I[987]	220			
				$PtCl_6$	XV[987]	190			
H₂NP	Ph	Ph	Ph	$Fe^{III}(NO)(CN)_5$	XV[43]	193			
				$Fe^{III}(CN)_6$	XV[43]	125			
				$OPh(NO_2)_3$-2,4,6	XV[43]	120			
				O_3S- (anthraquinone-2-sulfonate)	XV[43]	214			
				Cl	I[992]	230			
				ClO_4	XV[992]	172			
				$PtCl_6$	XV[992]	190			
				PF_6	XV[992]	185			
				JO_4	XV[992]	163			
				Br	I[174]	247	CHCl₃/Et₂O	160	IR, HMR
	NEt₂	NEt₂	NEt₂	Cl	I[422]	85	Petroleum ether	59	IR, HMR, PMR -63[673a]
	NPr₂	NPr₂	NPr₂	PF_6	XV[422]	219	H₂O		IR
				Cl	I[422]	153	Petroleum ether	71	IR
	NBu₂	NBu₂	NBu₂	Cl	I[422]	159	Hex	72	
	NEt₂	Ph	NEt₂	Cl	I[422]	175	Et₂O	48	
	NPr₂	Ph	NPr₂	Cl	I[422]	162	Et₂O	79	
	NBu₂	Ph	NBu₂	Cl	I[422]	82	Et₂O	79	

NMe_2	Me	Me	Cl	I[173]	190	$CHCl_3$/Hex	61	IR, HMR / PMR -39[673a]
NH-t-Bu	Ph	Ph	Cl	I[422]	180	PhH	60	
			PF_6	XV[422]	170	H_2O		PMR -54[673a]
NHEt	Bu	Bu	Cl	I[422]	201	PhH	64	
NH-i-Pr	Ph	Ph	Cl	I[422]	161	Me_2CO/Et_2O		
			PF_6	XV[422]				
NEt_2	Ph	Ph	Cl	I[422]	175	PhH	54	
			PF_6	XV[422]	169	H_2O		
NPr_2	Ph	Ph	Cl	I[422]	162	Et_2O	67	
			PF_6	XV[422]	129	H_2O		
$N=PEt_2$ / NH_2	Et	Et	Cl	I[989]	58	PhH	75	IR, HMR
$N=PPh_2$ / NH_2	Ph	Ph	O_3S (anthraquinone-9,10-dione structure)	XV[988]	218			
			$OPh(NO_2)_3$-2,4,6					
			Br	XV[988,85]	155[85] / 224	MeOH	65[444]	IR,[988] HMR,[988] PMR -20.3 ppm[988]
			Cl	I[85]	245[85]	MeOH[85]		
$NHNMe_2$	$NHNMe_2$	Ph	PF_6	[988]	153			
$N-NMe_2$ / Me	Ph	Ph	Cl	I[996]	196	$CHCl_3$		
			Cl	I[991]	195[991]	Et_2O[991]	90	
			Cl	I[766]	174	Et_2O	86[776]	
$N-NMe_2$ / Et	Ph	Ph	Cl	I[766]	196	EtOH/Et_2O	85	IR

Table 16 (Continued)

Cation			Anion X	Method	m.	Solvent	Yield (%)	Other Data and Remarks
H₂NP NNMe₂ $\overset{\mid}{\text{H}_2\text{N}=\text{PPh}_2}$	Ph	Ph	Cl	I[766]	204	CHCl₃/petroleum ether		IR
(MeNH)₄P	Me	Ph	Cl	I[880]	239[266]	MeOH[266]	17[880]	
MeNHP	Ph	Ph	J	I[299]	108	Et₂O		
				ΞV[992]	118			
			OPh(NO₂)₃-2,4,6 BF₄	X=V[1114]	125[1114]	CHCl₃/EtOAc[1114]	88[1114]	IR[985]
			Br	≡⁻[1112]	197[1112]	CHCl₃/EtOAc[1112]	77[1112]	IR,[985] HMR[579]
								IR[985]
			Cl	X=V[1114]	216[1114]			
			ClO₄	ΞV[992]	175			
			PtCl₆	ΞV[992]	185			
			PF₆	ΞV[992]	185			
			J	XIV[1112]	184[1112]	CHCl₃/EtOAc[1112]		IR[985]
H₂NCNHP $\overset{\text{O}}{\underset{\parallel}{}}$	Ph	Ph	Cl	I[1061]	194	EtOH		
(EtNH)₄P			Cl	I[880]	170[266]	MeOH[266]	40[880]	
			J	₂[266]	109			
EtNHP	Ph	Ph	BF₄	XIV[1114]	129[1114]	CHCl₃/EtOAc[1114]	92[1114]	IR[985]
			Br	I[1112]	246[1112]	CHCl₃/EtOAc[1112]	78[1112]	IR[985]
			Cl	XI/[1108]	268[1108]			IR[985]
			J	XI/[1112]	170[1112]	CHCl₃/EtOAc[1112]		IR[985]

Compound				Anion	Method / mp	Solvent	Yield (%)	Ref
(CH₂=CHCHNH)₄P				Cl	XXII[880] 133		24	
CH₂=CHCHNHP	NH-t-Bu	NH-t-Bu	NH-t-Bu	Cl	XVI[880] >300		50	
(PrNH)₄P				Cl	122	MeOH		
				J	120			
PrNHP	Ph	Ph	Ph	Br	XXII[1112] 190	CHCl₃/EtOAc	81	
				Cl	XIV[1108] 190	CHCl₃/EtOAc		
				J	XIV[1112] 154	CHCl₃/EtOAc		
(i-PrNH)₄P				Cl	XXII[880] 178		43	
i-PrNHP	Ph	Ph		BF₄	XIV[1114] 151[1114]	CHCl₃/EtOAc[1114]	86[1114]	IR[985]
				Br	XXII[1112] 242[1112]	CHCl₃/EtOAc[1112]	70[1112]	IR[985]
				Cl	XIV[1108] 230[1108]	CHCl₃/EtOAc		IR[985]
				J	XIV[1112] 224[1112]			IR[985]
(BuNH)₄P				Cl	XXII[880] 97[880]	MeOH[266]	23[880]	
				J	117			
BuNHP	Ph	Ph	Ph	Br	439 141	PhH	65	
	NH-t-Bu	NH-t-Bu	NH-t-Bu	Cl	880 278		67	
(i-BuNH)₄P				Cl	266 150	MeOH		
				J	266 135			
i-BuNHP	Ph	Ph		Br	XXII[1112] 136	CHCl₃/EtOAc	83	
				J	XIV[1112] 154	CHCl₃/EtOAc		
(t-BuNH)₄P				Cl	XXII[880] 300		25	
				J	XV[880] 300		86	
(t-BuNH)₂P	Me	Ph		J	I[1000] 176[1000]	i-Pr₂O[1000]	88[1000]	
	Bu	Ph		J	I[1064] 226			
	CH₂Ph-Cl₂-2,4	Ph		Cl	I[1064] 215		48	

Table 16 (Continued)

Cation		Anion X	Method	m.	Solvent	Yield (%)	Other Data and Remarks
(t-BuNH)₂P CH₂Ph-Cl₂-2,5	Ph	Cl	I[1064]	244			
t-BuNHP CH₂Ph	Ph	Cl	I[1000]	232 dec.		69	
Me	Ph	J	I[993]	199			
(pyrimidine ring: N, N, Cl, Cl)	Ph	Cl	I[995]	45 dec.			
CH₂Ph-Cl₂-2,4	Ph	Cl	I[995]	271 dec.[995]	EtOH/Et₂O[994]	50[994]	
CH₂Ph-F-4	Ph	Cl	I[995]	227[995]	DMF/PhH/Et₂O[1001]	41[1001]	
CH₂Ph	Ph	BF₄	XV[1001]	167 dec.[995]		100[1001]	IR[1001]
		OPh(NO₂)₃-2,4,6	XV[995]	104 dec.			
		Cl	I[995]	247 dec.			
		PF₆	XV[1001]	139	EtOH/Et₂O	48	
		Cl	I[994]	247 dec.	EtOH/Et₂O	48	
Ph	Ph	BF₄	XIV[1114]	169[1114]	CHCl₃/EtOAc[1114]	75[1114]	IR[985]
		Br	XXII[1112]	223[1112]	CHCl₃/EtOAc[1112]	92[1112]	IR[985]
		Cl	XIV[1108]	202[1108]			IR[985]
		J	XIV[1112]	197[1112]	CHCl₃/EtOAc[1112]		HMR[579] IR[985]

Compound	R	R'	Group	Ref.	m.p.	Yield
(t-BuNH)$_3$P–N(ring)–Me			Cl	[880]	267 dec.	90
(t-BuNH)$_3$P–NHCCH$_2$CMe$_3$ (Me)			Cl	[880]	252	63
[pyridine]–NHP	Ph	Ph	J	[1101]	202	
(H–N)$_3$P	Me	Ph	J	I[211]	251	
	Et	Ph	J	I[211]	178	
	i-Bu	Ph	J	I[211]	172	
	Me	Ph	Br	I[212]	97	
			Cl	[212]	130	
			PtCl$_6$	[212]	178	
			J	[212]	167	
(H–N)$_2$P	Me	Ph–Me–4	J	I[203]	186	
	Et	Ph	J	I[212]	174	
	Pr	Ph–Me–4	J	I[203]	191	
	i-Bu	Ph–Me–4	J	I[203]	197	
	CH$_2$Ph	Ph–Me–4	J	I[203]	204	
		Ph	PtCl$_6$	I[212]	204	
		Ph–Me–4	J	I[203]	125	
(PhNH)$_4$P			O$_2$CMe	XXII[181]	206	
			O$_2$CEt	[181]	240	
			Cl	[150]	275	
			SO$_4$	[150]	312	
(PhNH)$_3$P		Ph	OH	XXII[209]	216	
			Br	[209]	235	
			Cl	[209]	250 dec.	
			J	[209]	165	
			NO$_3$	[209]	160	
		Ph–Me–4	OH	XXII[209]	240	
			Br	[209]	238	

Table 16 (Continued)

Cation	Anion X	Method	m.	Solvent	Yield (%)	Other Data and Remarks
(PhNH)₃P Ph-Me-4	Cl	209	245			
	J	209	235			
	NO₃	209	180			
Ph-Me₃-2,4,5	OH	XXII[182]	204			
	Br	182	259			
	Cl	182	247			
	J	182	220			
	NO₃	182	224			
PhNHP Ph Ph Ph	Br	I[466]	218 dec.[496]	PhCl[496]	80[496]	
	Cl	XIV[789]				PMR −34[673a]
	ClO₄	XV[789]	188	CHCl₃/EtOAc		
(c-HexNH)₄P	J	1101	194			
	Cl	XXII[880]	264			
(PhCH₂NH)₄P	Cl	XXII[266]	206	MeOH	44	
	J	266	137			
(2-Me-PhNH)₄P	O₂CMe	XXII[181]	221			
	OEt	161	114			
	O₂ClEt	181	203			
	Cl	177	254 dec.			
	NO₃	177	250			
3-Me-PhNHP Ph Ph	Br	XXII[496]	168	PhH	60	
(2,4-Me₂-PhNH)₄P	OMe	XXII[161]	98			
	O₂CMe	181	210			
	OEt	161	107			

R	R′	R″	X	Ref.	M.p.	Solvent	Yield	Method
$(2,6\text{-}Me_2\text{-}PhNH)P$	Ph	Ph	Cl	[177]	264 dec.	H_2O	50	
			NO_3	[177]	246			
			Br	I[466]	220		73	
$(Me_3CCH_2C(O)NH)_4P$			C=O	XV[830]	113			
			CO_2H	XV[880]	159		33	
			O_2CMe	XXII[880]	252		63	
			Cl	XV[880]	156		52	
			NO_3	XV[880]	125		75	
			SO_4	XXII[880]	67		100	
$(OctNH)_4P$			Cl	XXII[202]	148			
			$PtCl_6$	[202]	230			
			J	[202]	188			
$\left(\text{⟨N-⟩}\right)_3P$	Me		J	XXII[202]	136			
$\left(\text{⟨N-⟩}\right)_2P$	Me	Ph	SCN	[1101]	164			
(1-naphthyl-NH)P	Ph	Ph	J	[1101]	230			
			SeCN	[1101]	136			
(2-naphthyl-NH)P	Ph	Ph	Cl	XXII[880]	75		25	
$Ph_2CHC(O)NHP$	Ph	Ph	Cl	XIV[790]	185	MeCN		IR, HMR
$(OctadecNH)_4P$	Ph	Ph	Cl	XXII[880]	91		5	
$Ph_2CHC=NP$, $NHPh\text{-}Cl_2\text{-}3,4$	Ph	Ph	Cl	XIV[789]	210	MeCN/PrCN	62	HMR

Table 16 (Continued)

Cation			Anion X	Method	m.	Solvent	Yield (%)	Other Data and Remarks
$Ph_2CHC=NP$–$NHPh-NO_2-4$	Ph	Ph	Cl	XIV[789]	213	MeCN/EtOAc		
$Ph_2CHC=NP$–$NHPh$	Ph	Ph	Cl	XIV[789]	230	$CHCl_3$/EtOAc	78	HMR
$Ph_2CHC=NP$–$NHPh-NMe_2-4$	Ph	Ph	Cl	XIV[789]	231	i-PrOH/EtOAc	69	HMR
$(Me_2N)_3P$	CCl_3		$OPh(NO_2)_3-2,4,6$	I + XV[685]	285	H_2O	61	
			ClO_4	XV[874]	290 dec.			
	$CHCl_2$		$OPh(NO_2)_3-2,4,6$	XV[874]	220	H_2O	85	
			ClO_4	XV[874]	285	H_2O	61	
	$CHClF$		Cl	[1054]				PMR -44 ppm E1/2 1.642, 2.098
			Br	[482]				
	Me		J	I[195]	>360[768]	Et_2O[195]	100[770]	HMR,[195] PMR -58.6[673a]
	Pr		Br	[482]				E1/2 1.653, 2.111
	Bu		Br	[1054]				PMR -62 ppm
	$Ph-OH-2-Br-3-Cl-5$		ClO_4	XV[875]	160 dec.			Red-brown
	$Ph-OH-2$		ClO_4	XV[875]	128 dec.			
	$CH_2Ph-Cl_2-2,4$		Cl	I[288]	203	$CHCl_3$/dioxane		
	$CH_2Ph-Cl-2$		Cl	I[287]	233	Dioxane	74	
	$CH_2Ph-Cl-4$		Cl	I[287]	211	Dioxane		PMR -55.5 ppm

Structure	Anion	Ref.	mp	Solvent	Yield	Color / Spectra
(1-Me-naphthol, OH)	Cl	I[287]	209	CHCl$_3$/dioxane		
	ClO$_4$	XV[875]	162 dec.			Orange-yellow
CH$_2$CH=CCH=CHCH=CHCH=CMe; Dec; CO$_2$Et; Me	Br	I[289]	72			
	Br	I[941]	174	Me$_2$CO/EtOAc	71	
(9-anthrol, OH, Me)	ClO$_4$	XV[875]	202 dec.			Orange-yellow
—CHCPh / —CH$_2$CPh, O= ‖O	Cl·HCl	XIV[836]	129	CHCl$_3$/Et$_2$O		IR, HMR
Hexadec	Br	I[289]	72	Dioxane/Et$_2$O		
Octadec; CPh$_3$; C=COCH$_2$Ph	Br	I[254]	178		13	
	BF$_4$	I[250]	166	MeOH	100	
H$_2$C—Ph; O=CPh	Br	XIV[836]	143	CH$_2$Cl$_2$/Et$_2$O	93	IR, HMR
NHCHEt$_2$	Cl	[266]	80			
N=NPh-Cl$_4$-2,3,4,5-OH-6	ClO$_4$	XV[875]	160			Red
NHN⋯S, =O	ClO$_4$	XV[875]	196 dec.			Yellow

437

Table 16 (Continued)

Cation		Anion X	Method	m.	Solvent	Yield (%)	Other Data and Remarks
$(Me_2N)_3P$	NHN=CCPh(=O)Ph	ClO_4	ΞV[875]	150 dec.			Yellow
	OCH=CCl₂	Cl	I + XVI[673]				HMR, PMR −34.8 ppm HMR[195]
$(Me_2N)_2P$	Me, Me	J	—[647]	270[647]	EtOH/Et₂O[647]		
	Me, Ph	HgJ₃	ΞV[647]	295 dec.	MeOH		Greenish
	Et, (CH₂)₂OH	HgJ₃	—[300]	132			
	CH₂Ph, Ph	HSO₄	I—I[368]				
		Br	—[512]	179	THF	65	
		Cl	—[512]	180		40	IR
Me/(N)₂P/Ph (cyclic)	Et, Et	J	[364]	192			
Me₂NP	Me, Me	J	—[136]	315 dec.	EtOH/Et₂O		
	Bu, Bu	Br	—[174]	131		100	HMR
	Ph, Ph	Br	—[174]	145	Hex	100	
	Ph, Ph	J	X—V[1112]	185	CHCl₃/EtOAc	98	IR, HMR
Me\Et NP	Ph, Ph	J	X—V[1112]	177	CHCl₃/EtOAc	82	
Me\Pr NP	Ph, Ph	J	X—V[1112]	205	CHCl₃/EtOAc	79	

R₁	R₂	R₃	X	Method	mp (°C)	Solvent	Yield (%)	Characterization
Me\i-Pr—NP	Ph	Ph	J	XIV[1112]	211	CHCl₃/EtOAc	80	
Me\i-Bu—NP	Ph	Ph	J	XIV[1112]	206	CHCl₃/EtOAc	77	
Me\t-Bu—NP	Ph	Ph	J	XIV[1112]	214[1112]	CHCl₃/EtOAc[1112]	94[1112]	HR[579]
Me\Ph—NP	Ph	Ph	Br	XIV[1075]	236 dec.	H₂O	85	
(Et₂N)₃P			J	XIV[1075]	240	PhH	80	
Ph-OH-2			Cl	XVI[234]	184	CH₂Cl₂/THF	73	HMR
Ph-OH-4			PF₆	XV[234]	157	H₂O	96	HMR
CH₂Ph-Cl₂-3,4			Cl	XIV[234]	155	THF/Et₂O		HMR
			Cl	I[287]	145	Dioxane		
Ph-OMe-2-OH-5			PF₆	XV[234]	194	EtOH		IR, HMR
Ph-OMe-4			J	XIV[234]	116	THF/diglyme	95	HMR
CH₂Ph-OMe-2-Cl-5			PF₆	XV[234]	195	EtOH	18	HMR
CH₂Ph-NO₂-3-OMe-4			Cl	I[707]	165		24	
CH₂Ph-OMe-4			Cl	I[707]	149		40	$n_D^{20} = 1.5430,\ d_4^0 = 1.1204$
			Cl	I[707]				
Dec			Cl	I[254]	217		16	
CPh₃			BF₄	I[250]	101 dec.	MeOH	33	
Octadec			Br	I[254]	224	Et₂O	87	HMR, PMR −34.5 ppm
OPh			Cl	II[234]				
OPh-OMe-4			PF₆	XV[234]	88	EtOH	90	HMR
			Cl	II[234]				HMR

Table 16. (Continued)

Cation			Anion X	Method	m.	Solvent	Yield (%)	Other Data and Remarks
$(Et_2N)_3P$		OPh-OMe-4	PF_6	XV[234]	88	EtOH		HMR
$(Et_2N)_2P$	Me	Ph	HgJ_3	XV[300]	90	EtOH		Greenish
			J	I[300]			100	Λ_m, viscous
CPh_3		OEt	BF_4	I[250]	145 dec.	CH_2Cl_2/Et_2O	100	Λ_c
Et_2N–P–Pr_2N	Me	Ph	HgJ_3	XV[300]	82	MeOH/Et_2O		Greenish
Et_2NP	Me	Ph	J	I[300]	107			
			HgJ_3	XV[300]	116			Greenish
	Et	c-Hex	HgJ_3	XV[943]	76		82	
	Et	c-Hex	J	I[943]	132	EtOH/Me_2CO	65	
	c-Hex	Ph	J	I[539]	83 dec.	MeOH/Et_2O	74	
	Ph	Ph	J	XIV[1112]	174	$CHCl_3$/EtOAc	71	
CPh_3	OEt	OEt	BF_4	I[250]	152 dec.	Et_2O	98	Λ_c
Et–NP–i-Pr	Ph	Ph	J	XIV[1112]	202	$CHCl_3$/EtOAc	60	
Et–NP–i-Bu	Ph	Ph	J	XIV[1112]	184	$CHCl_3$/EtOAc	68	
Et–NP–t-Bu	Ph	Ph	J	XIV[1112]	220	$CHCl_3$/EtOAc	84	

Compound	R	X	Type[ref]	m.p. (°C)	Solvent	Yield %	Remarks
Et(Ph)N–P, Ph	Ph, Ph, Ph	J	XIV[1110]	220		30	n_D^{20} = 1.5109, d_4^{20} = 1.0639; IR
$(Pr_2N)_3P$	CH_2Ph-OMe-4	Cl	I[707]		MeOH		
$(Pr_2N)_2P$	Me, Ph	HgJ_3	XV[300]	59			
		J	I[300]	111			
(morpholino)$_3$P	$CHCl_2$	$OPh(NO_2)_3$-2,4,6	XV[874]	208	$EtOH/H_2O$	45	Brown-red
	Me	Ph_4B	XV[1086]	239	$Me_2CO/EtOH$		Brown
		J	I[1086]	328 dec.	$EtOH/H_2O$		Yellow
	CH_2OH	Ph_4B	XV[1086]	272	MeOH		
		J	XIV[1086]	268	MeOH/EtOH		
	N=NPh-OH-2-Cl_2-3,5	ClO_4	XV[875]	158 dec.			
	N=NPh-OH-2	ClO_4	XV[875]	167 dec.			
	NHN=CHPh	ClO_4	XV[875]	234 dec.			
	NHN=CCPh (C=O)	ClO_4	XV[875]	217 dec.			
(morpholino)$_2$P	CH_2CO_2Et, Ph	Br	I[512]	115		85	IR; $E_{1/2}$ 1.627, 2.206
	CH_2Ph, CH_2Ph	Br	[482]	85			
	CH_2CPh-NO_2-4 (C=O), Ph	Br	I[512]	100	Et_2O		IR, unstable
(piperidino)$_3$P	$CHCl_2$	$OPh(NO_2)_3$-2,4,6	XV[874]	149	$EtOH/H_2O$	85	
	Me	J	I[137]	242			
(piperidino)$_2$P	CH_2CO_2Et, Ph	Br	I[512]	113		60	IR

Table 16 (Continued)

Cation			Anion X	Method	m.	Solvent	Yield (%)	Other Data and Remarks
pyridinium (=N)$_2$P, N–Et	OEt	OEt	BF$_4$	XVI[566]	159			
pyridinium =N–P Me, N–Et	OEt	OEt	BF$_4$	XV[566]	85			
			J	I[566]	87			
indolyl N–P	Ph	Ph	Br	[1067]				IR
bicyclic (CH$_2$–CH$_2$)$_2$ N–P Ph		CH$_2$Ph	AuCl$_4$	XV[1000]	166		88	Yellow
			Cl	I[1000]	273		44	
			J	I[1000]	236 dec.		43	
bicyclic (CH$_2$–CH$_2$)$_2$ N–P Me	Ph	Ph	AuCl$_4$	XV[1000]	166	Me$_2$CO	88	
			J	I[1000]	206 dec.	PhH/Et$_2$O	55	
Ph$_2$CHC=N–P, SMe	Ph	Ph	Cl	XIV[789]		Xylene	46	HMR
Ph$_3$P=N–P	Ph	Ph	Cl	[322]	272	H$_2$O		
			NO$_3$	XV[322]	232			
Ph$_3$P	HNNH$_2$		Br	XXII[1113]	215	CHCl$_3$/EtOAc	83	

Substituent / Structure	Phosphine	X	Method[ref]	mp (°C)	Solvent	Yield (%)	Notes
N–NH₂ (N–Me)		J	XIV[1113]	181 dec.		100	HMR
HNHNPh		Br	XIV[483]	225 dec.	EtOAc	22	
HNHNPh-OH-4		J	XIV[501]	186 dec.	EtOAc	100	Pale yellow
HNHN– (naphthalenol, OH)		Cl	XIV[501]	226	MeOH/Et₂O	65	
HNHNPh-SNHPh-3-OH-4 (O=S=O)		Cl	XIV[501]	175 dec.	MeOH/Et₂O	82	
		J	I[766]	156[766]	Et₂O[991]		IR[766]
HNNMe₂	Ph₂P–Me	Br	XXII[1113]	190 dec.	CHCl₃/EtOAc	90	HMR
HNN(Ph)(Me)	Ph₃P	Br	XXII[1113]	213	CHCl₃/EtOAc	87	HMR
HNNPh₂		Br	XXII[1113]	249	CHCl₃/EtOAc	99	Light yellow
–NNMe₂ (Me, Ph)	Ph₂P–Me	J	I[1113]	143 dec.[1113]	CHCl₃/EtOAc[1113]	98[1113]	IR[766] HMR[1113]
NN(Ph)(Me) (Me)	Ph₃P	J	XIV[1113]		CHCl₃/EtOAc		HMR
NNPh₂ (Me)		J	XIV[1113]	179 dec.		80	HMR, light yellow
–NNPh₂ (Et)		J	XIV[1113]	162 dec.		53	Light yellow

Table 16 (Continued)

Cation	Anion X	Method	m.	Solvent	Yield (%)	Other Data and Remarks
Ph_3P $-NN{=}CH_2$ | Me	Ph_4B	XV[71]	205 dec.	i-PrOH	99	
	ClO_4	XV[71]	166 dec.	i-PrOH	100	
	J	[71]	143 dec.	i-PrOH	43	
$NN{=}CH_2$ $-C{=}O$ Ph	SCN	XV[71]	248 dec.	H_2O	81	
	Cl	[71]	207 dec.	i-PrOH	93	
	ClO_3	XV[71]	192 dec.	CH_2Cl_2/EtOAc		
	CrO_4	XV[71]	dec.	H_2O	72	Dark yellow
	J	XV[71]	253 dec.	CH_2Cl_2/EtOAc	71	
	NO_2	XV[71]	90 dec.	H_2O	94	
	NO_3	XV[71]	206 dec.	H_2O	76	
$-NN{=}CH_2$ | CH_2Ph	J	[71]	170 dec.	i-PrOH	94	
$-HNN{=}CMe_2$	J	[71]	187 dec.	i-PrOH	74	
$-NN{=}CMe_2$ | $CH_2{-}CH{=}CH_2$	Br	[71]	165	i-PrOH	85	
	J	[71]	151 dec.	i-PrOH/Et_3N	80	
$-NN{=}CMe_2$ | CH_2Ph	J	[71]	203 dec.	i-PrOH	92	
$-HNN{=}CH{-}$ (2-furyl)	Br	XXII[986]	189 dec.	$CHCl_3$/EtOAc	74	Brownish
	J	XXII[986]	215 dec.	$CHCl_3$/EtOAc	93	HMR, brownish

R group	X	Ref.	mp (°C)	Solvent	Yield (%)	Notes
[furan]–NN=C(—Et)(H)	J	XIV[986]	219 dec.	CHCl₃/EtOAc	90	HMR, brownish
H–NN=CPh–Cl–4 (–Me)	J	XIV[986]	204 dec.	CHCl₃/EtOAc	76	HMR
H–HNN=CPh–NO₂–4	Br	XXII[986]	162	CHCl₃/EtOAc	88	Yellow
H–NN=CPh–NO₂–4 (–Me)	J	XIV[986]	223	CHCl₃/EtOAc	93	HMR, yellow
–NN=CPh–NO₂–4 (–Et)	J	XIV[986]	193	CHCl₃/EtOAc	90	Yellow
–HNN=CPh	Br	XXII[986]	225	CHCl₃/EtOAc	93	
–NN=CPh (–Me)	J		180 dec.	i-PrOH	97	
	J	XIV[986]	229 dec.	CHCl₃/EtOAc	100	HMR
–NN=CPh (–Et) (C=O)	J	XIV[986]	210 dec.	CHCl₃/EtOAc	93	HMR
–HNN=CHCPh–NO₂–4 (C=O)	Cl	XIV[71]	153	i-PrOH	77	IR, dark yellow
–HNN=CHCPh (C=O)	Cl	[71]	140	i-PrOH	91	IR, light yellow

Table 16 (Continued)

Cation	Anion X	Method	m.	Solvent	Yield (%)	Other Data and Remarks
c-Hex₃P	$-NN=CHCPh-NO_2-4$ (C=O), Me	J	71	234 dec.	i-PrOH/MeOH	76	Orange-yellow
	$-NN=CHCPh$ (C=O), Me	J	71	246 dec.	MeOH	77	Light yellow
Ph₃P	$-NN=CPh$, Me, Me Me	J	71	199 dec.	i-PrOH	90	
c-Hex₃P	$-NN=CCPh$ (C=O), Me Me	J	71	199 dec.	i-PrOH	99	Light yellow
Ph₃P	$HNN=CHCCH_2Ph$	Cl	71	154	i-PrOH	85	IR, gold yellow
	NHN, OMe₂, Me ring (C=O)	Cl	71	176	i-PrOH	76	IR
	NN–Me, OMe₂, Me ring (C=O)	J	71	209 dec.	EtOAc	82	

Structure	Halide	Method	mp (°C)	Solvent	Yield (%)	Remarks
Me, NHN=CMe₂ (bicyclic)	Br	XIV[986]	173	CHCl₃/EtOAc	99	
Me, NN=CMe₂, –Me	J	XIV[986]	199	CHCl₃/EtOAc	71	HMR
Me, NN=CMe₂, –Et	J	XIV[986]	192	CHCl₃/EtOAc	62	HMR
NN=CH₂, Me, Me Me	Br	XIV[986]	167	CHCl₃/EtOAc	74	
	J	XIV[986]	238	CHCl₃/EtOAc	93	HMR
Me, NN=CH₂, –Et, Me Me	J	XIV[986]	248 dec.	CHCl₃/EtOAc	88	HMR
NHN=CHC(=O)–naphthyl	Cl	71	147	i-PrOH	100	IR, light yellow
fluorenyl –HNN	Br	XXII[986]	226	CHCl₃/EtOAc	98	Yellow

447

Table 16 (Continued)

Cation		Anion X	Method	m.	Solvent	Yield (%)	Other Data and Remarks
Ph₃P	−HNN	Cl	[71]	dec.	i-PrOH	84	IR, light yellow
	NN–Me	J	XIV[986]	180 dec.	CHCl₃/EtOAc	85	HMR, yellow
	NN–Et	J	XIV[986]	209 dec.	CHCl₃/EtOAc	58	Yellow
	−HNN=CHPh₂	Br	XXII[986]	200	CHCl₃/EtOAc	94	
	−NN=CHPh₂ Me	J	XIV[986]	230 dec.[71]	i-PrOH/Et₃N[71]	88[71]	HMR[986]
	NN=CHPh₂ Et	J	XIV[986]	190 dec.	CHCl₃/EtOAc	85	

			mp	Solvent	Yield	Color
Me₃P	N_3					
	Cl	XXI^{286}	160 dec.	EtOH/Et₂O		Yellow
	$SbCl_5N_3$	XV^{919}	80 dec.		68	
	$SbCl_6$	$\overline{X}V^{919}$	251 dec.	PhNO₂/CCl₄	86	Yellow

Table 17. Salts with P-O Bonds

Cation				Anion X	Method	m.	Solvent	Yield (%)	Other Data and Remarks
HOP	Ph	Ph	Ph	HSO₄·Ph₃PO	[53]	177			
(MeO)₄P				SbCl₆	I[178]	139 dec.	CH₂Cl₂/Et₂O	38	
(MeO)₃P	Me			BF₄	I[761]		CH₂Cl₂		HMR
	O=								
	CH₂CMe			ClO₄	I[166]				
	CPh₃			BF₄	I[250]	104	CH₂Cl₂/Et₂O	97	IR
	SMe			SbCl₆	I[1027]	127 dec.	CH₂Cl₂/MeNO₂/Et₂O	100	
(MeO)₂P	SEt			SbCl₆	I[451]	125 dec.		55	
MeOP	OPh	OPh		BF₄	I[761]				HMR
	OPh	OPh		BF₄	I[671]				HMR
(EtO)₄P				BF₄	(EtO)₃PO + Et₃O⁺BF₄⁻ [763]		CH₂Cl₂/Et₂O	72	
	Me			BF₄	XVI[760]	-25			
(EtO)₃P	Et			BF₄	I[250]	-14	CH₂Cl₂/Et₂O		IR
				F	EtP(OEt)₂ + Et₃O⁺BF⁻ ; I[598]	8, b$_{0.03}$ 60		100	n$_D^{20}$ = 1.4020, d$_{20}$ = 1.3293

Phosphorus	R	R'	X	Method	m.p./b.p. (°)	Yield (%)	Solvent	Notes
(EtO)₂P	Et	$-C(F)=C(CF_3)_2$	F	I[598]	10, $b_{0.02}$ 58			$n_D^{20} = 1.3730$, $d_{20} = 1.3341$
		$-CPh_3$	BF₄	I[250]	145	67	CH₂Cl₂/Et₂O	
		$-SMe$	SbCl₆	I[451]	120 dec.	63	CH₂Cl₂/Et₂O	
		$-SEt$	SbCl₆	I[451]	109	83	CH₂Cl₂/Et₂O	
EtOP	Me	Et	BF₄	Et₂POEt + Et₃O⁺BF₄⁻ [763]	3		CH₂Cl₂/Et₂O	$[\alpha]_D = 19$ Me₂CO
	t-Bu	Ph	SbCl₆					
EtOP	Et	Et	BF₄	Et₃PO + Et₃O⁺BF₄⁻; I[853]	118	85	CH₂Cl₂/Et₂O	
	Et	CH₂CO₂Et	J	I[1096]	33		Et₂O	
	i-Pr	CH₂CO₂Et	J	I[1096]	56		Et₂O	
	Bu	CH₂CO₂Et	J	I[951]	47			
EtOP	Ph	Ph	Br		123 dec.	62	EtOH/Et₂O	
(PrO)₃P	Me	(cyclopentadiene ring: Cl, Cl, Cl, Cl, CPh₃)	BF₄	XVI[760]	−36			
(BuO)₃P			Cl	I[130]	$b_{0.776}$			$n_D^{25} = 1.4765$
			BF₄	I	99	96	CH₂Cl₂/Et₂O	
(i-BuO)₃P	Me		BF₄, O₃SPh-Me-4	XVI[760], XV[235]	−40			Half-life 180'/20° CH₂Cl₂; Half-life 40'/20°
(Me₃CCH₂O)₄P			O₂Me	XV[235]				

Table 17 (Continued)

Cation	Anion X	Method	m.	Solvent	Yield (%)	Other Data and Remarks
$(Me_3CCH_2O)_4P$	Cl	I[235]				HMR, half-life 3'/20°
$(Me_3CCH_2O)_3P$ Me Me_3CCH_2OP Ph Ph	BF_4 Cl	XV[760] XXI[235]	124 62 dec.	Hex	85	IR Half-life 23h
(chloro-substituted phenyl)—OP Ph Ph	OH	XI[490]	122	PhH/petroleum ether/H_2O		
$(4\text{-Cl-PhO})_3P$ Me	J OPh	I[208] XXII[34]	71 46			Very hygroscopic, Half-life 0.11h
$(PhO)_4P$	J	[762]				
$(PhO)_3P$ Me	BF_4 J	I[1086] I[1086]	69 140[371]	Et_2O $MeNO_2$/Et_2O[1086]		HMR[761]
Ph	Cl	II[762]	60	PhCl	100	Quarter-life 1.07h
	J	XV[762] VI[814]	130[762]	CH_2Cl_2/Et_2O[762]	8[814]	
$(CH_2)_2Ph$	J	I[351]	154	Me_2CO/Et_2O		
CPh_3	BF_4	I[250]	121	CH_2Cl_2/Et_2O	100	

R¹O–	R²	OC=CH₂ / Ph	X	Products (ref.)	120 dec.	PhMe	95	Notes
$(PhO)_2P$	Ph	Ph	Br	I[513]	120 dec.			Quarter-life 2.4ʰ
	Ph	Ph		J[762]	198	PhMe	95	
2,6-Cl_2PhOP	Ph	Ph	OH	Ph_3P + $ArOH$[145]	97			
2-Cl-PhOP	Ph	Ph	OH	Ph_3P + $ArOH$[145]	105			
4-Cl-PhOP	Ph	Ph	OH	Ph_3P + $ArOH$[145]	80			
4-O_2N-PhOP	Ph	Ph	OH	Ph_3P + $ArOH$[145]	107			
(PhO)P	Me	CH_2Ph	Cl	I[183]	193			
PhOP	Me	Ph	J	I[210]	134			
PhOP	Ph	Ph	OH / J	[145]	106			
c-HexOP	CH_2Ph	Ph	Cl	I[66]	267			No decomposition in 60 hr. IR
4-OHC-PhOP	Ph	Ph	OH	Ph_3P + $ArOH$[145]	66			IR
$(PhCH_2O)_3P$	Me	Ph	BF_4	XVI[760]	89			
2-Me-PhOP	Ph	Ph	OH	Ph_3P + $ArOH$[145]	89			
3-Me-PhOP	Ph	Ph	OH	Ph_3P + $ArOH$[145]	67			
4-Me-PhOP	Ph	Ph	OH	Ph_3P + $ArOH$[145]	44			
⬡–CH_2–OP (cyclohexylmethyl)	Ph	Ph	Cl	I[233]	193	Me_2CO/Hex	80	
endo- ⬡–CH_2–OP	Ph	Ph	Br	Ph_3P + ROH + Br_2[900]	82	$CHCl_3$/Et_2O	100	

Table 17 (Continued)

Cation			Anion X	Method	m.	Solvent	Yield (%)	Other Data and Remarks
endo- ⬡–CH₂–OP⟨Ph Ph	Ph	Ph	Br	Ph₃P + ROH + Br₂[900]	161		87	IR, UV, HMR
Cl₂C=COP–Ph (Ph)	Ph	Ph	Cl	I[790]		PhH	78	IR, hygroscopic
[Ph(CH₂)₂O]₃P	Et		BF₄	I[250]	35	CH₂Cl₂/CCl₄		
4-EtO₂C-PhOP	Ph	Ph	OH	Ph₃P + ArOH[145]	104			
4-t- ⟨Bu–⬡–H⟩–OP	CH₂Ph	Ph	Cl	I[66]	268 dec.	H₂O		IR
ClC=COP–PhPh	Ph	Ph	Cl	I[790]		PhH	62	IR, hygroscopic

Table 18. Salts with P-O, P-S, P-Sb, and P-Si Bonds

Cation				Anion X	Method	m.	Solvent	Yield (%)	Other Data and Remarks
$Ph_2C=COP$(–Ph)	Ph	Ph	Ph	Cl	I[790]		Xylene	68	IR, HMR, PMR –63 ppm, hygroscopic Half-life 10'
Ph_2CCHOP(–Ph), $OPPh_3$	Ph	Ph		Cl	Ph_3P + ROH + Cl_2[235]	50 dec.			
(tetrachloro-1,4-phenylene-dioxy, $OPPh_3$ / $OPPh_3$, Cl substituents)				(tetrachloro aromatic dioxy structure, Cl, O)	XI[829]				UV
(chloro-phenylene-dioxy, $OPPh_3$, Cl)				(dichloro aromatic dioxy structure, O, J)	XI[829]				UV
$(CH_3S)_4P$				J	I[35]				Vapor pressure[19] 20mm Hg
Ph_2SbP	Me	Ph		Cl	I[175]	138	EtOH		
$(Me_3Si)_3P$	Me			J	I[952]	175 dec.	EtOH/ Et_2O		

Table 19. Salts with Phosphorus in the Ring with P-Cl Bonds

Cation	Anion X	Method	m.	Solvent	Yield (%)	Other Data and Remarks
Cl₂P⟨(O)(O) (benzo ring)	Cl	XXI[34]			61	
Cl₂P⟨(O)(O) (methyl benzo ring, Me)	Cl	XX[120]	$b_{11}158$			

Table 20. Salts with P-N and P-O Bonds with Phosphorus in the Ring

Cation	Anion X	Method	m.	Solvent	Yield (%)	Other Data and Remarks
(cyclopentane) H₂NP-Ph	OPh(NO₂)₃-2,4,6	XV[992]	171			
	O₃S (anthraquinone)	XV[992]	207			
	PF₆	XV[992]	76	EtOH		
(cyclohexane) H₂NP-Ph	OPh(NO₂)₃-2,4,6	XV[992]	150			
	O₃S (anthraquinone)	XV[992]	243 dec.			
	PF₆	XV[992]	120	EtOH		
(fluorene) Et₂NP-i-Me	OPh(NO₂)₃-2,4,6	XV[17]	115	EtOH/H₂O		
	J	I[17]				HMR

457

Table 20 (Continued)

Cation Cyclo Quasi P-O	Anion X	Method	m.	Solvent	Yield (%)	Other Data and Remarks
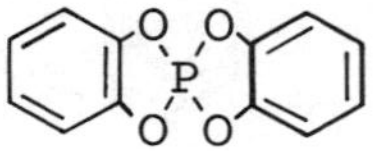 (catechol phosphorus structure)	OPh	XXII[34]	192			
	Cl	XXII[34]	b₁₀ 245			
(benzene 1,2-diolate structure)	(o-phenylene O,O)	XXII[34]	166 b₁₁ 194			
(bis-catechol P, OPh OPh structure)	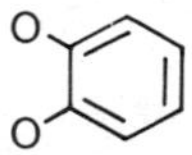 OPh	XXII[34]	200 dec.		95	
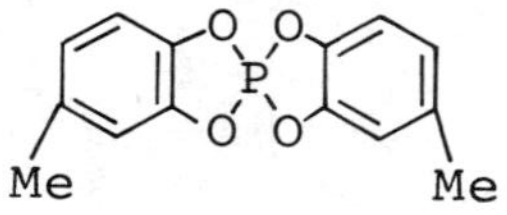 (dimethyl catechol P structure)	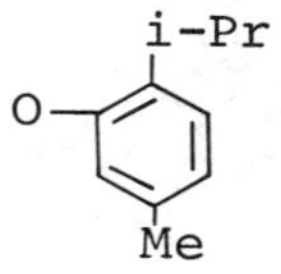 i-Pr, Me	XXII[120]			65	

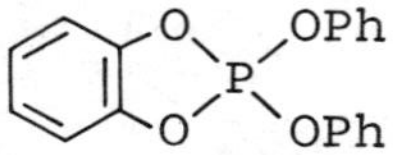

458

Table 21. Diquaternary Salts with P-N Bonds

Cation RPR'PR"	Anion X	Method	m.	Solvent	Yield (%)	Other Data and Remarks
$(t\text{-BuNH})_2\overset{\text{Ph}}{P}CH_2$—⟨$C_6H_4$⟩—$CH_2\overset{\text{Ph}}{P}(NH\text{-}t\text{-B})_2$	Cl	I[1000]	>335	EtOH	28	
$t\text{-Bu-N}\overset{\text{Ph}}{H}P\text{-}CH_2\text{-}\overset{\text{Ph}}{P}NH\text{-}t\text{-Bu}$ (Ph, Ph)	J	I[995]	252			
$-CH_2CH=CHCH_2-$	Br	I[995]	205 dec.[995]		54[1001]	
$-(CH_2)_2O(CH_2)_2-$	Br	I[995]	278 dec.[995]	EtOH/Et₂O[994]	50[994]	
	OPh(NO₂)₃-2,4,6	[1001]	164 dec.			
$-CH_2\overset{\text{Me}}{\underset{\text{Me}}{Si}}\text{-O-}\overset{\text{Me}}{\underset{\text{Me}}{Si}}CH_2-$	Br	I[995]	205 dec.			
$-CH_2-$⟨C_6H_4⟩$-CH_2-$	Cl	I[995]	269 dec.[995]	PhH[995]	80[994]	IR[994]
$-CH_2-$⟨C_6Me_4⟩$-CH_2-$ (Me Me / Me Me)	Cl·2 H₂O	I[994]	295 dec.		52	

Table 21 (Continued)

Cation RPR'PR''	Anion X	Method	m.	Solvent	Yield (%)	Other Data and Remarks
t-BuNHP–CH₂ (anthracene) –CH₂–PNH–t-Bu (Ph, Ph substituents)	OPh(NO₂)₃-2,4,6	[1001]	240 dec.			
	Cl	I[995]	279 dec.			
	PF₆	[1001]	259 dec.		85	
Me,Bu,Ph–NPCH₂–⟨C₆H₄⟩–CH₂P–N (Ph, Me, Ph, Bu substituents)	Cl	I[995]	322			
Ph₃P–NHNHPPh₃	Cl	XXII[125]	265 dec.			
(EtNH)₃P–N–P(NHEt)₃, Et / NHBu	J	XXII[266]	172		80	
(EtNH)₃P–N–P(NHEt)₂, Bu	Cl	XXII[266]	196			
(PrNH)₃P–N–P(NHPr)₃, Pr	Cl	XXII[266]	212			

$\begin{array}{c} \text{NHBu} \\ | \\ (\text{PrNH})_3\text{P}-\text{N}-\text{P}(\text{NHPr})_2 \\ | \\ \text{Bu} \end{array}$	Cl	XXII[266]	207
$\begin{array}{c} (\text{BuNH})_3\text{P}-\text{N}-\text{P}(\text{NHBu})_3 \\ | \\ \text{Bu} \end{array}$	Cl	XXII[266]	206
$\begin{array}{c} (\text{i-BuNH})_3\text{P}-\text{NP}(\text{NHi-Bu})_3 \\ | \\ \text{i-Bu} \end{array}$	J	XXII[266]	180
	Cl	XXII[266]	198
	J	XXII[266]	162

Table 22. Betaines with P-N, P-O, and P-P Bonds

Structure	Method	m.	Solvent	Yield (%)	Other Data and Remarks
(PhO)$_3$P$^{(+)}$–NC–CF$_3$ with $^{(-)}$OCl and F	359				IR, quadrupole resonance
(Me$_2$N)$_3$P$^{(+)}$–O–C with OEt, C=O; $^{(-)}$O–C–OEt, CO$_2$Et	I[841]	119[839]	Hex/PhH[841]	90[841]	IR, HMR, PMR –38.0 ppm[841]
–O–C–CO$_2$Et, $^{(-)}$O–C–CO$_2$Et, CO$_2$Et	I[839]				IR, HMR, PMR –22 ppm
(Me$_2$N)$_3$P$^{(+)}$ phenanthrene $^{(-)}$	I[841]	101	PhH/Hex		IR, HMR, PMR –38.6 ppm
–O–C–Ph, $^{(-)}$O–C–Ph	I[840]				IR, HMR, PMR –30.2 ppm

Compound	Preparation	M.p. (°C)	Solvent	Yield (%)	Remarks
O=C–Ph / –O–C / (–) O–C–Ph (–)	I[841]	119			IR, HMR, PMR –35.9 ppm
(Me$_2$N)$_2$P(+)–Ph, –O(–) ... OH	630				ESR
naphthol–O(–) ... OH (–)	630				ESR
(Et$_2$N)$_3$P– –Ph–O(–)–2 (+)	XV + XVI[234]	119	Petroleum ether	79	HMR
–PhO(–)–4	XV + XVI[234]	120		93	HMR
–O–C$_6$H$_3$(CH$_3$)–OH (–)	I[234]	222	THF/Me$_2$CO	95	ESR
(Et$_2$N)$_2$P–Ph (+), –O ... OH (–)	630				ESR
Ph$_3$P– –N=NPh–O(–)–4–Cl$_2$–3,5 (+)	XXIII[873]	183	CHCl$_3$/Et$_2$O	93	
–N=NPh–O(–)–2–(NO$_2$)$_2$–3,5	XXIII[872]	184 dec.	DMF/Et$_2$O		Orange

Table 22 (Continued)

Cation	Anion X	Method	m.	Solvent	Yield (%)	Other Data and Remarks
Ph_3P- (+)	$-N=N-Ph-CO_2H-3-O^{(-)}-4-NO_2-5$	XXIII[873]	205	MeOH		
	$-N=N-Ph-CO_2H-3-O^{(-)}-4$	XXIII[873]	173[873]	$CHCl_3/Et_2O$[873]		Red[872]
	$-N=N-Ph-Me-3-O^{(-)}-4-CO_2H-5$	XXIII[873]	180	$MeOH/Et_2O$		
		XXIII[872]	177	$CHCl_3/EtOAc$	84	Red
$Ph-CHNPh-3-O^{(-)}-4$		XXIII[873]	174	$CHCl_3/Et_2O$	80	
Ph_3P- (+)		XXIII[872]	202 dec.	$ClCH_2CH_2Cl/Et_2O$		Orange-red
		XXIII[873]	202	Dioxane	91	
		XXIII[872]	175	$CHCl_3/Et_2O$		Red-orange
		XXII[873]	164 dec.[872]	DMF/Et_2O[872]	76[873]	Red[872]

Structure	Method	m.p.	Solvent	Yield	Color
(structure: $O^{(-)}$, SO_2NPh/Me naphthalene, $-N=N-$)	XXIII[873]	146	MeOH/Et$_2$O	77	
$-N=N-Ph-SO_2HN-OPh-3-O^{(-)}-4$	XXIII[873]	187 dec.[872]	Et$_2$O/CHCl$_3$[872]	89[873]	Red-orange[872]
$-N=N-Ph-SO_2HN-Ph-3-O^{(-)}-4$	XXIII[873]	187	EtOAc	89	
(structure $-N=N$, polycyclic)	XXIII[873]	204	Dioxane	93	
(structure $-N=N$, polycyclic)	XXIII[873]	204 dec.	ClCH$_2$CH$_2$Cl		Red
(structure $O_2S-OPh-CMe_3-4$, $O^{(-)}$, $-N=N$)	XXIII[873]	130 dec.[873]	PhH/petroleum ether[872]	63[873]	Red-orange[872]

Table 22 (Continued)

Cation	Anion X	Method	m.	Solvent	Yield (%)	Other Data and Remarks
(+) Ph$_3$P− ![pentachlorophenoxide structure] −O Cl Cl Cl Cl (−) 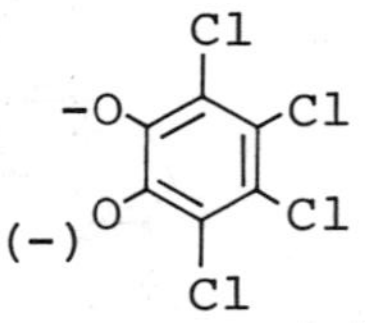		XXIII^{490}	217	CHCl$_3$/petroleum ether		
−O−Ph−O$^{(-)}$−2		XXIII^{490}	90 dec.		95	
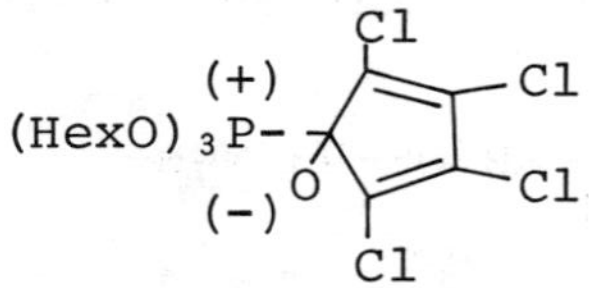 (+) (HexO)$_3$P− tetrachlorocyclopentadiene−O (−)		130				$n_D^{25} = 1.4700$
Bu$_3$P S‖−P−Ph (−)S		I^{320}	97			

466

S
‖
−P−PhOMe−4   I^{320}   87
(−)S
S
‖
−P−PhOEt−4   I^{320}   85
(−)S   O (+)
          ‖
(−) O−S−PPh₃   I^{798}   181
          ‖
          O

(received January 15, 1971)

467

REFERENCES

1. Abbiss, T., A. Soloway, and V. Mark, J. Med. Chem.,
    7, 763 (1964).
2. Acker, D., and D. Blomstrom (E.I. du Pont de Nemours
    & Co.), U.S. Pat. 3,162,641 (Dec. 22, 1964).
3. Agiuar, A., and H. Agiuar, J. Amer. Chem. Soc., 88,
    4090 (1966).
4. Agiuar, A., H. Agiuar, and T. Archibald, Tetrahedron
    Lett., 1966, 3197.
5. Agiuar, A., and J. Brisler, J. Org. Chem., 27, 1001
    (1962).
6. Akamsin, V., and N. Rizpolozhenski, Dokl. Akad. Nauk
    SSSR, 168, 807 (1966).
7. Akhtar, M., T. Richards, and B. Weedon, J. Chem. Soc.,
    1959, 933.
8. Aksnes, G., Acta Chim. Scand., 15, 692 (1961).
9. Aksnes, G., Acta Chim, Scand., 15, 438 (1961).
10. Aksnes, G., and L. Brudvik, Acta. Chim. Scand., 17,
    1616 (1963).
11. Aksnes, G., and J. Songstad, Acta Chim. Scand., 16,
    1426 (1962).
12. Aksnes, G., and J. Songstad, Acta. Chim. Scand., 18,
    655 (1964).
13. Albright & Wilson Ltd. Belg. Pat. 626,626 (April 16,
    1963).
14. Albright & Wilson Ltd. Belg. Pat. 1,264,384 (March 28,
    1968).
15. Allen, D., F. Mann, and J. Millar, Chem. Ind. 1966,
    196.
16. Allen, D., and J. Millar, J. Chem. Soc. B, 1969, 263.
17. Allen, D., and J. Millar, J. Chem. Soc. C, 1969, 252.
18. Allen, D., J. Millar, and J. Tebby, Tetrahedron Lett.,
    1968, 745.
19. Alt, G., and A. Speziale, J. Org. Chem., 30, 1407
    (1965).
20. American Cyanamid Co., Brit. Pat. 898,759 (June 14,
    1962).
21. Anderson, C., and R. Keeler, Anal. Chem., 26, 213
    (1954).
22. Ansell, M., and P. Thomas, J. Chem. Soc., 1961, 539.
23. Antosz, M., Roczniki Chem., 36, 979 (1962).
24. Antosz, M., Roczniki Chem., 39, 501 (1965).
25. Appel, R., and R. Schoellhorn, Angew. Chem., 76, 991
    (1964).
26. Appleyard, G., and C. Stiling, J. Chem. Soc. C, 1969,
    1904.
27. Arbuzov, A., J. Russ. Phys. Chem. Soc., 42, 549 (1910).
28. Arbuzov, A., G. Kamai, and C. Nesterov., Trudy Kazan.
    Khim. Tekhnol. Inst. Inh. S.M. Kirova, No. 16, 17
    (1951).

29. Arbuzov, B., N. Rizpolozhenski, and M. Zvereva, Izvest Akad. Nauk, SSSR Otdel Khim Nauk, 1958, 706.
30. Asahi Chemical Industry Co. Ltd., Japan Pat. 3917, 3918, 3919 (June 9, 1960).
31. Asher, W., Nature, 200, 912 (1963).
32. Atherton, N., J. Locke, and J. McCleverty, Chem. Ind., 1965, 1300.
33. Atkinson, J., M. Fischer, D. Horley, and A. Morse, Can. J. Chem., 43, 1614 (1965).
34. Auger, V., and M. Billy, Compt. Rend., 139, 597 (1904).
35. Aylett, B., H. Eméleus, A. Maddock, J. Inorg. Nuclear Chem., 1, 187 (1955).
36. Bailey, W., and S. Buckler, J. Am. Chem. Soc., 79, 3567 (1957).
37. Bailey, W., S. Buckler, and F. Marktscheffel, J. Org. Chem., 25, 1996 (1960).
38. Baird, W., J. Owen, and A. Parkinson (Imperial Chemical Industries Ltd.), Brit. Pat. 939,607 (Oct. 16, 1963).
39. Bajpai, L., C. Whewell, and J. Woodhouse, J. Soc. Dyers Colourists, 77, 193 (1961).
40. Balzer, W., Chem. Ber., 102, 3546 (1969).
41. Baranauckas, C., and J. Gordon, U.S. Pat. 3,248,429 (April 26, 1966).
42. Barber, M., J. Davis, L. Jackman, and B. Weedon, J. Chem. Soc., 1960, 2870.
43. Barrett, W., (W.R. Grace & Co.) Fr. Pat. 1,339,351 (Oct. 4, 1963).
44. Barton, D., Y. Chow, A. Cox, and G. Kirby, J. Chem. Soc., 1965, 3571.
45. BASF, Neth. Appl. 6,405,660 (Nov. 25, 1964).
46. BASF, Neth. Appl. 6,413,900 (May 31, 1965).
47. BASF, Brit. Pat. 812,522 (April 29, 1959).
48. BASF, Brit. Pat. 813,539 (May 21, 1959).
49. Baumgaertner, R., W. Sawodny, and J. Goubeau, Z. Anorg. Allgem. Chem., 333, 171 (1964).
49a. Bayer (Farbenfabrik), Neth. Appl. 6,401,031 (Aug. 10, 1964).
50. Bayer (Farbenfabrik), Neth. Appl. 6,600,826 (July 22, 1966).
51. Becher, R., W. Weyerts, and W. Salminen (Kodak Soc. Anon.), Belg. Pat. 610,707 (May 23, 1962).
52. Becke-Goering, M., and P. Hormuth, Z. Anorg. Allg. Chem., 369, 105 (1969).
53. Becke-Goering, M., and H. Thielemann, Z. Anorg. Allg. Chem., 308, 33 (1961).
54. Beeby, M., and F. Mann, J. Chem. Soc., 1951, 411.
55. Beg, M., and H. Clark, Can. J. Chem., 40, 283 (1962).
56. Behrends, K., Z. Anal. Chem., 226, 1 (1967).
57. Behrends, K., and H. Klein, Z. Anal. Chem., 499, 165 (1970).

58. Beninate, J., G. Drake, and W. Reeves, (United States Dept. of Agriculture), U.S. Pat. 3,268,360 (Aug. 23, 1966).
59. Benson, R. (E.I. du Pont de Nemours & Co.), U.S. Pat. 3,255,195 (June 7, 1966).
60. Berenson, H., A. Dornbush, and D. Wehner (American Cyanamid Co.,), U.S. Pat. 3,364,141 (Jan. 16, 1968).
61. Bergel'son, L., V. Solodovnik, and M. Shemyakin, Izv. Akad. Nauk. SSSR, Ser. Khim., 1966, 499.
62. Bergel'son, L., V. Vaver, L. Barsukov, and M. Shemyakin, Izv. Akad. Nauk SSSR Otd. Khim. Nauk, 1963, 1053.
63. Bergel'son, L., V. Vaver, L. Barsukov, and M. Shemyakin, Izv. Akad. Nauk SSSR Otd. Khim. Nauk, 1963, 1134.
64. Bergel'son, L., V. Vaver, V. Kowtun, L. Senyavina, and M. Shemyakin, Zh. Obshch. Khim., 32, 1802 (1962).
65. Bergmann, E., and J. Dusza, J. Org. Chem., 23, 1245 (1958).
66. Berlin, K., D. Hellwege, M. Nagabhushanam, and E. Gaudy, Tetrahedron, 22, 2191 (1966).
67. Bestmann, H., Chem. Ber., 95, 58 (1962).
68. Bestmann, H., and B. Arnason, Chem. Ber., 95, 1513 (1962).
69. Bestmann, H., R. Engler, and H. Hartung, Angew. Chem., Int. Ed. Engl., 5, 1040 (1966).
70. Bestmann, H., R. Engler, and H. Hartung, Ger. Pat. 1,287,570 (Jan. 23, 1969).
71. Bestmann, H., and L. Goethlich, Ann. 655, 1 (1962).
72. Bestmann, H., H. Haeberlein, and W. Eisele, Chem. Ber., 99, 28 (1966).
73. Bestmann, H., H. Haeberlein, and J. Pils, Tetrahedron, 20, 2079 (1964).
74. Bestmann, H., H. Haeberlein, H. Wagner, and O. Kratzer, Chem. Ber., 99, 2848 (1966).
75. Bestmann, H., and H. Hartung, Chem. Ber., 99, 1188 (1966).
76. Bestmann, H., G. Hoffman, Ann., 716, 98 (1968).
77. Bestmann, H., and O. Kratzer, Chem. Ber., 95, 1894 (1962).
78. Bestmann, H., and O. Kratzer, Chem. Ber., 96, 1899 (1963).
79. Bestmann, H., and K. Schnabel, Ann., 698, 106 (1966).
80. Bestmann, H., and H. Schulz, Tetrahedron Lett., 1960, 5.
81. Bestmann, H., and H. Schulz, Chem. Ber., 95, 2921 (1962).
82. Bestmann, H., and F. Seng, Tetrahedron, 21, 1373 (1965).
83. Bestmann, H., F. Seng, and H. Schulz, Chem. Ber., 96, 465 (1963).
84. Beveridge, A., G. Harris, and F. Inglis, J. Chem. Soc. A., 1966, 520.

85. Bezmann, I., and J. Smally (Armstrong Cork. Co.), U.S. Pat. 3,080,422 (March 5, 1963).
86. Bieber, T., and E. Eisman, J. Org. Chem., 27, 678 (1962).
87. Bigley, D., N. Rogers, and J. Barltrop, J. Chem. Soc., 1960, 4613.
88. Biilman, E., and K. Jensen, Bull. Soc. Chim., 3, 2306 (1936).
89. Birum, G., and J. Dever, Abstr. Am. Chem. Soc., 134th meeting, Chicago, 1958, p. 101P.
90. Birum, G., and C. Matthews, J. Am. Chem. Soc., 88, 4198 (1966).
91. Birum, G., and C. Matthews, Chem. Commun., 1967, 137.
92. Birum, G., and C. Matthews (Monsanto Co.), U.S. Pat. 3,390,165 (June 25, 1968); U.S. Pat. 3,390,166 (June 25, 1968).
93. Birum, G., and C. Matthews (Monsanto Co.), U.S. Pat. 3,445,570 (May 20, 1969).
94. Bissing, D., J. Org. Chem., 30, 1296 (1965).
95. Bladé-Font, A., C. VanderWerf, and W. McEwen, J. Am. Chem. Soc., 82, 2396 (1960).
96. Blandamer, M., T. Gough, M. Symons, Symposium Reprints, 1, 10 (1963).
97. Blicke, F., and S. Raines, J. Org. Chem., 29, 204 (1964).
98. Blomquist, A., and V. Hruby, J. Am. Chem. Soc., 86, 5051 (1964).
99. Bock, R., and J. Jainz, Z. Anal. Chem., 198, 315 (1963).
100. Boehme, H., and M. Haake, Chem. Ber., 100, 3609 (1967).
101. Boehringer, C.H. u. Sohn, Neth. Appl. 6,414,757 (June 21, 1965).
102. Bohlmann, F., C. Arndt, H. Bornowski, and P. Herbst, Chem. Ber., 93, 981 (1960).
102a. Bohlmann, F., D. Bohm, and C. Rybak, Chem. Ber., 98, 3087 (1965).
103. Bohlmann, F., H. Bornowski, and P. Herbst, Chem. Ber., 93, 1931 (1960).
104. Bohlmann, F., H. Bornowski, and P. Herbst, Chem. Ber., 94, 3189 (1961).
105. Bohlmann, F., and P. Herbst, Chem. Ber., 92, 1319 (1959).
106. Bohlmann, F., E. Inhoffen, and P. Herbst, Chem. Ber., 90, 1661 (1957).
107. Bohlmann, F., and H. Mannhardt, Chem. Ber., 89, 1307 (1956).
108. Bohlmann, F., and J. Politt, Chem. Ber., 90, 130 (1957).
109. Bohlmann, F., and J. Ruhnke, Chem. Ber., 93, 1945 (1960).

110. Bohlmann, F., and H. Viehe, Chem. Ber., **88**, 1245 (1955).
111. Bonnett, R., A. Spark, and B. Weedon, Acta. Chim. Scand., **18**, 1739 (1964).
112. Borisov, A., A. Abramova, and A. Nesmeyanov, Izv. Akad. Nauk, SSSR Otd. Khim. Nauk, **1962**, 1258.
113. Borowitz, J., K. Kirby, and R. Virkhaus, J. Org. Chem., **31**, 4031 (1966).
114. Bott, K., Angew. Chem., **77**, 683 (1965).
115. Bott, K., B. Dowden, and C. Eaborn, Intern. Symp. Organosilicon, Chem. Sci. Commun., Prague, **1965**, 290.
116. Bott, R., B. Dowden, and C. Eaborn, J. Chem. Soc., **1965**, 4994.
117. Bott, R., B. Dowden, and C. Eaborn, J. Organometal. Chem., **4**, 291 (1965).
118. Braddock, J., and G. Coates, J. Chem. Soc., **1961**, 3208.
119. Brannon, J. (Bakelite Comp.), U.S. Pat. 2,475,005 (July 5, 1949).
120. Bredig, G., Z. Physik, Chem., **13**, 289 (1894).
121. Brett, C., M. Murphy, R. Reinking, and J. Kansas, Entomol. Soc., **24**, 112 (1951).
122. Brewis, S., W. Dent, and R. Smith, J. Chem. Soc., **1965**, 1539.
123. British Hat & Allied Feltmakers Research Assoc., Fr. Pat. 1,400,248 (May 21, 1965).
124. Brooks, E., F. Glockling, and K. Hooton, J. Chem. Soc., **1965**, 4283.
125. Brophy, J., K. Freemann, and M. Gallagher, J. Chem. Soc. C, **1968**, 2760.
126. Brophy, J., and M. Gallagher, Aust. J. Chem., **20**, 503 (1967).
127. Brophy, J., and M. Gallagher, Aust. J. Chem., **22**, 1385 (1969).
128. Brown, C., and M. Sargent, J. Chem. Soc. C, **1969**, 1818.
129. Bruker, A., E. Grinshtein, and L. Soborovskii, Zh. Obshch. Khim., **36**, 484 (1966).
130. Bruson, H., and T. O'Day (Olin Mathieson Chemical Corp.), U.S. Pat. 3,037,044 (May 29, 1962).
131. Buchta, E., and F. Andree, Chem. Ber., **92**, 3111 (1959).
132. Buchta, E., and F. Andree, Ann., **640**, 29 (1961).
133. Buckler, S. (American Cyanamid Co.), U.S. Pat. 2,969,398 (Jan. 24, 1961).
134. Buckler, S. (American Cyanamid Co.), U.S. Pat. 3,013,085 (Dec. 12, 1961).
134a. Buckler, S., and M. Epstein, J. Org. Chem., **27**, 1090 (1962).
135. Buckler, S., V. Wystrach, J. Am. Chem. Soc., **83**,

168 (1961).

136. Burg, A., and P. Slota, J. Am. Chem. Soc., 80, 1107 (1958).
137. Burgada, R., Ann. Chim., 8, 347 (1963).
138. Burn, A., and J. Cadogan, J. Chem. Soc., 1963, 5788.
139. Burns, D., A. Fogg, and C. Higgens, J. Chromatog., 32, 793 (1968).
140. Butenandt, A., E. Hecker, M. Hopp, and W. Koch, Ann., 658, 39 (1962).
141. Butenandt, A., E. Truscheit, K. Eiter, and E. Hecker (Bayer, Farbenfabrik), Ger. Pat. 1,096,345 (Jan. 5, 1961).
142. Cabiddu, S., A. Maccioni, and M. Secci, Ann. Chim. (Rome), 54, 1153 (1964).
143. Caesar, F., and W. Balzer, Chem. Ber., 102, 1665 (1969).
144. Cahours, A., and A. Hofmann, Ann., 104, 1 (1857).
145. Caldwell, J., J. Chem. Soc., 109, 283 (1916).
146. Campbell, I., G. Fowles, and L. Nixon, J. Chem. Soc., 1964, 1389.
147. Campbell, I., R. McDonald, J. Org. Chem., 24, 730 (1959).
148. Campbell, T., R. McDonald, J. Org. Chem., 24, 1246 (1959).
149. Cannelongo, J. (American Cyanamid Co.), U.S. Pat. 3,422,048 (Jan. 14, 1969).
150. Carrol, B., and C. Allen (Eastman Kodak Co.), U.S. Pat. 2,271,622 (Feb. 3, 1942).
151. Cass, R., G. Coates, and R. Hayter, J. Chem. Soc., 1955, 4007.
152. Castro, B., R. Burgada, G. Laville, and J. Villiéras, Compt. Rend., 268, 1067 (1969).
153. Cathey, M., Phyton, 21, 203 (1964).
154. Cathey, M., and A. Diringer, Hort. Sci., 77, 608 (1961).
155. Chadha, R. (Stauffer Chemical Co.), Belg. Pat. 661,912 (Oct. 1, 1965).
156. Chamberland, B., and E. Muetterties, Inorg. Chem., 3, 1450 (1964).
157. Chance, L., D. Daigle, and G. Drake, J. Chem. Eng. Data, 12, 282 (1967).
158. Charrier, C., W. Chodkievicz, and P. Cadiot, Bull. Soc. Chim. France, 1966, 1002.
159. Chas. Pfizer & Co., Inc., Neth. Appl. 6,411,861 (Apr. 15, 1965).
160. Chatt, J., and F. Hart, J. Chem. Soc., 1960, 1378.
161. Chatt, J., and F. Mann, J. Chem. Soc., 1940, 1192.
162. Chechak, A., (Eastman Kodak Co.), U.S. Pat. 3,347,992 (Oct. 17, 1967).
163. Chiddiy, M., and E. Williams (General Aniline & Film Corp.), U.S. Pat. 2,946,824 (July 26, 1960).

164. Chopard, P., Chimia, 20, 172 (1966).
165. Chopard, P., J. Org. Chem., 31, 107 (1966).
166. Chopard, P., V. Clark, R. Hudson, and A. Kirby, Tetrahedron, 21, 1961 (1965).
167. Chopard, P., and R. Hudson, Z. Naturforsch., 18b, 509 (1963).
168. Chopard, P., and R. Hudson, J. Inorg. Nucl. Chem., 25, 801 (1963).
169. Chopard, P., R. Hudson, and G. Klopman, J. Chem. Soc., 1965, 1379.
170. Chopard, P., R. Searle, and F. Devitt, J. Org. Chem., 30, 1015 (1965).
171. CIBA Ltd., Fr. Pat. 1,535,554 (Aug. 9, 1968).
172. Clark, D., P. Holton, R. Meredith, A. Ritchie, T. Walker, and K. Whiting, J. Chem. Soc., 1962, 2479.
173. Clemens, D., and H. Sisler, Inorg. Chem., 4, 1222 (1965).
174. Clemens, D., W. Woodford, E. Dellinger, and Z. Tyndall, Inorg. Chem., 8, 998 (1969).
175. Coates, G., J. Livingstone, Chem. Ind., 1958, 1366.
176. Coates, H., and B. Chalkley, U.S. Pat. 3,236,676 (Feb. 22, 1966).
177. Coffman, D., and C. Marvel, J. Am. Chem. Soc., 51, 3496 (1929).
178. Cohen, J., Tetrahedron Lett., 1965, 3491.
179. Cohen, O., and M. Scott (Monsanto Chemical Co.), U.S. Pat. 2,772,189 (Nov. 27, 1956).
180. Colichmann, E., Anal. Chem., 26, 1204 (1954).
181. Collie, N., Phil. Mag., 24, 27 (1887).
182. Collie, N., J. Chem. Soc., 53, 636,714 (1888); 55, 223 (1889).
183. Collie, N., J. Chem. Soc., 127, 964 (1925).
184. Conen, J., Chem. Ber., 31, 2919 (1898).
185. Conix, A. (Gevaert Photo-Producten N.V.), Ger. Pat. 1,157,782 (Nov. 21, 1963).
186. Connelly, N., J. Locke, J. McCleverty, and A. Jon, Inorg. Chim. Acta, 2, 411 (1968).
187. Considine, W., J. Org. Chem., 27, 647 (1962).
188. Coombs, M., and R. Houghton, J. Chem. Soc., 1961, 5015.
189. Cooper, R., J. Davis, and B. Weedon, J. Chem. Soc. 1963, 5637.
190. Cottrell, W., and R. Morris, Chem. Commun., 1968, 409.
191. Cotton, F., O. Faut, and D. Goodgame, J. Am. Chem. Soc., 83, 344 (1961).
192. Cotton, F., D. Goodgame, and M. Goodgame, J. Am. Chem. Soc., 84, 167 (1962).
193. Coucouvanis, D., J. Am. Chem. Soc., 92, 707 (1970).
194. Coulson, D., Tetrahedron Lett., 1964, 3323.
195. Cowley, A., and R. Pinnell, J. Am. Chem. Soc., 87,

4454 (1965).
196. Craig, J., D. Hamon, H. Brewer, and H. Haerle, J. Org. Chem., 30, 907 (1965).
197. Cramer, R., R. Lindsey, Jr., C. Prewitt, and V. Stolberg, J. Am. Chem. Soc., 87, 638 (1965).
198. Creighton, J., G. Deacon, and J. Green, Aust. J. Chem., 20, 583 (1967).
199. Croatto, V., and V. Scatturin, Ricerca Sci., 18, 113 (1948).
200. Cross, R., A. Lauder, and G. Coates, Chem. Ind., 1962, 2013.
201. Cunnigham, J., and R. Gigg, J. Chem. Soc., 1965, 2968.
202. Czimatis, L., Chem. Ber., 15, 2014 (1882).
202a. Czimatis, L., and A. Michaelis, Chem. Ber., 15, 2018 (1882).
203. Czimatis, L., Jahresber., 1883, 1306.
204. Diagle, D., L. Chance, and G. Drake, J. Chem. Eng. Data, 13, 585 (1968).
205. Dallacker, F., F. Kornfeld, and M. Lipp, Monatsh. Chem., 91, 688 (1940).
206. Dallacker, F., K. Ulrichs, and M. Lipp, Ann., 667, 50 (1963).
207. Davidson, R., R. Sheldon, and S. Trippet, J. Chem. Soc. (C), (1966), 722.
208. Davies, W., J. Chem. Soc., 1933, 1043.
209. Davies, W., J. Chem. Soc., 1935, 462.
210. Davies, W., and W. Jones, J. Chem. Soc., 1929, 33.
211. Davies, W., and W. Lewis, J. Chem. Soc., 1934, 1599.
212. Davies, W., and F. Mann, J. Chem. Soc., 1944, 276.
213. Davies, W., and C. Morris, J. Chem. Soc., 1932, 2880.
214. Davies, W., P. Pearse, and W. Jones, J. Chem. Soc., 1929, 1262.
215. Davis, J., L. Jackmann, P. Siddons, and B. Weedon, J. Chem. Soc. (C), 1966, 2154.
216. Davis, M., and F. Mann, Chem. Ind., 1962, 1539.
217. Davis, M., and F. Mann, J. Chem. Soc., 1964, 3770.
218. Davis, M., F. Mann, J. Chem. Soc., 1964, 3786.
219. Davison, A., N. Edelstein, R. Holm, and A. Maki, Inorg. Chem., 3, 814 (1964).
220. Day, J., and L. Venanzi, J. Chem. Soc. (A), 1966, 197.
221. Day, J., and L. Venanzi, J. Chem. Soc. (A), 1966, 1363.
222. Deacon, G., J. Inorg. Nucl. Chem., 24, 1221 (1962).
223. Deacon, G., J. Green, and R. Nyholm, J. Chem. Soc., 1965, 3411.
224. Deacon, G., and R. Jones, Aust. J. Chem., 15, 555 (1962).
225. Deacon, G., and R. Jones, Aust. J. Chem., 16, 499 (1963).

226. Deacon, G., R. Jones, and P. Rogasch, Aust. J. Chem., 16, 360 (1963).
227. Deacon, G., and B. West, J. Chem. Soc., 1961, 3929.
228. Deacon, G., and B. West, J. Inorg. Nucl. Chem., 24, 169 (1962).
229. Deacon, G., and B. West, Aust. J. Chem., 16, 1132 (1963).
230. Dearborn, R., U.S. Pat. 3,087,836 (April 30, 1963).
231. De'ath, N., and S. Trippett, Chem. Commun., 1969, 172.
232. Dehn, W., and R. Conner, J. Am. Chem. Soc., 34, 1409 (1912).
233. Denney, D., and R. Dileone, J. Chem. Soc., 84, 4737 (1962).
234. Denney, D., and S. Felton, J. Am. Chem. Soc., 90, 183 (1968).
235. Denney, D., and H. Relles, Tetrahedron Lett., 1964, 573.
236. Denney, D., C. Rossi, and J. Vill, J. Am. Chem. Soc., 83, 3336 (1961).
237. Denney, D., C. Rossi, and J. Vill, J. Org. Chem., 29, 1003 (1964).
238. Denney, D., and L. Smith, J. Chem., 27, 3404 (1962).
239. Denney, D., and J. Song, J. Org. Chem., 29, 495 (1964).
240. Depoorter, H., J. Nys, and A. van Dormael, Bull. Chem. Soc. Belges, 73, 921 (1964).
241. Depoorter, H., J. Nys, and A. van Dormael, Bull. Chem. Soc. Belges, 74, 12 (1965).
242. Depoorter, H., J. Nys, and A. van Dormael (Gevaert Photo-Producten N.V.), Brit. Pat. 1,022,211 (March 9, 1966).
243. Derkach, G., and L. Samarai, Zh. Obshch. Khim., 34, 1161 (1964).
244. Dershowitz, S. (Polaroid Corp.), U.S. Pat. 3,019,108 (Mar. 18, 1959).
245. Deutsche Gold- und Silber-Scheideanstalt vorm. Roessler, Brit. Pat. 888,958 (Feb. 7, 1962).
246. Dickoré, K., Ann., 671, 135 (1964).
247. Dickoré, K., K. Sasse, R. Wegler, and L. Eue (Bayer Farbenfabrik), Belg. Pat. 628,113 (Aug. 7, 1963).
248. Diekmann, J., W. Hertler, and R. Benson, J. Org. Chem., 28, 2719 (1963).
249. Dimroth, K., H. Follmann, and G. Pohl, Chem. Ber., 99, 642 (1966).
250. Dimroth, K., and A. Nuerrenbach, Chem. Ber., 93, 1649 (1960).
251. Dimroth, K., and G. Pohl, Angew. Chem., 73, 436 (1961).
252. Dimroth, K., G. Pohl, and H. Follmann, Chem. Ber., 99, 634 (1966).

253. Dittmer, D., H. Simmons, and R. Vest, J. Org. Chem., 29, 497 (1964).
254. Divinskaya, L., V. Limanov, E. Skvortsova, G. Putyatina, A. Starkov, N. Grinshtein, and E. Nifant'ev. Zh. Obshch. Khim. 36, 1244 (1966).
255. Dodonov, J., and H. Medoks, Chem. Ber., 61, 907 (1928).
256. Doering, W.v.E., and A. Hoffmann, J. Am. Chem. Soc., 77, 521 (1955).
257. Dombrovskii, A., V. Listvan, A. Grigorenko, and M. Shevchuk, Zh. Obshch. Khim., 36, 1421 (1966).
258. Dombrovskii, A., and M. Shevchuk, Zh. Obshch. Khim., 33, 1263 (1963).
259. Dombrovskii, V., M. Shevchuk, and A. Dombrovskii, Zh. Obshch. Khim., 34, 3741 (1964).
260. Dormael, A. van, J. Nys, H. Depoorter (Gavaert Photo-Producten N.V.), Belg. Pat. 583,922 (Feb. 16, 1960).
261. Dormael, A. van, J. Nys, and H. Depoorter, Sci. Ind. Phot., 31, 389 (1961).
262. Dornfeld, C. (G.D. Searle & CO.), U.S. Pat 2,867,665 (Jan. 6, 1959).
263. Dornfeld, C., U.S. Pat. 2,882,321 (April 14, 1959).
264. Dornfeld, C., and A. Clinton, U.S. Pat. 2,884,461 (April 8, 1959).
265. Dornfeld, C., and L. Thielen (G.D. Searle & CO.), U.S. Pat. 2,865,964 (Dec. 23, 1958).
266. Drach, B., I. Zhmurova, and A. Kirsanov, Zh. Obshch. Khim., 37, 2524 (1967).
267. Drawe, D., and G. Caspari, Angew. Chem., 78, 331 (1966).
268. Dreeskamp, H., H. Elser, and C. Schumann, Ber. Bunsenges. Physik. Chem., 70, 751 (1966).
269. Drefahl, G., D. Lorenz, and R. Breng, J. Prakt. Chem., 18, 297 (1962).
270. Drefahl, G., and D. Lorenz, J. Prakt. Chem., 24, 312 (1964).
271. Drefahl, G., and G. Plötner, Chem. Ber., 93, 990 (1960).
272. Drefahl, G., and G. Plötner, Chem. Ber., 93, 998 (1960).
273. Drefahl, G., and G. Plötner, Chem. Ber., 94, 907 (1961).
274. Drefahl. G., G. Plötner, W. Hartrodt, and R. Kühmstedt, Chem. Ber., 93, 1799 (1960).
275. Drefahl, G., G. Plötner, and R. Scholz, Z. Chem., 1, 93 (1961).
276. Drefahl, G., G. Plötner, and K. Winnefeld, Chem. Ber., 94, 2002 (1961).
277. Drefahl, G., and W. Schermer, J. Prakt. Chem., 23, 225 (1964).
278. Drefahl, G., and K. Thalmann, J. Prakt. Chem., 20,

56 (1963).

279. Drefahl, G., and K. Winnefeld, J. Prakt. Chem., _28_, 242 (1965).

280. Drefahl, G., and K. Winnefeld, J. Prakt. Chem., _29_, 72 (1965).

281. Driscoll, J., D. Grisley, J. Pustinger, J. Harris, and C. Matthews, J. Org. Chem., _29_, 2427 (1964).

282. Driscoll, J., and C. Matthews, Chem. Ind., _1963_, 1282.

283. Driscoll, J., and C. Matthews, U.S. Pat. 3,374,256 (March 19, 1968).

284. Driver, G., and M. Gallagher, J. Chem. Soc. (D), _1970_, 150.

285. Dupont, J., and M. Hawthorne, Chem. Ind., _1962_, 405.

286. Dupont, J., and M. Hawthorne, U.S. Pat. 3,154,528 (Oct. 27, 1964).

287. Dye, W. (Monsanto Chemical Co.), U.S. Pat. 2,703,814 (March 8, 1955).

288. Dye, W., U.S. Pat. 2,730,547 (Jan. 10, 1956).

289. Dye, W., (Monsanto Chemical Co.), U.S. Pat. 2,774,658 (Dec. 18, 1956).

290. Eastman Kodak Co., Brit. Pat. 1,000,348 (Aug. 4, 1965).

291. Eberhardt, G., and W. Griffen, U.S. Pat. 3,472,911 (Oct. 14, 1969).

291a. Eiter, K., H. Oediger (Bayer Farbenfabrik), Ger. Pat. 1,160,855 (Jan. 9, 1964).

292. El Hewehi, Z., and D. Hempel, J. Prakt. Chem., _22_, 1 (1963).

293. Ellis, D., J. Orr, and B. Thomas (British Petroleum Co., Ltd.), Brit. Pat. 932,641 (July 31, 1963).

294. Ellzey, S., U.S. Pat. 3,359,321 (Dec. 19, 1967).

295. Ellzey, S., Can. J. Chem., _47_, 1251 (1969).

296. Emrys, R., H. Jones, and F. Mann, J. Chem. Soc., _1955_, 4472.

297. Epstein, M., and S. Buckler, U.S. Pat. 3,026,327 (March 20, 1962).

298. Epstein, M., and S. Buckler, Tetrahedron, _18_, 1231 (1962).

299. Ewart, G., A. Lane, J. McKechnie, and D. Payne, J. Chem. Soc., _1964_, 1543.

300. Ewart, G., D. Payne, A. Porte, and A. Lane, J. Chem., _1962_, 3984.

301. Eyles, C., and S. Trippet, J. Chem. Soc. (C), _1966_, 67.

302. Ezzell, B., and L. Freedman, J. Org. Chem., _34_, 1777 (1969).

303. Fackler, J., G. Dolbear, and D. Coucouvanis, J. Inorg. Nucl. Chem., _26_, 2035 (1964).

304. Fackler, J., D. Coucouvanis, Chem. Commun., _1965_, 556.

305. Fackler, J., and D. Coucouvanis, J. Am. Chem. Soc., 88, 3913 (1966).
306. Faithful, B., R. Gillard, D. Tuck, and R. Ugo, J. Chem. Soc. (A), 1966, 1185.
307. Falkowski, L., T. Ziminski, W. Mechlinski, and E. Borowski, Roczniki Chem., 39, 225 (1965).
308. Farrar, T., and J. Rush, Nat. Bur. Stand. (U.S.) Spec. Publ. No. 301,245 (1967).
309. Farlane, W.Mc., and J. Nash, J. Chem. Soc. (D), 1968, 524.
310. Fatuzzo, E., R. Nitscho, H. Routschi, and S. Zingg, Phys. Rev., 115, 514 (1962).
311. Fedorova, G., and A. Kirsanov, Zh. Obshch. Khim., 30, 4044 (1960).
312. Fedorova, G., Y. Shaturski, and A. Kirsanov, Probl. Organ. Sintesa. Akad. Nauk SSSR Otd. Obshch i. Tekhn. Khim., 1965, 258.
313. Fenton, G., and C. Ingold, J. Chem. Soc., 1929, 2342.
314. Ferrania, Societa per Azioni, Fr. Pat. 1,477,002 (April 14, 1967).
315. Filipescu, N., L. Kindley, H. Podall, and F. Seratin, Can. J. Chem., 41, 821 (1963).
316. Finkelstein, M., J. Org. Chem., 27, 4076 (1962).
317. Fischer, H., and H. Fischer, Chem. Ber., 99, 658 (1966).
318. Fishwick, B. (Imperial Chemical Industries Ltd.) Brit. Pat. 926,998 (May 22, 1963).
319. Fliszár, S., R. Hudson, and G. Salvadori, Helv. Chim. Acta, 46, 1580 (1963).
320. Fluck, E., and H. Binder, Angew. Chem., Intern. Ed. Engl., 5, 666 (1966).
321. Fluck, E., and K. Issleib, Chem. Ber., 98, 2674 (1965).
322. Fluck, E., and R. Reinisch, Ber., 96, 3085 (1963).
323. Ford, J., and C. Wilson, J. Org. Chem., 26, 1433 (1961).
324. Ford, J., and C. Wilson, J. Org. Chem., 28, 875 (1963).
325. Forster, D., and D. Goodgame, Inorg. Chem., 4, 715 (1965).
326. Forster, D., and D. Goodgame, Inorg. Chem., 4, 823 (1965).
327. Forster, D., and D. Goodgame, Inorg. Chem., 4, 1712 (1965).
328. Forster, D., and D. Goodgame, J. Chem. Soc., 1965, 268.
329. Franco, S. (Ferrania Societa per Azioni), U.S. Pat. 3,429,839 (Feb. 25, 1969).
330. Frank, A., and J. Gordon, Can. J. Chem., 44, 2593 (1966).
331. Frank, R., and E. Rothstein, J. Chem. Soc., 1964, 3872.

332. Frazier, S., and H. Sisler, Inorg. Chem., 5, 925 (1966).

333. Freemann, J., Chem. Ind., 1959, 1254.

334. Freemann, J., J. Org. Chem., 31, 538 (1966).

335. Freyschlag, H., N. Grassner, A. Nuerrenbach, H. Pommer, W. Reif, and W. Sarnecki, Angew. Chem., 77, 277 (1965).

336. Freyschlag, H., W. Reif, and A. Nuerrenbach (BASF), Ger. Pat. 1,216,862 (May 18, 1966).

337. Friedrich, K., and H. Hennig, Chem. Ber., 92, 2756 (1959).

338. Friederich, H., and K. Sepp (BASF), Ger. Pat. 948,056 (Aug. 30, 1956).

339. Frieser, E., Spinner, Weber Textilveredl., 79, 1140 (1961); 79, 1144 (1961); 80, 122 (1962).

340. Fritchie, C., Acta Cryst., 20, 107 (1966).

341. Fritzsche, H., U. Hasserodt, F. Korte, G. Friese, K. Adrian, and H. Arenz, Chem. Ber., 97, 1988 (1964).

342. Fuerst, H., G. Wetzke, W. Berger, and W. Schubert, J. Prakt. Chem., 17, 299 (1962).

343. Fussgaenger, R., and G. Taeuber (Farbwerke Hoechst AG), Ger. Pat. 1,145,292 (March 14, 1963).

344. Gaertner, R. van (Monsanto Chemical Co.), U.S. Pat. 2,828,332 (March 25, 1958).

345. Gallagher, M., E. Kirby, and F. Mann, J. Chem. Soc., 1963, 4846.

346. Gallagher, M., and F. Mann, J. Chem. Soc., 1962, 5110.

347. Gans, P., B. Hathaway, and B. Smith, Spectrochim. Acta, 21, 1589 (1965).

348. Gans, P., and B. Smith, J. Chem. Soc., 1964, 4172.

349. Ganushchak, N., M. Yukhomenko, M. Stadnichuk, and M. Shevchuk, Zh. Obshch. Khim., 36, 1150 (1966).

350. Garner, A., J. Abramo, and E. Chapin (Monsanto Chemical Co.), U.S. Pat. 3,065,272 (Nov. 20, 1962).

351. Garner, A., E. Chapin, and P. Scanlon, J. Org. Chem., 24, 532 (1959).

352. Gee, W., R. Shaw, and B. Smith, J. Chem. Soc., 1965, 3354.

353. Geerts, J., and R. Martin, Bull. Soc. Chim. Belges, 69, 563 (1960).

354. Geigy, J.R., A.G., Swiss. Pat. 246,911 (Feb. 15, 1947).

355. Geigy, J.R., A.G., Brit. Pat. 726,936 (March 23, 1955).

356. Gerecke, M., G. Ryser, and P. Zeller (Hofmann-LaRoche, Inc.), U.S. Pat. 2,912,467 (Nov. 19, 1959).

357. Gevaert Photo-Producten, N.V., Belg. Pat. 588,783 (Sept. 19, 1960).

358. Gevaert Photo-Producten, N.V., Belg. Pat. 650,289 (Nov. 3, 1964).

359. Gevorkyan, A., B. Dyatkin, and J. Knunyants, Izv. Akad. Nauk SSSR Ser. Khim., 1965, 1599.
360. Gigg, J., R. Gigg, and C. Warren, J. Chem. Soc. (C), 1966, 1872.
361. Gillard, R., and G. Wilkinson, J. Chem. Soc., 1963, 5885.
362. Gillette, Co., Belg. Pat. 639,015 (April 22, 1964).
363. Gillham, H., and A. Sherr (American Cyanamid Co.), U.S. Pat. 3,322,861 (May 30, 1967).
364. Gilyarov, V., and M. Kabachnik, Zh. Obshch. Khim., 36, 282 (1966).
365. Ginzel, K., H. Klupp, and O. Kraupp, Arch. Exptl. Pathol. Pharmakol., 221, 336 (1954).
366. Ginzel, K., H. Klupp, O. Kraupp, and G. Werner, Arch. Exptl. Pathol. Pharmokol., 217, 173 (1953).
367. Girard, A., Ann. Chim., 2, 11 (1884).
368. Glabich, D., G. Oertel (Bayer Farbenfabrik), Ger. Pat. 1,210,836 (Feb. 17, 1966).
369. Gleichmann, L., Chem. Ber., 15, 198 (1882).
370. Gleysteen, L., and C. Kraus, J. Am. Chem. Soc., 69, 451 (1947).
371. Gonikberg, M., J. Fainshtein, Izv. Akad. Nauk SSSR Ser. Khim., 1965, 1469.
372. Goodfellow, R., and L. Venanzi, J. Chem. Soc. (A), 1966, 784.
373. Gordon, M., and C. Griffin, J. Chem. Phys., 41, 2570 (1964).
374. Gosselck, J., H. Schenk, and H. Ahlbrecht, Angew. Chem., Int. Ed. Engl., 6, 249 (1967).
375. Goubeau, J., and G. Wenzel, Z. Physik. Chem., 45, 31 (1965).
376. Gough, S., and S. Trippett, J. Chem. Soc., 1961, 4263.
377. Graybill, B., A. Pitochelli, and M. Hawthorne, Inorg. Chem., 1, 626 (1962).
378. Grayson, M., J. Am. Chem. Soc., 85, 79 (1963).
379. Grayson, M., P. Keough (American Cyanamid Co.), U.S. Pat. Appl. 3,005,013 (Oct. 30, 1958).
380. Grayson, M., and P. Keough, J. Am. Chem. Soc., 82, 3919 (1960).
381. Grayson, M., and P. Keough (American Cyanamid Co.), U.S. Pat. 3,116,317 (Dec. 31, 1963).
382. Grayson, M., and P. Keough (American Cyanamid Co.), U.S. Pat. 3,148,205 (Sept. 8, 1964).
383. Grayson, M., P. Keough, and G. Johnson, J. Am. Chem. Soc., 81, 4803 (1959).
384. Grayson, M., P. Keough, and M. Rauhut (American Cyanamid Co.), U.S. Pat. 3,214,434 (Oct. 26, 1965).
385. Grayson, M., P. Keough, and M. Rauhut (American Cyanamid Co.), U.S. Pat. 3,299,143 (Jan. 17, 1967).
386. Grayson, M., P. Keough, and M. Rauhut (American

Cyanamid Co.), U.S. Pat. 3,364,245 (Jan. 16, 1968).
387. Griffin, G., and M. Kaufmann, Tetrahedron Lett.,
      1965, 773.
388. Griffin, C., K. Martin, and B. Douglas, J. Org. Chem.,
      27, 1627 (1962).
389. Griffin, C., and M. Gorden, J. Organometall. Chem.,
      3, 414 (1965).
390. Griffin, C., and J. Peters, J. Org. Chem., 28, 1715
      (1963).
391. Griffin, C., and G. Witschard, J. Org. Chem., 27,
      3334 (1962).
392. Griffin, C., and G. Witschard, J. Org. Chem., 29,
      1001 (1964).
393. Griffiths, T., J. Chem. Eng. Data, 8, 568 (1963).
394. Grigorenko, A., M. Shevchuk, and A. Dombrovskii, Zh.
      Obshch. Khim., 35, 1227 (1965).
395. Grigorenko, A., M. Shevchuk, and A. Dombrovskii, Zh.
      Obshch. Khim., 36, 506 (1966).
396. Grigorenko, A., M. Shevchuk, and A. Dombrovskii, Zh.
      Obshch. Khim., 36, 1121 (1966).
397. Grim, S., W. McFarlane, E. Davidoff, and J. Marts,
      J. Phys. Chem., 70, 581 (1966).
398. Grim, S., and D. Seyferth, Chem. Ind., 1959, 849.
399. Grimshaw, J., and J. Ramsey, J. Chem. Soc. (B),
      1968, 63.
400. Grinshtein, E., A. Bruker, and L. Soborovskii, Zh.
      Obshch. Khim., 36, 302 (1966).
401. Grinshtein, E., A. Bruker, and L. Soborovskii, Zh.
      Obshch. Khim., 36, 1138 (1966).
402. Grisley, D. (Monsanto Research Corp.), U.S. Pat.
      3,330,868 (July 11, 1967).
403. Grisley, D. (Monsanto Research Corp.), U.S. Pat.
      3,334,145 (Aug. 1, 1967).
404. Grisley, D. (Monsanto Research Corp.), U.S. Pat.
      3,341,605 (Sept. 12, 1967).
405. Grisley, D., J. Alm, and C. Matthews, Tetrahedron,
      21, 5 (1965).
406. Gruettner, G., and E. Krause, Ber., 49, 437 (1916).
407. Gruettner, G., and M. Wiernik, Ber., 48, 1473 (1915).
408. Gryszkiewiez-Trochimowski, E., C. Monard, J. Quin-
      chon, and M. LeSech, Bull. Soc. Chim. France, 1961,
      2408.
409. Gupta, A., and H. Peters, Fette, Seifen, Anstrichmit.,
      68, 349 (1966).
410. Haas, H. (Böhme Fettchemie GmbH), Ger. Pat. 937,949
      (Jan. 19, 1956).
411. Hall, D., and B. Prakobsantisukh, J. Chem. Soc.,
      1965, 6311.
412. Hamid, A., and S. Trippett, J. Chem. Soc. (C), 1967,
      2625.
413. Hands, A., and A. Mercer, J. Chem. Soc., 1965, 6055.

414. Hands, A., and A. Mercer, J. Chem. Soc. (C), 1968, 1331.
415. Harris, J., and C. Matthews (Monsanto Research Corp.), U.S. Pat. 3,098,878 (July 23, 1963).
416. Harris, G., and D. Payne, J. Chem. Soc., 1956, 3038.
417. Harrison, I., and B. Lythgoe, J. Chem. Soc., 1958, 843.
418. Harrison, I., B. Lythgoe, and B. Trippett, J. Chem. Soc., 1955, 507.
419. Hart, F., J. Chem. Soc., 1960, 3324.
420. Hart, F., and F. Mann, J. Chem. Soc., 1955, 4107.
421. Hart, F., and F. Mann, J. Chem. Soc., 1957, 3939.
422. Hart, F., and H. Sisler, Inorg. Chem., 3, 617 (1964).
423. Harting, H., Pharm. Ind., 25, 57 (1963).
424. Hartmann, H., C. Beermann, and H. Czempik, Z. Anorg. Allgem. Chem., 287, 261 (1956).
425. Hantzsch, A., Chem. Ber., 52, 1544 (1919).
426. Hauser, M., J. Org. Chem., 27, 43 (1962).
427. Hauser, C., T. Brocks, M. Miles, M. Raymond, and G. Butler, J. Org. Chem., 28, 372 (1963).
428. Hawes, W., and S. Trippett, Chem. Commun., 1968, 295.
428a. Hawthorne, M., J. Am. Chem. Soc., 80, 3480 (1958).
429. Hawthorne, M., A. Pitochelli, R. Strahm, and J. Miller, J. Am. Chem. Soc., 82, 1825 (1900).
430. Hay, A. (General Electric Co.), U.S. Pat. 3,378,508 (April 16, 1968).
431. Hays, H., J. Org. Chem., 31, 3817 (1966).
432. Hayton, B., and B. Smith, J. Inorg. Nucl. Chem., 31, 1369 (1969).
433. Heal, H., J. Inorg. Nucl. Chem., 16, 208 (1961).
434. Hecker, H., and F. Hein, Z. Anal. Chem., 174, 354 (1960).
435. Hellmann, H., J. Bader, H. Birkner, and O. Schumacher, Ann., 659, 49 (1962).
436. Hellmann, H., and O. Schumacher, Ann., 640, 79 (1961).
437. Hellmann, H., and O. Schumacher (Bayer Farbenfabrik AG), Ger. Pat. 1,176,657 (Aug. 27, 1964).
438. Hellwinkel, D., Angew. Chem., 77, 378 (1965).
439. Hellwinkel, D., Chem. Ber., 98, 576 (1965).
440. Hellwinkel, D., Chem. Ber., 102, 548 (1969).
440a. Henderson, Jr., W.A., and S.A. Buckler, J. Am. Chem. Soc., 82, 5794 (1960).
441. Hendrickson, J., C. Hall, R. Rees, and J. Templeton, J. Org. Chem., 30, 3312 (1965).
442. Hendrickson, J., M. Maddox, J. Sims, and H. Kaesz, Tetrahedron, 20, 449 (1964).
443. Herring, D., J. Org. Chem., 26, 3998 (1961).
444. Herring D., and C. Douglas, Inorg. Chem., 3, 428 (1964).
445. Herrmann, K., and H. Rudolph (Bayer Farbenfabrik),

Ger. Pat. 1,261,668 (Feb. 22, 1968).

446. Hertler, W., U.S. Pat. 3,226,388 (Dec. 28, 1965).

447. Hewertson, W., R. Shaw, and B. Smith, J. Chem. Soc., 1964, 1020.

448. Hewertson, W., and H. Watson, J. Chem. Soc., 1962, 1490.

449. Heying, T., and C. Naar-Colin, Inorg. Chem., 3, 282 (1964).

450. Hiestand, A., and A. Maeder (Ciba Ltd.), S. African Pat. 6,707,257 (April 24, 1968).

451. Hilgetag, G., and H. Teichmann, Chem. Ber., 96, 1465 (1963).

452. Hinton, R., and F. Mann, J. Chem. Soc., 1959, 2835.

453. Hinton, R., F. Mann, and D. Todd, J. Chem. Soc., 1961, 5454.

454. Hitchcock, C., and F. Mann, J. Chem. Soc., 1958, 2081.

455. Hoeniger, A., and H. Schindlbauer, Ber. Bunsenges. Physik. Chem., 69, 138 (1965).

456. Hofmann, A., Ann. Suppl., 1, 1, 275 (1861); Phil. Trans., 150, 409 (1860).

457. Hofmann, A., Chem. Ber., 6, 292, 301 (1873).

458. Hoffmann, A., J. Am. Chem. Soc., 43, 1684 (1921); 52, 2995 (1930).

459. Hoffmann, H., Ann., 634, 1 (1959).

460. Hoffmann, H., Chem. Ber., 94, 1331 (1961).

461. Hoffmann, H., Chem. Ber., 95, 2563 (1962).

462. Hoffmann, H., and H. Diehr, Chem. Ber., 98, 363 (1965).

463. Hoffmann, H., and H. Foerster, Tetrahedron Lett., 1964, 983.

464. Hoffmann, H., and R. Gruenewald, Chem. Ber., 94, 186 (1961).

465. Hoffmann, H., R. Gruenewald, and L. Horner, Chem. Ber., 93, 861 (1960).

466. Hoffmann, H., L. Horner, H. Wippel, and D. Michael, Chem. Ber., 95, 523 (1962).

467. Hoffmann, H., and D. Michael, Chem. Ber., 95, 528 (1962).

468. Hoffmann, H., and D. Schellenbeck, Chem. Ber., 99, 1134 (1966).

469. Hoffmann, H., and D. Schellenbeck, Chem. Ber., 100, 692 (1967).

470. Hoffmann-LaRoche & Co. AG., Fr. Pat. 1,383,944 (Jan. 4, 1965).

471. Holliman, F., and F. Mann, J. Chem. Soc., 1947, 1634.

472. Holliman, F., and F. Mann, Nature, 159, 438 (1947).

473. Holliman, F., F. Mann, and A. Thornton, J. Chem. Soc., 1960, 9.

474. Hook, W., G. Howarth, W. Hoyle, and G. Roberts, Chem. Ind., 1965, 1630.

475. Hooker Chemical Corp., Ger. Pat. 1,221,605 (July 28, 1966).

476. Horn, P., and E. Rothstein, J. Chem. Soc., 1963, 1036.
477. Horner, L. (Farbwerke Hoechst), Ger. Pat. 937,587 (Jan. 12, 1956).
478. Horner, L., P. Beck, and H. Hoffmann, Chem. Ber., 92, 2088 (1959).
479. Horner, L., P. Beck, and F. Roettger (Farbwerke Hoechst), Ger. Pat. 1,183,338 (Dec. 10, 1964).
480. Horner, L., H. Fuchs, H. Winkler, and A. Rapp, Tetrahedron Lett., 1963, 965.
481. Horner, L., and J. Haufe, Chem. Ber., 101, 2903 (1968).
482. Horner, L., and J. Haufe, Electroanal. Chem., Interfacial Electrochem., 20, 245 (1969).
483. Horner, L., and H. Hoffmann, Chem. Ber., 91, 45 (1958).
484. Horner, L., and H. Hoffmann, Chem. Ber., 91, 50 (1958).
485. Horner, L., H. Hoffmann, and G. Hassel, Ber., 91, 58 (1958).
486. Horner, L., H. Hoffmann, W. Klink, H. Ertel, and V. Toscano, Chem. Ber., 95, 581 (1962).
487. Horner, L., H. Hoffmann, H. Wippel, and G. Hassel, Chem. Ber., 91, 50 (1958).
488. Horner, L., W. Jurgeleit, and K. Kluepfel, Ann., 591, 108 (1955).
489. Horner, L., W. Klink, and H. Hoffmann, Chem. Ber., 96, 3133 (1963).
490. Horner, L., and K. Kluepfel, Ann., 591, 69 (1955).
491. Horner, L., and E. Lingnau, Ann., 591, 135 (1955).
492. Horner, L., and A. Mentrup, Ann., 646, 65 (1961).
493. Horner, L., and H. Moser, Chem. Ber., 99, 2789 (1966).
494. Horner, L., G. Mummenthey, H. Moser, and P. Beck, Chem. Ber., 99, 2782 (1966).
495. Horner, L., and B. Nippe, Chem. Ber., 91, 67 (1958).
496. Horner, L., and H. Oediger, Ann., 627, 142 (1959).
497. Horner, L., and F. Roettger, Korrosion, 16, 57 (1963).
498. Horner, L., and F. Roettger, Werkstoffe Korrosion, 15, 125 (1964).
499. Horner, L., F. Roettger, and H. Fuchs, Ber., 96, 3141 (1963).
500. Horner, L., F. Schedlbauer, and P. Beck, Tetrahedron Lett., 1964, 1421.
501. Horner, L., and H. Schmelzer, Chem. Ber., 94, 1326 (1961).
502. Horner, L., and H. Winkler, Tetrahedron Lett., 1964, 175.
503. Horner, L., and H. Winkler, Tetrahedron Lett., 1964, 455.
504. Horner, L., and H. Winkler, Ann., 685, 1 (1965).
505. Horner, L., H. Winkler, A. Rapp, A. Mentrup, H. Hoffmann, and P. Beck, Tetrahedron Lett., 1961, 161.

506. House, H., and H. Babad, J. Org. Chem., 28, 90 (1963).

507. House, H., V. Jones, and G. Frank, J. Org. Chem., 29, 3327 (1964).

508. Houghton, R., J. Chem. Soc, 1963, 6075.

509. Hsia Chen, C., and D. Grabar, J. Polymer Sci. (C), No. 4, 869 (1964).

510. Hudson, R., and P. Chopard, Helv. Chim. Acta, 46, 2178 (1963).

511. Hudson, R., and P. Chopard, J. Org. Chem., 28, 2446 (1963).

512. Hudson, R., P. Chopard, and G. Salvadori, Helv. Chim. Acta, 47, 632 (1964).

513. Imaev, M., and A. Shakirova, Dokl. Akad. Nauk SSSR, 163, 656 (1965).

514. Ingold, C., F. Shaw, and J. Wilson, J. Chem. Soc., 1928, 1280.

515. Inhoffen, H., K. Brueckner, G. Domagk, and H. Erdmann, Chem. Ber., 88, 1415 (1955).

516. Inhoffen, H., K. Brueckner, and H. Hess, Chem. Ber., 88, 1850 (1955).

517. Inhoffen, H., K. Irmscher, G. Friedrich, D. Kampe, and O. Berges, Chem. Ber., 92, 1772 (1959).

518. Isler, O., L. Chopard-dit-Jean, M. Montavon, R. Rüegg, and P. Zeller, Helv. Chem. Acta, 40, 1256 (1957).

519. Isler, O., H. Gutmann, H. Lindlar, M. Montavon, R. Rüegg, G. Ryser, and P. Zeller, Helv. Chim. Acta, 39, 463 (1956).

520. Isler, O., H. Gutmann, M. Montavon, R. Rüegg, G. Ryser, and P. Zeller, Helv. Chim. Acta, 40, 1242 (1957).

521. Isler, O., M. Montavon, R. Rüegg, and P. Zeller (Hoffmann-LaRoche & Co., AG.), U.S. Pat. 2,842,599 (1958).

522. Issleib, K., and H. Anhoeck, Z. Naturforsch., 16b, 837 (1961).

523. Issleib, K., and B. Biermann, Z. Anorg. Allgem. Chem., 347, 39 (1966).

524. Issleib, K., and R. Bleck, Z. Anorg. Allgem. Chem., 336, 234 (1965).

525. Issleib, K., and H. Haeckert, Z. Naturforsch., 21b, 519 (1966).

526. Issleib, K., and B. Hamann, Z. Anorg. Allgem. Chem., 339, 289 (1965).

527. Issleib, K., and G. Harzfeld, Chem. Ber., 97, 3430 (1964).

528. Issleib, K., K. Jasche, Chem. Ber., 100, 412 (1967).

529. Issleib, K., F. Krech, Chem. Ber., 94, 2656 (1961).

530. Issleib, K., F. Krech, Chem. Ber., 98, 2545 (1965).

531. Issleib, K., F. Krech, Z. Anorg. Allgem. Chem., 328,

21 (1964).

532. Issleib, K., and R. Lindner, Ann., 699, 40 (1966).
533. Issleib, K., and R. Lindner, Ann., 707, 112 (1967).
534. Issleib, K., and R. Lindner, Ann., 707, 120 (1967).
535. Issleib, K., and O. Loew, Z. Anorg. Allgem. Chem., 346, 241 (1966).
536. Issleib, K., and H. Moebius, Chem. Ber., 94, 102 (1961).
537. Issleib, K., and E. Priebe, Chem. Ber., 92, 3183 (1959).
538. Issleib, K., and E. Priebe, Chem. Ber., 92, 3175 (1959).
539. Issleib, K., and R. Rieschel, Chem. Ber., 98, 2086 (1965).
540. Issleib, K., and K. Rockstroh, Chem. Ber., 96, 407 (1963).
541. Issleib, K., and H. Roloff, Chem. Ber., 98, 2091 (1965).
542. Issleib, K., and G. Thomas, Chem. Ber., 93, 803 (1960).
543. Issleib, K., and G. Thomas, Chem. Ber., 94, 2244 (1961).
544. Issleib, K., and A. Tzschach, Chem. Ber., 92, 1118 (1959).
545. Issleib, K., and A. Tzschach, Chem. Ber., 93, 1852 (1960).
546. Issleib, K., and H. Voelker, Chem. Ber., 94, 392 (1961).
547. Issleib, K., and B. Walther, Chem. Ber., 97, 3424 (1964).
548. Ivashchenko, S., I. Sarycheva, and N. Preobrazbenski, Zh. Org. Kh., 2, 2181 (1966).
549. Ivashchenko, S., I. Sarycheva, and N. Preobrazbenski, Zh. Obshch. Khim., 36, 1380 (1966).
550. Jackson, J., W. Davies, and W. Jones, J. Chem. Soc., 1930, 2298.
551. Jackson, J., W. Davies, and W. Jones, J. Chem. Soc., 1931, 2109.
552. Jackson, J., and W. Jones, J. Chem. Soc., 1931, 575.
553. Jensen, K., and P. Nielsen, Acta Chim. Scand., 17, 547 (1963).
554. Jerchel, D., Chem. Ber., 76, 600 (1943).
555. Jerchel, D., and J. Kimmig, Chem. Ber., 83, 277 (1950).
556. Johnson, A., and R. Amel, Can. J. Chem., 46, 461 (1968).
557. Johnson, A., and V. Kyllingstad, J. Org. Chem., 31, 334 (1966).
558. Johnson, A., and R. LaCount, Tetrahedron, 9, 130 (1960).
559. Johnson, A., S. Lee, R. Swor, and L. Royer, J. Am.

Chem. Soc., 88, 1953 (1966).
560. Jones, A., and M. Talmet, Aust. J. Chem., 18, 903 (1965).
561. Jones, M., and D. Chesnut, J. Chem. Phys., 38, 1311 (1963).
562. Jones, M., and S. Trippett, J. Chem. Soc. (C), 1966, 1090.
563. Jones, R., U.S. Pat. 3,419,344 (Dec. 31, 1968).
564. Jones, W., W. Davies, S. Bowden, C. Edwards, V. Davis, and L. Thomas, J. Chem. Soc., 1947, 1446.
565. Jong, G. de (Stanis Carbon N.V.), U.S. Pat. 2,775,564 (Dec. 25, 1956).
566. Kabachnik, M., V. Gilyarov, and M. Yusupov, Dokl. Akad. Nauk SSSR, 164, 812 (1965).
567. Kalovoulos, J., and J. Giesching, Trans. Int. Congr. Soil Sci., 8th Bukarest, 3, 417 (1964).
568. Kamai, G., Zh. Obshch. Khim., 2, 524 (1932).
569. Kamai, G., and O. Belorossova, Bull. Acad. Sci. URSS, Classe Sci. Chim., 1947, 191.
570. Kamai, G., and L. Khismatullina, Dokl. Akad. Nauk SSSR, 92, 69 (1953).
571. Kamai, G., and L. Khismatullina, Zh. Obshch. Khim., 28, 3426 (1956).
572. Kamai, G., and L. Khismatullina, Ser. Khim. Nauk, 4, 79 (1957).
573. Kamai, G., and G. Rusetskaya, Zh. Obshch. Khim., 32, 2848 (1962).
574. Kamai, G., and G. Usacheva, Zh. Obshch. Khim., 34, 785 (1964).
575. Kamai, G., and G. Usacheva, Zh. Obshch. Khim., 34, 3606 (1964).
576. Kanai, K., T. Hashimoto, H. Kitano, and K. Fukui, Nippon Kagaku Zasshi, 86, 534 (1965).
577. Kantor, S., and A. Gilbert (General Electric Co.), U.S. Pat. 2,883,366 (April 21, 1959).
578. Kaplan, E., Diss. Abstr., 28, 107 (1967).
579. Kaplan, E., G. Singh, and H. Zimmer, J. Phys. Chem., 67, 2509 (1963).
580. Kawamori, A., J. Chem. Phys., 47, 3091 (1967).
581. Keaveney, W., and D. Hennessy, J. Org. Chem., 27, 1057 (1962).
582. Kelbe, W., Chem. Ber., 11, 1499 (1878).
583. Kennedy, J., E. Lane, and J. Willans, J. Chem. Soc., 1956, 4670.
584. Keough, P., and M. Grayson, J. Org. Chem., 29, 631 (1964).
585. Ketcham, R., D. Jambotkar, and L. Martinelli, J. Org. Chem., 27, 4666 (1962).
586. Khaleeluddin, K., and J. Scott, Chem. Commun., 1966, 511.
587. Khaskin, B., N. Tuturina, and N. Mel'nikov, Zh.

Obshch. Khim., 38, 2652 (1968).
588. Kimmig, J., and D. Jerchel, Klin. Wschr., 28, 429 (1950).
588a. Klamann, D., and P. Weyerstahl, Angew. Chem., 75, 89 (1963).
589. Klamann, D., and P. Weyerstahl, Chem. Ber., 97, 2534 (1964).
590. Kleine-Weischede, K. (Bayer Farbenfabrik), Ger. Pat. 1,203,772 (Oct. 28, 1965).
591. Kleine-Weischede, K. (Bayer Farbenfabrik), Ger. Pat. 1,212,529 (March 17, 1966).
592. Kline, E., and C. Kraus, J. Am. Chem. Soc., 69, 814 (1947).
593. Knott, E., J. Chem. Soc., 1965, 3793.
594. Knypl, J., Naturwissenschaften, 54, 146 (1967).
595. Kochetkov, N., A. Vasil'ev, and S. Levchenko, Zh. Obshch. Khim., 35, 190 (1965).
596. Koebrich, G., H. Trapp, K. Flory, and W. Drischel, Chem. Ber., 99, 689 (1966).
597. Kolitowska, I., Roczniki Chem., 8, 568 (1928).
598. Knunyants, I., V. Tyuleneva, E. Pervova, and R. Sterlin, Izv. Akad. Nauk SSSR Serb. Khim., 1964, 1797.
599. Kolmerten, J., and J. Epstein, Anal. Chem., 30, 1536 (1958).
600. Komkov, I., K. Karavanov, and S. Ivin, Metody Polucheniya Khim. Reaktivovi Preparatov, 12, 73 (1965).
601. Kosmin, M. (Monsanto Chemical Co.), U.S. Pat. 2,818,368 (Dec. 31, 1957).
602. Kottler, A., H. Scheffler, and G. Werner (Dr. K. Thomae GmbH), Ger. Pat. 941,193 (April 5, 1956).
603. Kramer, G. (Esso Research and Engineering Co.), U.S. Pat. 3,364,280 (Jan. 16, 1968).
604. Krivun, S., Dokl. Akad. Nauk SSSR, 182, 347 (1968).
605. Kroehnke, F., Chem. Ber., 83, 291 (1950).
606. Krubiner, A., and E. Oliveto, J. Org. Chem., 31, 24 (1966).
607. Kuchen, W., and H. Buchwald, Angew. Chem., 69, 307 (1957).
608. Kukhtin, V., A. Kazymov, and T. Voskoboeva, Dokl. Akad. Nauk SSSR, 140, 601 (1961).
609. Kumli, K., W. McEwen, and C. Vanderwerf, J. Am. Chem. Soc., 81, 248 (1959).
610. Kuzmin, K., Z. Panfilovich, and L. Pavlova, Tr. Kazan Khim. Tekhnol. Inst., 34, 392 (1965).
611. Kuznetsov, E., T. Sorokina, and R. Valetdinov, Zh. Obshch. Khim., 33, 2631 (1963).
612. LaLancette, E., J. Org. Chem., 29, 2957 (1964).
613. Lamza, L., J. Prakt. Chem., 25, 294 (1964).
614. Lane, E., Analyst, 80, 675 (1955).
615. Lane, E., J. Chem. Soc., 1956, 2006.
616. Larsen, D. (National Lead Co.), Ger. Pat. 1,103,336

(March 30, 1961).
617. Laughlin, R., J. Org. Chem., 30, 1322 (1965).
618. Ledermann, B., Chem. Ber., 21, 405 (1888).
619. LeGoff, E., J. Am. Chem. Soc., 84, 1505 (1962).
620. Leland, F., A. Kraus, and F. Gleysteen, J. Am. Chem. Soc., 69, 451 (1947).
621. Letsinger, R., J. Nazy, and A. Hussey, J. Org. Chem., 23, 1806 (1958).
622. Letts, E., and R. Blake, Proc. Roy, Soc. (Edinburgh), 16, 193 (1867); Trans. Roy. Soc., 35, 527 (1889).
623. Letts, E., and N. Collie, Phil. Mag., 22, 183 (1886).
624. Levine, N., V. Ivens, M. Kleckner, and J. Sonder, Ann. J. Vet. Research, 17, 117 (1956).
625. Lewis, R., K. Naumann, K. DeBruin, and K. Mislow, J. Chem. Soc. (D), 1969, 1010.
626. Listvan, V., and A. Dombrovskii, Zh. Obshch. Khim., 38, 601 (1968).
627. Locke, J., and J. McCleverty, Chem. Commun., 1965, 102.
628. Locke, J., and J. McCleverty, Inorg. Chem., 5, 1157 (1966).
629. Lorenz, W., and H. Fischer, Ber. Bunsenges. Physik. Chem., 69, 689 (1965).
630. Lucken, E., J. Chem. Soc., 1963, 5123.
631. Lucken, E., Z. Naturforsch., 18b, 166 (1963).
632. Lucken, E., and C. Mazeline, J. Chem. Soc. (A), 1966, 1074.
633. Luettringhaus, A., Am. Dyestuff Reptr., 37, 57 (1948).
634. Lyon, D., and F. Mann, J. Chem. Soc., 1942, 666.
635. Machleidt, H., and W. Grell, Ann., 690, 79 (1965).
636. McDonald, R., and T. Campbell, J. Org. Chem., 24, 1969 (1959).
637. McDonald, R., and R. Campbell, J. Am. Chem. Soc., 82, 4669 (1960).
638. McEwen, W., G. Axelrad, M. Zanger, and C. VanderWerf, J. Am. Chem. Soc., 87, 3948 (1965).
639. McEwen, W., A. Bladé-Font, and C. VanderWerf, J. Am. Chem. Soc., 84, 677 (1962).
640. McEwen, W., K. Kumli, A. Bladé-Font, M. Zanger, and C. VanderWerf, J. Am. Chem. Soc., 86, 2378 (1964).
641. Maerkl, G., Chem. Ber., 95, 3003 (1962).
642. Maerkl, G., Tetrahedron Lett., 1962, 1027.
643. Maerkl, G., Z. Naturforsch., 17b, 782 (1962).
644. Maerkl, G., Angew. Chem., 75, 168 (1963).
645. Maerkl, G., Z. Naturforsch., 18b, 84 (1963).
646. Maerkl, G., Z. Naturforsch., 18b, 1136 (1963).
647. Maier, L., Helv. Chim. Act, 46, 2667 (1963).
648. Maier, L., Helv. Chim. Acta, 49, 1119 (1966).
648a. Maier, L., Helv. Chim. Acta, 49, 2458 (1966).
649. Maier, L., D. Seyferth, F. Stone, and E. Rochow, J. Am. Chem. Soc., 79, 5884 (1957).

650. Makarova, L., and A. Nesmeyanov, Akad. Nauk SSSR, Inst. Org. Khim. Sintezy Org. Soedinenii Sbornik, 1, 142 (1950).
651. Makin, S., Dokl. Akad. Nauk. SSSR, 138, 387 (1961).
652. Makin, S., Zh. Obschch. Khim., 32, 3159 (1962).
653. Makin, S., G. Lapitskii, and A. Kolunova, Zh. Vses. Khim. Obshch. D.J. Mendelleva, 8, 708 (1963).
654. Malatesta, L., Gazz. Chim. Ital., 77, 518 (1947).
655. Mallion, K., and F. Mann, J. Chem. Soc. Suppl., 1964, 6121.
656. Mallion, K., and F. Mann, J. Chem. Soc., 1964, 5716.
657. Mallion, K., and F. Mann, J. Chem. Soc., 1965, 4115
658. Mallion, K., and F. Mann, Chem. Ind., 1963, 654.
659. Mallion, K., F. Mann, B. Tong, and V. Wystrach, J. Chem. Soc., 1963, 1327.
660. Manchand, P., R. Rüegg, U. Schwieter, P. Siddons, and B. Weedon, J. Chem. Soc., 1965, 2019.
661. Mann, F., and F. Hart, Nature, 175, 952 (1955).
662. Mann, F., and R. Hinton, J. Chem. Soc., 1959, 2835.
663. Mann, F., and J. Millar, J. Chem. Soc., 1951, 2205.
664. Mann, F., and J. Millar, J. Chem. Soc., 1952, 4453.
665. Mann, F., J. Millar, and B. Smith, J. Chem. Soc., 1953, 1130.
666. Mann, F., J. Millar, and F. Stewart, J. Chem. Soc., 1954, 2832.
667. Mann, F., J. Millar, and H. Watson, J. Chem. Soc., 1958, 2516.
668. Mann, F., and M. Pragnell, J. Chem. Soc. (C), 1966, 916.
669. Mann, F., B. Tong, and V. Wystrach, J. Chem. Soc., 1963, 1155.
670. Mann, F., and J. Watson, J. Org. Chem., 13, 502 (1948).
671. Mann, F., and H. Watson, J. Chem. Soc., 1957, 3945.
672. Margulis, T., and D. Templeton, J. Chem. Phys., 36, 2311 (1962).
673. Mark, V., J. Am. Chem. Soc., 85, 1884 (1963).
673a. Mark, V., C. Dungan, M. Crutchfield, and J. Van Wazer, in Top. Phosphorus Chem., 5, 227.
674. Marquardt, C., J. Chem. Phys., 48, 994 (1968).
675. Marsi, K., and G. Homer, J. Org. Chem., 28, 2150 (1963).
676. Marvel, C., and E. Gall, J. Org. Chem., 24, 1494 (1959).
677. Martin, D., and C. Griffin, J. Org. Chem., 30, 4034 (1965).
678. Martin, G., and G. Mavel, Compt. Rend., 253, 644 (1961).
679. Massey, A., E. Randall, and D. Shaw, Spectrochim. Acta, 21, 263 (1965).
680. Massy-Westropp, R., and G. Reynolds, Aust. J.

Chem., 19, 891 (1966).
681. Matschiner, H., and K. Issleib, Z. Anorg. Allgem. Chem., 354, 60 (1967).
682. Matsui, K., and T. Kimijima, J. Soc. Org. Synthetic Chem. (Japan), 7, 239 (1949).
683. Matsumoto, M., M. Kawanishi, K. Enomoto, and M. Sasamoto, Ann. Rept. G. Tanabe Co., Ltd., 1, 29 (1956).
684. Matsuo, H., and S. Chaki, Bunseki Kagaku, 11, 762 (1962).
685. Matteson, D., J. Org. Chem., 29, 3399 (1964).
686. Matthews, G. (Monsanto Research Corp.), U.S. Pat. 3,262,971 (July 26, 1966).
687. Matthews, G., and G. Birum (Monsanto Co.), U.S. Pat. 3,426,073 (Feb. 4, 1969).
687a. Matthews, G., J. Driscoll, and G. Birum, Chem. Commun., 1966, 736.
688. Matthews, G., J. Driscoll, and J. Harris, U.S. Dept. Com., Office Tech. Serv. A.D., 278, 229 (1962).
689. Matthews, G., J. Driscoll, J. Harris, and R. Winemann, J. Am. Chem. Soc., 84, 4349 (1962).
690. Maxwell, K. (Shell Development Co.), U.S. Pat. 2,418,652 (April 8, 1947).
691. Mayer, H., P. Schudel, R. Rüegg, and O. Isler, Helv. Chim. Acta, 46, 650 (1963).
692. Medoks, G., Zh. Obshch. Khim., 8, 289 (1938).
693. Medoks, G., Zh. Obshch. Khim., 26, 382 (1956).
694. Medoks, G., and V. Andronova, Zh. Obshch. Khim., 22, 2058 (1952).
695. Medoks, G., and H. Sakharova, Dokl. Akad. Nauk. SSSR, 73, 1201 (1950).
696. Medoks, G., and E. Soshestvenskaya, Zh. Obshch. Khim., 27, 271 (1957).
697. Medoks, G., and E. Soshestvenskaya, Zh. Obshch. Khim., 27, 271 (1957).
698. Medoks, G., E. Soshestvenskaya, and W. Sakharova, Zh. Priklad. Khim., 25, 1111 (1952).
699. Meisenheimer, J., and L. Lichtenstadt, Ann., 449, 213 (1926).
700. Melby, L., R. Harder, W. Hertler, W. Mahler, R. Benson, and W. Mochel, J. Am. Chem. Soc., 84, 3374 (1962).
701. Mel'nikov, N., B. Khaskin, N. Petruchenko, L. Stonov, L. Bakumenko, and V. Kazakova, All-Union Scientific Research Institute of Chemicals for Plant Protection. USSR, 181,442, April 15, 1966.
702. Mel'nikov, N., B. Khaskin, and N. Tuturina, Zh. Obshch. Khim., 36, 645 (1966).
703. Mel'nikov, N., B. Khaskin, and N. Tuturina, All-Union Scientific Research Institute of Chemicals for Plant Protection. USSR, 201,401, Sept. 7, 1967.

704. Mel'nikov, N., B. Khaskin, N. Tuturina, C. Pershin, and S. Milovanova, Khim.-Farm. Zh., 2, 11 (1968).
705. Mel'nikov, N., B. Khaskin, N. Tuturina, L. Stonov, N. Usacheva, L. Bakumenko, and V. Kazakova, All-Union Scientific Research Institute of Chemicals for Plant Protection. USSR, 181,444, April 15, 1966.
706. Mel'nikov, N., and V. Kraft, Zh. Obshch. Khim., 30, 1918 (1960).
707. Mel'nikov, N., N. Petrina, A. Prokofwa, N. Popovkina, I. Vladimirova, and B. Khaskin, Zh. Obshch. Khim., 39, 1244 (1969).
708. Mel'nikov. N., N. Tuturina, and B. Khaskin, Zh. Obshch. Khim., 36, 1082 (1966).
709. Melton, T. (Virginia-Carolina Chemical Co.), U.S. Pat. 3,050,543 (Aug. 21, 1962).
710. Messinger, J., and C. Engels, Chem. Ber., 21, 326, 2919 (1888).
711. Michaelis, A., Ann., 293, 193 (1896); 294, 1 (1896).
712. Michaelis, A., Ann., 315, 43 (1901).
713. Michaelis, A., and H. Gimborn, Chem. Ber., 27, 272 (1894).
714. Michaelis, A., and L. Gleichmann, Chem. Ber., 15, 801 (1882).
715. Michaelis, A., and R. Kaehne, Chem. Ber., 31, 1048 (1898).
716. Michaelis, A., and H. Koehler, Chem. Ber., 32, 1566 (1899).
717. Michaelis, A., and A. Link, Ann., 207, 193 (1881).
718. Michaelis, A., and A. Schenk, Ann., 260, 1 (1890).
719. Michaelis, A., and H.v. Soden, Ann., 229, 295 (1885).
719a. Mingoia, Q., Gazz. Chim. Ital., 60, 144 (1930).
720. Milas, N., F. Serratosa, L. Pohmer, Y. Fellion, H. Pendse, and E. Ghera, Compt. Rend., 248, 3455 (1959).
721. Miles, T., F. Hoffmann, and A. Delasanta (to U.S. Dept. of the Army), U.S. Pat. 2,993,746 (July 25, 1961).
722. Miller, N., J. Am. Chem. Soc., 87, 390 (1965).
723. Miller, N., Inorg. Chem., 4, 1458 (1965).
724. Misumi, S., and M. Nakagawa, Bull. Chem. Soc. Japan, 36, 399 (1963).
725. Mitsunobu, O., and M. Yamada, Bull. Chem. Soc. Japan, 40, 2380 (1967).
726. Modro, T., Bull. Acad. Pol. Sai, Ser. Sci. Chim., 16, 585 (1968).
727. Moedritzer, K., U.S. Pat. 3,281,365 (Oct. 25, 1966).
728. Mondon, A., Ann., 603, 115 (1957).
729. Monsanto Chemical Co., Fr. Pat. 1,347,066 (Dec. 27, 1963).
730. Monsanto Chemical Co., Fr. Pat. 1,482,337 (May 26, 1967).
731. Moocia, K., J. Chem. Phys., 37, 910 (1962).

732. Morgan, P., and B. Herr, J. Am. Chem. Soc., 74, 4526
     (1952).
733. Moura De Campos, M., L. do Amaral, Arch. Pharm.,
     298, 92 (1965).
733a. Mueller, A., and F. Bollmann, Z. Naturforsch., 236,
     1539 (1968).
734. Mueller, E., H. Eggensperger, and K. Scheffler, Ann.,
     658, 103 (1962).
735. Mueller, W., E. Bennett, and J. Fuller, J. Dairy
     Sci., 29, 751 (1946).
736. Muetterties, E., H. Balthis, Y. Chia, W. Knoth, and
     H. Miller, Inorg. Chem., 3, 444 (1964).
737. Murayama, K., S. Morimura, Y. Nakamura, and G.
     Sunagawa, Yakugaku Zasshi, 85, 757 (1965).
738. Muxfeldt, H., G. Grethe, K. Uhlig, and H. Zeugner,
     Chem. Ber., 96, 2943 (1963).
739. Nakaya, T., and M. Imoto, Bull. Chem. Soc. Japan,
     39, 1547 (1966).
740. Nakaya, T., T. Tomomoto, and M. Imoto, Bull. Chem.
     Soc. Japan, 39, 1551 (1966).
741. Naldini, L., Gazz. Chim. Ital., 90, 1231 (1960).
742. Nasipuri, D., G. Pyne, D. Roy, R. Bhattacharya, and
     P. Dutt, J. Chem. Soc., 1964, 2146.
743. Nast, R., and K. Kaeb, J. Organometal. Chem., 6, 456
     (1966).
744. Neeb, R., Z. Anal. Chem., 152, 158 (1956).
745. Neeb, R., Z. Anal. Chem., 177, 420 (1960).
746. Neeb, R., Z. Anal. Chem., 179, 21 (1961).
747. Neeb, R., Z. Anal. Chem., 182, 10 (1961).
748. Neilands, O., and S. Kalnina, Zh. Org. Khim., 4, 140
     (1968).
749. Nemours E.I. du Pont de, & Co., Brit. Pat. 793,673
     (April 23, 1958).
750. Nesmeyanov, N., S. Berman, L. Ashkinadze, L. Kazityna,
     and O. Reutov, Zh. Org. Khim., 4, 1685 (1968).
751. Nesmeyanov, N., and L. Makarova, Org. Khim., 7, 109
     (1956).
752. Nesmeyanov, N., and V. Novikov, Dokl. Akad. Nauk
     SSSR, 162, 350 (1965).
753. Nesmeyanov, N., V. Novikov, and O. Reutov, J. Organo-
     metal. Chem., 4, 202 (1965).
754. Nesmeyanov, N., V. Novikov, and O. Reutov, Zh. Organ.
     Khim., 2, 942 (1966).
755. Nesmeyanov, N., and O. Reutov, Dokl. Akad. Nauk.
     SSSR, 171, 111 (1966).
756. Nesmeyanov, N., and O. Reutov, Zh. Organ. Khim., 2,
     1716 (1966).
757. Nesmeyanov, N., Y. Vasyak, V. Kalyavin, and O.
     Reutov, Zh. Org. Khim., 4, 385 (1968).
758. Nesmeyanov, N., S. Zhuzhlikova, and O. Reutov, Dokl.
     Akad. Nauk SSSR, 151, 856 (1963).

759. Nesmeyanov, N., S. Zhuzhlokova, and O. Reutov, Izv. Akad. Nauk SSSR Ser. Khim., 1965, 194.
760. Nesterov, L., A. Kessel, and L. Maklakov, Zh. Obshch. Khim., 38, 318 (1968).
761. Nesterov, L., A. Kessel, Y. Samitov, and A. Musina, Zh. Obshch. Khim., 39, 1179 (1969).
762. Nesterov, L., and R. Mutalapova, Zh. Obshch. Khim., 37, 1843 (1967).
763. Nesterov, L., and R. Mutalapova, Tetrahedron Lett., 1968, 51.
764. Neunhoeffer, O., and L. Lamza, Chem. Ber., 94, 2514 (1961).
765. Nicholson, D., E. Rothstein, R. Saville, and R. Whiteley, J. Chem. Soc., 1953, 4019.
766. Nielson, R., J. Vincent, and H. Sisler, Inorg. Chem., 2, 760 (1963).
767. Nifant'ev, E., M. Grachev, L. Bakinovskii, S. Kara-Murza, and N. Kochetkov, Zh. Prikl. Khim., 36, 676 (1963).
768. Noeth, H., and H. Vetter, Chem. Ber., 94, 1505 (1961).
769. Noeth, H., and H. Vetter, Chem. Ber., 96, 1109 (1963).
770. Noeth, H., and H. Vetter, Chem. Ber., 98, 1981 (1965).
771. Noether, D., Diss. Abstr., 25, 2237 (1964).
772. Normant, H., T. Cuvigny, J. Normant, and B. Angelo, Bull. Soc. Chim. France, 1965, 3446.
773. Novikov, Y., P. Palli, and S. Kapurshina, At. Energ., 26, 420 (1969).
774. Nuerrenbach, A., W. Sarmecki, and W. Reif (BASF), Fr. Pat. 1,395,458 (April 9, 1965).
775. Obol'nikova, E., and G. Samokhvalov, Zh. Obshch. Khim., 33, 1860 (1963).
776. Obol'nikova, E., M. Yanotovskii, and G. Samokhvalov, Zh. Obshch. Khim., 34, 1499 (1964).
777. Olaj, O., J. Breitenbach, and B. Buchberger, Angew. Makromol. Chem., 3, 160 (1968).
778. Oppegard, A. (E.I. du Pont de Nemours & Co.), U.S. Pat. 2,687,437 (Aug. 24, 1954).
779. Pagilagan, R., Diss. Abstr., 26, 5733 (1966).
780. Pagilagan, R., and W. McEwen, Chem. Commun., 1966, 652.
781. Pappas, J., and E. Gancher, J. Org. Chem., 31, 3877 (1966).
782. Parisek, C., Diss. Abstr., 24, 520 (1963).
783. Parisek, C., W. McEwen, and C. VanderWerf, J. Am. Chem. Soc., 82, 5503 (1960).
784. Parshall, G., J. Am. Chem. Soc., 86, 361 (1964).
785. Parshall, G., D. England, and R. Lindsey, J. Am. Chem. Soc., 81, 4801 (1959).
786. Partheil, A., and A. Gronover, Chem. Ber., 33, 606 (1900).
787. Partheil, A., and A. Gronover, Arch. Pharm., 241,

411 (1903).

788. Partheil, A., and A. Haaren, Arch. Pharm., 238, 35
      (1900).

789. Partos, R., and K. Ratts, J. Am. Chem. Soc., 88,
      4996 (1966).

790. Partos, R., and A. Speziale, J. Am. Chem. Soc., 87,
      5068 (1965).

791. Paschoal, S., J. Am. Chem. Soc., 81, 4169 (1959).

792. Pattenden, G., and B. Walker, J. Chem. Soc. (C),
      1969, 531.

793. Paul, B., and D. Ramana, Indian J. Chem., 6, 395
      (1968).

794. Pauson, D., and W. Watts, J. Chem. Soc., 1963, 2990.

795. Peerdeman, A., J. Holst, L. Horner, and H. Winkler,
      Tetrahedron Lett., 1965, 8111.

796. Pershin, G., and N. Begdanova, Farmakol. i. Toksikol.,
      25, 209 (1962).

797. Petragnani, N., and M. de Moura Campos, Chem. Ind.,
      1964, 1461.

798. Petrov, K., A. Gavrilova, V. Nam, and V. Chuchkanova,
      Zh. Obshch. Khim., 32, 3711 (1962).

799. Petrov, K., and V. Parshina, Zh. Obshch. Khim., 31,
      2729 (1961).

800. Petrov, K., and V. Parshina, Zh. Obshch. Khim., 31,
      3417 (1961).

801. Petrov, K., V. Parshina, and V. Gaidamak, Zh. Obshch.
      Khim., 31, 3411 (1961).

802. Petrov, K., V. Parshina, and V. Gaidamak, Zh. Obshch.
      Khim., 31, 3421 (1961).

803. Petrov, K., V. Parshina, and M. Luzanova, Zh. Obshch.
      Khim., 32, 553 (1962).

804. Petrov, K., V. Parshina, and A. Manuilov, Zh. Obshch.
      Khim., 35, 1602 (1965).

805. Petrov, K., V. Parshina, and A. Manuilov, Zh. Obshch.
      Khim., 35, 2062 (1965).

806. Petrov, K., V. Parshina, and G. Petrova, Zh. Obshch.
      Khim., 39, 1247 (1969).

807. Petrovich, J., Elektrochim. Acta, 12, 1429 (1967).

808. Petry, G., and W. Scheele, Kautschuk, Gummi, Kunst-
      stoffe, 19, 526 (1966).

809. Philips, N.V., Gloeilampenfabrieken, Neth. Appl.
      6,606,914 (Nov. 20, 1967).

810. Philips, N.V., Gloeilampenfabrieken, Neth. Appl.
      6,606,916 (Nov. 20, 1967).

811. Pinck, L., and G. Hilbert, J. Am. Chem. Soc., 69,
      723 (1947).

812. Plieninger, H., M. Hoebel, and V. Liede, Chem. Ber.,
      96, 1618 (1963).

813. Plumb, J., and C. Griffin, J. Org. Chem., 27, 4711
      (1962).

814. Plumb, J., R. Obrycki, and C. Griffin, J. Org. Chem.,

$\underline{31}$, 2455 (1966).

815. Pommer, H., G. Wittig, and U. Schöllkopf, Ger. Pat. 1,048,568 (Jan. 15, 1959).

816. Pope, W., and C. Gibson, J. Chem. Soc., $\underline{101}$, 736 (1912).

817. Preston, W. (United States Government), U.S. Pat. 3,230,069 (Jan. 18, 1966).

818. Protiva, M., and O. Exner, Chem. Listy, $\underline{48}$, 1370 (1954).

819. Protopopov, I., and M. Kraft, Med. Prom. SSSR, $\underline{13}$, 5 (1959).

820. Protopopov, I., and M. Kraft, Zh. Obshch. Khim., $\underline{33}$, 3050 (1963).

821. Protopopov, I., and M. Kraft, Zh. Obshch. Khim., $\underline{34}$, 1446 (1964).

822. Ptitsyna, O., M. Pudeeva, N. Bel'kevich, and O. Reutov, Dokl. Akad. Nauk. SSSR, $\underline{163}$, 383 (1965).

823. Pyykkö, P., Chem. Phys. Lett., $\underline{2}$, 559 (1968).

824. Querfurth, W. (Deutsche Gold- und Silberscheideanstalt vorm. Roessler), U.S. Pat. 3,080,351 (March 5, 1963).

825. Quin, L., and D. Mathews, Chem. Ind., $\underline{1963}$, 210.

826. Rabinowitz, R., A. Henry, and R. Marcus, J. Polymer Sci., Pt. A, $\underline{3}$, 2055 (1965).

827. Rabinowitz, R., and J. Pellon, J. Org. Chem., $\underline{26}$, 4623 (1961).

828. Ramirez, F., and S. Dershowitz, J. Am. Chem. Soc., $\underline{78}$, 5614 (1956).

829. Ramirez, F., and S. Dershowitz, Chem. Ind., $\underline{1956}$, 665.

830. Ramirez, F., and S. Dershowitz, J. Org. Chem., $\underline{22}$, 41 (1957).

831. Ramirez, F., N. Desai, B. Hansen, and N. McKelvie, J. Am. Chem. Soc., $\underline{83}$, 3539 (1961).

832. Ramirez, F., N. Desai, and N. McKelvie, J. Am. Chem. Soc., $\underline{84}$, 1745 (1962).

833. Ramirez, F., and S. Levy, J. Org. Chem., $\underline{21}$, 1333 (1956).

833a. Ramirez, F., and S. Levy, J. Org. Chem., $\underline{23}$, 2035 (1958).

834. Ramirez, F., and S. Levy, J. Am. Chem. Soc., $\underline{79}$, 67 (1957).

835. Ramirez, F., and S. Levy, J. Am. Chem. Soc., $\underline{79}$, 6167 (1957).

836. Ramirez, F., O. Madan, and C. Smith, Tetrahedron Lett., $\underline{22}$, 567 (1966).

837. Ramirez, F., and N. McKelvie, J. Am. Chem. Soc., $\underline{79}$, 5829 (1957).

838. Ramirez, F., N. McKelvie, and N. Desai, Preprints Papers Intern. Symp. Free Radicals 5th, Uppsala, $\underline{1961}$, p. 57.

839. Ramirez, F., A. Patwardhan, H. Kugler, and C. Smith,

J. Am. Chem. Soc., 89, 6276 (1967).

840. Ramirez, F., A. Patwardhan, H. Kugler, and C. Smith, Tetrahedron, 24, 2275 (1968).

841. Ramirez, F., A. Patwardhan, and C. Smith, J. Am. Chem. Soc., 87, 4973 (1965).

842. Ramirez, F., D. Rhum, and C. Smith, Tetrahedron Lett., 21, 1941 (1965).

843. Raschig, F., Ger. Pat. 223,684 (1910); 233,631 (March 27, 1911).

844. Ratcliffe, G. (National Lead Co.), U.S. Pat. 2,622,987 (Dec. 23, 1952).

845. Rauhut, M., G. Borowitz, and H. Gillham, J. Org. Chem., 28, 2565 (1963).

846. Rauhut, M., G. Borowitz, and M. Grayson, U.S. Pat. 3,422,149 (Jan. 14, 1969).

847. Rauhut, M., J. Hechenbleikner, H. Currier, F. Schaefer, and V. Wystrach, J. Am. Chem. Soc., 81, 1103 (1959).

848. Rauhut, M., and A. Semsel, J. Org. Chem., 28, 473 (1963).

849. Rauhut, M., and A. Semsel (American Cyanamid Co.), U.S. Pat. 3,099,690 (July 30, 1963).

850. Rauhut, M., and A. Semsel (American Cyanamid Co.), U.S. Pat. 3,251,883 (May 17, 1966).

851. Rave, T. (Procter and Gamble Co.), Fr. Pat. 1,506,964 (Dec. 22, 1967).

852. Rave, T., and H. Hays, J. Org. Chem., 31, 2894 (1966).

853. Razumov, A., and N. Zabusova, Zh. Obshch. Khim., 32, 2688 (1962).

854. Razuvaev, G., and N. Osanova, Dokl. Akad. Nauk. SSSR, 104, 552 (1955).

855. Reddy, G., and R. Schmutzler, Inorg. Chem., 5, 164 (1966).

856. Reeves, W., and J. Guthrie, U.S. Bur. Agr. and Ind. Chem., Mimeographed Circ. Ser. A.J.C. 364, 1953.

857. Reeves, W., and J. Guthrie (United States of America, as represented by the Secy of Agr.), U.S. Pat. 2,772,188 (Nov. 27, 1956).

858. Reicheneder, F., and F. Stolp (BASF), Ger. Pat. 1,157,629 (Nov. 21, 1963).

859. Reid, D., J. Chem. Soc., 1965, 5920.

860. Reilly, G., and W. McEwen, Tetrahedron Lett., 1968, 1231.

861. Renshaw, R., and R. Bishop, J. Am. Chem. Soc., 60, 946 (1938).

862. Reppe, W., and K. Friedrich, U.S. Pat. 2,730,546 (Jan. 10, 1956).

863. Reppe, W., H. v. Kutepow, and W. Morsch (BASF), Brit. Pat. 742,740 (Jan. 4, 1956).

864. Reppe, W., W. Schweckendiek, and H. Friedrich (BASF),

U.S. Pat. 2,738,364 (March 13, 1956).

865. Reuter, M. (Farbwerke Hoechst), U.S. Pat. 2,912,466 (Nov. 10, 1959).

866. Reuter, M., and L. Orthner (Farbwerke Hoechst), Ger. Pat. 1,041,957 (Dec. 4, 1958).

867. Reuter, M., L. Orthner, F. Jacob, and E. Wolf (Farbwerke Hoechst), U.S. Pat. 2,937,207 (May 17, 1960).

868. Reuter, M., L. Orthner, and E. Wolf (Farbwerke Hoechst), Ger. Pat. 1,067,812 (Oct. 29, 1959).

869. Reuter, M., E. Wolf, L. Orthner, and F. Jacob (Farbwerke Hoechst), Ger. Pat. 1,042,583 (Nov. 6, 1958).

870. Reuter, M., E. Wolf, L. Orthner, and F. Jacob (Farbwerke Hoechst), Ger. Pat. 1,045,401 (Dec. 4, 1958).

871. Richards, H., D. Cooper, J. Manesfield, P. Palmer, A. Patel, W. Sherren, K. Slater, and H. Zimmermann, Text. Chem. Color, $\underline{1}$, 54 (1969).

872. Ried, W., and H. Appel, Z. Naturforsch., $\underline{15b}$, 684 (1960).

873. Ried, W., and H. Appel, Ann., $\underline{646}$, 82 (1961).

874. Ried, W., and H. Appel, Ann., $\underline{679}$, 51 (1964).

875. Ried, W., and H. Appel, Ann., $\underline{679}$, 56 (1964).

876. Riley, F., and E. Rothstein, J. Chem. Soc., $\underline{1964}$, 3872.

877. Robinson, B., and J. Fergusson, J. Chem. Soc., $\underline{1964}$, 5683.

878. Roesky, H., Chem. Ber., $\underline{101}$, 636 (1968).

879. Roesky, H., Chem. Ber., $\underline{101}$, 2977 (1968).

880. Röhm & Haas Co., Neth. Appl. 6,407,501 (Jan. 11, 1965).

881. Roitburd, T., and R. Valetdinov, Material 1-bi (Pervoi) Konf. Molodykh Nauchn. Rabotn. Kazani, Sektsiya Khim. Kazan Sb., $\underline{1959}$, 91.

882. Ross, T., and D. Denney, Anal. Chem., $\underline{32}$, 1896 (1966).

883. Rothstein, E., S. Rowland, and P. Horn, J. Chem. Soc., $\underline{1953}$, 3994.

884. Rothberg, I., and E. Thornton, J. Am. Chem. Soc., $\underline{85}$, 1704 (1963).

885. Rothberg, I., and E. Thornton, J. Am. Chem. Soc., $\underline{86}$, 3296 (1964).

886. Rudich, S., and J. Lind, J. Chem. Phys., $\underline{50}$, 3055 (1969).

887. Rüegg, R., U. Schwieder, G. Ryser, P. Schudel, and O. Isler, Helv. Chim. Acta, $\underline{44}$, 985 (1961).

888. Rüegg, R., U. Schwieder, G. Ryser, P. Schudel, and O. Isler, Helv. Chim. Acta, $\underline{44}$, 994 (1961).

889. Ruff, J., Inorg. Chem., $\underline{2}$, 813 (1963).

890a. Rush, J., A. Melveger, T. Farrar, and T. Tsang, Chem. Phys. Lett., $\underline{1968}$, 2.

890. Rush, J., A. Melveger, and E. Lippincott, Chem. Phys., $\underline{51}$, 2947 (1969).

891. Rydon, H., and B. Tonge, J. Chem. Soc., 1956, 3043.
892. Ryl'tsev, E., I. Boldeskul, N. Feshchenko, J. Mako-
     veelskii, and Y. Egoroo, Teor. Eksp. Khim., 5, 563
     (1969).
893. Sachs, H., Chem. Ber., 29, 1514 (1892).
894. Saikachi, H., Y. Taniguchi, and H. Ogawa, Yakugaku
     Zasshi, 82, 1262 (1962).
895. Saikachi, H., Y. Taniguchi, and H. Ogawa, Yakugaku
     Zasshi, 83, 582 (1963).
896. Sakharova, N., Dokl. Akad. Nauk. SSSR, 77, 73 (1951).
896a. Sarnecki, W., A. Nuerrenbach, and W. Reif (BASF),
     Ger. Pat. 1,155,126 (Oct. 3, 1963).
897. Sarnecki, W., and H. Pommer (BASF), U.S. Pat.
     2,950,321 (Aug. 23, 1960).
898. Sastri, M., and T. Rao, J. Inorg. Nucl. Chem., 30,
     1727 (1968).
899. Schaefer, F., and J. Ross, J. Org. Chem., 29, 1527
     (1964).
900. Schaefer, J., and D. Weinberg, J. Org. Chem., 30,
     2635 (1965).
901. Schiemenz, G., Tetrahedron Lett., 1964, 2729.
902. Schiemenz, G., Angew. Chem., 78, 777 (1966).
903. Schiemenz, G., Angew. Chem., Int. Ed. Engl., 6, 564
     (1967).
904. Schiemenz, G., Chem. Ber., 98, 65 (1965).
905. Schiemenz, G., Chem. Ber., 99, 504 (1966).
906. Schiemenz, G., Chem. Ber., 99, 514 (1966).
907. Shciemenz, G., and J. Engelhard, Chem. Ber., 94,
     578 (1961).
908. Schiemenz, G., and H. Rast, Tetrahedron Lett., 1969,
     2165.
909. Schiemenz, G., and J. Thobe, Chem. Ber., 99, 2663
     (1966).
910. Schindlbauer, H., Chem. Ber., 96, 2109 (1963).
911. Schindlbauer, H., Monatsh. Chem., 96, 1793 (1965).
912. Schindlbauer, H., Spectrochim. Acta, 20, 1143 (1964).
913. Schindlbauer, H., L. Golser, and V. Hilzensauer,
     Chem. Ber., 97, 1150 (1964).
914. Schindlbauer, H., and F. Mitterhofer, Z. Anal. Chem.,
     221, 394 (1966).
915. Schloegel, K., and H. Egger, Ann., 676, 76 (1964).
916. Schlosser, M., Chem. Ber., 97, 3219 (1964).
917. Schmidbaur, H., and W. Tronich, Chem. Ber., 100,
     1032 (1967).
918. Schmidpeter, A., and G. Weingand, Angew. Chem., Int.
     Ed. Engl., 7, 210 (1968).
919. Schmidt, A., Chem. Ber., 101, 4015 (1968).
920. Schmidt, L., Arch. Exptl. Pathol. Pharmakol., 241,
     538 (1961).
921. Schmutzler, R., J. Am. Chem. Soc., 86, 4500 (1964).
922. Schmutzler, R., J. Chem. Soc., 1965, 5630.

923. Schneider, D., and C. Garbers, J. Chem. Soc., <u>1964</u>, 2465.
924. Schoeller, C. (BASF), Ger. Pat. 806,992 (June 21, 1951).
925. Schoenberg, A., K. Brosowski, and E. Singer, Chem. Ber., <u>95</u>, 2144 (1962).
926. Schroeder, H., U.S. Dept. Com., Office Tech. Serv. A.D., 267, 991 (1961).
927. Schroeder, H., T. Heying, and J. Reiner, Inorg. Chem., <u>2</u>, 1092 (1963).
928. Schutt, J., and S. Trippett, J. Chem. Soc. (C), <u>1969</u>, 2038.
929. Schweckendiek, W., and K. Sepp (BASF), Ger. Pat. 831,693 (Feb. 18, 1952).
930. Schweizer, E., J. Am. Chem. Soc., <u>86</u>, 2744 (1964).
931. Schweizer, E., and R. Bach, J. Org. Chem., <u>29</u>, 1746 (1964).
932. Schweizer, E., M. El Bakoush, K. Light, and K. Oberle, J. Org. Chem., <u>33</u>, 2590 (1968).
933. Schweizer, E., C. Berninger, D. Crouse, R. Davis, and R. Logothetis, J. Org. Chem., <u>34</u>, 207 (1969).
934. Schweizer, E., W. Creasy, K. Light, and E. Schaffer, J. Org. Chem., <u>34</u>, 212 (1969).
935. Schweizer, E., and K. Light, J. Org. Chem., <u>31</u>, 870 (1966).
936. Schweizer, E., and K. Light, J. Org. Chem., <u>31</u>, 2912 (1966).
937. Schweizer, E., and R. Schepers, Tetrahedron Lett., <u>1963</u>, 979.
938. Schweizer, E., E. Shaffer, C. Hughes, and G. Berninger, J. Org. Chem., <u>31</u>, 2907 (1960).
939. Schweizer, E., and J. Thompson, Chem. Commun., <u>1966</u>, 666.
940. Schweizer, E., J. Thompson, and T. Ulrich, J. Org. Chem., <u>33</u>, 3082 (1968).
941. Schwieter, U., H. Gutmann, H. Lindlar, R. Marbet, N. Rigassi, R. Rüegg, S. Schaeren, and O. Isler, Helv. Chim. Acta, <u>49</u>, 369 (1966).
942. Scoggins, L., and T. Yokley (Phillips Petroleum Co.), U.S. Pat. 3,367,989 (Feb. 6, 1968).
943. Seidel, W., Z. Anorg. Allgem. Chem., <u>330</u>, 141 (1964).
944. Semmens, P. (Imperial Chemical Industries Ltd.), Brit. Pat. 786,902 (Nov. 27, 1957).
945. Senise, P., Anal. Chem. Proc. Intern. Symp., Birmingham, Univ., Birmingham, England, 1962, p. 171.
946. Senise, P., and L. Pitombo, Anais Assoc. Brasil. Quim., <u>20</u>, 93 (1961).
947. Senise, P., and L. Pitombo, Anal. Chim. Acta, <u>26</u>, 85 (1962).
948. Senise, P., and L. Pitombo, Talanta, <u>11</u>, 1185 (1964).
949. Senyavina, L., E. Dyatlovitskaya, Y. Sheinker, and

502   Quaternary Phosphonium Compounds

L. Bergel'son, Izv. Akad. Nauk SSSR Ser. Khim., **1964**, 1979.

950. Senyavina, L., Y. Sheinker, V. Zheltova, A. Dombrovskii, and M. Shevchuk, Izv. Akad. Nauk SSSR Ser. Khim., **1965**, 895.
951. Serratosa, F., and E. Sole, Anales Real. Soc. Espan. Fis. Khim., Ser. B, **62**, 431 (1966).
952. Seyferth, D., J. Am. Chem. Soc., **80**, 1336 (1958).
953. Seyferth, D., and K. Braendle, J. Am. Chem. Soc., **83**, 2055 (1961).
954. Seyferth, D., M. Eisert, and J. Heeren, J. Organometal. Chem., **2**, 101 (1964).
955. Seyferth, D., and J. Fogel, J. Organometal Chem., **6**, 205 (1966).
956. Seyferth, D., J. Fogel, and J. Heeren, J. Am. Chem. Soc., **88**, 2207 (1966).
957. Seyferth, D., and S. Grim, J. Am. Chem. Soc., **83**, 1610 (1961).
958. Seyferth, D., and S. Grim, J. Am. Chem. Soc., **83**, 1613 (1961).
959. Seyferth, D., S. Grim, and T. Read, J. Am. Chem. Soc., **82**, 1510 (1960).
960. Seyferth, D., S. Grim, and T. Read, J. Am. Chem. Soc., **83**, 1617 (1961).
961. Seyferth, D., J. Heeren, G. Singh, S. Grim, and W. Hughes, J. Organometal. Chem., **5**, 267 (1966).
962. Seyferth, D., W. Hughes, and J. Heeren, J. Am. Chem. Soc., **87**, 2847 (1965).
963. Seyferth, D., W. Hughes, and J. Heeren, J. Am. Chem. Soc., **87**, 3467 (1965).
964. Seyferth, D., G. Raab, and S. Grim, J. Org. Chem., **26**, 3034 (1961).
965. Seyferth, D., and G. Singh, J. Am. Chem. Soc., **87**, 4156 (1965).
966. Seyferth, D., G. Singh, and R. Suzuki, Pure Appl. Chem., **13**, 1594 (1966).
967. Seyferth, D., and T. Wada, Inorg. Chem., **1**, 78 (1962).
968. Seyferth, D., and M. Weiner, J. Org. Chem., **26**, 4797 (1961).
969. Shell International Research Maatschappij N.V., Neth. Appl. 6,414,352 (June 11, 1965).
970. Shevchenko, V., A. Pinchuk, and N. Kozlova, Zh. Obshch. Khim., **34**, 3955 (1964).
971. Shevchuk, M., A. Antonyuk, and A. Dombrovskii, Zh. Obshch. Khim., **39**, 860 (1969).
972. Shevchuk, M., and A. Dombrovskii, Zh. Obshch. Khim., **34**, 916 (1964).
973. Shevchuk, M., and A. Dombrovskii, Zh. Obshch. Khim., **34**, 1473 (1964).
974. Shinagawa, M., Kagaku no Ryoiki, **10**, 111 (1956).
975. Shinagawa, M., and H. Matsuo, Japan Analyst, **5**, 20 (1956).

976. Shinagawa, M., H. Matsuo, and R. Khoara, Japan
      Analyst, 5, 29 (1956).
977. Shokol, V., L. Molyavko, and G. Derkach, Zh. Obshch.
      Khim., 36, 930 (1966).
978. Shubina, L., L. Malkes, B. Zadorozhnyi, and I.
      Ishchenko, Zh. Obshch. Khim., 36, 1991 (1966).
979. Siemiatycki, M., Compt. Rend., 248, 817 (1959).
980. Siemiatycki, M., and H. Strzelecka, Compt. Rend.,
      250, 3489 (1960).
981. Siemons, W., P. Bierstedt, and R. Kepler, J. Chem.
      Phys., 39, 3523 (1963).
982. Simalty, M., and H. Chakine (Centre Nationale de
      la Recherche Scientifique), Fr. Pat. 1,532,290
      (July 12, 1968).
983. Simalty, M., M. Siemiatycki, J. Caretto, and F.
      Malbec, Bull. Soc. Chim. France, 1962, 125.
984. Simonnin, M., J. Organometal. Chem., 5, 155 (1966).
985. Singh, G., and H. Zimmer, J. Org. Chem., 30, 313
      (1965).
986. Singh, G., and H. Zimmer, J. Org. Chem., 30, 417
      (1965).
987. Sisler, H., H. Ahuja, and N. Smith, J. Org. Chem.,
      26, 1819 (1961).
988. Sisler, H., H. Ahuja, and N. Smith, Inorg. Chem.,
      1, 84 (1962).
989. Sisler, H., and S. Frazier, Inorg. Chem., 4, 1204
      (1965).
990. Sisler, H., S. Frazier, R. Rize, and M. Sanchez,
      Inorg. Chem., 5, 326 (1966).
991. Sisler, H., and R. Nielsen, Inorg. Syn., 8, 74
      (1966).
992. Sisler, H., A. Sarkis, H. Ahuja, R. Drago, and N.
      Smith, J. Am. Chem. Soc., 81, 2982 (1959).
993. Sisler, H., and N. Smith, J. Org. Chem., 26, 611
      (1961).
994. Sisler, H., and N. Smith, J. Org. Chem., 26, 4733
      (1961).
995. Sisler, H., and N. Smith (W.R. Grace & Co.), Fr.
      Pat. 1,330,978 (June 28, 1963).
996. Sisler, H., and J. Weiss, Inorg. Chem., 4, 1514
      (1965).
997. Skinner, D., Diss. Abstr., 22, 80 (1961).
998. Smalley, A., Diss. Abstr., 26, 5727 (1966).
999. Smith, L., and J. Baldwin, J. Org. Chem., 27, 1770
      (1962).
1000. Smith, N., J. Org. Chem., 28, 863 (1963).
1001. Smith, N., and H. Sisler, J. Org. Chem., 28, 272
      (1963).
1002. Smucker, L., Diss. Abstr., 29, 120 (1968).
1003. Soerensen, J., and N. Soerensen, Acta Chim. Scand.,
      20, 992 (1966).

1004. Sokolov, V., and O. Reutov, Izv. Akad. Nauk SSSR Ser. Khim., 1964, 394.

1005. Sollott, G., and E. Howard, J. Org. Chem., 27, 4034 (1967).

1006. Sollott, G., H. Mertwoy, S. Portnoy, and J. Snead, J. Org. Chem., 28, 1090 (1963).

1007. Sollott, G., J. Snead, S. Portnoy, W. Peterson, and H. Mertwoy, U.S. Dept. Com., Office Tech. Serv. A.D. 611, 869 Vol. II, 441 (1965).

1008. Songstad, J., L. Stangeland, and T. Ausfad, Acta. Chim. Scand., 24, 355 (1970).

1009. Speziale, A., and D. Bissing, J. Am. Chem. Soc., 85, 3879 (1963).

1010. Speziale, A., and K. Ratts, J. Am. Chem. Soc., 84, 854 (1962).

1011. Speziale, A., and K. Ratts, J. Am. Chem. Soc., 85, 2790 (1963).

1012. Speziale, A., and K. Ratts, J. Am. Chem. Soc., 87, 5603 (1965).

1013. Speziale, A., and K. Ratts, J. Org. Chem., 28, 465 (1963).

1014. Speziale, A., and L. Smith, J. Am. Chem. Soc., 84, 1868 (1962).

1015. Staab, H., F. Graf, and B. Junge, Tetrahedron Lett., 1966, 743.

1016. Stamicarbon, N.V., Dutch Pat. 75,705 (Aug. 16, 1954).

1017. Staudinger, H., and J. Meyer, Helv. Chim. Acta, 2, 635 (1919).

1018. Steinkopf, W., and G. Schwen, Chem. Ber., 54, 2969 (1921).

1019. Stepanov, B., and A. Bikanov, Zh. Obshch. Khim., 34, 3849 (1964).

1020. Stern, M. (Eastman Kodak Co.), U.S. Pat. 2,945,069 (July 12, 1960).

1021. Stewart, L., Plant. Desease Rept., 46, 469 (1962).

1022. Surmatis, J., and A. Ofner, J. Org. Chem., 26, 1171 (1961).

1023. Surmatis, J., and A. Ofner, J. Org. Chem., 28, 2735 (1963).

1024. Takayuki, O., and M. Matsui, Agr. Biol. Chem. (Tokyo), 30, 759 (1966).

1025. Tate, F., and J. Wild (Imperial Chemical Industries Ltd.), Brit. Pat. 837,120 (June 9, 1960).

1026. Tayler, E., K. Lenard, and B. Loev, Tetrahedron, 23, 77 (1967).

1027. Teichmann, H., and G. Hilgetag, Chem. Ber., 96, 1454 (1963).

1028. Thielen, L. (G.D. Searle & Co.), U.S. Pat. 2,862,970 (Dec. 2, 1958).

1029. Thielen, L., and C. Dornfeld (G.D. Searle & Co.), U.S. Pat. 2,862,971 (Dec. 2, 1958).

1030. Titov, A., and P. Gitel, Dokl. Akad. Nauk SSSR, 158, 1380 (1964).

1031. Tomaschewski, G., J. Prakt. Chem., 33, 168 (1966).

1032. Tomaschewski, G., and G. Giessler, Z. Chem., 6, 26 (1966).

1033. Toyo Rayon Co., Ltd., Fr. Pat. 1,377,361 (Nov. 6, 1964).

1034. Tribalat, S., Anal. Chim. Acta, 5, 115 (1951).

1035. Trippett, S., J. Chem. Soc., 1961, 2813.

1036. Trippett, S., Proc. Chem. Soc., 1963, 19.

1037. Trippett, S., and B. Walker, J. Chem. Soc., 1959, 3874.

1037a. Trippett, S., and B. Walker, J. Chem. Soc., 1960, 2976.

1038. Trippett, S., and B. Walker, J. Chem. Soc., 1961, 1266.

1039. Trippett, S., and B. Walker, J. Chem. Soc., 1961, 2130.

1040. Trippett, S., and B. Walker, J. Chem. Soc. (C), 1966, 887.

1041. Trippett, S., and B. Walker, Chem. Commun., 1965, 106.

1042. Trippett, S., B. Walker, and II. Hoffmann, J. Chem. Soc., 1965, 7140.

1043. Trostyanskaya, E., J. Losev, A. Tevlina, S. Makarova, G. Nefedova, and L. Hsiang-Jao, Mezhdunarod Simpozium po Makromol. Khim. Dokl., Moscow (1960), Sektsiya 3, 124.

1044. Trostyanskaya, E., I. Losev, A. Tevlina, S. Makarova, G. Nefedova, and L. Syan'Zhao, J. Polymer Sci., 59, 379 (1962).

1045. Trostyanskaya, E., and S. Makarova, Zh. Prikl. Khim., 39, 1754 (1966).

1046. Trostyanskaya, E., S. Makarova, and I. Losev, Vysokomolekul. Soedin, 5, 325 (1963).

1047. Truscheit, E., and K. Eiter, Ann., 658, 65 (1962).

1048. Tsivunin, V., G. Kamai, and V. Kormachey, Zh. Obshch. Khim., 35, 1819 (1965).

1049. Tsivunin, V., G. Kamai, R. Shagidullin, and R. Khisamutdinova, Zh. Obshch. Khim., 35, 1234 (1965).

1050. Tsuge, O., T. Tomita, and A. Torii, Nippon Kagaku Yasshi, 89, 1104 (1968).

1051. United States Rubber Co., Brit. Pat. 834,286 (May 4, 1960).

1052. Wagenknecht, J., and M. Baizer, J. Org. Chem., 31, 3885 (1966).

1053. Wagner, G. (Hooker Chemical Corp.), Belg. Pat. 623,724 (April 18, 1963).

1054. Wazer, J. van, C. Callis, J. Shoolery, and R. Jones, J. Am. Chem. Soc., 78, 5715 (1956).

1055. Webb, R. (Union Camp. Corp.), U.S. Pat. 3,293,286

(Dec. 20, 1966).

1056.   Wedekind, E., Chem. Ber., _45_, 2933 (1912).
1057.   Wegler, R., and E. Regel, Makromol. Chem., _9_, 1
        (1952).
1058.   Weiher, J., L. Melby, and R. Benson, J. Am. Chem.
        Soc., _86_, 4329 (1964).
1059.   Weis, C., J. Org. Chem., _27_, 3520 (1962).
1060.   Wellcome Foundation, Ltd., Neth. Appl. 6,414,305
        (June 10, 1965).
1061.   Wiesboeck, R., J. Org. Chem., _30_, 3161 (1965).
1062.   Willard, H., L. Perkins, and F. Blicke, J. Am. Chem.
        Soc., _70_, 737 (1948).
1063.   Wilson, B. (Eastman Kodak Co.), U.S. Pat. 3,345,177
        (Oct 3, 1967).
1064.   Wilson, H., and C. Glassick (Roehm u. Haas Co.),
        Fr. Pat. 1,366,248 (July 10, 1964).
1065.   Wilson, J. (Virginia-Carolina Chemical Corp.), U.S.
        Pat. 3,103,431 (Sept. 10, 1963).
1066.   Windus, W., and W. Happich, J. Am. Leather Chemists'
        Assoc., _58_, 646 (1963).
1067.   Witschard, G., and C. Griffin, Spectrochim. Acta,
        _19_, 1905 (1963).
1068.   Wittig, G., and W. Boell, Chem. Ber., _95_, 2526
        (1962).
1069.   Wittig, G., W. Boell, and K. Krueck, Chem. Ber.,
        _95_, 2514 (1962).
1069a. Wittig, G., and K. Clauss, Ann., _577_, 26 (1952).
1070.   Wittig, G., H. Eggers, and P. Duffner, Ann., _619_,
        10 (1958).
1071.   Wittig, G., and G. Geissler, Ann., _580_, 44 (1953).
1072.   Wittig, G., and A. Haag, Chem. Ber., _88_, 1654 (1955).
1073.   Wittig, G., and A. Haag, Chem. Ber., _96_, 1535 (1963).
1074.   Wittig, G., and O. Hellwinkel, Chem. Ber., _97_, 769
        (1964).
1075.   Wittig, G., and E. Kochendörfer, Chem. Ber., _97_,
        741 (1964).
1076.   Wittig, G., and H. Laib, Ann., _580_, 57 (1953).
1077.   Wittig, G., and A. Maercker, Chem. Ber., _97_, 747
        (1964).
1078.   Wittig, G., and H. Matzura, Angew. Chem., _76_, 187
        (1964).
1079.   Wittig, G., and H. Matzura, Ann., _732_, 97 (1970).
1080.   Wittig, G., H. Pommer, and E. Hartwig (BASF), Ger.
        Pat. 957,942 (1957).
1081.   Wittig, G., and M. Rieber, Ann., _562_, 177 (1949).
1082.   Wittig, G., and M. Schlosser, Chem. Ber., _94_, 1373
        (1961).
1083.   Wittig, G., and M. Schlosser, Tetrahedron, _18_, 1023
        (1962).
1084.   Wittig, G., and U. Schoellkopf, Chem. Ber., _87_,
        1318 (1954).

1085. Wittig, G., and K. Schwarzenbach, Ann., 650, 1 (1961).
1086. Wittig, G., H. Weigmann, and M. Schlosser, Chem. Ber., 94, 676 (1961).
1087. Wittig, G., and D. Wittenberg, Ann., 606, 1 (1957).
1088. Wolff, R., and L. Pichat, Compt. Rend., 246, 1868 (1958).
1089. Wolinsky, J., and K. Erickson, J. Org. Chem., 30, 2208 (1965).
1090. Worral, D., J. Am. Chem. Soc., 52, 2933 (1930).
1091. Yamamoto, K., and M. Oku (Mitsu Chemical Industries Ltd.), Japan Pat. 9067 (Dec. 14, 1955).
1092. Yomo, H., and H. Iinuma, Planta, 71, 113 (1966).
1093. Young, D., W. McEwen, D. Velez, J. Johnson, and C. VanderWerf, Tetrahedron Lett., 1964, 359.
1094. Young, S., J. Turner, and D. Tarbell, J. Org. Chem., 28, 928 (1963).
1095. Yur'ev, Y., and D. Eckhardt, Zh. Obshch. Khim., 31, 3536 (1961).
1096. Zabusova, N., A. Razumov, and T. Kazansk. Khim.-Tekhnol. Inst., 33, 161 (1964).
1097. Zander, M., and W. Franke, Chem. Bcr., 94, 446 (1961).
1098. Zbiral, E., Monatsh. Chem., 91, 1144 (1960).
1099. Zbiral, E., Monatsh. Chem., 95, 1759 (1964).
1100. Zbiral, E., Monatsh. Chem., 97, 180 (1966).
1101. Zbiral, E., Tetrahedron Lett., 1966, 2005.
1102. Zbiral, E., and L. Berner-Fenz, Tetrahedron, 24, 1363 (1968).
1103. Zbiral, E., and L. Fenz, Monatsh. Chem., 96, 1983 (1965).
1104. Zbiral, E., and H. Hengstberger, Monatsh. Chem., 99, 412 (1968).
1105. Zbiral, E., and H. Hengstberger, Monatsh. Chem., 99, 429 (1968).
1106. Zbiral, E., M. Rosberger, and H. Hengstberger, Ann., 725, 22 (1969).
1107. Zbiral, E., and E. Werner, Ann., 707, 130 (1967).
1108. Ziegler, M., and O. Glemser, Angew. Chem., 75, 574 (1963).
1109. Zhdanov, Y., L. Uzlova, and G. Dorofeenko, Zh. Vscs. Khim. Obshch. D.J. Mendeleeva, 10, 600 (1965).
1110. Zhmurova, I., A. Kisilenko, and A. Kirsanov, Zh. Obshch. Khim., 32, 2580 (1962).
1111. Zhmurova, I., and I. Voitsekhovskaya, Zh. Obshch. Khim., 34, 1171 (1964).
1112. Zimmer, H., and G. Singh., J. Org. Chem., 28, 483 (1963).
1113. Zimmer, H., and G. Singh., J. Org. Chem., 29, 1579 (1964).
1114. Zimmer, H., and G. Singh, J. Org. Chem., 29, 3412 (1964).

1115.   Zorn, H., H. Schindlbauer, and H. Hagen, Chem. Ber., 98, 2431 (1965).
1116.   Zosel, D., H. Ritschel, and H. Haensel, Phys. Status Solidi, 38, 177 (1970).